Fossil Algae

Recent Results and Developments

Edited by Erik Flügel

With 119 Figures and 32 Plates

Springer-Verlag
Berlin Heidelberg New York 1977

Prof. Dr. Erik Flügel, Institut für Paläontologie,
Universität Erlangen-Nürnberg, Loewenichstraße 28,
8520 Erlangen/W.-Germany

Cover photo: Well-preserved structure of the alga *Solenopora cassiana* Flügel, together with calcareous ooids. Lowermost Upper Triassic (Cordevol), Southern Alps. Thin-section, ×25.

ISBN 3-540-07974-2 Springer-Verlag Berlin Heidelberg New York
ISBN 0-387-07974-2 Springer-Verlag New York Heidelberg Berlin

Library of Congress Cataloging in Publication Data. International Symposium on Fossil Algae, 1st, University Erlangen-Nürnberg, 1975. Fossil algae. Organized by the Institute of Paleontology of the University Erlangen-Nürnberg. Bibliography: p. Includes index. 1. Algae, Fossil-Congresses. I. Flügel, Erik. II. Erlangen. Universität Erlangen-Nürnberg. Institut für Paläontologie. III. Title. QE955.I59 1975 561'.93 76-46461.

Printed in Germany.

Printing and bookbinding: Konrad Triltsch, Graphischer Betrieb,
D-8700 Würzburg

Preface

Algae and Algal Carbonates are of interest to geo- and bioscientists because of the significant interactions between algal life and environment. Much information about systematical, morphological and ecological data of fossil algae has been accumulated during the last two decades. Facies interpretation, sedimentology, and biostratigraphic correlations, as well as speculations about Precambrian ecosystems have received new impulses from the detailed study of Recent algae and their fossil counterparts.

During the First International Symposium on Fossil Algae, organized by the Institute of Paleontology of the University Erlangen-Nürnberg in October 1975, about 130 scientists were able to discuss the present status of knowledge; 55 lectures were given, 37 of which now have been chosen for publication.

Some of these articles comprehend fundamental reviews on algal groups and algal constructions, such as Monty on the stromatolites, Herak et al. and the "French Group" on the dasyclads, and Wray and Poignant on the Paleozoic and Mesozoic red algae. Other articles deal with the biology and morphology of blue-green algae, with the arduous problem of the affinities of many algae and "pseudo-algae", with biometrical methods and with the ultrastructure of calcareous algae. A broad section of this book is reserved for some stimulating papers about the role of algae as facies indicators and sedimentological factors in Recent and Ancient environments.

Thanks to the gratefully accepted co-operation of many collegues, we hope this book may offer a good cross-section through main trends of recent algal research.

I am indebted to all contributors of the book for their confidence, to Mrs. U. Scholl of the Paleontological Institute Erlangen for all her work during and after the Symposium, and to the staff of the Springer-Verlag, Heidelberg, for the assistance during the production of the book.

Erik Flügel

Contents

Blue-Green Algae and Stromatolites

Algae and Sedimentary Environments

List of Contributors

Abate, B., Istituto di Geologia dell'Università di Palermo, Corso Tuköry 131, Palermo/Italy

Adolphe, J.-P., Géologie 1, Université Paris VI, Tours 14–16, 4, place Jussieu, 75230 Paris, Cedex 05/France

Alexandersson, T., Paleontological Institute, Department of Historical Geology and Paleontology, Box 558, 75122 Uppsala 1/Sweden

Anderson, M.M., Department of Geology, Memorial University of Newfoundland, St. John's, Newfoundland/Canada

D'Argenio, B., Istituto di Geologia e Geofisica dell'Università di Napoli, Largo S. Marcellino 10, Napoli/Italy

Barghoorn, E.S., Palaeobotanical Laboratories, Harvard University, 16 Divinity Avenue, Cambridge, Mass. 02138/USA

Bassoullet, J.P., CNRS, Laboratoire de Micropaléontologie, Université P. & M. Curie, 4, place Jussieu, 75230 Paris, Cedex 05/France

Bernier, P., Department Science de la Terre, Université Claude Bernhard, 15–43 Bd. du 11 Nov. 1918, Villeurbanne/France

Bosence, D., Department of Geology, University of London Goldsmiths' College, New Cross, London SE 14 6 NW/England

Buchbinder, B., Geological Survey of Israel, 30 Malkhe Yisrael St., 95501 Jerusalem/Israel

Catalano, R., Istituto di Geologia dell'Università di Palermo, Corso Tuköry 131, Palermo/Italy

Cohen, Y., Department of Microbiological Chemistry, Hebrew University, Jerusalem/Israel

Conrad, M.A., Departement de Géologie et Paléontologie, Université Genève/Suisse

Deloffre, R., Société Nationale des Pétroles d'Aquitaine, Centre de Recherches de Pau, BP. 65, 64001 Pau/France

Dricot, E., Université Catholique de Louvain, Institut de Géologie et de Géographie, Laboratoire de Paléontologie, Place Louis Pasteur, 3, 1348 Louvain-la-Neuve/Belgique

Edhorn, A.-St., Memorial University, St. John's, Newfoundland/Canada

Flajs, G., Institut für Paläontologie der Universität Bonn, Nußallee 8, 5300 Bonn/W. Germany

Flügel, E., Institut für Paläontologie, Universität Erlagen-Nürnberg, Loewenichstr. 28, 8520 Erlangen/W.-Germany

Förster, R., Bayerische Staatssammlung für Paläontologie und historische Geologie, Richard-Wagner-Str. 10, 8000 München/W.-Germany

Genot, P., Laboratoire de Micropaléontologie, Université P. & M. Curie, 75230 Paris, Cedex 05/France

Golubic, St., Department of Biology, Boston University, 2 Cummington Street, Boston, Mass. 02215/USA

Gunatilaka, A., School of Mines, University of Zambia, Lusaka/Zambia

Gusic, I., Geološko-paleontološki zavod, Prir.-mat. fak., Soc. revolucije 8, 41000 Zagreb/Yugoslavia

Herak, M., Geološko-paleontološki zavod, Prir.-mat. fak., Soc. revolucije 8, 41000 Zagreb/Yugoslavia

Jaffrezo, M., Département de Géologie structurale, Université P. & M. Curie, associé au CNRS no. 145, Géologie méditerranéenne, Paris/France

Kochansky-Devide, V., Geološko-paleontološki zavod, Prir.-mat. fak., Soc. revolucije 8, 41000 Zagreb/Yugoslavia

Krumbein, W.E., Carl-von-Ossietzky-Universität, Postfach 943, 2900 Oldenburg/W.-Germany

Lemone, D., Department of Geology, University of Texas at El Paso, P.O. Box 3, El Paso, Texas/USA

Massieux, M., Laboratoire de Paléobotanique, USTL, Place E. Bataillon, 34060 Montpellier, Cedex/France

Maurin, A.F., Compagnie Française des Pétroles, 39 Quai Citroen, 75739 Paris/France

Milliman, J.D., Woods Hole Oceanographic Institution, Woods Hole, Mass. 02543/USA

Monninger, W., Paläontologisches Institut der Universität, Pleicherwall, 8700 Würzburg/W.-Germany

Monty, C., Université de Liège, Laboratoire de Paléontologie animale, 7, Place du Vingt-Aout, 4000 Liège/Belgique

Noel, D., Musée d'Histoire Naturelle, Paris/France

Orszag-Sperber, F., Laboratoire de Géologie structurale, Université Paris XI, Orsay/France

Peryt, T.M., Instytut Geologiczny, Rakowicka 4, 00–975 Warsaw/Poland

Piatkowski, T.S., Instytut Geologiczny, Rakowicka 4, 00–975 Warsaw/Poland

Poignant, A.-F., Laboratoire de Géologie des Bassins Sédimentaires, Université P. & M. Curie, place Jussieu, 75230 Paris, Cedex 05/France

Poisson, A., Equipe de Recherche Associée au C.N.R.S., Asie Alpine, Université Paris XI, Orsay/France

Riccobono, R., Istituto di Geologia dell'Universita di Palermo, Corso Tuköry 131, Palermo/Italy

Riding, R., Department of Geology, University College, Radiff CF1 1XL/England

Rietschel, S., Forschungsinstitut Senckenberg, Senckenberganlage 25, 6000 Frankfurt a.M./W.-Germany

Schmitt, M., Paläontologisches Institut der Universität, Pleicherwall, 8700 Würzburg/W.-Germany

Schneider, J., Geologisch-Paläontologisches Institut der Universität, Goldschmidtstr. 3, 3400 Göttingen/W.-Germany

Schnitzer, W.A., Geologisches Institut der Universität, Pleicherwall, 8700 Würzburg/W.-Germany

Segonzac, G., 14, Place Dupuy, 31000 Toulouse/France

Sinha, A.K., Wadia Institute of Himalayan Geology, 36, Teg Bahadur Marg, Dalanwala, 248001 Dehra Dun/India

Srivastava, N.K., Geologisches Institut der Universität, Pleicherwall, 8700 Würzburg/W.-Germany

Di Stefano, P., Istituto di Geologia dell'Università di Palermo, Corso Tuköry 131, Palermo/Italy

Termier, G., Laboratoire de Géologie structurale, Université P. & M. Curie, 4 place Jussieu, 75230 Paris, Cedex 05/France

Termier, H., Laboratoire de Géologie structurale, Université P. & M. Curie, 4 place Jussieu, 75230 Paris, Cedex 05/France

Toomey, D.F., University of Texas of the Permian Basin, Odessa, Texas/USA

Tsien, H.H., Université Catholique de Louvain, Institut de Géologie et de Géographie, Laboratoire de Paléontologie, Place Louis Pasteur, 3, 1348 Louvain-la-Neuve/Belgique

Vachard, D., Institut de Géologie IGAL, 1 rue d'Assas, 75270 Paris, Cedex 06/France

Wachendorf, H., Institut für Geologie und Paläontologie der Technischen Universität, 3300 Braunschweig/W.-Germany

Wray, J.L., Marathon Oil Company, Denver Research Center, Littleton, Colorado 80120/USA and Colorado School of Mines, Golden, Colorado 80401/USA

Zamarreno, I., Departamento de Geologie, Faculdad de Ciencias, Universidad de Oviedo, Oviedo/Spain

Zorn, H., Geologisch-Paläontologisches Institut der Universität, Senckenberganlage 32, 6000 Frankfurt a.M./W.-Germany

1 Interpretation of Microbial Fossils with Special Reference to the Precambrian

St. GOLUBIĆ, Boston, Mass., USA, and E.S. BARGHOORN, Cambridge, Mass., USA

1.1. Introduction

During the last two decades the occurrence of microbial fossils in silicified Precambrian rocks has become firmly established. The significance of these early microorganisms in the bio-geo-chemical evolution has been recognized, and their frequent affiliation with stromatolitic structures, common in Precambrian strata, has been documented (Reviewed by Schopf, 1970b; Licari and Cloud, 1972). A large number of Precambrian fossil microorganisms has been formally described (Pflug, 1965, 1966; Schopf, 1968; Schopf and Blacic, 1971). Based on these descriptions and identifications of taxonomic affinities of Precambrian microorganisms, dates have been proposed for important early evolutionary events, such as photosynthesis (Schopf et al., 1971), N_2-fixation (Cloud, 1974), Eukaryotic cell organization (Schopf, 1968, 1970b; Cloud et al., 1969) and sex (Schopf and Blacic, 1971). However, the criteria used in identification of fossil microorganisms remain in many cases dubious, leaving the already generally accepted conclusions on shaky grounds.

Studies of post-mortem degradation of extant mat-forming cyanophytes (Awramik et al., 1972) have shown that early diagenesis may be responsible for significant morphological modification of fossilized forms. Comparisons with degraded extant microorganisms have been successfully applied in the interpretation of Precambrian microbial fossils (Knoll et al., 1975; Knoll and Barghoorn, 1975; Golubić and Hofmann, 1976).

It is now possible to summarize the principles and approaches used in those and the continuing studies, in order to refine the means of identification and interpretation of microbial fossils: 1. A comparison with extant counterparts is sought in comparable modern environments. 2. Post-mortem changes and early diagenesis are studied in modern natural populations, and experimentally induced in cultured unispecific populations of selected taxa. 3. Long-term changes are followed by studying the preservation of microorganisms in Holocene sediments. 4. Whenever possible, entire populations are studied and their morphological variability, including cell division patterns and colony formation, is evaluated. 5. Maximal attention is paid to the few well-preserved fossil taxa which offer the most complete record, in order to establish and verify the criteria for interpretation of less complete or partially obliterated specimens.

The following questions appear to be important in the interpretation of fossil microorganisms: 1. What is the preservation state of a microbial fossil? 2. What are the nature, extent, and morphological consequences of its diagenesis? 3. Under what conditions is a preservation of a microorganism possible? 4. How selective is the preservation process? Is the fossil microbial assemblage biased by this selection, and, if so, which specific structures have a higher preservation potential than others?

1.2 The State of Preservation of Fossil Microorganisms

A fossil assemblage is usually biased toward organisms with hard tissues and skeletons. This general rule applies also for microscopic organisms, such as coccolithophorids, silicoflagellates, and diatoms. However, such microorganisms have not been found in Precambrian strata. Most of the Precambrian microbial fossils are comparable with extant "soft bodied" cyanophytes and bacteria.

The best-known microbial fossils are found embedded in silicious cherts. Optical properties of these cherts (calcedony) permit a convenient in situ study in pretrographic thin sections, and a control of HF-maceration, to ensure the autochthoneity of isolated specimens (Schopf, 1970a). The morphology of microfossils embedded in silica appears remarkably well preserved as compared with those embedded in carbonates (Hofmann, 1974). The size of the preserved structures range from fractions of a micron to about 20 μm in most cases, requiring the use of high power – high resolution microscopy.

Under high power transmission light microscope (TLM) and incident light microscope (ILM), the structures are usually sharply outlined, dark brown to red brown in color. Some forms show more or less distinct extracellular pigmentation marking individual cell envelopes and sheaths, or entire colonies. The original pigmentation of the microorganisms seems entirely lost and replaced by a uniform coloration, depending on the preservation conditions, rather than on the nature of the microorganisms. This pigment seems to be based on carbonaceous organic matter (Kerogen), possibly enriched with complexed or mineralized heavy metals (Pflug, 1967).

Structures macerated out of the silicious matrix by HF retain their coherence and external features, and can be observed by scanning electron microscope (SEM) (Schopf, 1970a). Under transmission electron microscope (TEM), the structures resolve into fibrous, often loosely reticulate, outlines without further structural details (Cloud and Hagen, 1965; Schopf, 1970a; Cloud and Licari, 1972), or they represent homogeneous masses with scalloped outlines and occasional internal holes (Oehler, in press). Outlines of globose cell-like structures, when packed in aggregates, often fuse on contact into similar homogeneous, electron-dense masses. Both reticulation and fusion of initially distinct membranous structures produced by *individual* cell entities are diagenetic in origin (Oehler, 1976). At least the coarser reticulations may originate from crystal growth within the matrix, as illustrated by Pflug (1967).

High resolution images under TEM definitely show that the preserved microfossils did suffer substantial diagenetic alteration, so that an unequivocal identification with the ultrastructure of extant microbial cells, and, thus, a distinction between prokaryotic and eukaryotic organization, is hardly possible. A study of early diagnenetic alterations of extant microbial cells, however, may provide relevant information for the interpretation of microbial cells.

1.3 Early Diagenesis of Organic Structures

Under normal post mortem circumstances, an organism undergoes several steps of morphological and chemical decomposition, which includes autolytic processes as well as enzymatic lysis by various decomposing microorganisms. The rate of this degradation depends, in the first place, on structural quality and composition of the organism in general, and its cell wall structures in particular.

Prokaryotes (cyanophytes and bacteria) have a reputation of being generally tough, and able to adjust and to survive extreme life conditions. Although this is true for the group as a whole, at the species level they represent a wide spectrum of specialized types, ranging from extremely fine and sensitive, to very robust and resistant to environmental changes. Marine and freshwater planktonic and permanently submerged benthic taxa often show little resistance to degradation. They lyse explosively at slight changes in osmotic pressure or temperature, leaving little or no detectible structural remains. At the other end of the spectrum are some intertidal forms, and such as live on periodically wetted bare rocks. These microorganisms can survive periods of complete desiccation, and they leave ample structural remains after death.

A similarly wide range of adaptational types and structures is encountered among the "soft-bodied" microscopic eukaryotes. Small flagellates, for example, belong to the most sensitive ones. They lyse explosively within minutes of collection, leaving unidentifiable spherical droplets of plasma. At the other extreme are many unicellular algae and fungi and their spores, with rigid cellular walls fortified by multiple layers of cellulose, chitin, or other resistant compounds. The correlation between the structural type and the environment, however, is less evident in eukaryotes than in prokaryotes. A selection for potential fossils, which favors more resistant structures among both prokaryotes and eukaryotes, takes place already at the level of occupancy of an environment by an adapted microbial assemblage. Some environments, as well as some groups of microorganisms, have better chances of being represented in the fossil record than others.

After death, one to several specialized decomposers may be required to convert the organic matter into simple inorganic compounds. A disappearance of all structures is then the ultimate consequence. However, the decomposition processes in nature are neither perfect nor complete. They may be arrested, terminated, or modified at various stages, thus preserving a choice of materials and corresponding morphological characters, adding to the variability of forms in the fossil assemblage.

Decomposition processes may by themselves alter the microenvironment and modify the conditions for their continuation. They may change the redox potential and pH, and thus change the solubility state of metals and other elements. An intensive aerobic decomposition, for example, which is usually very efficient and complete, becomes self-limiting by depleting the environmental oxygen. The conditions for an anaerobic decomposition are then introduced, and a different set of microorganisms takes over. A complete anaerobic decomposition of organic matter, however, requires several steps, performed by different specialized decomposers in a successive order. These conditions are not always met under natural conditions, so that anaerobic decomposition often remains incomplete. Other environmental factors, when expressed as extremes (e.g. high salinity, low pH) may further restrict the decomposing flora, and become responsible for preservation of organic structures. The brines beneath the Abu Dhabi sabkha (Persian Gulf), for example, preserve green pigmented cellular structures of cyanophytes for 8000 years (Golubić, 1973). Preservation of organic remains in bogs is well known.

It is generally assumed that the best conditions for microbial fossilization are met by early embedding in a mineral matrix. Incomplete degradation within soft sediments, however, may provide a time extension within which mineralization can occur, thus increasing the chance for microbial fossilization.

1.4 The Prokaryotic Organization

Today it is generally accepted that the most important step in the evolution of life since its origin is between the prokaryotic and eukaryotic cellular organization. It has also become established that the prokaryotes have dominated the Earth during the vast time spans of the Precambrian, prior to the evolution of eukaryotes, probably maintaining completely functional ecosystems. The main metabolic pathways evolved at the prokaryotic level of organization, and the principal modifications of the Earth as a habitat, can largely be attributed to these microorganisms. Yet, ancient and conservative as their organization may be, it has proven itself extremely successful in terms of the number, and functional importance, of ecological niches which the prokaryotes occupy in modern ecosystems. Their present metabolic diversity exceeds by far that of the more complex and advanced eukaryotes.

Partly because of their simplicity, but possibly also due to the enormous evolutionary distance, many early and mid-Precambrian microorganisms cannot be easily related to modern forms. Some mid-Precambrian forms, however, (e.g. *Eoentophysalis belcherensis* Hofmann, 1.9 x 10^9 years old) and many late Precambrian ones can be easily compared with extant cyanophyte genera (Schopf, 1968; Licari and Cloud, 1972). At the time of the presumed appearance of the first eukaryotes, the world was already inhabited with familiar, still extant prokaryotic types.

Differences between the prokaryotic and eukaryotic cellular organizations are well defined in modern microorganisms at the ultrastructural level. Due to diagenetic alterations, however, this approach cannot be satisfactorily applied for microbial fossils, so that identification rests primarily on comparisons of external features and cellular arrangements. It seems, therefore, reasonable to examine first the potential of prokaryotes to produce various organizational types, before a given morphology can be safely attributed to eukaryotes.

The most advanced prokaryotic organization, and a correspondingly high morphological diversity among the prokaryotes, has been achieved in the cyanophytes. As photosynthetic microorganisms and primary producers, they occupy the base position in a trophic pyramid, and their abundance in ancient prokaryotic ecosystems may be expected to be at least one order of magnitude higher than that of the contemporary heterotrophs. The modern cyanophytes are, therefore, the most likely candidates for a successful interpretational model of Precambrian microbial fossils.

Post mortem degradation of extant microorganisms has been studied in marine intertidal and shallow water algal mats (Awramik et al., 1972), an environment which in many respects resembles fossil stromatolitic strata (Hofmann, 1974; Golubić and Hofmann, 1976). The cyanophytes which have been studied are characterized by coccoid or filamentous cellular organization, and by a production of extracellular gelatinous matter, which is frequently layered and pigmented by a yellow-brown, light-dependent pigment ("scytonemine"). After death, the microorganisms are subjected through water loss and shrinkage. In all cyanophytes studied, the cellular structures deform more than their extracellular envelopes. Moreover, the envelopes with their pigmentation often remain preserved after the cells have completely disintegrated.

Entophysalis major Ercegović, which dominates recent stromatolites in Shark Bay, Australia, and in the Persian Gulf, illustrates a shrinkage pattern typical of many coccoid cyanophytes. A spectrum of forms originates from progressive shrinkage, superimposed on the cell division cycle (Fig. 1). A spherical cell, surrounded by multiple gelatinous envelopes caves in, with the loss of volume producing a polyhedral body with flat, and later concave, planes (Fig. 1a, I and II). Further shrinkage causes complete collapse and substantial size reduction of the cell. After complete disintegration of the cell structures, the extracellular envelopes often remain, conforming to the morphology of the cell that has produced them (Fig. 1a, III and IV). If a cell dies during the state of division, the shrinkage is modified by the presence of a connection between the two separating halves and of the newly synthesized cell wall portions. The result is a formation of a dumbbell-shaped body (Fig. 1b, I–III). The deformation may be transferred to the inner envelopes but with a dampening effect (Fig. 1b, IV). If the cell dies immediately after division, before the cell wall formation is completed, a cell may cave in asymmetrically by an invagination of the weaker side, producing a convex-concave body that appears semilunar in side view (Fig. 1c, I–III). In other cases, it may also continue a type b deformation producing a tear-shaped body (Fig. 1d, I–III). The external envelopes in all cases are less deformed (Fig. 1, column IV).

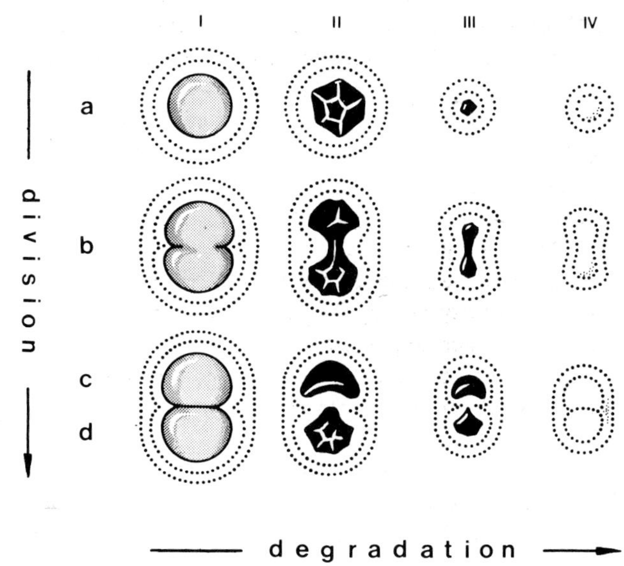

Fig. 1. Early diagenesis of a coccoid cyanophyte (*Entophysalis*) prograding degradation (I–IV), superimposed over division cycle, as four shrinkage sequences (a–d); column IV contains only empty envelopes

The shrinkage is strongest at the level of the maximum density difference between the cell content and the extracellular matter, delineated by the osmotically sensitive plasma membrane. The cell wall layers appear to be tightly bound to the plasma mem-

Plate 1. Comparison between selected modern microorganisms (*left*, Figs. 1–10, and ▶ Precambrian microbial fossils (*right*, Figs. 11–18). Scale bars are 10 μm long. The scale bar in Fig. 14 refers to Figs. 1, 14 and 15; that in Fig. 1 to all other photomicrographs.

Left: modern microorganisms
Plate 1/1. *Entophysalis major* Ercegović, pigmented colony surface, lower intertidal zone, Abu Dhabi, Persian Gulf; *arrow*: dumbbell-shaped shrinkage of a dividing cell and its envelopes. Plate 1/2. *Chrococcus turgidus* (Kütz). Naeg. culture; different shrinkage of dividing cells. Plate 1/3. *Entophysalis major*, tetrahedral arrangement of shrinking cells, the forth cell is behind; *arrow*: deformed inner envelope. Plate 1/4. *Gloeothece* sp., algal mats, intertidal zone, Andros Island, Bahamas, polyhedral packing of individual envelopes. Plate 1/5. Iron accumulation within cell excretions in *Cocconeis* sp. (sessile diatom). Plate 1/6. Iron accumulation within the envelope of a chlorophyte. Plate 1/7. Sessile bacteria dividing in one plane (Bacilli); Plate 1/5–7 are from an exposed slide at mud-water interface 3 m deep, Buzzard Bay, Massachusetts, USA. Plate 1/8. *Lyngbya aestuarii* (Mert.) Liebm. forming a left-screw sheath. Plate 1/9. Turgescent cells of a marine *Scytonema* with divergent sheath formation, intertidal zone, Andros Island, Bahamas. Plate 1/10. Polarized shrinkage of *Scytonema* (same as in Plate 1/9) with divergent sheaths (cf. Fig. 5b).

Right: microbial fossils, Figs. 11, 14, and 15 from the Belcher Island Formation, Canada, 1.9×10^9 years old; Figs. 12, 13, 16, 17, and 18 from the Bitter Springs Formation, Australia, 0.9×10^9 years old.
Plate 1/11. *Eoentophysalis belcherensis* Hofmann, shrunken cells in various division states surrounded by polygonally packed envelopes; *arrow*: dumbbell-shaped shrinkage of a dividing cell and its envelopes (cf. Plate 1/1, 4 and Fig. 1b). Plate 1/12. *Eozygion minutum* Schopf and Blacic, different shrinkage of dividing cells (cf. Plate 1/2 and Fig. 1c, d). Plate 1/13. *Eotetrahedron princeps* Schopf (cf. Plate 1/3 and Fig. 2b). Plate 1/14. *Eoentophysalis belcherensis*, detached colonies with polyhedral packing deformation (cf. Plate 1/4 and Fig. 1c). Plate 1/15. *Eosynechococcus moorei* Hofmann, *left*: spiral packing of cells into an aggregate; *arrow*: series of one dimensionally dividing cells (cf. Plate 1/7 and Fig. 4). Plate 1/16. *Heliconema australiense* Schopf, ribbon-like spiral structure, a probable oscillatoriacean sheath (cf. Plate 1/8 and Fig. 5a). Plate 1/17. *Palaeolyngbya barghoorniana* Schopf, separating trichome cells (cf. Plate 1/8 and 9). Plate 1/18. *Caudiculophycus rivularioides* Schopf, turgescent trichome cells attenuating at the end, a possible local shrinkage.

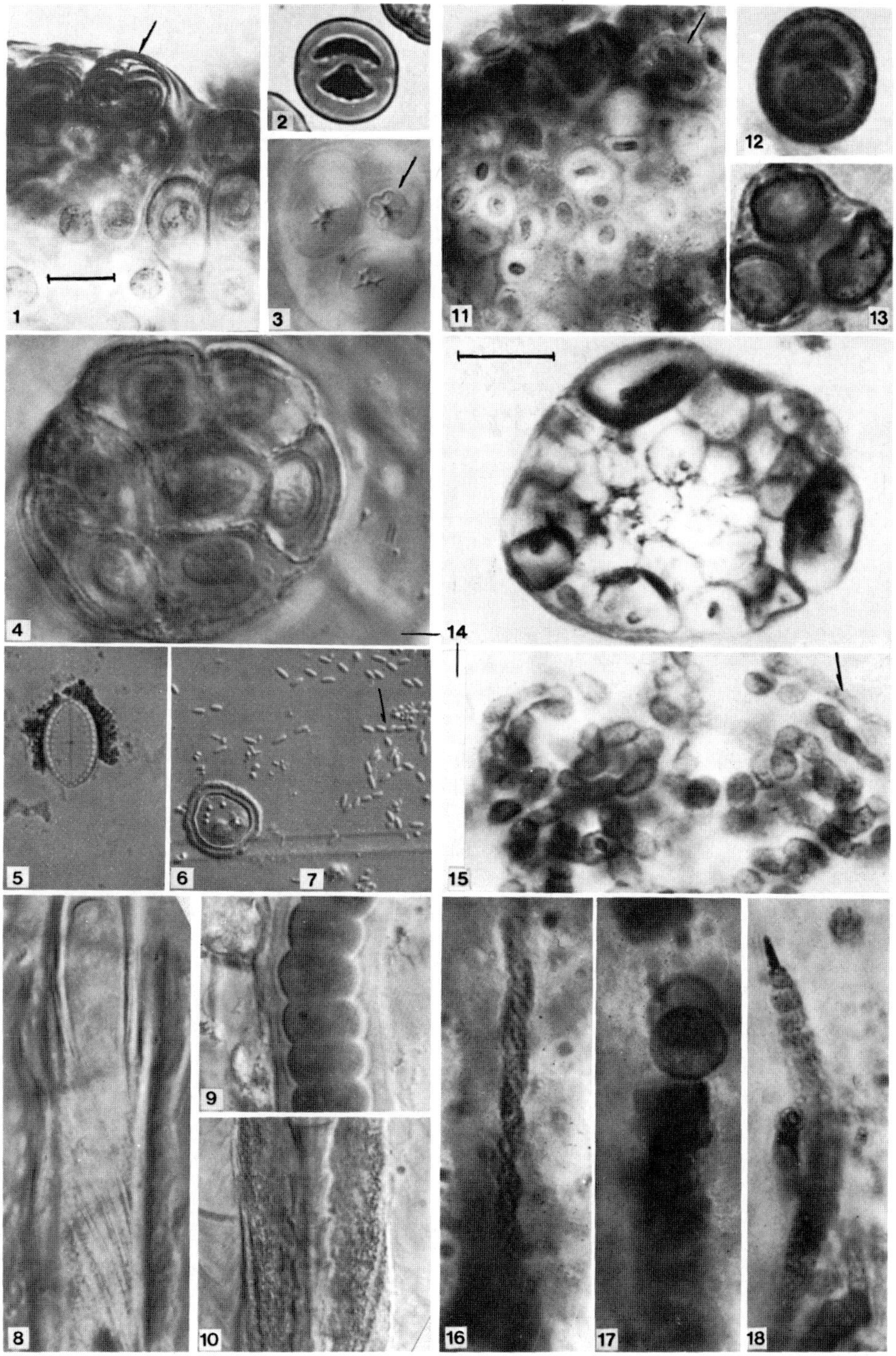

Plate 1. Legend see opposite page

brane and flexible enough to follow the deformation and collapse of the cell content. The layers of external envelopes, however, demark discontinuities between areas of similar density. With water loss, they shrink evenly, following a pattern of a deflating rubber ball. Only the innermost envelopes are pulled in by the collapsing cell (Plate 1/1, 3 and 11, arrows). Various shrinkage morphologies found in the modern *Entophysalis* (Plate 1/1, 3) have been recognized in the mid-Precambrian *Eoentophysalis belcherensis* Hofmann (Golubić and Hofmann, 1976). Comparable cases of dumbbell shrinkage morphology are marked with arrows in Plate 1/1 and 11 respectively. A counterpart to shrinkage introduced in cultures of *Chroococcus turgidus* (Kützing) Naegeli (Plate 1/2) has been found in many Precambrian taxa (Plate 1/12) (Knoll and Barghoorn, 1975).

The morphology of entire colonies (cell populations) is determined by the sequence of cell divisions and by their subsequent arrangement within a more or less cohesive gelatinous matrix. Recent prokaryotic genera characterized by such colony formation provide an instructive comparison material for the interpretation of microbial fossils.

Entophysalis divides predominantly into three perpendicular planes forming three-dimensional packets. Variants of this pattern are caused by displacement and compaction within common gelatinous envelopes (Golubić, 1976a). The following compaction morphologies are frequent in cyanophytes that divide into three perpendicular planes – tetrahedral tetrad formation, and polyhedral packing. After two cell divisions the resulting four cells are normally so arranged to form a plate. A three-dimensional, cubical arrangement of 8 cells is introduced after the third division as in the cyanophyte genus *Eucapsis* (Fig. 2a). A similar form was found by Licari and Cloud (1972) in silicified stromatolites of the 1.6 x 10^9 years old Paradise Creek formation, Australia. In large, tightly enveloped species of *Chroococcus*, however, the 4 cells after the second division are frequently displaced, so that the minimal surface to volume ratio of the packet is achieved: they form a tetrahedral tetrad. This mechanical "slippage" can be explained as a rotation of one half of the division plane with respect to the other for 90°, followed by a locking together of the cells along the rotation axis (Fig. 2b). Tetrahedral tetrads have been found in 0.9 x 10^9 years old Bitter Spring formation, Australia, and described as *Eotetrahedron princeps* Schopf and Blacic, a presumed cukaryote characterized by meiotic divison and sexual reproduction (Plate 1/13). Knoll and Barghoorn (1975) explained these tetrads as possible prokaryotes, following the cell division and compaction pattern of *Chroococcus* as its extant analog. The presence of triads in the fossil population supports this interpretation, since meiotic divisions are synchronous.

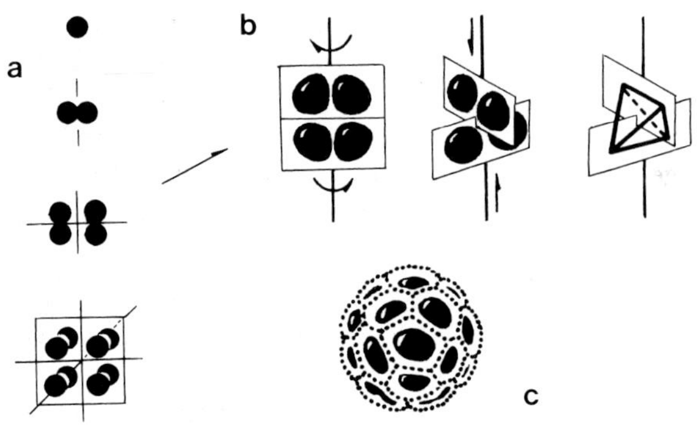

Fig. 2 a–c. Three-dimensional cell division in coccoid cyanophytes; (a) sequence of 3 divisions resulting in cubical arrangement of 8 cells (*Eucapsis*); (b) mechanical rearrangement of 4 cells (after 2 divisions) under compaction to form a tetraheder (*Chroococcus*); (c) polyhedral packing of envelopes within a tight cellular aggregate (*Gloeocapsa*)

Polyhedral packing morphology originates from mutual compaction of a large number of cells, or cell envelopes within a tight common envelope (Fig. 2c). It is common in recent cyanophytes (Plate 1/4) and in Precambrian fossils (Plate 1/14). Packing phenomena observed in Precambrian *Eoentophysalis* (Hofmann, 1976), *Palaeopleurocapsa* (Knoll et al., 1975) and others are similar to those in extant genera *Gloeocapsa, Gloeothece,* and *Entophysalis* (Chroococcales) and in *Hyella, Pleurocapsa,* and *Xenococcus* (Pleurocapsales).

In cyanophytes that divide in two planes, a normal consequence of multiple divisions is the formation of a plate which is held together by extracellular gelatinous matter (Fig. 3a and b), as in the genus *Merismopedia.* However, two-dimensional cell division associated with unidirectional gel production, perpendicular to the divisions, may result in the formation of hollow balls, as in genera *Gomphosphaeria* and *Coelosphaerium.* A recent interpretation of the mid-Precambrian fossil *Eosphaera tyleri* Barghoorn from the Gunflint Iron Formation, Canada (Barghoorn and Tyler, 1965), as a possible hollow sphere with peripheral cell arrangement, comparable with fossil and extant eukaryotic Volvocaceae, was considered by Kaźmierczak (1975). It should be noted here that the hollow sphere morphology has been achieved also at the level of coccoid prokaryotes.

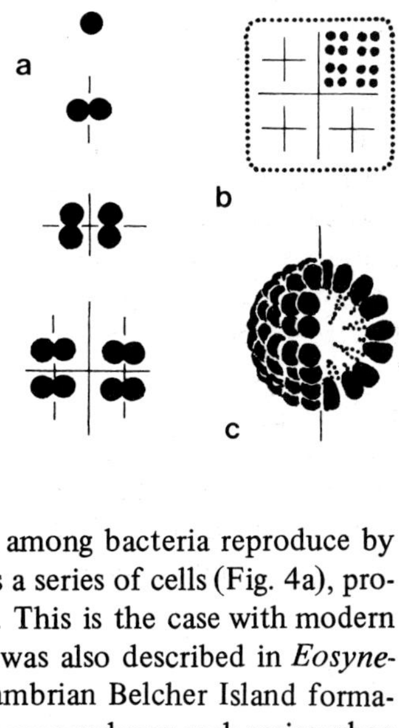

Fig. 3 a–c. Two-dimensional cell division in coccoid cyanophytes; (a) sequence of 3 divisions resulting in a plate arrangement of 8 cells; (b) plate-shaped colony of 64 cells (*Merismopedia*); (c) hollow sphere formation by two-dimensional cell division, when combined with unidirectional, centripetal gel production (*Gomphospheria*)

Several genera of coccoid cyanophytes and the bacilli among bacteria reproduce by cell division in one plane only. The normal consequence is a series of cells (Fig. 4a), providing that the cells are not dispersed after each division. This is the case with modern bacteria colonizing firm substrates (Plate 1/7, arrow). It was also described in *Eosynechococcus moorei* (Hofmann, 1976) from the mid-Precambrian Belcher Island formation (Plate 1/15, arrow). Modern cyanophyte *Synechococcus* produces such series when growing on agar; in nature it is usually dispersed. Spiral crowding and formation of globular aggregates (Fig. 4b) occur in genera which produce extracellular gelatinous envelopes: *Aphanothece* and *Gloeothece* (Plate 1/4). Initiation of crowding is shown in *Eosynechococcus* (Plate 1/15, left).

Trichome formation in filamentous cyanophytes is in principle based on one-dimensional cell division. True branching of trichomes, which occurs in the most advanced forms (Stigonematales), originates from a change of plane of division (Golubić, 1976b). Post mortem cell degradation and shrinkage morphology of filamentous cyanophytes

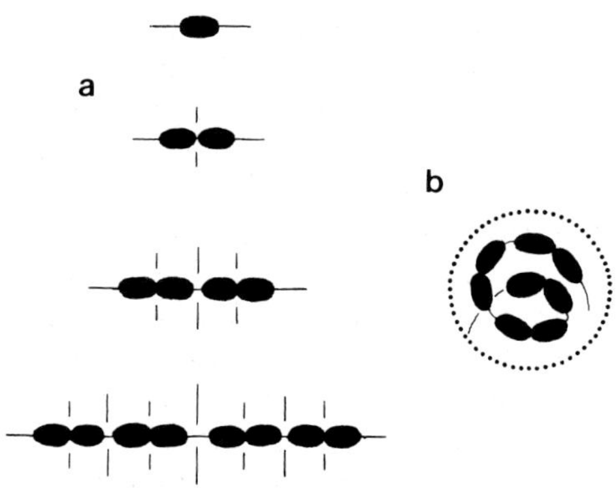

Fig. 4 a and b. One-dimensional cell division in coccoid cyanophytes; (a) sequence of 3 divisions resulting in a cell series (*Synechococcus*; (b) spherical colony resulting from a spiral deformation of the division axis in the presence of extracellular gel (*Aphanothece*)

is also in many respects specific to families and genera, and can be observed in modern algal mat communities, side by side with their coccoid members.

In the family Oscillatoriaceae, the basic cell shape is a cylinder, connected to neighboring cells at its circular base. Shrinkage of a cylindrical cell follows a pattern described by Thompson (1917), resulting in formation of a spool- or fish vertebra shaped body. A detachment of cells from each other usually accompanies the shrinkage (Fig. 5a, I–III). The extracellular gelatinous envelopes are usually called "sheaths" in filamentous cyanophytes. As in coccoid forms, they show a higher resistance to degradation and deform less than the cells. In addition, filamentous cyanophytes have the ability to glide out of their sheaths and to form new ones. This habit is one of the successful adaptations for building stromatolites. As a consequence, sheaths of filamentous cyanophytes can be expected to have both higher fossilization potential and a higher frequency of occurrence in the sediments than the cellular structures. It is, therefore, possible that many of the non-septate Precambrian filamentous forms in fact represent empty sheaths of cyanophytes (Schopf, 1968), rather than fungi or siphonous green algae.

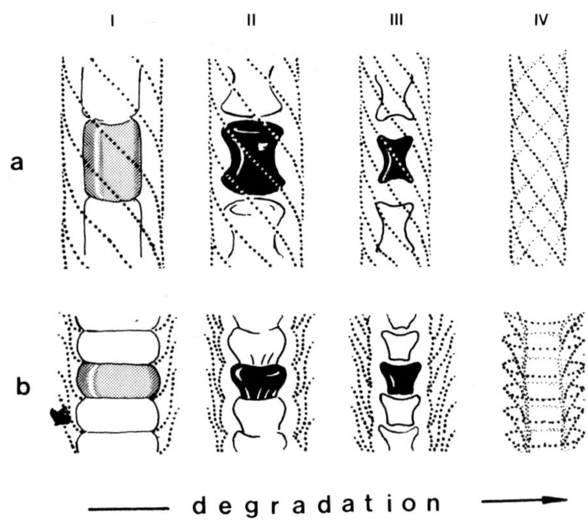

Fig. 5 a and b. Early diagenesis in filamentous cyanophytes; (a) prograding degradation (I–IV) of an oscillatoriacean form (*Lyngbya*): cells shrink into a spool-shaped body, the sheath is a progressing spiral (screw); (b) degradation of a scytonematacean form (*Scytonema*): cells shrink into a series of inverted, truncated cones, sheath is composed of inserted funnel-shaped units

The gliding movement of most members of the family Oscillatoriaceae is a progressing spiral (screw). Both right and left spirals occur, and represent a genetically fixed property. The sheaths are formed in the same fashion, and may identify a gliding type on dead or fossil material (Fig. 5a, IV). Plate 1/8 illustrates sheath formation in extant *Lyngbya aestuarii* (Mertens) Liebman from an algal mat of Melum Island near Wilhelmshaven, Germany. *Heliconema australiensis* Schopf shows a ribbon-like spiral, probably representing a sheath of a narrower oscillatoriacean taxon (Schopf, 1968) (Plate 1/16).

The family Scytonemataceae includes heterocystous cyanophytes which form strong and thick sheaths. These sheaths are never spirally wound, but are frequently divergent, being composed of successively produced funnel-shaped units which appear inserted into each other and stacked (Fig. 5b). The polarized asymmetry of the sheath reveals the growth direction of the filament and indicates the structure of the cell that has produced individual funnel unit. The cells show such asymmetry only after shrinking has occurred: a conical deformation of an initially biconvex, turgescent cylindrical cell, with the smaller base turned down (Fig. 5b, I–III). A similar asymmetry of shrunken cells is expressed in the late Precambrian *Halythrix nodosa* Schopf (Schopf, 1968, Plate 77/7). Diverging *Scytonema* sheaths (Fig. 5b, IV) have been found preserved in carbonate rocks of Cretaceous age, and described as *Aeolisaccus kotori* Radoicic (De Castro, 1975).

Shrinkage does not necessarily involve all cells of a population. Different degrees of shrinkage are shown in Plate 1/14. Parts of the same trichome, usually the ends, show stronger shrinkage (in modern *Hydrocoleum* and *Microcoleus*), so that the trichomes appear tapered. Some tapering trichomes from the Bitter Spring Formation described as a possible rivulariacean *Caudiculophycus* Schopf (1968) (Plate 1/18) may well represent terminal shrinkage, since the attenuated cells do not show the elongation typical of extant Rivulariaceae. Other cells in this trichome, as well as those of *Palaeolyngbya barghoorniana* Schopf (1968) (Plate 1/17) indicate a turgescent state in the moment of fossilization and an absence of shrinkage deformation. Some of these cells, however, are separating from the trichome, a phenomenon commonly observed in modern *Lyngbya aestuarii*. Degradation of some modern cyanophytes (*Pseudanabaena*) may be "explosive": individual cells within a trichome increase in volume, and sometimes burst. Such enlarged cells found in fossil forms have been interpreted as possible heterocysts, and their occurrence linked with N_2-fixation (Licari and Cloud, 1972). However, an unambiguous Precambrian cyst (Cloud et al., 1969) or heterocyst is yet to be found.

1.5. Conclusions

Early diagenesis of modern cellular structures will significantly alter morphology and reduce the size of structures. The same should apply to fossil microorganisms. The state of preservation of microbial fossils seems to corroborate this assumption. This condition imposes a limitation on the recognition of a microbial fossil and its taxonomic affinity, and calls for additional criteria.

Of particular importance is the observed preservation potential of cyanophycean extracellular envelopes and sheaths, as compared with collapsed cellular structures. Microbial fossils characterized by small granules within spherical, cell-like outlines are now recognized as shrunken prokaryotic cells within single or multiple envelopes, rather than nuclear residues or organelles within eukaryotic cells. This does not mean that all Precambrian microorganisms have to be considered to be prokaryotic. It does mean, however, that the distinction between prokaryotes and eukaryotes in the fossil record is probably more difficult than previously assumed.

The most convincing interpretation of a Precambrian eukaryote is the TEM work on *Glenobotrydion aenigmatis* Schopf, as an eukaryotic spherical cell, with preserved intracellular membranous residues and a residual pyrenoid body (Oehler, in press). This conclusion is not unequivocal, although several subtile features come to its support: e.g. peripheral position of the granules and their electron opacity, distinctive nature of the membranous structures comprising the cell outlines vs. those within them. Both prokaryotes and eukaryotes have intracellular inclusions which may show similar electron opacity, and both have intracellular membranous structures (e.g. thylakoids). Both may have extracellular gelatinous envelopes. The exclusive feature of an eukaryotic cell, an intracellular organelle within a double membrane packing, was not possible to demonstrate in the fossils. Little is known about the degradation and early diagenesis of eukaryotic microogranisms, and our knowledge about microbial fossils from Phanerozoic times, when the existence of eukaryotes is unquestioned, is equally meager.

The reason for the preferential preservability of envelopes over the cells is not clear, although metal chelation by extracellular products (Sorensen and Conover, 1962; Walsby, 1974) may be responsible. Spotty Fe-enrichment, up to 10% of dry weight, was found in the sheaths of modern *Scytonema* with electron microprobe (type ARL-EMX-SM, M. Tarkian, personal communication). Incorporation of ferric iron around photosynthesizing cells of prokaryotes and eukaryotes alike is often observed in low Eh and low pH environments, where iron is available in a soluble form (Plate 1/5 and 6). On the other hand, heavy metals have been detected as a part of microbial fossils (Pflug, 1967). These observations suggest that a number of microbial fossils may be preserved as, or fortified by, casts and incorporations of metals.

With only limited structural features available, a confusion of microbial fossils with chemical fossils and mineralogical phenomena is possible (Pflug, 1967). Cellular organization, and particularly cell division, is one of the distinctive properties of microorganisms as opposed to non-living fossilized matter. A variety of cell division patterns have evolved within various groups of microorganisms. We have tried to illustrate that these patterns, and the resulting colony formation, provide a valid tool for recognition and identification of microbial fossils.

New interpretations and alternative explanations of microbial fossils invite a reconsideration of a number of descriptions of presumed biological Precambrian taxa. Identifications and comparisons with extant counterparts are almost invariably restricted to morphology as only one of many biological identification criteria, and conclusions about biological functions and diversity of ancient microorganisms need to be treated with caution in view of these considerations.

Acknowledgments. This work was supported by the NSF Grant GA–43391 to S. Golubić and NSF Grant GA–13821 and NASA Grant NGL–22-007-069 to E.S. Barghoorn. E.T. Degens critically read the manuscript.

1.6 Bibliography

Awramik, S.M., Golubić, S., Barghoorn, E.S.: Blue-green algal cell degradation and its implication for the fossil record. Geol. Soc. Am., Abstr. Programs **4** (7), 438 (1972).

Barghoorn, E.S., Tyler, S.A.: Microorganisms from the Gunflint chert. Science, N.Y. **147**, 563–577 (1965).

Cloud, P.E. Jr.: Evolution of ecosystems. Am. Scientist **62**, 54–66 (1974).

Cloud, P.E. Jr., Hagen, H.: Electron microscopy of the Gunflint microflora: preliminary results. Proc. Natl. Acad. Sci. US **54**, 1–6 (1965).

Cloud, P.E. Jr., Licari, G.R.: Ultrastructure and geologic relations of some two-aeon old nostocacean algae from northeastern Minnesota. Am. J. Sci. **272**, 138–149 (1972).

Cloud, P.E. Jr., Licari, G.R., Wright, L.A., Troxel, B.W.: Proterozoic eucaryotes from eastern California. Proc. Natl. Acad. Sci. US **62**, 623–631 (1969).

De Castro, P.: Osservazioni su *Aeolisaccus kotori* Radoicic, 1959 (Cyanoschizophyta). Boll. Soc. Natur. in Napoli **84**, 1–44 (1975).

Golubić, S.: The relationship between blue-green algae and carbonate deposits. In: The Biology of Blue-green Algae. Carr, N.G., Whitton, B.A. (eds.). Oxford–London–Edinburgh–Melbourne: Blackwell Scientific Publications, 1973, pp. 434–472.

Golubić, S.: Organisms that build stromatolites. In: Stromatolites. Walter, M.R. (ed.). Amsterdam: Elsevier Publ. Co., 1976a, pp. 113–129.

Golubić, S.: Taxonomy of extant stromatolite-building cyanophytes. In: Stromatolites. Walter, M.R. (ed.). Amsterdam: Elsevier Publ. Co., 1976b, pp. 127–140.

Golubić, S., Hofmann, H.J.: Comparison of Holocene and mid-Precambrian Entophysalidaceae (Cyanophyta) in stromatolitic algal mats: cell division and degradation. J. Paleontol. **50**, 1074–1082 (1976).

Hofmann, H.J.: Mid-Precambrian prokaryotes (?) from the Belcher Islands, Canada. Nature **249**, 87–88 (1974).

Hofmann, H.J.: Precambrian microflora, Belcher Islands, Canada: significance and systematics. J. Paleontol. **50**, 1040–1074 (1976).

Kaźmierczak, J.: Colonial Volvocales (Chlorophyta) from the Upper Devonian of Poland and their palaeoenvironmental significance. Acta Palaeont. Pol. **20**, 73–85 (1975).

Knoll, A.H., Barghoorn, E.S.: Precambrian eukaryotic organisms: a reassesment of the evidence. Science, N.Y. **190**, 52–54 (1975).

Knoll, A.H., Barghoorn, E.S., Golubić, S.: *Palaeopleurocapsa wopfnerii gen. et sp. nov.:* A late Precambrian alga and its modern counterpart. Proc. Natl. Acad. Sci. US **72**, 2488–2492 (1975).

Licari, G.R., Cloud, P.E. Jr.: Prokaryotic algae associated with Australian Protorozoic stromatolites. Proc. Natl. Acad. Sci. US **69**, 2500–2504 (1972).

Oehler, D.Z.: Transmission electron microscopy of organic microfossils from the late Precambrian Bitter Springs Formation of Australia: Techniques and survey of preserved ultrastructure. J. Paleontol. **50**, 90–106 (1976).

Oehler, D.Z.: Pyrenoid-like structures in late Precambrian algae from the Bitter Springs Formation of Australia. J. Paleontol., in press.

Pflug, H.D.: Organische Reste aus der Belt-Serie (Algonkium) von Nordamerika. Paläontol. Z. **39**, 10–25 (1965).

Pflug, H.D.: Einige Reste niederer Pflanzen aus dem Algonkium. Palaeontographica **117** B, 59–74 (1966).

Pflug, H.D.: Structured organic remains from the Fig Tree Series (Precambrian) of the Barberton Mountain Land (South Africa). Rev. Palaeobot. Palynol. **5**, 9–29 (1967).

Schopf, J.W.: Microflora of the Bitter Springs Formation, Late Precambrian, Central Australia. J. Paleontol. **42**, 651–688 (1968).

Schopf, J.W.: Electron microscopy of organically preserved Precambrian microorganisms. J. Paleontol. **44**, 1–6 (1970a).

Schopf, J.W.: Precambrian micro-organisms and evolutionary events prior to the origin of vascular plants. Biol. Rev. **45**, 319–352 (1970b).

Schopf, J.W., Blacic, J.M.: New microorganisms from the Bitter Springs Formation (Late Precambrian) of the North-Central Amadeus Basin, Australia. J. Paleontol. **45**, 925–960 (1971).

Schopf, J.W., Oehler, D.Z., Horodyski, J., Kvenvolden, K.A.: Biogenicity and significance of the oldest-known stromatolites. J. Paleontol. **45**, 477–485 (1971).

Sorensen, L.O., Conover, J.T.: Algal mat communities of *Lyngbya confervoides* (C. Agardh) Gomont. Inst. Mar. Sci. Univ. Texas, Publ. **8**, 237–249 (1962).
Thompson, D.W.: On Growth and Form. 793 pp. Cambridge: Univ. Press 1917.
Walsby, A.E.: The extracellular products of *Anabaena cylindrica* Lem. I. Isolation of a macromolecular pigment-peptide complex and other components. Brit. Phycol. J. **9**, 371–381 (1974).

2 Evolving Concepts on the Nature and the Ecological Significance of Stromatolites

Cl. MONTY, Liège, Belgium

2.1 Pre-Walcott Times

2.1.1 Generalities

The German authors used to call "Mumien" ovoid to cylindrical, finely laminated, calcareous nodules resulting from periodical algal encrustations around shells, pieces of wood, etc. (Pia, 1933; Rutte, 1953). These deposits are more precisely called "Schneckelistein" (Schmidle, 1910), or "Schnegglisand" (Baumann, 1911), when the nucleus is a snail. These freshwater structures take us back to one of the first descriptions of stromatolite-related objects: in 1649, indeed, laminated calcareous nodules from what is now known as the Eocene 'Calcaire de Castre' Formation (France) were studied by Borel and called "Priapolithes". As may be guessed from the name, Borel provided one of the best descriptions of algal structures ever found in the literature. More than a century later, the name priapolithe persists and is defined as follows (Bernard's Dictionary of Fossils, 1763, p. 486): "Priapolithe: lapides membrum virile mentientes (Petri Borelli observationes de priapolithis Cent. III observ. LXXXVI, p. 260)". These freshwater Cenozoic deposits became widely known in the French and German literature of the XIXth and XXth centuries. The term "priapolithe" was, however, dropped and does not seem to appear in the geological literature of the early 20th century, where it is replaced by words such as "nodules", "balls", "Knollen", "Kugeln", "biscuits", "boulets", etc. From the very beginning, they have generally been treated as sedimentological objects, and although they belong to Pia's Oncolithi (1927, p. 37), i.e. the free-lying counterpart of Stromatolithi, none of these Cenozoic nodules ever receive any taxonomical name as had been given to the older oncolites such as *Pycnostroma, Osagia,* etc.

Laminated calcareous objects, that were to be called "stromatolites" later, were apparently first reported in 1825 by Steel from Upper Cambrian rocks outcropping near Saratoga Springs, New York. However, they remained obscure geological objects until 1883 when James Hall reconsidered them in detail and ascertained their biological origin. He described and reillustrated these reef builders which be called *Cryptozoon proliferum.* In Europe, part of Kalkowsky's Mesozoic material was first reported by Naumann (1862) as concentrical shell-like jointed structures, known in the Harz region under the name of "Napfstein" (rock in form of cup or bowl as, by alteration, these rocks would liberate the successivly stacked hemispheroids which were used as dishes).

Increased interest and recovery of similar structures generalized the application of a Linnean nomenclature necessary to name the various forms retrieved from Cambrian and Precambrian rocks. The example of Hall was soon followed by Matthew, who named the first Precambrian stromatolites (1890a, b) previously reported by Bailey and Matthew

(1872); he first called them *Eozoon*, then *Archeozoon acadiense* n. gen. n. sp. (1890b, p. 67).

The origin of these "fossils" remained, however, highly conjectural: they were "sea plants" for Hall (1847, p. 5; 1883), as well as for Kalkowsky (1908, p. 124) and some others who were convinced that they could be related to mosses, myxomycetes, or even red algae. On the other hand, they were protozoans for Matthew (1890a, b) and Gürich (1906), whereas Steinmann (1911), Seely (1906, p. 171), and Grabau (1909, vol. I, p. 46) related them to the stromatoporoids. Finally, many geologists claimed that stromatolites were inorganic concretions: Reis (1908) considered them as sorts of mega-ooids; Linck (1909) compared them with the concretions found in the caves of Karlsbad, whereas Voss (1928) interpreted them as marine sinters formed in deeper waters than the ooids.

2.1.2 Gürich and the Spongiostromids

In 1906, Gürich published a very detailed petrographical study of laminated encrusting organisms that were fairly common in the Belgian Middle and Upper Visean carbonates (Monty, 1963). These fossils had been identified as stromatoporoids by Belgian geologists, an interpretation that Gürich refutes by showing that they can neither be hydrozoans, nor sponges, nor algae, but protozoans. As these fossils have a unique layered fabric, made of alternating looser and denser laminae, which frequently look "spongious" in thin sections, he erected a new order of protozoans: the Spongiostromaceae with one single family: the Spongiostromidae. He introduces them as follows (Gürich, 1906, p. 7) ". . . the unquestionable relationships I have discovered between the various thin sections show that we deal here with a family of fossil encrusting organisms for which I have chosen the name Spongiostromidae, after the typical genus and the prominent particularity of all the forms of this group". He was probably the first to make a systematic study of stromatolitic microstructures, to classify them – granular, vermicular, nodular (that is pelletal), vesicular, spongy, bacilliferous, sinuous, etc. – and to use them for characterizing five genera and 14 species. These were launched in literature by Rothpletz (1913), then by Pia (1927), and later by Johnson (1961). In spite of this publicity, the word spongiostromid did not receive the favor of geologists. The reason probably lies in the fact that Gürich has neither described nor illustrated the growth form and field relations of his spongiostromids, but limited his excellent illustrations to the microstructures. These sort of mythic fossils did not appeal to field geologists, who evidently felt safer when referring to Kalkowsky's stromatolites, as he had clearly described and illustrated their overall features and variety or forms.

2.1.3 What Kalkowsky Really Did Say

Kalkowsky is repeatedly quoted in recent literature as the initiator of the word stromatolite, but Hofmann (1973, p. 341) was perhaps the only modern author to call our attention back to the fact that Kalkowsky had, in fact, defined two words: *stromatolites* and *stromatoids*. However, as will be shown below, Hofmann misunderstood Kalkowsky when he writes (Hofmann, 1973) that "He, however, applied the term [stromatolite] to the whole bed (bioherm, biostrome) using the word 'stromatoid' for individual forms, which the current literature refers to as stromatolites, . . .". Although his German is sometimes difficult, I believe that Kalkowsky deserves more than being referred to as the man who coined a name. That is why I think a short review of his 1908 paper may be of interest.

One of Kalkowsky's major contributions, besides deciphering their microorganic origin, was to unify a variety of deposits, ranging from flat-lying and stratiform banks to discrete columns and variously shaped laminated objects, under the name of stromatolite (Stromatolith); he separated them, microstructurally and genetically, from the ooids-pisolites type which were found in very close association in the field.

What Are Kalkowsky's Stromatolites? The term is introduced as follows (p. 68, § 3): "The new name *Stromatolith* is applied to limestone masses with particular structure and particular constructional pattern, which occur together with the Rogenstein in the Buntsandstein of Northern Germany". (This and following translations have been translated into modern English and checked by Flügel.) He adds (p. 69): "the name stromatolite designs limestone masses showing a fine, more or less planar, layered structure in contrast to concentric structures of the oolite grains". This "prominent layered structure [Lagenstruktur] [. . .] appears clearly and distinctly on weathered surfaces particularly . . . " (p. 102, § 67). "Whereas the most striking feature of oolites is the form of the [constitutive] ooids, stromatolites stand apart by their larger dimensions. If oolites are always made of ooids and a cement which binds them in a rock, allothigenous constituents, ooids and spar do not play any significant role in the build up of stromatolites (;) the essential components of the true stromatolites and of stromatolite-rocks are the thin, more or less planar, layers of carbonate with a particular structure" (p. 101, § 65). Accordingly, Kalkowsky introduces the problem by describing what we would call *stratifera*, or cryptalgal laminite to-day (ebene Lagenstruktur). However, on page 11, § 84, under the title "Free stromatolitic heads" (Stock) he broadens the scope: "These [the stromatolites] can occur not only as simple regular and laterally continuous banks, but also as better individualized masses, perfectly delimited heads, which tower up the enclosing sediments with, at least, their upper half free. They must simply be called 'free stromatolite heads'. Their growth forms can be massive, branched (astig), columnar (röhrig), cauliflower-like etc. . . . In many cases, stromatolites start from protuberances in the underlying oolite; the upper surface of these bulges is first encrusted by stromatolitic layers which, as growth proceeds, differenciate the various reported morphologies"; As far as the growth pattern of stromatolites is concerned, he writes (p. 117, § 86) "The peculiar structure of the individual thicker layers of the head shows that a stromatolite has a periodic growth, i.e., that it has 'periods of sprout'. For instance, the short branches in a layer stop at the same time and die, but enough germs are left to allow a new layer to start growing again with eventually another constructional pattern. This is not only true for the large individualized structures but also for the small nodules, as well as for the flat-lying stromatolites".

The Stromatoid. Kalkowsky introduces the notion of stromatoid as follows (p. 101, § 65): "Stromatolites generally lack the individualized habit yielded by the organic builders of ooids, habit which is characterized by a proper form limited on all sides (;) nevertheless these thin carbonate layers [laminae] with their proper structure can be introduced here as *stromatoid*, because, in a similar way, they owe their origin to an organic builder which seems to be related to the ooid builders". Accordingly, the word stromatoid refers to the component laminae of stromatolites, not to the stromatolitic head as understood by Hofmann (1973). Let us try to explain Kalkowsky's logic. An ooid is a closed architecture made of concentrical spherical layers or shells. A stromatoid is an open architecture made of piling up of somewhat flat-lying or hemispherical layers. The concept is nicely illustrated by the upper portion of his Fig. 1, Plate 6, which shows ooids growing on a stromatoid. Accordingly, an accumulation of ooids will yield an oolite, whereas a piling of stromatoids will yield a stromatolite. So the term stromatolite refers to the final resulting structure, which is a bed in case of stratiform deposit, or variously shaped heads in case of discrete bodies, whereas *stromatoid* refers to what constitutes it, i.e. sets of laminae. I would believe that his stromatoid is nothing else that what we call a laminated algal mat. That is why Kalkowsky does not put dimensional limits between both structures (how many laminae are required for passing from a stromatoid to a stromatolite) as is shown by the following references: delimited stromatolite banks can reach thicknesses of 1.20 m and "be composed either of pure layers of stromatoid or of thicker or thinner layers of stromatoids interspersed with numerous interstratifications of thin layers, or with very limited pockets, of ooids and variable amounts of detritus" (p. 113–114, § 82). Elsewhere (p. 112, § 79) he reports oolitic bulges (ripples?), which "are overgrown by a millimeter-sized layer of stromatolite made of layers of stromatoid only". Describing the features of stromatoids, what we would now call microstructures, he says (p. 103, § 67): "the layered structure of stromatoids (and not only the overall layered structure of the stromatolite which may yet result from other factors) can be recognized on polished surfaces and in thin sections [. . .]". This remark is quite correct, as the laminated fabric of stromatolite can first result from an alternation of algal mat phases (stromatoid) and detrital laminae, whereas the algal mats show their own – first-order – lamination. "In its elementary layers, the stromatoid is built up by fine threads (or filaments or fibers – in the 1908 Muret and Sanders Enzyklopädisches Wörterbuch, the meaning filament is given first; fiber appears in a secondary acception) which frequently show a poor fan-like arrangement as if there were a tendency for a radial growth pattern. The 'fibrosity' of the stromatoid is often less distinct [than that of ooids] and [appears] very vague when using higher magnification" (p. 103, § 68). "Decalcification of a thin section of stromatoid, leaves but fine clayey material and scattered sand grains which replicate – although in a poor fashion – the layered fabric. Accordingly, in the whole stromatolite, the stromatoid, in a similar way to ooids, looks relatively poor in allothigenous components, which have been incorporated in very small quantities during growth".

Origin of Stromatolites. p. 124, § 95: "Among stromatolites, these which show a marked planar development could be tentatively considered as simple laminated and thinly bedded limestone provided one remains blind and discards their fine structural features. The individualized bodies, the heads, the reefs, the nodules, the wrinkled knobs are unique structures that nobody could explain in terms of plain inorganic constructions. Nobody will want to accept [the proposal that] in the case of branching stromatolites, the calcium carbonate deposited itself mechanically on the tops of the branches only, while all the detritus, all the ooids would carefully accumulate out of the way, in the interstices [between the branches]". He then notes the similarity between many stromatolitic growth forms and that of corals and sponges. Yet, the growth pattern of stromatolites, the superposition of convex layers of stromatoid, the tendency to grow divergent columns, etc. . . . are many evidences which fit well with organisms in search of light and food. Having eliminated the possibilities of coral or spongal affinities, he concludes (p. 125): "we must then admit that they have been lower vegetal organisms which initiated the precipitation of calcium carbonate. In these conditions, then, we may, as far as the conditions of formation are concerned, think of the frequent occurrence, among the plants, of forms of vegetation represented by the crust-forming genera and species which are found in the phanerogames, as well as in mosses and myxomycetes. I cannot say here, how the plant builders of the stromatoid were constituted, any more than I know the way to classify the ooids builders in the plant kingdom".

What preceeds shows that by pure field and paleontological methods, Kalkowsky, as a fine observer, had avoided all the traps and by 1908 was hinting at the very nature of stromatolites. From the study of thin sections, he inferred the fundamental role of the stromatoid – our algal mat – in the construction of stromatolites, and this concurrently with the environmental sedimentological factors. In fact, "stromatoid" is much better than "algal mat", as it synthesizes a concept, a structure, a behavior, a community, in a nonspecific way, for we all know that an algal mat is not always (entirely) algal. It is thus right that geological literature has retained "stromatolite" and discarded "spongiostromid", but Kalkowsky's fame should not be restricted to the fact that he coined the name. By the way, the Group Kalkowskyphyton is still open!

2.2. Walcott and the Era of Freshwater Stromatolites

Very little change occurred in Europe after Kalkowsky's contribution; in 1913, Rothpletz kept on interpreting the spongiostromids as hydrozoans and considered the possibility of symbiosis between these spongiostromids and the tabulate corals, or various calcareous algae (*Solenopora, Hedströmia*, etc. . .) which they embodied during growth. Furthermore, Kalkowsky's paper was severely criticized by people who thought that stromatolites were nothing else than inorganic concretions (Reis, 1908; Linck, 1909; etc. . .).

Beside Rothpletz (Lower Paleozoic), Gürich (Upper Paleozoic) and Kalkowsky (Lower Mesozoic), the bulk of the European literature concerned with stromatolite-related objects seems to concentrate on Cenozoic freshwater nodules and tufas. In spite of a very abundant literature on recent freshwater analogs (Bornemann, 1887; Barber, 1890; Linden, 1890; Thiselton-Dyer, 1891; Murray, 1895; Schröter and Kirchner, 1896; Kirchner, 1898; Forel, 1904; Früh and Schröter, 1904; Le Roux, 1908; Schmidle, 1910; Baumann, 1911, 1912; Wesenberg-Lund, 1914, etc. . .), the Cenozoic laminated concretions are still interpreted as inorganic structures by most geologists (Burger, 1911; Berz, 1915; Sommermeier, 1915, etc. . .), although an algal origin is proposed at times (Berz, 1915, p. 341; Sommermeier, 1914a, b, 1915, etc. . .); no direct relationships with Kalkowsky's stromatolites are, however, elaborated.

In America, on the contrary, concepts began to evolve rather rapidly. By that time, many calcareous concretions showing a concentric lamination had also been described from lakes (Murray, 1895; Penhallow, 1896; Clarke, 1900; Davis, 1900; Blatchley and Ashley, 1901; Hale et al., 1903; Powell, 1903), rivers (Tilden, 1910; Roddy, 1915), hot springs (Cohn, 1862; Weed, 1889), or salt lakes (Jones, 1914; Russel, 1893), but no serious connections had yet been made between these botanical investigations and the paleontology of stromatolites. This will be the task of Walcott. Three months before him, however, Wieland (April 1914), in a review of the problem, concludes that "it is now believed that at least all those forms once included amongst the stromatoporoids, which lack a tubule system with corresponding surface pustulations and are in greater part of characteristically laminate linear, or much branched *Lithothamnium* form, are all primitive algae which form the abundant record of a far more luxuriant seaweed growth than has hitherto been understood to have characterized the Paleozoic" (p. 246). He adds that the Precambrian, Cambrian, and Ordovician times can be characterized by the "reign of algae". But, what kind of algae were they?

Walcott (1914) was the first to compare the overall features of the proterozoic stromatolites from the Belt Mountains of Montana with the freshwater calcareous concretions that had been described from streams, lakes, and ponds in various parts of the States. He concluded that the *Collenia, Cryptozoon* etc. were biological constructions resulting from the life activity of lime-depositing blue-green algae. Walcott's uniformitarianism led him to the statement that all proterozoic stromatolites had formed in freshwater lakes or basins, which fitted his former ideas that Precambrian sedimentary rocks were entirely non-marine.

Walcott's views are soon supported by Roddy (1915, p. 255), who stressed the similitude between the structure of his recent freshwater algal concretions and that of *Cryptozoon.*

Another important step was realized in 1918, when Moore described silicified concretionary limestones associated with the now classical Iron Formation of the Belchers Island: without any hesitation, he concluded that the laminated heads resulted from the activity of blue-green algae; he almost proved it, as in haematitic concretions, he finds and illustrates (Fig. 14, p. 427) cells and filaments that Hill assimilates to coccoid and filamentous cyanophytes. This is the first reported Precambrian microflora, 50 years before Tyler and Barghoorn (1954) and 60 years before Hofmann's paper on the Belchers (1974, 1975).

At this time, Kalkowsky's contribution had not yet reached the States, as in 1919 Twenhofel introduced the term "Coenoplase" as a common name for describing these algal laminated structures. He recognized two main types of coenoplases: (1) the columnar dome-shaped coenoplases (such as *Collenia, Cryptozoon*) and the nodular, biscuit-like, coenoplases (such as *Ottonosia*). As far as the growth environment of precambrian stromatolites was concerned, Twenhofel disagreed with Walcott, but was not convincing when merely stating that "The algae do not even prove that the waters were salty as lime-precipitating algae thrive in both fresh and salt waters but the tremendous growth of the Kona algae suggests marine conditions" (p. 345). Twenhofel's doubts and the eventual impact of descriptions of blue-green algal encrustations from *salt*water lakes (Russel, 1893; Andrussow, 1897; Stahl, 1905; Jones, 1914, 1925, etc. . .) will be hindered by the conclusions of Bradley's superb study of Eocene River Formation (1929): the imposing component algal reefs have been built in freshwater lakes.

By that time, despite some geologists who still doubted the organic origin of stromatolite (Holtedahl, 1919, 1921) or even rejected it (Seward, 1931; Voss, 1928), it was firmly established that the *Cryptozoon* and the like were biological constructions resulting from the in situ precipitation of calcium carbonate in blue-green algal mat communities.

In Europe, this theory was strongly supported by Pia (1928a, 1933), who repeatedly outlined the similarities between the recent non-marine "Algenknollen", tufa, etc. . . and the fossil *Pycnostroma, Archeozoon, Gymnosolen*, etc. . . (Pia, 1933, pp. 143, 171, 174, 178, 185). The cenotic complexity of stromatolite is also well known; for instance, Pia (1933) recalls that in algal nodules from the Bodensee (which, he says, compare very well with *Pycnostroma*) Schmidle (1910) had found 12 species in seven genera of blue-green and green algae, several species of diatoms, and one fungus. In spite of such a cenotic complexity, Pia (1933, p. 173) really feels the uniqueness of many stromatolites when he writes "It is just wonderful, the way such a colony behaves in many aspects as one single organism, characterized by a given typical form, a regulated architecture and a proper growth pattern". On the other hand, Fenton and Fenton (1933) were trying to simplify the taxonomy by describing the morphological plasticity of Precambrian stromatolites. As will be discussed elsewhere, both were right: some stenobiontic stromatolites have a unique and proper growth form, while some eurybiontic ones may exhibit different morpholgies.

In late twenties – early thirties, the postulated non-marine origin of stromatolites became more and more questioned by stratal evidences as, in Paleozoic formations, fossil stromatolites were not infrequently found in association with marine organisms, whereas traces of saline environments (salt casts, etc. . .) were reported from otherwise unfossiliferous stromatolitic beds (see discussion in Fenton and Fenton, 1931, p. 679–680). Times were ready for a change.

It is interesting, at this stage, to point out that, although Pia (1927) had formalized the term Stromatolithi to group all the structures called *Wedia, Collenia, Cryptozoon, Archeozoon, Gymnosolem,* etc. . . , Kalkowsky's term will painfully infiltrate into the geological literature of the 1930s; indeed it was adopted neither by Bradley, although he refers to Kalkowsky (1929, p. 207, 214), nor by the Fentons (1931), Johnson (1936, 1937) nor Goldring (1938) who keep on speaking of "algal limestones", algal reefs, etc. . . ; it occurs, however, in Høeg (1929), Young (1932, 1934), and Maslov (1937); Cloud (1942) will probably make the final decision by writing: "Although organic in origin these structures are themselves neither organisms nor parts of organisms. Nevertheless they must be called something and the term *stromatolite* . . . seems most apropriate for the simply laminated forms" (p. 363).

2.3. Black and the Era of Tidal Flat Stromatolites

The publication of Black's preliminary report on the algal sediments of Andros Island, Bahamas (1933), can be termed a geological event. First of all, Black reported the formation of blue-green algal laminated structures in intertidal marine and brackish waters. As stated above, the recovery of recent analogues of fossil stromatolites in association with marine conditions was badly needed. This is illustrated by Young's relief (1934, p. 158). "In a previous paper in which the small branched type of stromatolite was briefly referred to, I adopted the algal theory of their origin. The circumstance, however, that no modern instance could at the time be pointed to in which evenly bedded algal deposits were being formed in marine surroundings on a substratum of loose calcareous sediment made this explanation highly speculative... By Mr. Maurice Black's discoveries on Andros Island, the position is now completely changed . . . ".

Secondly, Black put the accent on an alternative mode of formation of stromatolites: he did not notice any in-situ precipitation of carbonate in the algal mats, but claimed that all the particles, from mud to silt and sand were detrital in origin; these particles, he said, are washed by the tides and the storms over the algal mats which trap and bind them into laminated algal structures. Finally, Black described four types of mats forming in different environments and tried to show the impact of the overall environmental conditions on the shaping and eventually the microstructure of the algal deposits.

In the 1950s, Ginsburg took over and started a far-reaching investigation on the algal mats covering the intertidal flats of Florida and the Bahamas (Ginsburg et al., 1954; Ginsburg, 1955; Ginsburg and Lowenstam, 1958). Elaborating on Black's attempts, he tried to relate the properties and the microstructures of the mats to the physical environmental parameters (sediment influx, degree of exposure, etc. . .). His studies were soon extended to a variety of tidal flat settings from Europe to the Persian Gulf and to Australia (Hommeril and Rioult, 1965; Kinsman, 1964; Kendall and Skipwith, 1966, 1968, 1969; Davies, 1970, etc. . .).

Although Ginsburg himself had first reported the occurrence of subtidal stromatolites (1960), the intertidal myth progressively settled, and culminated when Logan (1961) described large cemented algal heads crowding intertidal platforms and headlands in Shark Bay, Australia.

The accumulation of the resulting literature, focussing on this sole blue-green algal setting, while ignoring all the others, led to the definitive crystallization of a threefold dogma, as much extremist as Walcott's theory, of which it took all the counterparts: (1) stromatolites form on intertidal flats or eventually on very shallow banks (2) by trapping and binding detrital particles, whereas (3) their overall morphology is ruled by environmental factors. The algal control is nil, or nearly so. What became characteristic of the Western geological literature of the 1960s – early 1970s is the physical, mechanistic interpretation of the stromatolites features and dynamics; the task of blue-green algae – whoever they are – is to bind detrital grains, whereas the overall characteristics of the resulting stromatolites, from their microstructure, to their linkage and growth form, were accounted for in terms of wave splashes, scouring effects, tidal movements, sediment influx, dessication cracks, etc. . . Furthermore, the processes of in situ precipitation was definitely discarded, as can be seen in all of the definitions of stromatolites given during that decade; they invariably ran like "laminated structures composed of particulate sand, silt and clay size sediment which have been formed by the trapping and binding of detrital sediment particles by an algal film" (Logan et al., 1962). A convincing illustration of the pollution brought about by the dogma in geological studies is found in Davis' study (1966, pp. 917, 918, 921, point 5), among others.

Such biased views were at the opposite extreme of those of the Russian school, which appear perhaps more realistic: Maslov was indeed less restrictive when he considered the processes of mineralization of stromatolites: "In the process of deposition of 'mixed carbonate' or 'stromatolitic constructions', intervene chemical, biochemical, and, in particular cases, physiological processes of carbonate deposition, together with the intervention of mechanical sedimentation of various amounts of clastic material (1961, p. 84, author's translation). Furthermore, he insisted on the algal, rather than purely environmental, control of stromatolitic features: "It appears that the diversity of carbonate-accumulating algal cenoses determines various morphological and structural types of stromatolites; accordingly a considerable diversification of these concretions is observed" (1953, p. 112, author's translation).

2.4 The Reaction

One of the most straightforward reactions against the three points of the dogma can be read in Serebryakov and Semikhatov (1974, English). As will be shown below, they are the spokemen of an ever-increasing number of geologists who, beside the Russian School, try to stick to stratal and paleontological evidence before transferring Shark Bay or the Bahamas, as such, to Precambrian or Lower Paleozoic stromatolitic fields. Accordingly, a sound actualism appears to be, hopefully, in view.

From the beginning of my research on recent Bahamian "stromatolites", I have always been impressed by the strong impact of life processes on the internal as well as superficial features of algal mats and heads; these appeared to me as unique microcosms, unique biological entities, at every level of which something is always happening. It soon became clear that the composition, the structure and the behavior of the algal cenoses were determinant in the development of particular microstructures and, in some cases, of particular growth forms. These observations forced me to conclude that "stromatolites are *algally controled* structures built by *coordinated societies* of primitive organisms (blue-green algae and bacteria) . . . " (Monty, 1965a, p. 373; 1967, p. 98) incomplete (as this statement may be), and to oppose the threefold dogma reported here above (Monty, 1965a, p. 317–380, a discussion which is to be found scattered through later papers).

2.4.1 Reaction Against the Stated Processes of Mineralization of Stromatolites

Many paleontologists soon disagreed with the idea that the "trap sediment" mechanism was the only, or even the main, mode of formation of stromatolites (Maslov, 1960, 1961; Donaldson, 1963; Korolyuk, 1963; Krylov, 1963; Komar, 1966, etc. . . as well as most of the authors working on Cenozoic stromatolites); they were convinced that many ancient stromatolites had formed by way of physiological, or more generally biological, precipitation. In fact, Black's types C and D stromatolites, which contributed so much to the evolution of concepts, were shown to result *not* from the trapping of detrital muds as postulated by him, but from processes of in situ precipitation (Monty, 1965a, 1967). Further observations have shown that this was applicable to all of the freshwater algal mats of Andros Island, and the Everglades, Florida (Gleason, 1972; Monty, 1972, 1975; Gleason and Spackman, 1974; Monty and Hardie, 1976), with the exception of Western Andros Id. where detrital pelletal layers are periodically deposited by storms over the normally autocalcified mats (Monty, 1975; Monty and Hardie, 1976).

It also appeared that if the bulk of the carbonate content of the modern freshwater cryptalgal deposits is precipitated in situ, as known since the beginning of the century, the trapping and binding of detrital particles takes over and becomes the leading process as we move toward marine waters; here, it seems that no, or little, precipitation – other than beach-rock and abiogenic submarine cementation processes – is observed in the mats (Monty, 1972; Gebelein, 1974). This is the present-day situation resulting from a long ecological evolution of stromatolitic communities (Monty, 1973b); it is *by no means* a model for interpreting pre-Cenozoic stromatolites, when blue-green algal communities had still the faculty of precipitating calcium carbonate in sea water. It just shows that both processes may have been equally important in ancient stromatolites: some of them, like those reported by Szulczewski (1967, Jurassic), Hoffman (1967, Aphebian) or Davis (1968, Ordovician) definitely formed by trapping and binding detrital particles (ooids, shell fragments, quartz grains), whereas many others, like the cases reported by Korolyuk (1963), Komar (1966), Serebryakov (1968), Semikhakov et al. (1970), Krylov et al. (1971), for instance definitely formed by in situ precipitation of carbonate.

Hofmann (1969, p. 6) was among the first Anglo-Saxon author to reintroduce, in a formal definition of stromatolites, the eventual intervention of both processes, amending in that way the second point of the dogma: "... The sedimentary material is accumulated by trapping or agglutination of particles from suspension on the organic film, or by direct or indirect precipitation resulting from the metabolic activity of elements of the microbiota". This finally put an end to the old disagreement that had separated the partisans of Walcott's views and those of Black's. To be complete, it should be added that the mineralization of stromatolites can yet result from the cristallization of impregneting gels of silica or yet from the precipitation of various iron oxydes, as will be shown elsewhere.

2.4.2 Reaction Against the Overgeneralization of Intertidal Stromatolites

As stated above, that part of the dogma which established the intertidal zone as the typical setting for stromatolitic growth, really polluted the geological literature for years; the prejudice can be measured by Aitken's embarassment in his paper on thrombolites (1967, p. 1173–1174): "The hazard in naming structures stromatolites (although they clearly intergrade with digitate algal stromatolites), is that such a name leads to an interpretation of intertidal origin, which, as shown above, is not necessarily justified".

This shows how deep the intertidal doctrine had became rooted in the geologists' minds. Its almightiness eventually annihilated any personal geological logic and reasoning, as shown by the way in which Davis (1966), after such a severe brainwashing, is led to define, classify, and interpret an Ordovician stromatolitic suite. It also led to the delineation of incredibly large intertidal zones, reaching widths of 50 miles at least in Davis (1966, Fig. 5) and up to 600–1200 km in Lochman-Balk (1970), where the intertidal significance of stromatolite "reefs" is taken for granted. The word "intertidal" was so much distorted that it lost all its meaning, as it was finally used in the sense of "supratidal to shallow subtidal", although these are three particular ecological belts characterized by specific environmental conditions, organic communities, sedimentary structures, and diagenetic processes.

It is astonishing to see how strongly the myth grew and how long it lasted in spite of the available opposing evidence; the geological literature clearly showed that stromatolites had not been confined to the one and only intertidal setting; on the contrary, even if, during the Phanerozoic, stromatolites were progressively displaced toward perimarine environments, as discussed in Monty (1973b), it is clear that, during Paleozoic and Mesozoic times, they also grew in various subtidal marine environments, such as lagoons, reefs, shelves where they could be associated with rich marine fauna and eventually with other encrusting communities (Rothpletz, 1913; Wolfenden, 1958; Smith and Anderson, 1958; Hadding, 1960; Szulczewsky, 1967, 1968; Radwanski, 1968; Playford and Cockbain, 1969; Maanten, 1971; Tsien, 1971; Elloy, 1972; Golonka and Rajchel, 1972; Marcinowski and Szulczewski, 1972, etc. . .). In the absence of fauna, stratigraphic and facietal relationships allowed the recognition of subtidal Precambrian stromatolites as well (Trompette, 1969; Walter, 1970, 1972; Serebryakov, 1971; Awramik, 1971; Bertrand-Sarfati, 1972; Larsson, 1973; Hoffman, 1974; Truswell and Erikson, 1973, 1975, etc. . .). In fact, the evolution of stromatolitic communities, as traced in Monty (1973b), allows us to think that the Precambrian was the time of the maximal colonization of all the subtidal environments, as all of them were free of competitive animal communities.

The Late Mesozoic and the complete Cenozoic record shows, almost exclusively, an intense development of fluviolacustrine freshwater, or more generally non-marine, stromatolitic structures and bioherms (Bradley, 1929; Johnson, 1936, 1937; Rutte, 1953a, b, 1954; Colom, 1961; Freytet, 1965; Freytat and Plaziat, 1965, 1972; Bertrand-Sarfati et al., 1966; Mouline, 1966; Gigot and Rioult, 1972; Ott and Volkheimer, 1972; Stapf, 1973; Donsimoni, 1975). In fact, freshwater stromatolites have also been reported in North Africa Paleozoic limestone (Bertrand-Sarfati, 1976).

Finally, the present day occurrence, beside the well-known prevalence of stromatolitic and various cryptalgal structures in non-marine settings, and the development of intertidal leathery algal mats and laminated sediments, have also provided several instances of shallow-water subtidal biscuits and heads (Florida: Ginsburg, 1960; The Bahamas: Monty, 1965, 1967; Bermuda: Gebelein, 1969; The Great Barrier Reefs: Monty personal observations, 1969; Australia: Logan et al., 1974; Playford, 1976; etc. . .).

In front of such compelling evidence, one can easily understand the exasperation of Achauer and Johnson (1969, p. 1479): "Obviously, stromatolites can not be regarded as indicating exclusively intertidal areas of deposition"; or yet Serebryakov and Semikhatov's disapointment (1974, p. 560): "The arguments for the subtidal origin of some stromatolites published in some recent articles (. . .) have not yet caused a revision of the viewpoint prevailing in the Western literature on the intertidal nature of the vast majority of these structures".

2.4.3 The Compromise: If not Always Intertidal, Stromatolites Are Very Shallow-Water Forms at the Most

The intertidal myth, joined to a poor knowledge of the physiology of blue-green algae, had another repercussion: when an intertidal origin was strongly opposed by stratal evidence, stromatolites were automatically said to be very shallow-water forms, as if people were afraid to go too far away from the tidal line. This was not only apparently justified by the present-day subtidal occurrences reported here above, but also by various subjective factors which unconsciously enhanced such an interpretation; one of them was the association of the word "stromatolite" with the word "reef"; as everybody knows that a "coral reef" is a very shallow-water to intertidal structure, why should a stromatolitic reef not be so too? Furthermore, as stromatolites are built by photosynthetic algae, it was easy to argue that these needed the full illumination of the surficial waters to grow profusely. These tendences are illustrated in the following statement (Hutchinson's written communication in Cloud, 1942, p. 371): "The building of an algal reef, however, presumably would require considerably more light than is available at the lower limit of photosynthesis; and in view of the probability that a considerable amount of suspended calcium carbonate now represented by the matrix, would have been present in the water, a maximum depth of 10 m might well be an overestimate". On the basis of this information, Cloud (1942) concludes: "It is undoubtedly conservative to say that stromatolite-forming algae will not ordinarily flourish below a depth of 30 m in marine waters". Althoug Cloud was, at the time, much more realistic than the authors of most of the recent studies, where maximal depths of a few meters to 10 m are generally admitted, such theoretical assumptions could not resist the verdict of nature.

One more time indeed, stratal evidence opposed the intertidal to very shallow subtidal *limitation* of stromatolites. In 1969, Playford and Cockbain described columnar stromatolites that grew on the slope of Devonian reefs in *at least* 45 m of water as shown by paleo-topographical data. The next year, Jenkyns, studying manganese nodules associated with stromatolites on Jurassic seamounts of Sicily, conservatively optimized between stratal evidence and algal requirements by noting (1970, p. 769): "The topographic highs probably reached to within a hundred meters or so of the surface, since the mineral concretions are often associated with algal stromatolites and thus a depth limit must be imposed by the extent of the photic zone". Finally, Hoffman (1974) has nicely documented a series of stromatolitic Precambrian deposits ranging from platform-shoaled facies to basin floor facies, where stromatolites are associated with graywacke-turbidite sequences.

2.4.4 From Shallow to "Deep" Water Stromatolites

So facts are available to show that fossil subtidal stromatolites may have ranged from very shallow to fairly deep waters. Our knowledge of the metabolisms of recent blue-green algae supports these paleontological observations and shows that to restrict stromatolites to very shallow waters for photosynthetic purposes is nonsense. Let us first make a revealing comparison: fossil calcareous algae have always been considered as indicative of shallow well-oxygenated (lagoonal) waters; the richer the accumulation of skeletal debris, the shallower the waters, as prolific growth of dasyclad or codiacean calcarous algae supposedly required bright illumination. In fact, to-day's calcareous green algae such as *Halimeda* can profusely grow very large thalli which cover reefs drop off at depths of 100 m (personal observation by scuba diving, off Andros Id., 1970), or even more (Wray, personal communication, 1975, observation by submersible). What about the blue-green algae then, which can actively grow in twilight to no light conditions, as discussed in Monty (1971)? The present-day depth distribution of benthic blue-green algae is poorly known and relies on the generalization of observations by deep scuba diving and submersible; up to now species of *Lyngbya* have been traced to depth of 30 m (Feldman, 1959; Donze, 1968), whereas rich growths of Oscillatoriaceae, like various species of *Schizothrix*, including *S. calcicola, Microcoleus*, etc. . . have been found at depths of 20–25 m off South Eastern Florida and the Bahamas. In fresh water lakes, Forel has dredged blue-green algal calcareous crusts from depths of 60 m in the Genfer See (in Pia, 1933, p. 146); transposing this last occurrence in the sea could imply that marine crusts might form in waters deeper than at least 100 m. More is known about the depth distribution of planktonic blue-greens, as they are more easily recovered in plankton nets. For instance, Drouet (1963) reports growths of *Schizothrix calcicola* around 400 m in the Dead Sea; Bernard and Leçal (1960) state that species of *Nostoc* are found abundantly in the Indian Ocean, the Arabian Sea, the Mediterranean Sea, and the Atlantic Ocean at depths of 600–1000 m. The different life habit between benthic and planktonic blue-green algae is not important to our problem, which is to estimate the range of depths spanned by blue-green algal life. The reported data show that marine blue-green algae can actively grow from 0 to at least 1000 m. In first approximation, the depth-distribution of blue-green algal communities relies on three main energetic processes: (1) in shallow, brightly illuminated waters: syntheses rely on the direct activity of the chlorophyll *a*; (2) in shaded to twilight conditions, chromatic adaptation triggers the enhanced production of the red pigment (Phycoerythrin), which absorbs the remaining blue radiation and transmits the energy to the chlorophyll *a*; (3) in complete darkness, chemolithotrophic (Iron, Manganese, Sulfur) or chemoorganotrophic metabolism insure the required synthesis (Monty, 1971; Krumbein and Cohen, 1976, etc.).

What preceeds shows that subtidal fossil cryptalgal structures may have formed at depths of 1, 10, 50, 100, 500, 1000 . . . m? In most cases, the associated sediments and fossil debris, as well as the lateral and vertical facies relationships, will yield the major clues as to the approximate depth of formation; in some cases, the material trapped by the successive algal films may be indicative (coccoliths, or various pelagic organisms, for instance).

In the aphotic zone, the laminated fabric of stromatolites could, among other things, result from periodic heavy rains of pelagic tests (after period of bloom, or periodic shift in movements of oceanic waters), such as found in the Callovian columnar stromatolites described by Radwanski and Szulscewsky (1965) where lamine are made of coccoliths; it could also result from growth cycles of the algo-bacterial mats in relation to vertical movements of nurients, or yet from intrinsic periodic mechanisms ruling microbiota dynamics. In other environments no laminated pattern may show up and massive mud mounds – sorts of massive aphanitic bioherms – would form; this seems, for instance, to be the case of some middle Devonian Mud mounds of Southern Belgium, which I have found loaded with algal filaments.

It should be finally pointed out, that stromatolites are generally built by communities of blue-green algae *and* of bacteria. In many cases, blue-green algae play the dominant role, but situations can, however, be found where the prevailing environmental conditions originate a shift toward the bacterial pole. This is, for instance, what occurs in the stromatolites described in the hot springs of Yellowstone National Park (Walter et al., 1972, 1976); in the pools where the water temperature is above the uppermost limit tolerated by blue-green algae, the stromatolitic community becomes restricted to filamentous bacteria and plain bacterial stromatolites are formed. On the basis of personal microscopic observations (Monty, 1973b) and laboratory experiments (Ehrlich, 1963, 1966, 1972, 1974, 1975; Schweisfurth, 1971) it has also been proposed that colloform ferromanganic laminated nodules and crusts should be considered as oceanic bacterial stromatolites (Monty, 1973b). These occurrences still enlarge the environmental range of stromatolites.

2.4.5 Remark on the Inference of Growth Environments from Associated Sedimentary Features

When using sedimentary features to infer the growth environment of stromatolites, the *effective* time and space relationships between these features and the stromatolites themselves should be critically analyzed and clearly presented, which is not always the case. For instance, as noted by Serebryakov and Semikhatov (1974, p. 561), it is not uncommon that authors refer to the presence, in the section studied, of dessication cracks, flat pebble conglomerates, pseudomorphs after evaporites etc. . . as a clues to an inter- or supratidal origin of stromatolitic beds, although the reported features occur in the *over-* or *underlying* deposits and not in direct connection with the stromatolites themselves.

Also, the time-space relationships between stromatolitic bodies and the interstitial sediments (ooids, bioclasts, terrigenous material, etc. . .) must be unequivocally defined before using these sediments as indicators of the growth environment; this mean that a clear discussion should discriminate between the following three possibilities: (1) Stromatolite and interstitial material formed contemporaneously and concurrently in the same environment. (2) Stromatolites and interstitial sediment formed contemporaneously but in different environments, the sediment being introduced into the stromatolite field by periodic (sub) marine current, storms, etc. An example of such very misleading situations is illustrated by the stromatolitic deposits of Western Andros Id. (Monty and Hardie, 1976). Here, algal heads and mats grow in coastal marches which constitute terrestrial freshwater settings for the main part of the year; they are, however, flooded by episodic or yearly storms which redeposit subtidal sediments from the Great Bahama Bank on the algal growths. As a result, laminae of pellets, tests of foraminifera, etc. alternate with phases of algal growth or accumulate in between the heads, whereas some of the detrital grains infiltrate into the bundles of filaments; furtermore, the evaporation of seawater yields a very complex marine mineralogy with aragonites and high magnesian calcites. Consequently, this very punctual, episodic marine accident imprints the most striking evidence in the deposit and obliterates the clues to the prevailing year-long environmental conditions. Similar situations may occur in subtidal environments by seasonally shifting currents.

In such situations, the contemporaneity of the stromatolitic growth with the sediment deposition will be shown by (a) the incorporation, in the stromatolite laminae, of (generally finer) sedimentary grains of similar source as the interstitial sediment; (b) the presence of stromatolite laminae extending from one head to the other, over the depositional surface of the interstitial material (see for instance, Monty, 1976, Fig. 23). The original space relationships between the stromatolite growth-environment and the area where the interstitial sediment normally accumulates, or forms, will be elucidated by mapping the facies across the depositional strike to the source area where pure oolitic, bioclastic, pelletal, or yet quartzitic beds are found with their classical sedimentary structures.

(3) Stromatolites and the interstitial material are heterochronous, as may occur when the latter is brought about during a regressive or transgressive oscillation. In this case, few of the interstitial material is present in the stromatolites themselves; furthermore, particular features such as erosional traces, organic encrustations, caliches etc. draw a clear-cut physical boundary between stromatolites and the sediment.

2.5 Bibliography

Achauer, C.W., Johnson, H.: Algal stromatolites in the James Reef Complex (lower Cretaceous) Fairfield, Texas. J. Sedimentol. Petrol. **39**, 1466–1472 (1969).

Aitken, J.D.: Classification and environmental significance of cryptalgal limestones and dolomites, with illustrations from the Cambrian and Ordovician of Southwestern Alberta. J. Sedimentol. Petrol. **37**, 1163–1178 (1967).

Andrussow, N.: Der Adschi-darja oder Karabugas-Busen. Petermanns Mitt. **43**, 23 (1897).

Awramik, S.M.: Precambrian columnar stromatolite diversity: reflection of Metazoan appearance. Science **174**, 825–827 (1971).

Bailey, L.W., Matthew, G.F.: Preliminary report on the geology of Southern New Brunswick. Geol. Surv. Can. Rep. Progr. **1870–81**, 15–57 (1872).

Barber, C.A.: The structure of *Pachytheca.* Ann Botany **3**, 1889–1890 (1890).

Bauman, E.: Die Vegetation des Untersees (Bodensee). Archiv für Hydrobiol. und Planktonkunde. Suppl. Bd. 1, Stuttgart, 26–48 (1911).

Bauman, E.: Vegetation des Untersees (Bodensee). Eine pflanzenbiologische Skizze. Vegetationsbilder. Karsten, v.G., Schenk, H. (eds.), R. 9, H. 3, Jena, 1912.

Bernard, E.: Dictionnaire Universel des Fossiles Propres et des Fossiles Accidentels. Chambeau, L. (ed.). Avignon (France), 1763.

Bernard, F., Lecal, J.: Plancton unicellulaire récolté dans l'Océan Indien par le Charcot (1950) et le Norsel (1955–1956). Bull. Inst. Océanogr. Monaco **1166**, 1–59 (1960).

Bertrand-Sarfati, J.: Les stromatolites columnaires du Précambrien supérieur du Sahara Nord occidental; inventaire, morphologic et microstructure des laminations, corrélations stratigraphiques. Cent. Rech. Zones Arides, sér. Géol. **14**, Paris, 254 p. (1972).

Bertrand–Sarfati, J., Fabre, J.: Les stromatolithes des formations lacustres post-moscoviennes du Sahara septentrional (Algérie). In press (1976).

Bertrand-Sarfati, J., Freytet, P., Plaziat, J.-C.: Les calcaires concrétionnés de la limite Oligocène-Miocène des environ de Saint Pourçain sur Sioule (Limagne d'Allier): rôle des algues dans leur édification; analogie avec les stromatolithes et rapports avec la sédimentation. Bull. Soc. Géol. France, Sér. 7, **VIII**, 652–662 (1966).

Berz, K.C.: Petrographisch-stratigraphische Studien im oberschwäbischen Molassegebiet. Jahresh. Vaterländ. Naturk. Württemberg **71**, Stuttgart (1915).

Black, M.: The algal sediments of Andros Island, Bahamas. Philos. Trans. Ser. B, **CCXXII**, 165–192 (1933).

Blatchley, W.S., Ashley, G.H.: The lakes of Northern Indiana and their associated marl deposits. Ann. Rept. Indiana Dept. Geol. Nat. Resources **25**, 31–32 (1901).

Borel, P.: Les Antiquitez de Castres (1649).
Bornemann, J.G.: Geologische Algenstudien. Jb. Preuß. Geol. Landesanst. Berlin 1886 (1887).
Bradley, W.H.: Algae reefs and oolites of the Green River Formation. U.S.G.S. prof. Paper 154–G, 203–223 (1929).
Burger, O.: Über schwäbische Kalktuffe insbesondere des Echaztales. Inaug.-Dissert. Tübingen, 1911.
Clarke, J.M.: The water biscuit of Squaw Id. Canandaigua Lake, N.Y. Bull. New York State Mus. **39**, 8, Albany (1900).
Cloud, P.E.: Notes on stromatolites. Am. J. Sci. **240**, 363–379 (1942).
Cohn, F.: Über die Algen des Carlsbad Sprudels, mit Rücksicht auf die Bildung des Sprudelsinters. Schlesische Gesell. Vaterl. Cult. Abh., Abt. Naturwiss. u. Medicin, Heft 2, 35–55 (1862).
Colom, G.: La paléoécologie des lacs du Ludien-Stampein inférieur de l'île de Majorque. Rev. Micropalol. **4**, 17–29 (1961).
Davies, G.R.: Algal laminated sediments, Gladstone embayment, Shark Bay, Western Australia. In: Carbonate Sedimentation and Environments, Shark Bay, Western Australia. Logan, B.W., Davies, G.R., Read, J.F., Cebueski, D.E. (eds.). AAPG Memoir **13**, 169–205 (1970).
Davis, Ch.A.: A contribution to the natural history of marl. J. Geol. **8**, Chicago (1900).
Davis, R.A.: Willow River dolomite: Ordovician analogue of Modern algal stromatolite environments. J. Geol. **74**, 908–923 (1966).
Davis, R.A.: Algal stromatolites composed of quartz sandstone. J. Sedimentol. Petrol. **38**, 953–955 (1968).
Donaldson, J.A.: Stromatolites in the Denault Formation, Marion Lake, Coast of Labrador, Newfoundland. Geol. Surv. Can. Bull. **102**, 33 (1963).
Donsimoni, M.: Etude des calcaires concrétionnés lacustres de l'Oligocène supérieur et de l'Aquitanien du Bassin de Limagne (Massif Central, France). Unpublished thesis, Paris, 197 p. (1975).
Donze, M.: The algal vegetation of the Ria de Arosa (N.W. Spain). Blumea **16**, 159–192 (1968).
Drouet, F.: Ecophenes of *Schizothrix calcicola*. Proc. Acad. Nat. Sci. Philadelphia **115**, 9, 261–281 (1963).
Ehrlich, H.L.: Bacteriology of manganese nodules. I. Bacterial action on manganese in nodule enrichments. Appl. Microbiol. **11**, 15–19 (1963).
Ehrlich, H.L.: Reactions with manganese by bacteria from marine ferromanganese nodules. In: Developments in Industrial Microbiology **7**, 279–286 (1966).
Ehrlich, H.L.: The role of microbes in manganese nodule genesis and degradation. In: Ferromanganese deposits on the ocean floor. Horn, D. (ed). The office of the Intern. Decade of Ocean Exploration. Nat. Sci. Found., Washington D.C., 63–70 (1972).
Ehrlich, H.L.: Response of some activities of ferromanganese nodule bacteria to hydrostatic pressure. In: Effect of the Ocean Environment on Microbial Activities. Colwell, R.R., Morita, R.Y. (eds.). Baltimore–London: Univ. Park Press 1974, p. 208–221.
Ehrlich, H.L.: The formation of ores in the sedimentary environment of the deep sea with microbial participation: the case for ferromanganese concretions. Soil Sci. **119**, 36–41 (1975).
Elloy, R.: Réflexion sur quelques environments récifaux du Paléozoique. Bull. Cent. Rech. Paris S.N.P.A. **6**, no. 1, 105 p. (1972).
Feldman, J.: Sur l'ecologie des Rhodophycées et des Cyanophycées marines dan ses rapports avec les problèmes de leur photosynthèse. Bull. Soc, Fr. Physiol. Vég. **5**, 161–167 (1959).
Fenton, C.L., Fenton, M.A.: Algae and algal beds in the Belt Series of Glacier National Park. J. Geol. **39**, 670–686 (1931).
Fenton, C.L., Fenton, M.A.: Algal reefs or Bioherms in the Belt Series of Montana. Geol. Soc. Am. Bull. **44**, 1135–1142 (1933).
Forel, F.A.: Le Leman. Monographie limnologique. T. I–III, Lausanne (1904).
Freytet, P.: Sédimentation microcyclothémique avec croûtes zonaires à algues dans le calcaire de Beauce de Chauffour-Etrechy (Seine et Oise). Bull. Soc. Géol. de France (7), **VII**, 309–313 (1965).
Freytet, P., Plaziat, J.C.: Importance des constructions algaires dues a des Cyanophycées dans les formations continentales du Crétacé supérieur et de l'Eocène du Languedoc. Bull. Soc. Géol. France, 7ème série **7**, 679–694 (1965).

Freytet, P., Plaziat, J.C.: Les constructions algaires continentales stromatolithiques. Exemples pris dans le Crétacé supérieur et le Tertiaire de France et d'Espagne du Nord. 24th IGC, section 7, 524–534 (1972).
Früh, J., Schröter, C.: Die Moore der Schweiz mit Berücksichtigung der gesamten Moorfrage. Beitr. Geol. Schweiz., Geotech. Ser., Lfg. 3, Bern 1904.
Gebelein, C.D.: Distribution, morphology, and accretion rate of recent subtidal algal stromatolites, Bermuda. J. Sedimentol.. Petrol. **39**, 49–69 (1969).
Gebelein, C.D.: Biologic control of stromatolite microstructure: implication for Precambrian time stratigraphy. Am. J. Sci. **274**, 575–598 (1974).
Gebelein, C.D.: Holocene sedimentation and stratigraphy Southwest Andros Island, Bahamas. Mangin, J.Ph. (ed.). IXth Intern. Congr. Sedimentol. (I.A.S.), Nice 1975; Theme V, 1975, p. 191–197.
Gigot, P., Rioult, M.: Importance stratigraphique des concrétions calcaires d'origine algaire dans la série miocène de Châteauredon (Alpes de Haute Provence). C.R. Acad. Sci. Paris **274**, 3183–3185 (1972).
Ginsburg, R.N.: Recent stromatolitic sediments from South Florida. Abstr. J. Paleont. **29**, 723 (1955).
Ginsburg, R.N.: Ancient analogues of Recent stromatolites. Intern. Geol. Congress, XXI session, Norden, 1960, part XXII, Copenhagen 1960, 26–35 (1960).
Ginsburg, R.N., Isham, L.B., Bein, S.J., Kupferberg, J.: Laminated algal sediments of South Florida and their recognition in the fossil record. The marine laboratory, Univ. of Miami, unpublished (1954).
Ginsburg, R.N., Lowenstam, H.A.: The influence of marine bottom communities on the depositional environment of sediments. J. Geol. **66**, 310–318 (1958).
Gleason, P.J.: The origin, sedimentation and stratigraphy of a calcitic mud located in the Southern freshwater Everglades. Ph. D. thesis, Pennsylvania State Univ. (1972).
Gleason, P.J., Spackman, W.: Calcareous periphyton and water chemistry in the Everglades. In: Environments of South Florida: Present and Past. Gleason, P.D. (ed.). Miami Geol. Soc. Memoir 2, 1974, p. 146–181.
Goldring, W.: Algal Barrier Reefs in the Lower Ozarkian of New York with a chapter on the importance of coralline algae through the ages. N.Y. State mus. Bull. **315**, 75 p. (1938).
Golonka, J., Rajchel, J.: Upper Cretaceous stromatolites in the vincinity of Crakow. Kwart Geologiczny **16**, no. 3, 652–667 (1972). (Polish – English summary).
Grabau, A.W.: North American Index Fossils **I**, 853 p. (1909).
Gürich, G.: Les spongiostromides du Viséen de la Province de Namur. Mus. Hist. Natur. de Belgique, mem. 3 (4), 1–55 (1906).
Hadding, A.: Silurian algal limestones of Gothland. Lunds. Univ. Arssk. N.F. Avd. 2, Bd. 56, no. 7, 1–25 (1960).
Hale, D.J., Davis, C.A., Lane, A.C., Humphrey, R.L.: Marl (Bog lime) and its application to the manufacture of Portland cement. Geol. Surv. Michigan **VIII**, Pt. 3, Lansing (1903).
Hall, J.D.: *Cryptozoon (proliferum)* N.G. (an sp). New York State Mus. Ann. Rept. **36**, (1883).
Høeg, O.A.: A postglacial marine stromatolite from Southeastern Norway. Studies in Stromatolites. I.K. Norsk. Videnskap Selskabs. Skrift **1**, 1–60 (1929).
Hoffman, P.E.: Algal stromatolites use in stratigraphic correclation and paleocurrent determination. Science **157**, 1043–1045 (1967).
Hoffman, P.E.: Shallow and deepwater stromatolites in Lower Proterozoic platform – to basin facies change, Great Slave Lake, Canada. A.A.P.G. Bull. **58**, 856–867 (1974).
Hofmann, H.J.: Attributes of stromatolites. Geol. Surv. Can. **69–39**, 43 pp. (1969).
Hofmann, H.J.: Stromatolites: Characteristics and utility. Earth Sci. Rev. **9**, 339–373 (1973).
Hofmann, H.J.: Mid-Precambrian prokaryotes (?) from the Belchers Islands, Canada. Nature (Lond.) **249**, 87–88 (1974).
Hofmann, H.J.: Stratiform Precambrian Stromatolites, Belchers Islands, Canada: relations between silicified microfossils and microstructure. Am. J. Sci. **275**, 1121–1132 (1975).
Holtedahl, O.: On the Paleozoic formations of Finmarken in Northern Norway. Am. J. Sci., Ser. 4, **47**, 85–107 (1919).

Holtedahl, O.: On the occurrence of structures like Walcott's Algonkian algae in the Permian of England. Am. J. Sci., Ser. 5, **1**, 195–206 (1921).
Hommeril, P., Rioult, M.: Etude de la fixation des sédiments meubles par deux algues marines: *Rhodothamniella floridula* (Dillwyn) F. Feldm. et *Microcoleus chtonoplastes* Thur. Marine Geol. **3**, 131–155 (1965).
Jenkyns, H.C.: Fossil manganese nodules from the West Sicilian Jurassic. Eclogae Geol. Helv. **63**, 741–744 (1970).
Johnson, J.H.: Algae as rock builders, with notes on some algal limestones from Colorado. Univ. Colorado Studies **23**, no. 3, 217–222 (1936).
Johnson, J.H.: Algae and algal limestone from the Oligocene of South Park, Colorado. Geol. Soc. Am. Bull. **48**, 1227–1236 (1937).
Johnson, J.H.: Limestone building algae and algal limestones. Colorado School Mines, Spec. Publ., 297 p (1961).
Jones, J.C.: The geologic history of lake Lahontan. Science N.S. 40, New York (1914).
Jones, J.C.: The geologic history of Lake Lahontan. – Quaternary climates. Carnegie Inst. Washington Nr. 352, S.I. (1925).
Kalkowsky, E.: Oolith and Stromatolith im norddeutschen Buntsandstein. Z. Deut. Geol. Ges. **60**, 68–125 (1908).
Kendall, C.G.St.C., Skipwith, Sir Patrick A. d'E.: Recent algal stromatolites of the Khor al Bazam, Abu Dhabi, the Southwest Persian Gulf (abs.). Geol. Soc. Am. special papers for 1966, p. 108 (1966).
Kendall, C.G.St.C., Skipwith, Sir Patrick A. d'E.: Recent algal mats of a Persian Gulf lagoon. J. Sedimentrol. Petrol. **38**, 1040–1058 (1968).
Kendall, C.G.St.C., Skipwith, Sir Patrick A. d'E.: Holocene shallow-water carbonate and evaporite sediments of Khor al Bazam, Abu Dhabi, Southwest Persian Gulf. A.A.P.G. **53**, 841–869 (1969).
Kinsman, D.J.J.: Recent carbonate sedimentation near Abu Dhabi, Trucial coast, Persian Gulf. Ph. D. thesis, Univ. London, 315 p. (1964).
Kirchner, O.: Schizophyceae. In: Die natürlichen Pflanzenfamilien nebst ihren Gattungen und wichtigeren Arten. Engler, A., Prantl, K. (Hrsg.). T. 1, I, Abt. 1a. Leipzig 1898.
Komar, V.A.: Upper Precambrian stromatolites in the Northern part of the Siberian platform and their stratigraphical significance. Akad. Nauk SSSR, Geol. Inst. Tr. **154**, 122 p. (Russian) (1966).
Korolyuk, I.K.: Stromatolites of the Upper Precambrian. In: Upper Precambrian stratigraphy of USSR. Keller, B.M. (ed.). Moscow, Gosgeoltekhizdat **2**, 479–498 (Russian) (1963).
Krumbein, W., Cohen, Y.: Primary production, mat formation and lithification chances of oxygenic and facultative anoxygenic cyanophytes (Cyanobacteria). This book (1976).
Krylov, I.N.: Columnar branching stromatolites of Riphean beds of the Southern Urals and their significance for the stratigraphy of the Upper Precambrian. Akad. Nauk. SSSR. Geol. Inst., Tr. **69**, 133 (Russian) (1963).
Krylov, I.N., Shapovalova, I.G., Kolosov, P.N., Fedonkin, A.M.: Riphean deposits of the lower part of River Lena. Sov. Geol. **7**, 85–95 (1971).
Larsson, K.: The Lower Viruan in the autochtonous Ordovician sequence of Jämtland. Sveriges Geol. Unders. Ser. C NR 63, Arsbok **67**, 2, 82 p (1973).
Le Roux, M.: Recherches biologiques sur le lac d'Annecy. Ann. Biol. Lacustre Bruxelles **2**, 1907–1908 (1908).
Linck, G.: Über die Bildung der Oolithe und Rogensteine. Jenaische Z. Naturwiss. **45** (1909).
Linden, M.: Über Kalkabscheidungen aus wässeriger Lösung B. Die Indusienkalke der Hürbe. Ber. 23. Vers. Oberrhein. Geol. Vereins zu Sigmaringen (1890).
Lochman-Balk, K.: Upper Cambrian faunal pattern on the Craton. Geol. Soc. Am. Bull. **81**, 3197–3224 (1970).
Logan, B.W.: Cryptozoon and associated stromatolites from the Recent, Shark Bay, Western Australia. J. Geol. **69**, 517–533 (1961).
Logan, B.W., Rezak, R., Ginsburg, R.N.: Classification and environmental significance of stromatolites. The A and M College of Texas. Dept. Oceanogr. (1962), *also* J. Geol. **72**, 68–83 (1964).
Logan, B.W., Hoffman, P., Gebelein, C.: Algal mats, cryptalgal fabrics and structures, Hamelin Pool, Western Australia. A.A.P.G. Memoir **22**, 140–194 (1974).

Maanten, A.A.: Silurian reefs of Gotland. Dev. Sedimentol. **13**. Amsterdam: Elsevier Publ. Co. 1971.
Marcinowski, R., Szulczewski, M.: Condensed cretaceous sequence with stromatolites in the Polish Jura Chain. Acta Geol. Pol. **22**, no. 3, 517–539 (1972).
Maslov, V.P.: Algal deposits of the Lower Paleozoic of Siberia – Problems of Paleontology. Publications of the Laboratory of Paleontology. Moscow Univ. 249–348 (Russian) (1937).
Maslov, V.P.: Printzipi nomenklatyri i sistematiki stromatolitov. Acad. Nauk. SSSR, Ser. Geol. **4**, 105–112 (1953).
Maslov, V.P.: Algae and carbonate deposition. Izv. Akad. Nauk. SSSR, Ser. Geol. **12**, 81–86 (Russian) (1961).
Matthew, G.F.: President's annual address. 3. On the existence of organisms in the Pre-Cambrian rocks. Nat. Hist. Soc. New Bruswick, Bull. **2**, no. 9, 28–33 (1890a).
Matthew, G.F.: Eozoon and other low organisms in Laurentian rocks at St. John. Nat. Hist. Soc. New Brunswick Bull. **2**, no. 9, 36–41 (1890b).
Monty, C.L.V.: Biostromes stromatolithiques dans le Viséen moyen de la Belgique. C.R. Acad. Sci. Paris **256**, 5603–5606 (1963).
Monty, C.L.V.: Geological and environmental significance of Cyanophyta. Ph. D. thesis, 429 p., Princeton Univ. New Jersey, U.S.A., Microfilm 66–5003, Univ. Microf. Inc. Ann Arbor, Michigan (1965a).
Monty, C.L.V.: Recent algal stromatolites in the Windward Lagoon, Andros Island, Bahamas. Ann. Soc. Géol. Belg. Bull. **88**, no. 6, B 269–276 (1965b).
Monty, C.L.V.: Distribution and structure of Recent stromatolitic algal mats, Eastern Andros Id, Bahamas. Ann. Soc. Géol. Belg. **90**, 55–100 (1967).
Monty, C.L.V.: An autoecological approach of intertidal and deep water stromatolites. Ann. Soc. Géol. Belg. **94**, Fasc. III, 265–276 (1971).
Monty, C.L.V.: Recent algal stromatolitic deposits, Andros Island, Bahamas, Preliminary report. Geol. Rundschau **61**, 2, 742–783 (1972).
Monty, C.L.V.: Les nodules de manganèse sont des stromatolithes océaniques. C.R. Acad. Sci. Paris, Ser. D, **276**, no. 25, 3285–3288, 3 pl (1973a).
Monty, C.L.V.: Precambrian background and phanerozoic history of stromatolitic communities, an overview. Ann. Soc. Géol. Belg. **96**, fasc. III, 585–624 (1973b).
Monty, C.L.V.: An important but overlooked geological setting: the fresh water blue-green algal calcareous marsh. In: Publications du IX Congrès International de Sédimentologie, Nice, 1975. Mangin, J.Ph. (ed.); vol. , Thème 1: Indicateurs sédimentologiques, 1975, p. 225–235.
Monty, C.L.V.: The origin and development of Cryptalgal fabrics in "Stromatolites". Walter, M. (ed.). Amsterdam: Elsevier, in press, 1976.
Monty, C.L.V., Hardie, L.: The geological significance of the fresh water blue-green algal calcareous marsh. In: Stromatolites. Walter, M. (ed.). Amsterdam: Elsevier 1976, pp. 447–488.
Moore, E.S.: The iron formationon Belcher Islands, Hudson Bay, with special reference to its origin and its associated algal limestones. J. Geol. **26**, 412–438 (1918).
Mouline, M.P.: Importance et répartition paléogéographiques des cyanophycées dans le complexe fluvio-lacustre dit des "calcaires de Castres" (Tarn). Actes Soc. Linéenne de Bordeaux **103**, Sér. B, no. 8, 13 p. (1966).
Murray, G.: Calcareous pebbles formed by algae. Phycolog. Mem. London (1895).
Naumann, C.F.: Text book of Geognosy, Vol. II, p. 741 (1862).
Ott, E., Volkheimer, W.: Paleospongilla chubutensis n.g. et n.sp. – ein Süßwasserschwamm aus der Kreide Patagoniens. N. Jb. Geol. Paläontol. Abh. **140**, 1, 49–63 (1972).
Penhallow, D.P.: Note on calcareous algae from Michigan. Botan. Gazette **21**, 215–217 Madison (1896).
Pia, J.: Thallophyta. In: Handbuch der Paläobotanik. Hirmer (ed.), Bd. 1. München–Berlin: R. Oldenburg 1927, p. 31–136.
Pia, J.V.: Die Anpassungsformen der Kalkalgen. Palaebiologica. Festschr. Dollo. Wien–Leipzig: 1928, Vol. I, p. 211–224.
Pia, J.: Die rezenten Kalksteine. Z. Kristallogr. Mineral. Petrogr., Abt. B, Mineral. Petrogr. Mitteilungen, Ergänzungsband 12–13, 142–192 (1933).
Playford, P.E., Cockbain, A.E.: Algal stromatolites: deep water forms in the Devonian of Western Australia. Science **165**, 1008–1010 (1969).

Playford, P.E., Cockbain, A.E.: Modern algal stromatolites in Hamelin Pool, a hypersaline barred basins in Shark Bay, Western Australia. In: Stromatolites. Walter, M. (ed.). Amsterdam: Elsevier 1976, pp. 389–411.

Powell, Ch.: Observations on some calcareous pebbles. Minnesota Botan. Studies, Minneapolis **3**, H 1, 73–75 (1903).

Radwanski, A.: Petrographical and sedimentological studies of the high-tatric rhaetic in the Tatra Mountains. Studia Geol. Pol. **XXV**, 145 p. (1968).

Radwanski, A., Szulczewski, M.: Jurassic stromatolites in the Villary Mountains southern Hungaria). Ann. Univ. Sci. Budapest, Sect. Geol. **9**, 87–107 (1965).

Reis, O.M.: (Discussion of paper by) Kalkowsky: Über Oolith and Stromatolith im Norddeutschen Buntsandstein. N. Jb. Mineral., Geol., Paläontol. **2**, 114–138 (1908).

Roddy, H.J.: Concretions in streams formed by the agency of blue-green algae and related plants. Proc. Am. Philos. Soc. **54**, Philadelphia, 256–258 (1915).

Rothpletz, A.: Über die Kalkalgen, Spongiostromen und einige andere Fossilien. Sverig Geol. unders. Ser. Ca, **10**, 35 p. (1913).

Russel, J.C.: Geological history of Lake Lahontan. Mon. 11, U.S.G.S., 189–222 (1885).

Russel, J.C.: A geological reconaissance in Central Washington. U.S. Geol. Surv. Bull. **108** (1893).

Rutte, E.: Süßwasserkalke aus dem Kaiserstuhl und Breisgau. Berichte d. Naturforsch. Ges. Freiburg im Breisgau **43**, 1–38 (1953).

Rutte, E.: Die Algenkalke aus dem Miozän von Engelswies in Baden. Neues Jb. Geol. Paläontol. Abh. **98**, 149–174 (1953a).

Rutte, E.: Gesteinsbildende Algen aus dem Eozän von Kleinkems aus Isteiner Klotz in Baden. Neues Jb. Geol. Paläontol. Mh. **11**, 498–506 (1953b).

Rutte, E.: Süßwasserkalke und Kalkalgenbildungen in der chattischen unteren Süßwassermolasse von Hoppetenzell, nördlich Stockach/Baden. Geol. Jb. **69**, 517–536 (1954).

Schmidle, W.: Postglaziale Ablagerungen im nordwestlichen Bodenseegebiet. N. Jb. Mineral. **II**, 104 (1910).

Schröter, C., Kirchner, O.: Die Vegetation des Bodensees. „Die Bodenseeforschungen". 9. Abschn. Aus dem 25. Heft d. Schriften d. Verein für Gesch. d. Bodensees und seiner Umgebung, 1. Abt., 122 p. (1896).

Schweisfurth, R.: Manganknollen im Meer. Naturwissenschaften **58**, 344–347 (1971).

Seely, H.M.: Cryptozoa of the Early Champlain Sea. Vt. State Geol. Rep. **5**, 156–173 (1906).

Semikhatov, M.A., Komar, Vl.A., Serebryakov, S.N.: Judomian complex in stratotypical area. Akad. Nauk. SSSR, Geol. Inst. Tr. **210**, 207 p. (1970).

Serebryakov, S.N.: On the composition of Riphean stromatolite bioherms from the Eastern part of Siberia. Izv Akad. Nauk SSSR, Ser. Geol. **9**, 130–135 (1968).

Serebryakov, S.N.: Stromatolites in the rhythmic Riphean beds. USSR Acad. Sci. Proc. Geol. Ser. **10**, 127–134 (1971).

Serebryakov, S.N., Semikhatov, M.A.: Riphean and Recent stromatolites: a comparison. Am. J. Sci. **274**, 556–574 (1974).

Seward, A.C.: Plant life through the ages. New York–Cambridge: 1931, 601 p. (see p. 86–87).

Smith, B., Anderson, F.W.: Observations on the Magnesian limestone reefs of North-Eastern Durham. Bull. Geol. Survey G.B. **15**, 71–85 (1958).

Sommermeier, L.: Zur Geologie des Kartsteins. Z. Deut. Geol. Ges. **65** (1915); Monatsber. S. 341, Berlin (1914a).

Sommermeier, L.: Der Karstein und der Kalktuff von Dreimühlen bei Eiserfey in der Eifel. Verh. Naturhist. Ver. Preuß. Rheinlande Westfalens. Bonn, **70** (1914b).

Sommermeier, L.: Neue Ooide. Z. Deut. Geol. Ges. **66**, Abhandl. (1915).

Stahl, W.: Der Karabugas als Bildungsstätte eines marinen Salzlagers. Naturwiss. Wschr. **20**, 689 (1905).

Stapf, K.R.: Limnische Stromatolithen aus dem pfälzischen Rotliegenden. Mitt. Pollichia, III R. **20**, 103–112 (1973).

Steel, J.H.: A description of the oolitic formation lately discovered in the County of Saratoga and State of New York. Am. J. Sci. **9**, 16–19 (1825).

Steinmann, G.: Über *Gymnosolen ramsayi*, eine Coelenterate von der Halbinsel Kanin. Bull. Soc. Geogr. Finlande **3**, 18–23 (1911).

Szulczewski, M.: Stromatolitic structures within Middle Jurassic transgressive deposits at Wola Morawicka (Southern margins of the Holy Cross Mountains, Central Poland). Ann. Soc. Geol. Pol. **XXXVII**, fasc. 4, 515–528 (1967).

Thiselton-Dyer, W.T.: Note on Mr. Barbers' paper on *Pachytheca*. Ann. Botan. **5**, London (1891).

Tilden, J.: Minnesota Algae. I. Myxophyceae. Minneapolis (1910).

Trompette, R.: Les stromatolites du "Précambrien supérieur" de l'Adrar de Mauritanie, Sahara occidental. Sedimentology **13**, 123–154 (1969).

Truswell, J.F., Erikson, K.A.: Stromatolitic associations and their paleo-environmental significance: A reappraisal of a lower Proterozoic locality from the Northern Cape Province, S.A. Sedimentol. Geol. **10**, 1–23 (1973).

Truswell, J.F., Erikson, K.A.: A palaeoenvironmental interpretation of the Early Proterozoic Malmani dolomite from Zwartkops, South Africa. Prec. Res. **2**, 277–303 (1975).

Tsien, H.H.: The middle and upper Devonian reef complexes of Belgium. Petrol. Geol. Taiwan **8**, 119–173 (1971).

Twenhofel, W.H.: Pre-Cambrian and Carboniferous algal deposits. Am. J. Sci. **48**, 339–352 (1919).

Tyler, S.A., Barghoorn, E.S.: Occurrence of structurally preserved plants in the pre-Cambrian rocks of the Canadian shield. Science **119**, 606–608 (1954).

Voss, A.: Die Paläogeographische Verbreitung des Rogensteins im deutschen Unteren Buntsandstein. Abh. Preuß. Geol. Landesanst. **107** (1928).

Walcott, Ch.D.: Precambrian algonkian flora. Smithsonian Misc. Coll. **64**, 77–156 (1914).

Walter, M.R.: Stromatolites used to determine the time of nearest approach of Earth and Moon. Science **170**, 1331–1332 (1970).

Walter, M.R.: Stromatolites and the biostratigraphy of the Australian Precambrian and Cambrian. Paleol. Assoc. Lond. Spec. public. **11**, 189 p. (1972).

Walter, M.R., Bauld, J., Brock, T.D.: Siliceous algal and bacterial stromatolites in hot springs and Geyser effluents of Yellowstone National Park. Science **178**, 402–405 (1972).

Walter, M.R., Bauld, J., Brock, T.D.: Microbiology and Morphogenesis of Columnar Stromatolites (*Conophyton, Vacerrila*) from hot springs in Yellowstone National Park. In: Stromatolites. Walter, M. (ed.). Amsterdam: Elsevier 1976, pp. 273–310.

Weed, W.H.: Formation of Travertine and Siliceous Sinter by the vegetation of Hot Springs. U.S.G.S. Geol. Surv. 9th. Ann. Rept. 613–676 (1889).

Wesenberg-Lund, C.: Studier over Sökalk, Bonnemalm, eg Sogytje i danske Indsoer. Medd. fra Dansk Geol. Foren. **7**, Ser. 1, Kobenhavn (1901).

Wesenberg-Lund, C.: Studier over Søkalk, Bønnemalm, og Søgytje i danske Indsøer. Medd. fra Dansk Geol. Foren. **7**, Ser. 1 (1914).

Wieland, G.R.: Further notes on Ozarkian seaweeds and oolites. Am. Mus. Nat. Hist. Bull. **33**, 237–260 (1914).

Wolfenden, E.B.: Paeloecology of the Carboniferous Reef Complex and shelf limestone in Northwest Derbyshire, England. Geol. Soc. Am. Bull. **69**, 871–898 (1958).

Young, R.B.: The occurrence of stromatolitic or algal limestones in the Campbell Rand Series, Griqualand West. Trans. Geol. Coc., S.A. **35**, 29–36 (1932).

Young, R.B.: A comparison of certain stromatolitic rocks in the Dolomite Series of South Africa with Modern algal limestone in the Bahamas (1934).

3 Primary Production, Mat Formation and Lithification: Contribution of Oxygenic and Facultative Anoxygenic Cyanobacteria

W.E. KRUMBEIN, Oldenburg, W. Germany, and Y. COHEN, Jerusalem, Israel

3.1 Introduction

Recent algal stromatolites have been described by many authors (Gebelein, 1969; Golubic, 1973; Logan, 1961, 1964; Monty, 1967, 1972; Walter et al., 1973). Thermal algal mats have been described by Brock and Brock (1967), Castenholz (1973), Doemel and Brock (1974), Krumbein and Cohen (1974).

A new environment for the development of benthic blue-green algal mats in a sea-marginal lake on the shores of the Gulf of Aqaba (Friedman et al., 1973; Krumbein and Cohen, 1974) gave the opportunity to study algal mats developing without the sedimentation of marine clastic sediments. Productivity studies in the lake and the benthic algal mats, together with degradation studies within the mats, were conducted by bacteriological, chemical, and microscopic methods. In this contribution we want to describe a specific type of blue-green algal mat growing below two dense layers of photosynthetic bacteria, the specific type of photosynthesis adopted here and the lithification processes taking place in the Solar Lake environment. Possible consequences for the interpretation of precambrian stromatolites and fossilization processes will then be discussed.

Cyanobacteria generally do not excrete carbonates in or on their cell walls but they have a great importance for the study of fossil algal deposits, mainly for two reasons: (1) They belong to the oldest-known procaryotic organisms found in precambrian rocks and (2) they produce oncoidal and stromatolitic structures of different types, which are typical for intertidal deposits and shallow-water carbonates. They thus have a great significance as environmental indicators. Very rarely are the algal structures themselves preserved, but the structures produced by their "wax and wane" are to typical that it is easy to attribute them to cyanobacteria. It was interesting to study the processes in recent environments which lead to the destruction of the algae themselves but at the same time to the development of a stromatolitic rock.

3.2 Materials and Methods

3.2.1 Field Methods

The algal mats and sediments of Solar Lake, a small sea-marginal pond 18 km south of Elat, were observed and sampled by SCUBA-diving during 1971–1974. Objects had been placed in the lake to observe the development of mats and evaporitic gypsum-carbonate crusts. The samples were taken by inserting plastic corers (water-pipes) of 5 cm diameter and up to 350 cm length into the sediments. Surface samples were received by inserting sterile Petri-dishes and bacteriological syringes into the sediment. These samples were then taken for bacteriological and primary productivity determinations. The samples for Scanning electron microscopy (SEM) and electron dispersive X-ray analyses (EDAX) were treated immediately after sampling with 6% glutaraldehyde diluted in Solar Lake water from 25% solutions. This treatment avoided osmotic shock and artefacts.

3.2.2 Microscopic Methods

The layering of the shallow water algal mats was analysed and numbers of layers were counted by using a dissecting microscope. Light microscopy (phase contrast and Nomarsky interference) was used for identification and determination of species distribution. It was also used to examine interior parts of slime aggregates and capsules. It was frequently observed that samples which were extremely rich in diatoms, when examined by SEM did not yield any, for the diatoms were completely slime-covered (see Fig. 7a). SEM was used to study the mat ultramorphology, composition, and degree of lithification. Electron dispersive X-ray analyses using both EDAX and ORTEC equipment were used to determine degree of calcification and iron sulfide development on several typical bacterial and algal cells. The comparison of light microscopy using Nomarsky interference and of SEM methods enabled us to distinguish between carbonate spherules and algal colonies (see Krumbein, 1975).

3.2.3 Bacteriological Methods

Heterotroph bacteria were determined by using selective media as described by Krumbein (1971).

Productivity of the mats was measured using the ^{14}C-method described by Brock and Brock (1967) and the oxygen evolution method described by Schramm (1968). Pure cultures of *Oscillatoria limnetica, O. salina*, and *Microcoleus sp.*, were obtained by using their capability of actively gliding towards a light source. Bacteria were sieved off by the gliding movement through 0.5–1% agar. Anoxygenic photosynthesis of *Oscillatoria* was tested by measuring CO_2 photoassimilation at various sulfide concentrations. CO_2 photoassimilation at specific wavelengths were obtained by using interference filters peaking at 580 nm and 700 nm. The light intensity was measured by a "Yellow Springs radiometer" photometer. Light absorption by the algae was measured in a Cary 14 spectrophotometer. Determinations of sulfide ions in the medium were carried out by using Orion Res. pH-meter with sulfide-silver electrode. Standard curves were obtained by volumetric titration of $AgNO_3$. An experimental set-up allowed us to measure consumption of sulfide and CO_2-photoassimilation simultaneously. The experimental procedure ensured the complete inhibition of photosystem II and the exact determination of stoichiometric ratios between H_2S oxidized and CO_2-photoassimilation in the absence of O_2 evolution (Cohen et al., 1975a,b).

3.3 Description of the Four Different Algal Mat Types from Solar Lake

3.3.1 Flat Shallow-water Mat

Solar Lake is a small sea-marginal lake of 140 x 60 m and a maximal depth of 5 m. Its water level changes by evaporation in summer and a change in the water volume seeping into the lake through a 60 m wide bar separating Solar Lake from the open sea (Krumbein and Cohen, 1974). The water level changes reach 50–60 cm. Parts of the shallow-water mat are thus uncovered in summer and directly exposed to the sun. Mud cracks develop (Friedman et al., 1973). In this shallow-water region flat stromatolitic algal mats develop. They are finely laminated and reach maximal thickness of 120 cm. In summer a layer of coccoid blue-green algae (*Aphanothece* and *Aphanocapsa*) mixed with diatoms develops. These two cyanobacteria and the diatoms are less sensitive to photooxydation than the filamentous blue-green algae. In winter a layer of filamentous cyanobacteria (*Oscillatoria, Spirulina, Lyngbya, Microcoleus*) mixed with some Flexibacteria replaces the summer mat (Figs. 1 and 2). By shading the summer mat for 1–2 days it can easily be shown that the lamination is due to seasonal pattern and not to the usually described daily circadian rhythm. In this case the underlying filamentous algae penetrate through the summer mat of coccoid blue-green algae and diatoms.

Fig. 1. *Oscillatoria* sp. from filamentous winter mat of the shallow region (type I). Several bunches of trichomes are encoated by slime (— = 10 μ)

Primary production of the shallow water algal mat has been measured in the field and in the laboratory and values of 8–12 g C m^{-2} day^{-1} were reached. This would yield an annual growth rate of approx. 14 mm. Since counting of the laminae and ^{14}C radiocarbon dating (Krumbein and Cohen, 1974), has yielded in both only 0.5 to 1 mm accretion, it is concluded that about 95–99% of the originally produced organic matter decays and is

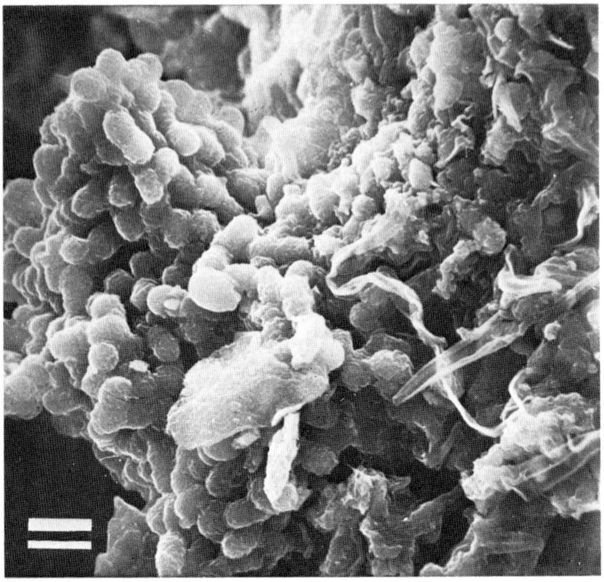

Fig. 2. Mass of relatively closely packed *Aphanothece* in the coccoid cyanobacteria diatom mat typical for the summer period of Solar Lake. Some slime-covered *Nitzschia* sp. can be detected (SEM-micrograph; ——— = 10 μ)

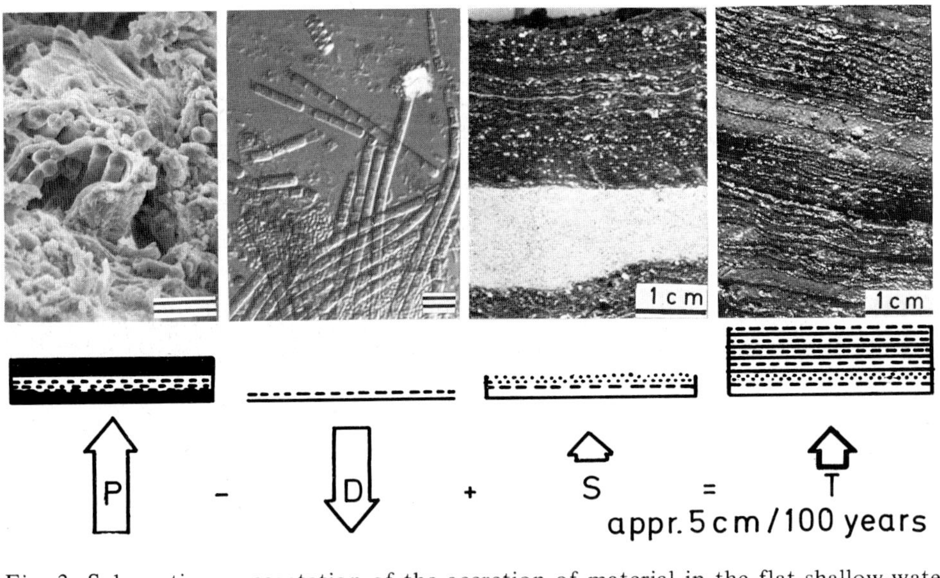

Fig. 3. Schematic representation of the accretion of material in the flat shallow-water mat (*D I*). ═══ = 10 μ; production (*P*) is relatively high and consists of summer and winter mat in alternating layers. Degradation (*D*) or decomposition is not complete due to establishment of reducing conditions and enormously high productivity. *Oscillatoria* and *Microcoleus* together with coccoid algae and diatoms form large aggregates in slime covers. Nomarsky-micrograph shows decay processes of *Oscillatoria* and one of the occasional *Spirulina* sp. Sedimentation (*S*) is low and occasional due to sporadic rain floods washing material into the lake from the surrounding metamorphic mountain ridges. Total accretion (*T*) has been found to be approx. 5 cm per 100 years

degraded immediately after production. We have, therefore, computed the overall accretion of the shallow-water algal mats using the schematic representation introduced by Golubić (1973), (Fig. 3). Primary production reduced by bacterial degradation, along with some clastic sedimentation mainly due to rain floods, thus yields a total accretion rate of 5 cm per 100 years. The organisms making up the shallow-water algal mat are shown in Figs. 1 and 2. Fig. 3 gives an impression of the summer and winter laminae.

3.3.2 Pinnacle Mats

At water depths between 1.2 m and 2.5 m the slope of Solar Lake is covered by pinnacle algal mats as described by Golubic (1973), (Fig. 4). At the Persian Gulf they occur above the shallow-water flat mat type. In Solar Lake they occur below these mats. In

Fig. 4. Pinnacle mat from Solar Lake (*D II*). Underwater photograph. The pinnacle height is between 1 and 3 cm

both cases the pinnacle pattern seems to be related to very fast decay processes, which in the case of Solar Lake are initiated by aeration of the mat from above *and* from below (aerated water is seeping into the lake at this water depth). During the summer, a few weeks before stratification builds up in the lake (Krumbein and Cohen, 1974), gypsum precipitates on these mats forming a "blister" appearance, as described by Golubić (1973). The gypsum is very soon redissolved, but the pinnacle pattern is preserved. As shown in Fig. 5, the pinnacle mat floats up in large patches, caused by oxygen bubbles and convectional mixing. During this time, gypsum precipitation continues and forms an algal-rich gypsum crust, which ultimately contributes most to the accretion of sediments in this part of the lake (Fig. 6).

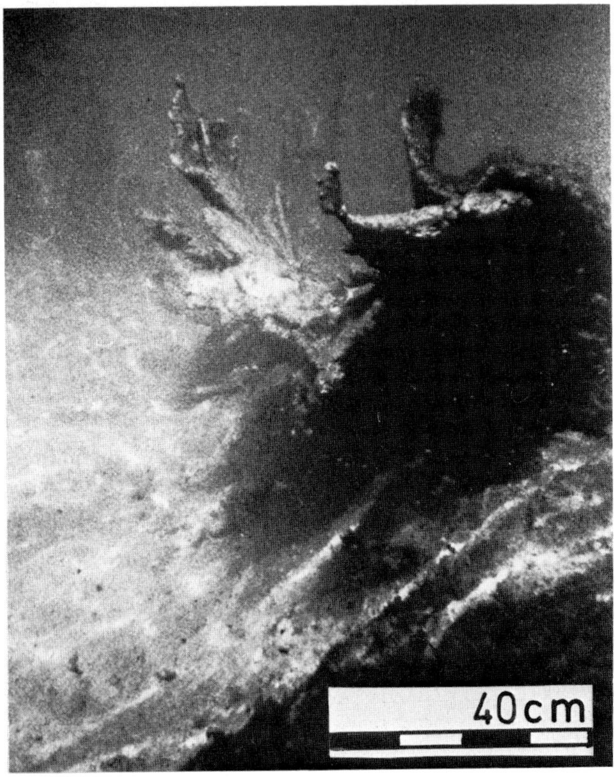

Fig. 5. Upfloating patches of the pinnacle mat (*D II*). This process of upfloating enhances considerably the growth of gypsum crusts below the pinnacle mat

3.3.3 Blue-Green Algal and Photosynthetic Bacterial Films

In deeper parts of the lake the hypersaline conditions prevail through the whole annual cycle. Temperatures reaching 50–60° C further enhance the extremely fast gypsum precipitation. Algal and bacterial growth on the rapidly growing gypsum crusts is continuously hindered by crystallization. Objects placed at this location were coated within 5 years by a 5–6 cm thick layer of gypsum, alternating with thin aragonite crusts. In summer, during holomixis, thin layers of filamentous algae and diatoms develop on top of the crust, while in winter photosynthetic bacteria and some diatoms produce a red film. Red and green films on the porous crust material can still be detected at 10–15 cm depth within the crust. Because the gypsum crust grows fast and *Artermia salina* feeds mainly in this water depth on algae and bacteria, very little organic matter is produced and preserved here. Gypsum precipitation, on the other hand, is maximal due to high salt concentration and temperatures. The overall accretion rate is high (Fig. 7). It seems that from this part of the lake gypsum deposits with thin aragonite layers grow towards

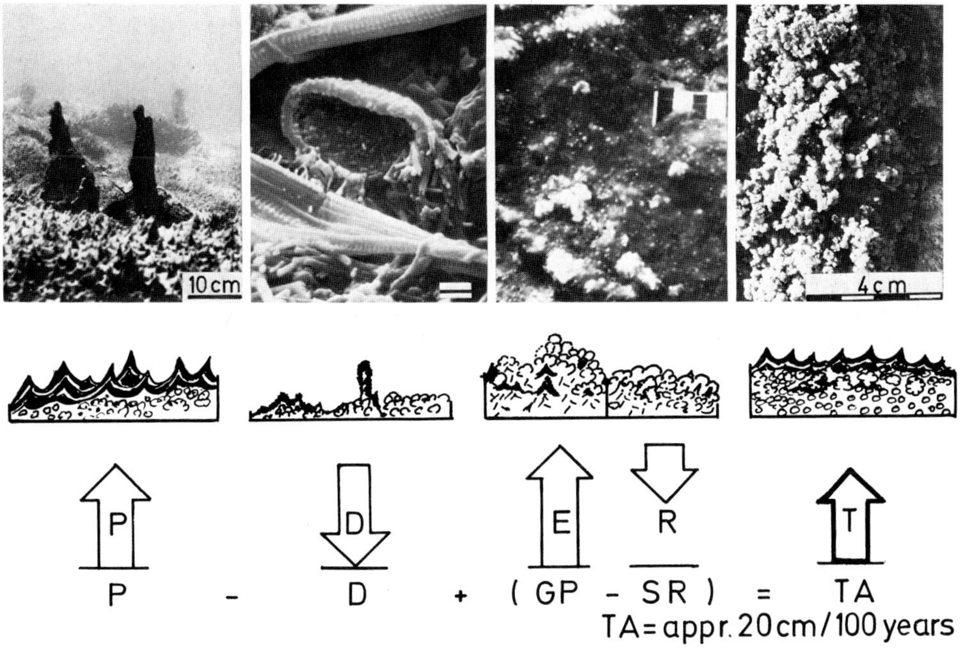

Fig. 6. 1–4. Schematic representation of the accretion rates of the pinnacle mat (*D II*). ——— = 1 μ. Production (*P*) is relatively high but immediately balanced by fast degradation (*D*). Pinnacle formation is partially due to degradation but seems also related to "blistering" by gypsum precipitation (3). A deflated *Aphanothece* cell shows some carbonate crystallization. Bacteria are dwelling on a diatom. A microcolony of heterotrophs is situated top right in (2) describing decay environment. Filamentous bacteria are active as well (bottom). In the late summer gypsum precipitation increases and is counterbalanced by some sulfate reduction (*R*) due to anaerobic decay. The resulting sediment (1) is a gypsum aggregate with some organic matter admixed

the center of the lake. The sediment accretion, thus, is more laterally orientated than vertically and the lake may get filled completely, more by this process than by the slow accumulation of sediment at the bottom (described below).

3.3.4 Flocculose Algal Mat

The flocculose algal mat of the bottom is not unusual as such. Golubic (1973) has described such mats extensively. The astonishing fact about these mats is that they thrive under completely anaerobic conditions, photosynthesize actively, and produce approx. 5 g organic C m^{-2} day^{-1} only in winter under stagnant conditions. In summer they either completely disintegrate at the bottom by myxobacteria activity or float up and decay under aerobic conditions. The bottom algal bloom was observed at H_2S concentrations of up to 39 ppm. It has been shown that the astonishing development of a flocculose *Oscillatoria* mat below two layers of photosynthetic sulfur bacteria, under extremely low light conditions (Krumbein et al., in press), is mainly due to the fact that

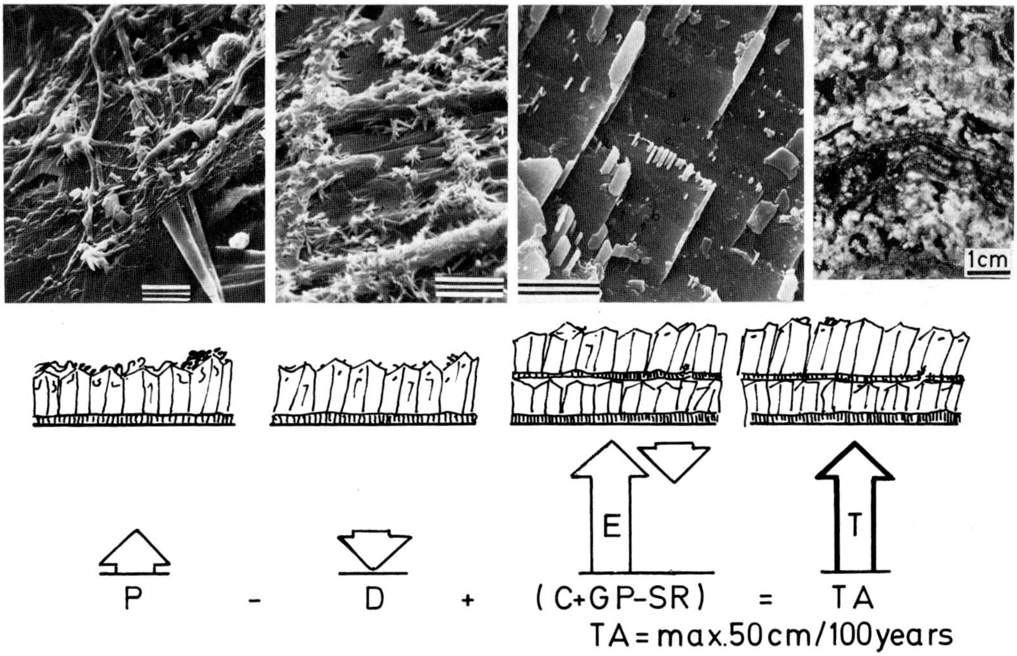

Fig. 7. 1–4. Schematic representation of algal and bacterial films (*D III*). ═══ = 10 μm. Production (*P*) is very low and only thin films are formed on crystals of gypsum and carbonate. (1) shows *Nitzschia* with filamentous cyanobacteria, filamentous bacteria, rods, and large bacteria. In (2) *Oscillatoria* is completely deformed by decay processes. Gypsum and carbonate crystals are abundant on the algal filaments. *Top right*: some bacteria settle on the mineral surface which is slime-covered. (3) shows growth pattern of gypsum with some bacteria adhering. (4) shows the final sediment with thin carbonate layers, some organic matter and thick gypsum layers. The total accretion is very high because the slope has highest salinities throughout the whole year and temperature reaches 50–60° C during stratification

O. limnetica in Solar Lake is capable of anoxygenic (H_2S dependent) bacterial type of photoassimilation under anaerobic, H_2S rich conditions while it switches to "normal" oxygenic photoassimilation, below a certain threshold concentration of H_2S (see D). Therefore, the algae try to reach as much light as possible and have the tendency to float up and distribute themselves as loosely as possible. On the other hand, they need relatively high H_2S concentrations. This keeps the population relatively close to the bottom, from where most of the H_2S is migrating.

Thus, high productivity of organic matter in winter under H_2S conditions and mat formation is counterbalanced in summer by aerobic and anaerobic decay of the mat. By sulfate reduction gypsum precipitates are partially redissolved. Thus, a mixture of decaying algae with a gypsum needle sediment forms which exhibits very little layering or lamination (Fig. 8). The overall accumulation is relatively high but still lower than at the hypersaline, hot, and aerobic slopes.

In this small lake four different types of algal mats occur, of which only one mat produces an algal laminite; one mat (i.e. film) is characterized by laminated aragonite gypsum layers and two mats are producing non-laminated gypsum sediments with a relatively low carbonate concentration.

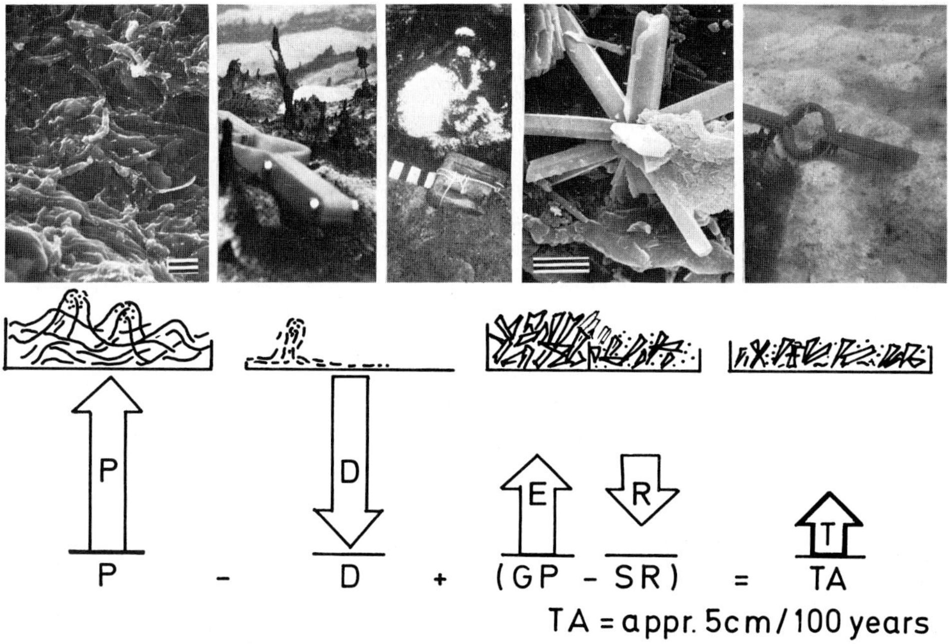

Fig. 8: 1–5. Schematic representation of the accretion rates of the flocculose mat (*D IV*). ═══ = 10 μm. (1) the *Oscillatoria* mat is characterized by typical (*bright*) sulfur granules on the cells. (2) Productivity is high 5 g (m^{-2} day^{-1}). During holomixis parts of the mat float up and decay, other parts (3) are decomposed by myxobacteria forming "macroplaques" on the mat. Gypsum precipitation (4) is partially balanced by sulfate reduction. The final sediment [surface view (5)] consists of gypsum aggregates in a light green mud of decaying algae. Organic matter concentration is approx. 5% organic C in the top 20 cm

3.4 Facultative Anoxygenic Photosynthesis of *Oscillatoria limnetica*

3.4.1 Field Observations

Usually blue-green algal mats occur only under aerobic or microaerophilic conditions. But is has been observed at many places that Cyanobacteria thrive better under low oxygen concentrations. The observation of these dense populations of *O. limnetica* under two discrete layers of photosynthetic sulfur bacteria was already indicative of the H_2S affinity. In addition, we observed sulfur granules on the surface of the filaments resembling those of sulfide oxidizing phototrophic bacteria. Filaments of *O. limnetica* collected in the field in winter were often whitish in color instead of the brilliant green usually observed. This lead us to analyse possibility of whether they have an unusual type of physiology.

3.4.2 Laboratory Experiments

The filamentous Cyanobacteria *O. limnetica* was isolated and purified from the bottom algal mat. It was tested by the above-described bacteriological methods. The results indicate that while this strain has the typical oxygenic photosynthesis, it can also photosynthesize under anaerobic conditions, using H_2S as the sole electron donor in photoassimilation of CO_2 with the operation of photosystem I alone. The experiments showed that when photosystem II is inhibited by the presence of DCMU, or not excited by using light of only 700 nm wave length, CO_2 was not photoassimilated under aerobic conditions. Under H_2S concentrations between 1.2–4 mM the photoassimilation of CO_2 proceeded in the presence of DCMU and also at 700 nm. The long wavelength light photosynthesis was as efficient as that of the short wavelength. As expected for an electron donor, the Na_2S concentration in the medium decreased at a constant rate as CO_2 photoassimilation proceeded. The stoichiometric relation was 2 for Na_2S consumed and CO_2 fixed, as should be expected in comparing this photosynthesis type to that of photosynthetic sulfur bacteria. CO_2 photoassimilation proceeds at the same rate with 3–4 mM Na_2S as under aerobic conditions with H_2O. Only when 0.5 and less mM H_2S were used under anaerobic conditions the photoassimilation decreased to about 30%. It is assumed that at low H_2S concentrations photosystem II is already inhibited, whereas the H_2S concentration is not yet sufficient to allow efficient electron flow. A higher threshold concentration is needed for efficient anoxygenic photoassimilation.

It is already known that certain cyanophytes and green algae can carry out photoreduction of CO_2 with H_2 in addition to their oxygenic photosynthesis. In these cases photosystem II was found to be partially active in addition to photosystem I. In this strain of *O. limnetica* both photosystems are fully developed and active. Yet photosystem I can support life efficiently and independently from photosystem II when reduced electron donors are available (Cohen et al., 1975a).

3.4.3 Environmental and Evolutionary Consequences

The type of *O. limnetica* isolated from the flocculose algal mats of Solar Lake presented here in its photosynthetic flexibility seems to be the first organism reported to grow with both types of photosynthesis. This is certainly a selective advantage for environments with extreme changes in environmental conditions such as occur in the monomictic Solar Lake. This *Oscillatoria* can grow under conditions where usually only sulfur photosynthetic bacteria, photosynthesizing on H_2S, survive. Under conditions when the short wavelength light that activates photosystem II is not limiting, it can also use H_2O as the electron donor with operation of both photosystems. These organisms are, thus, not restricted to special ecosystems where specific electron donors are available. They can thrive well under oxygen-free and H_2S-rich conditions, as well as under the conditions of holomixis with saturation of oxygen and high light intensity. Because *O. limnetica* displays both types of oxygenic and anoxygenic photosynthesis, it is tempting to speculate on its evolutionary position. It is possible that this cyanophyte represents a secondary specialisation, back towards anoxygenic photosynthesis typical for bacteria not possessing photosystem I and II, and which are generally regarded as the first organisms that were capable of using light energy. On the other hand, it is also possible that this alga represents a missing link between the two types of photosynthesis and has survived in special environments.

The possibility of isolation of photosystem II mutants may help to clarify the interconnection between system I and II, a subject still under dispute. The Solar Lake bottom mats may represent a model for the study of sulfur biotopes where sulfide plays the major role, as suggested with the term "sulfur biome" by Fenchel and Riedl (1970). Notions on *O. limnetica* and on the general evolutionary discussion are given by Cohen (1975) and Brock (1973). Though the evolutionary significance of our strain is highly suggestive, it has to be repeated that this organism is the first case reported of an oxygenic phototroph capable of growing anaerobically on H_2S.

Another feature, however, is the capability of cyanobacteria to be very sensitive to oxygen, which again brings them in close evolutionary relationship to the primitive photosynthetic sulfur bacteria. It has to be stated as well that primary productivity, for several reasons, in Solar Lake showed the highest values so far recorded for natural unpolluted waters. Astonishingly this high productivity took place under anaerobic conditions (Cohen, 1975). This type of productivity under anaerobic conditions may favor and enhance the capturing of larger amounts of organic matter under reducing conditions. Such conditions are generally regarded as favorable for the genesis of hydrocarbon deposits. Therefore, the described type of anaerobic and anoxygenic photosynthesis in algal mats may contribute to and explain the occurrence of larger amounts of fossil hydrocarbons.

3.5 Chances of Lithification of Cyanophyte Mats

3.5.1 Decomposition of Organic Matter

We have studied the decomposition of organic matter in the algal mats of Solar Lake by several methods. A comparison between our productivity measurements and the annual accretion rates proved that only 1–4% of the initially produced organic matter is accreted within the algal mat. The high water content (Fig. 9) of the recent stromatolitic mat helps to produce bacterial activity in considerable depth within the sediment. *Desulfovibrio* activity is recorded to a depth of 80 cm. Heterotroph bacteria, according to our bacteria determinations in a special case, are still active in 80 cm depth. Thus, after 95–99% of the initial production has been destroyed in the aerobic surface layers, biological decay processes continue to some extent under anaerobic conditions.

Therefore, the decomposition of organic matter is a fast and efficient process, which, in this environment, is largely controlled in the deeper sediment layers by anaerobic heterotroph bacteria.

Quantitative data on *Desulfovibrio* activity have been obtained by injecting $^{35}SO_4^{--}$ into the mats at specific depth and by sampling the radioactive S^{--} produced and the remaining radioactive sulfate (Jørgensen and Cohen, in press). It was thus found that not only *Desulfovibrio* contributes at least 10% of the decay activity in the upper layers, but that it is still active in depths within the algal mat where enough organic matter remains for such high reduction rates. It is inferred from this method, that organic compounds, mainly photosynthates produced at the upper mat region, must migrate approx. 5–10 cm down into the anaerobic parts of the recent stromatolite.

The degradation of organic matter in these laminites, therefore, is even more complicated by the transport of organic matter into deeper parts of the mats. The values for

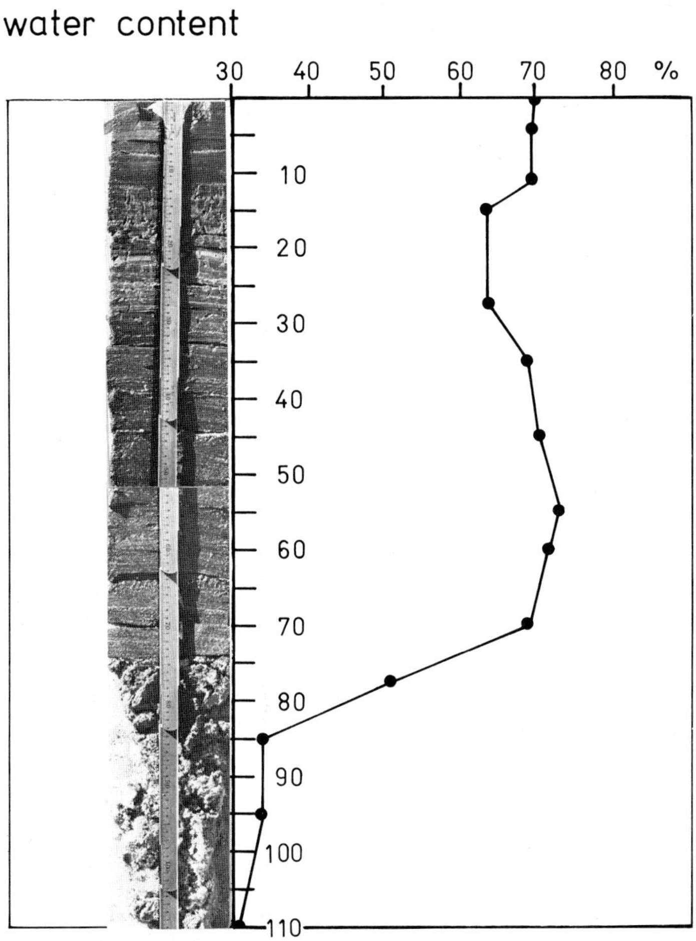

Fig. 9. Complete core of shallow-water algal mat with underlying gypsum sediments of different origin. Note borings by *Berosus*, increase in white carbonate particles with depth and fine lamination. At 78 cm, age determinations yielded 2400 years BP. The water content of the mat is still high, indicating ongoing processes of degradation and diagenesis, while the gypsum sediments are already dehydrated

---

Fig. 10. Upper parts of the flat shallow-water algal mat (*D I*) with sample locations of ▶ SEM-micrographs from different depths. *Oscillatoria, Lyngbya*, and *Microcoleus* bunches with adhering flexibacteria are still biologically active at the surface. In 3–5 cm depth *Microcoleus* bunches are already deforming and the algal cell sap is released to the environment. Hollow algal filaments later are either completely disintegrated (*bottom left* and Figs. 11–13) or encrusted with carbonates (*bottom right* and Fig. 15). Chitinous substances (*Artemia salina*) are better preserved. (═══ = 10 μm, ——— = 1 μm)

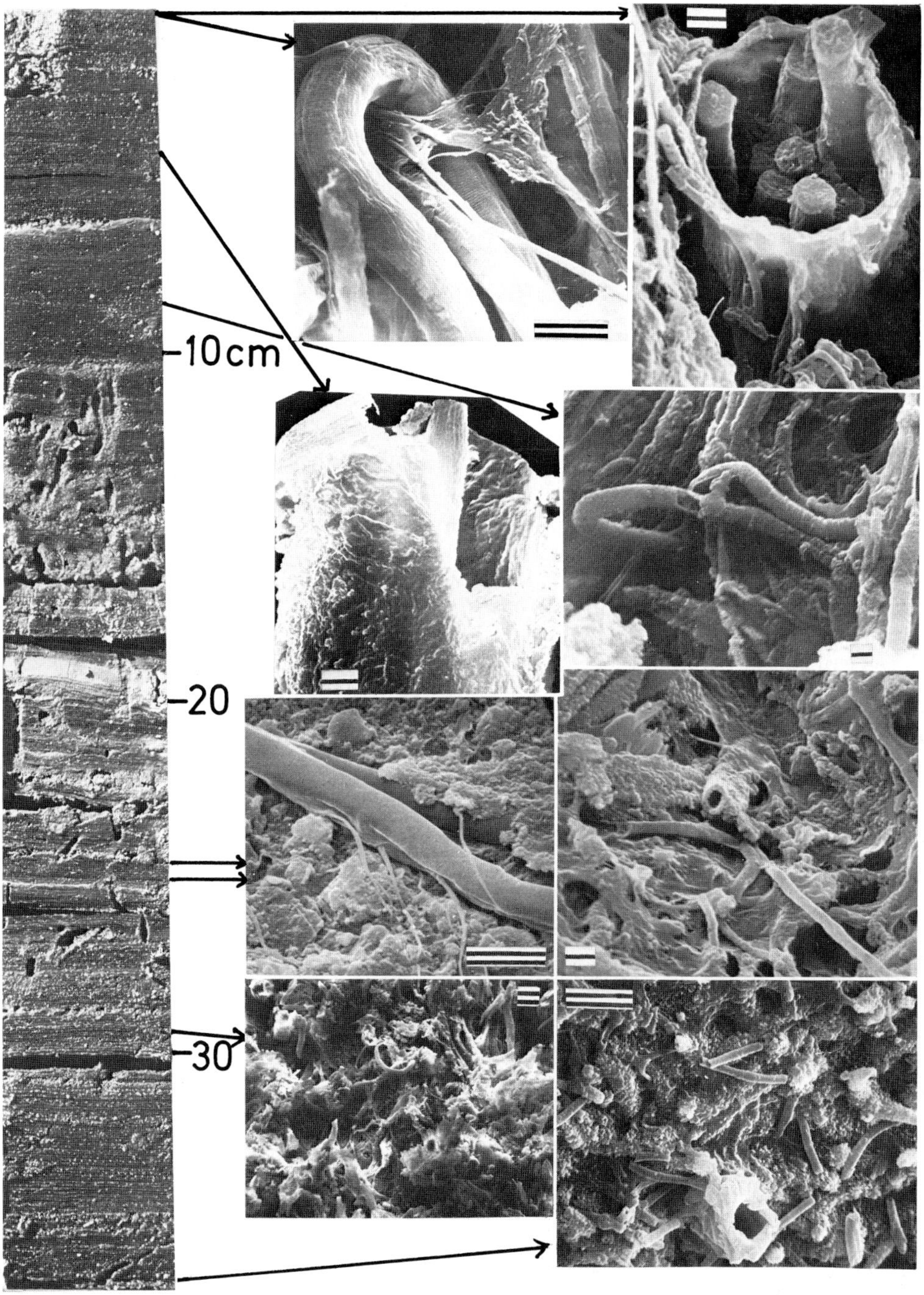

Fig. 10. Legend see opposite page

organic C, organic N, and the carbon nitroben ratios also show that decay processes are going on until approx. 30 cm depth within the algal mat (Krumbein et al., in press). This activity seems to be accelerated by boring organisms perforating the upper parts of the mat (a Coleoptera species, *Berosis* sp., is responsible for the borings; Por, pers. comm.) (Fig. 10).

Later, after the degradation by microorganisms, early diagenesis continues the decay processes. Then organic compounds may start to migrate to sediments close to this environment, which have higher porosities and lower concentrations of organic matter.

3.5.2 Destruction of Structures

In Fig. 10 we have collected some information on the pathway of gradual deformation and decomposition of organic structures with increasing depth within a laminated algal mat. Comparing the outlook of single trichomes and bunches of trichomes of e.g. *Microcoleus* sp. at various depths within the mat, one sees the rapid decay of structures within such an environment. Chitinous substances are better-preserved during these early stages of diagenesis. But most of the algal structures vanish relatively fast. Figs. 11, 12, and 13 demonstrate in some detail, by samples from different depths, the rapid transformation of algal filements into unrecognisable structures. This process is combined with the genesis of carbonate minerals and aggregates, replacing in many cases the original organic structures. The quantitative relationships of such processes have been calculated for recent marine sediments by Hartmann et al. (1973).

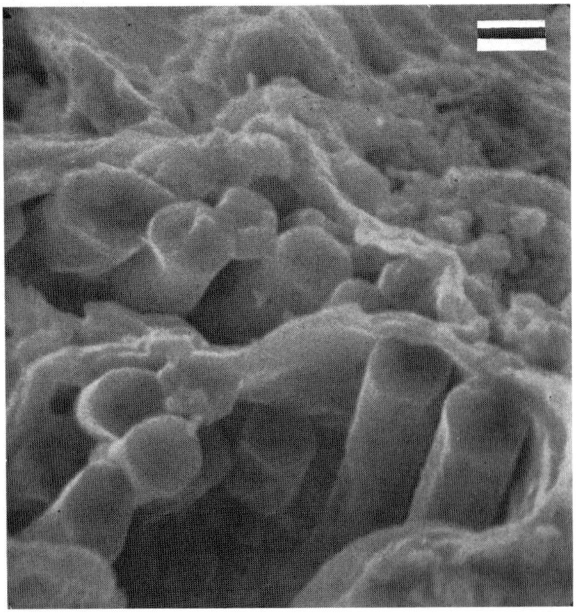

Fig. 11. SEM-micrograph of *Oscillatoria* filaments from surface layers of algal mat.
——— = 5 μm

Fig. 12. *Oscillatoria* filaments from deeper parts of the mat. Slime coatings are decayed. *O.* itself seems to disintegrate into newly forming mucous substance. ——— = 10 μm

Fig. 13. *Oscillatoria* from 40 cm depth. Relatively few isolated remainders of trichomes occur in micritic carbonate layers with the typical sandwich structure (see text). ——— = 10 μm

As can be seen from Figs. 10 and 14, filamentous bacteria, flexibacteria, rods, cocci, and possibly fungi are mainly involved in the decay of algal filaments and coccoid algae. Krumbein and Cohen (1974) have also shown that diatoms may be dissolved by biogeochemical reactions and direct bacterial contact, since the pH-values, ranging from 6.7–7.3 do not favor chemical silica dissolution.

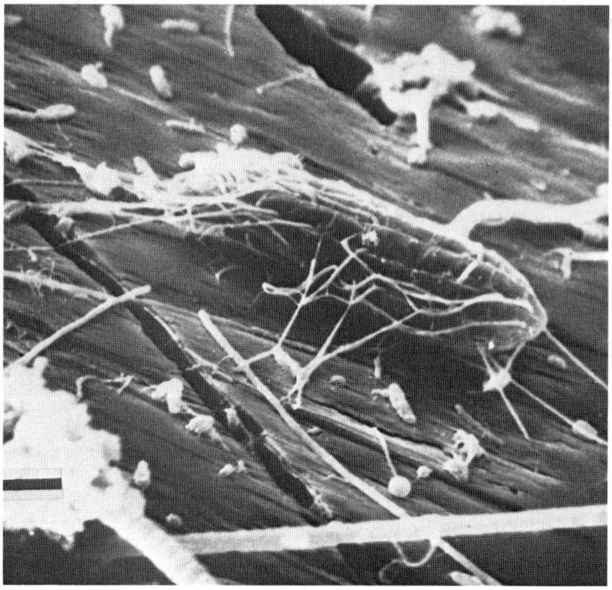

Fig. 14. Decay environment in algal mats. A large coccoid cyanobacterium (probably *Aphanothece*) is settling on the mucus sheath of a filamentous algal bunch. The coccoid alga is attacked and perforated by fungal or bacterial hyphae. Filamentous bacteria, rods, and cocci are dwelling within the decay environment. ——— = 2 μm

3.5.3 Preservation of Organisms

Although rapid and efficient decay processes are going on at large scale within recent algal mats (Figs. 10, 14), several organic structures may be fossilized by the precipitation of carbonates on the cell walls of algae. Fig. 15 demonstrates in a closer view that not only structureless carbonates form on the algal cell, but that some of the aragonite

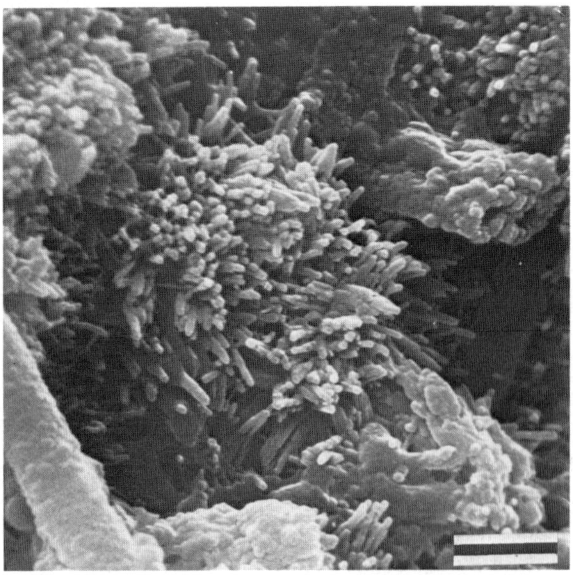

Fig. 15. Micritic coatings (*left*) and needle crystallites occur on empty trichomes of *Microcoleus* and *Oscillatoria*. ——— = 2 μm

needles which form on the algal filament are "fossilized" bacteria. Since these processes take place in depths below 5 cm within the algal mat (Fig. 10), it is concluded that not photosynthetic CO_2 uptake, but mainly complicated biogeochemical reactions during the bacterial degradation of algal material are responsable for eventual calcification of algal structures within this environment. Mineralogical analyses of this stromatolite have shown that, in such environments, extremely rich in organic matter, the physico-chemical conditions for carbonate formation are outweighed by biogeochemical reactions. Under the exclusion of fresh water (desert conditions) and marine detrital sediments (closed-off sea-marginal lake), with no gypsum precipitating within the mat, the presence of almost all forms of carbonates occurs in recent environments (aragonite, low and high Mg-calcite, and monohydrocalcite as well as dolomite). More information on the preservation of organic structures and of possible relation of the organic matter to calcification are given by Krumbein (1974, 1975) and Krumbein and Cohen (1974), and Krumbein et al. (in press).

3.4.5 Preservation of Structures

The preserved structures of organisms may be of sufficiently different size or shape, so that the original organism itself cannot be related to the preserved structure. In Fig. 16 we have taken an example where algal filaments of *Microcoleus* are transformed by decay processes into thick pellet-like elongated structures of Mg-calcite, which are orders of magnitude larger than the algal filament bunches. Since no tubes or smaller structures have been detected there by SEM methods, we can only deduce from the knowledge of the recent environment that such structures have been bunches of algal filaments before. The onkoids to the right of the elongated structures in some cases are

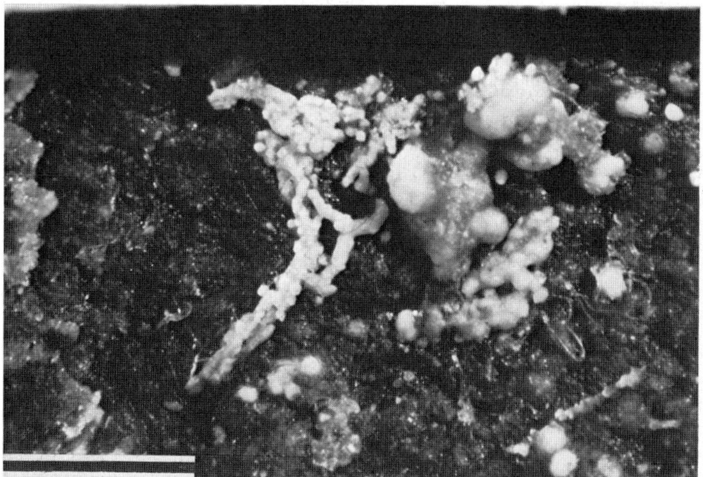

Fig. 16. Filament structures and onkoidal structures, both much larger than the original organisms, embedded in organic material of decaying algal mats. ——— = 2mm

caused by bacterial precipitation of carbonates, as is shown in Fig. 17. This bacterial carbonate spherulite has been produced in the laboratory within a pure culture of a rod-shaped heterotroph bacteria from Solar Lake algal mats, by adding 0.01% of lactic acid to seawater of 28‰ salinity. Sterile control did not yield any carbonates (Krumbein, 1973, 1974). Such spherules may as well be initiated on the mucous envelope of coccoid algae like *Aphanocapsa* and *Aphanothece*, occurring frequently in the summer shallow-water algal mat, a process which is demonstrated by Krumbein (1975).

Fig. 17. Bacterial spherulite (———— = 5 μm) from laboratory culture of a heterotroph rod isolated from algal mats. The spherulites form only in inoculated samples of seawater enriched with low nutrient concentrations.. Sterile control sets did not yield precipitates. Bacteria cells are smooth

Organisms are not the only structures which are preserved within these algal mat environments. Much more prominent and important is the preservation of the lamination of stromatolitic algal mats, be it a daily lamination as demonstrated by Monty (1967) and Gebelein (1969), or a seasonal lamination, as in the case of Solar Lake shallow-water algal mats. Due to different organism associations, and subsequently different calcification in the case of Solar Lake, or caused by different types of sediment-trapping in the case of Bermuda mats and many other stromatolitic environments, laminated sedimentary rocks are developing. Subrecent examples of lithified stromatolitic environments occur also along the Gulf of Aqaba, an example of which is shown in Figs. 18 and 19. Along the shores of the Gulf of Aqaba there are many elevated coral reefs and other pleistocene indication of sea level changes. Thus, it can be shown that the sea level has been higher, and also lower, than the present sea level. Therefore it is not astonishing that stromatolitic environments have been buried and preserved during an ancient higher sea level stage. In Fig. 18 we look into such an ancient closed-off lagoon or sea-marginal environment which has been uncovered and partially eroded by a recent Wadi outlet. The bright layers are more or less carbonate-gypsum mixtures of merely evaporitic sediments, while the dark layers and parts of the white layers at closer view are ancient algal laminites with the typical carbonate structures described previously.

Fig. 18. Outcrop of Pleistocene stromatolitic laminites south of Ras Burqa, approx. 40 km south of Elat (Gulf of Aqaba). The laminites witness higher sea levels with sea-marginal lagoons and algal mat formation. The sediments have been partially eroded by Wadi erosion. The outcrop is situated 150 m from the present shore

Fig. 19. Carbonate-gypsum-organic matter laminite from Fig. 18. The white layers are gypsum and carbonate. The brown layers are organic matter and carbonate. Algal material has been further condensed and compacted in comparison to recent examples from Solar Lake. Structures resembling algal filaments or bacteria have not yet been detected in these sediments. ——— = 1 cm

3.6 Bibliography

Brock, T.D.: Evolutionary and ecological aspects of cyanophytes. In: The Biology of Blue-green Algae. Carr, N.G., Whitton, B.A. (eds.). London: Blackwell 1973, pp. 487–500.

Brock, T.D., Brock, M.L.: The measurement of chlorophyll, primary productivity, photophosphorylation and macromolecules in benthic algal mats. Limnol. Oceanogr. **12**, 600–605 (1967).

Cohen, Y.: Dynamics of procaryotic photosynthetic communities of the Solar Lake. Ph. D. thesis, Jerusalem (1975), 125 p.

Cohen, Y., Padan, E., Shilo, M.: Facultative anoxygenic photosynthesis in the cyanobacterium Oscillatoria limnetica. J. Bacteriol. **123**, 855–861 (1975a).

Cohen, Y., Jørgensen, B.B., Padan, E., Shilo, M.: Sulphide dependent anoxygenic photosynthesis in the cyanobacterium *Oscillatoria limnetica*. Nature **257**, 489–492 (1975b).

Castenholz, R.W.: Ecology of blue-green algae in hot springs. In: The Biology of Blue-green Algae. Carr, N.G., Whittom, B.A. (eds.). Oxford: Blackwell 1973, p. 379–414.

Doemel, W.N., Brock, T.D.: Bacterial stromatolites: Origin of laminations. Science **184**, 1083–1085 (1974).

Fenchel, T., Riedl, R.J.: The sulfide system: A new biotic community underneath the oxidized layer of marine sand bottoms. Mar. Biol. **7**, 255–268 (1970).

Friedman, G.M., Amiel, A.G., Brown, M., Miller, D.S.: Generation of carbonate particles and laminites in algal mats, example from sea-marginal pool gulf of Aqaba, Red Sea. Am. Ass. Petrol. Geol. Bull. **57**, 541–557 (1973).

Gebelein, C.D.: Distribution, morphology and accretion rate of recent algal stromatolites, Bermuda. J. Sed. Petrol. **39**, 49–69 (1969).

Golubic, S.: The relationship between blue-green algae and carbonate deposition. In: The Biology of Blue-green Algae. Carr, N.G., Whitton, B.A. (eds.). Oxford: Blackwell 1973, p. 434–472.

Hartmann, M., Müller, P., Suess, E., Van der Weijden, C.H.: Oxidation of organic matter in recent sediments. Meteor Forsch.-Ergebnisse, Reihe C, **12**, 74–86 (1973).

Jørgensen, B.B., Cohen, Y.: Solar Lake V Algal mats sulfur cycle. Limnol. Oceanogr. (in press).

Jørgensen, B.B., Fenchel, T.: The sulfur cycle of a marine sediment model system. Mar. Biol. **24**, 189–201 (1974).

Krumbein, W.E.: Sediment microbiology and grain size distribution as related to tidal movement, during the first mission of the German underwater laboratory "Helgoland". Marine Biol. **10**, 101–102 (1971).

Krumbein, W.E.: Mikrobiologische Untersuchungen zur Fällung von Kalziumkarbonat aus Meerwasser. In: Bah (Ed.) Jahresberichte, Hamburg 1973, p. 50–54.

Krumbein, W.E.: On the precipitation of aragonite on the surface of marine bacteria. Naturwissenschaften **61**, 167 (1974).

Krumbein, W.E.: Biogenetic monohydrocalcite spherules in lake sediments of Lake Kivu (Africa) and the Solar Lake (Sinai). Sedimentology **22**, 631–634 (1975).

Krumbein, W.E., Cohen, Y.: Biogene, klastische und evaporitische Sedimentation in einem mesothermen, monomiktischen ufernahen See (Golf von Aqaba, Sinai). Geol. Rdsch. **63**, 1035–1065 (1974).

Krumbein, W.E., Cohen, Y., Shilo, M.: Solar Lake, Sinai: The stromatolotic algal mats. Limnol. Oceanogr. (in press).

Logan, B.W.: *Cryptozoon* and associated stromatolites from the recent Shark Bay, Australia. J. Geol. **69**, 517–533 (1961).

Logan, B.W., Rezak, R., Ginsburg, R.N.: Classification and environmental significance of algal stromatolites. J. Geol. **72**, 68–83 (1964).

Monty, C.L.V.: Distribution and structure of recent stromatolitic algal mats, eastern Andros islands, Bahamas. Ann. Soc. Geol. Belg. **90**, 55–99 (1967).

Monty, C.V.L.: Recent algal stromatolitic deposits, Andros Islands, Bahamas. Preliminary report. Geol. Rdsch. **61**, 742–783 (1972).

Schramm, W.: Kontinuierliche Messung der Assimilation und Atmung mariner Algen mittels der elektrochemischen Sauerstoffbestimmung. Helgol. Wiss. Meeresunters. **13**, 275–287 (1968).

Walter, M.R.S., Golubic, S., Preiss, W.V.: Recent stromatolites from hydromagnesite and aragonite depositing lakes, near Coorong Lagoon, South Australia. J. Sedimentol. Petrol. **43**, 1021–1030 (1973).

4 Skeletal Stromatolites

R. RIDING, Cardiff, Great Britain

4.1 Introduction

Stromatolites are usually regarded as organo-sedimentary structures seldom preserving the organisms responsible for their formation. The interpretation of fossil stromatolites as being produced by the trapping and binding of sediment by micro-organisms, particularly blue-green algae, is essentially based on examination of Recent living stromatolite surfaces. Yet some stromatolites do contain abundant calcified filaments and consequently show distinctive differences in internal structure from 'normal' stromatolites. I propose to term them *skeletal stromatolites* and *skeletal oncolites*, defined briefly as: stromatolites in which the organisms responsible for their formation are commonly preserved as calcified fossils. 'Normal' stromatolites, in which the micro-organisms are not calcified, are termed nonskeletal stromatolites.

4.2 Stromatolites, Oncolites, and Porostromata

Stromatolites with internal microstructures of fine tubes have been recognized before, but their importance has been obscured by terminological confusion. This stems from Pia's (1927, pp. 36–40) distinction between Spongiostromata (stromatolites and oncolites) and Porostromata (tubiform calcareous microfossils). At first sight this is a useful distinction to make because it separates external form from internal microstructure. But the distinction has been misconstrued and the Spongiostromata and Porostromata have come to be regarded as mutually exclusive groups of fossils, in which the Spongiostromata are laminated deposits with 'spongiostrome', i.e. fenestral micrite, microstructures, and the Porostromata are tubiform microfossils. Thus, Johnson (1963, p. 29) regarded Pennsylvanian Spongiostromata as having "little or no microstructure" but he placed oncolites such as *Osagia* Twenhofel and *Somphospongia* Beede, which contain internal tubules, in the Porostromata. Similarly, Johnson (1961, p. 1915) placed *Ottonosia* Twenhofel in the Porostromata. *Osagia* and *Ottonosia* are oncolites, as Pia (1927) recognized, but they are composed in part by calcified tubes, to some of which porostromate generic names can be applied. Hence, they are oncolites composed largely of calcareous tubes, mainly Porostromata, and are regarded here as skeletal oncolites.

Referral of skeletal stromatolite (including oncolite) genera to the Porostromata has not only been confusing, in that it has ranked genera such as *Ottonosia* along with *Girvanella* Nicholson and Etheridge, but it has also been inconsistent. Johnson (1963, p. 30) tabulated a variety of Pennsylvanian stromatolites, including *Cryptozoon* Hall,

Pycnostroma Gürich, and *Spongiostroma* Gürich, in which he noted the presence of fine tubes, but he nevertheless retained them in the Spongiostromata.

The terms skeletal and nonskeletal stromatolite recognize the distinction between external form and internal structure, but also emphasize their interrelationships and promote evaluation of the significance of calcification of organisms within stromatolites. The term 'skeletal" may be used as a prefix to either generic (*Collenia, Cryptozoon*) or geometric (SH, LLH) descriptions of stromatolites.

4.3 Stromatolites and Tufa

Blue-green algae are recorded in carbonate rocks by calcified filaments and microspherulites (which can both build skeletal stromatolites), by nonskeletal stromatolites, and by moulds and encrustations in tufa deposits. Two processes are involved, precipitation of $CaCO_3$ and trapping of loose sediment, both of which are probably operative in each case. Nevertheless, it is likely that one of the processes will dominate the other. Thus, the formation of nonskeletal stromatolites essentially involves sediment trapping, while tufa formation is dominated by $CaCO_3$ precipitation. Skeletal stromatolites are dominated by filaments and microspherulites preserved due to the precipitation of $CaCO_3$, but they also often contain fenestral micrite fabrics characteristic of the trapping of fine carbonate sediment. Consequently skeletal stromatolites can be regarded as intermediate between nonskeletal stromatolites and tufa. However, this is an oversimplification. Filaments in skeletal stromatolites are discrete, relatively thin-walled tubes. They are not usually thickly encrusted by $CaCO_3$ nor are they usually moulds within massive deposits of $CaCO_3$. The morphological constancy of these tubiform fossils suggests a degree of biological control over the calcification process (Riding, submitted). Furthermore, skeletal stromatolites do not show the massive crystalline dripstone fabrics characteristic of gratuitous physicochemical precipitation of $CaCO_3$. As a result, it is necessary to distinguish between predominantly physicochemical and predominantly biologically controlled precipitation of $CaCO_3$ in tufa (marine and nonmarine) and skeletal stromatolites respectively (Fig. 1).

4.4 Varieties of Skeletal Stromatolites

Algae which always calcify can produce laminated deposits such as the rhodolite nodules formed by red crustose corallines. But these are not usually regarded as a type of oncolite. Similarly, foraminifers can produce layered colonies (Wendt, 1969) which are stromatolite-like in form, but which are not likely to be considered as true stromatolites. Skeletal stromatolites differ from these in that they are formed by the calcification of organisms which are not obligate calcifiers and which, when uncalcified, are still capable of constructing stromatolites. This feature accounts for the intimate interrelationship between skeletal and nonskeletal stromatolites and it emphasizes the importance of being able to explain the mechanism of calcification involved. Nevertheless, organisms

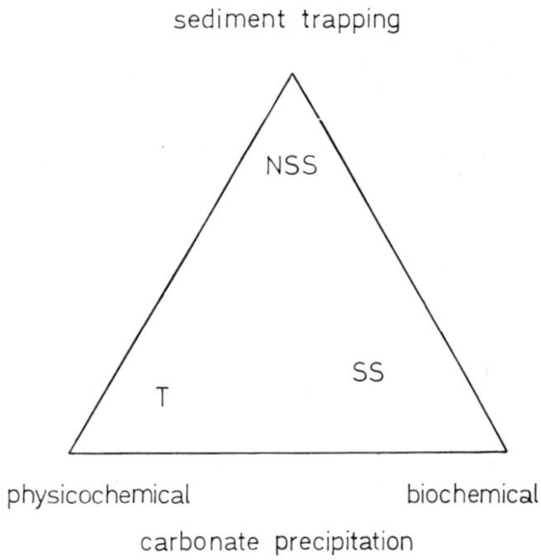

Fig. 1. Relationships between processes of carbonate deposition in the formation of nonskeletal stromatolites (*NSS*), tufa (*T*), and skeletal stromatolites (*SS*)

which always produce skeletons do occur within some skeletal stromatolites, as in the association of a foraminifer with *Girvanella* in *Osagia* reported by Johnson (1946). That skeletal stromatolites should act as substrates for encrusting organisms is reasonable, since the firmness engendered by calcification would promote settlement and attachment. But despite the addition of encrusters, a review of the geological literature indicates that skeletal stromatolites are formed essentially by porostromate fossils. In this respect skeletal stromatolites could equally well be regarded as porostromate stromatolites. However, it is likely that the status of the Porostromata will have to be scrutinized as the affinities of its component genera become better known, and the term skeletal stromatolite is considered preferable.

The form and appearance of many porostromate fossils suggests that calcification took place during life, but a distinctive group of skeletal stromatolites are formed by calcareous microspherulites which probably represent coccoid blue-green algae calcified during or after death (Krumbein and Cohen, 1974). Consequently it is possible that skeletal stromatolites can be constructed by organisms calcified during life or post mortem.

The internal structure of skeletal stromatolites is largely controlled by the form and mode of growth of the constituent organisms, together with the relative importance of calcification and sediment trapping effects. Calcified coccoid blue-greens produce a thrombolitic-spherulitic microfabric, while porostromate genera produce tubiform microfabrics. The orientation of the porostromate forms, whether prostrate or erect, exerts an additional important influence on the microstructure. *Ortonella* Garwood and *Garwoodia* Wood may form erect tufts separated by large sparite areas so that skeletal stromatolites produced by them commonly have an open, coarsely fenestral, microstructure, although in skeletal oncolites they can form denser growths. They contribute significantly to both skeletal stromatolites and oncolites in the British Lower Carboniferous (Garwood, 1931). Erect growths of *Cayeuxia* Frollo form comparable skeletal stromatolites in the Middle Jurassic of Scotland (Hudson, 1970). In contrast the prostrate filaments of *Sphaerocodium* Rothpletz emend. Wray produced more compact layered growth which form impressive skeletal stromatolites in the Upper Devonian of Western

Australia (Wray, 1967) and western Canada (Wray and Playford, 1970). The more tangled, but also normally prostrate, filaments of *Girvanella* contribute importantly to skeletal oncolites (Johnson, 1961, Pl. 86), but appear to be less conspicuous in skeletal stromatolites.

The control of calcification in the better-known porostomate genera, such as *Girvanella* and *Ortonella*, which represent calcified sheaths of filamentous blue-green algae (Riding, 1975), is believed to be to some degree biological, while remaining under an overall environmental control (Riding, submitted). In this respect it is interesting to note that skeletal stromatolites occur most commonly in marginal marine situations where evaporation or low salinity effects, or both, could promote calcification of the stromatolite-building micro-organisms.

4.5 Bibliography

Garwood, E.J.: The Tuedian Beds of northern Cumberland and Roxburghshire east of the Liddel Water. Quart. J. Geol. Soc. London **87**, 97–159 (1931).

Hudson, J.D.: Algal limestones with pseudomorphs after gypsum from the Middle Jurassic of Scotland. Lethaia **3**, 11–40 (1970).

Johnson, J.H.: Lime-secreting algae from the Pennsylvanian and Permian of Kansas. Geol. Soc. Am. Bull. **57**, 1087–1120 (1946).

Johnson, J.H.: Limestone-building algae and algal limestones. Boulder: Colorado School Mines 1961.

Johnson, J.H.: Pennsylvanian and Permian algae. Colorado School Mines Quart. **58**, 3, 1–211 (1963).

Krumbein, W.E., Cohen, Y.C.: Biogene, klastische und evaporitische Sedimentation in einem mesothermen monomiktischen ufernahen See (Golf von Aqaba). Geol. Rundschau **63**, 1035–1065 (1974).

Pia, J.: Thallophyta. In: Handbuch der Paläobotanik. Hirmer, M. (ed.). Pt 1. Munich: Oldenbourg 1927, pp. 31–136.

Riding, R.: *Girvanella* and other algae as depth indicators. Lethaia **8**, 173–179 (1975).

Riding, R.: Calcified *Plectonema* (blue-green algae), a Recent example of *Girvanella* from Aldabra atoll. Palaeontology (in press).

Wendt, J.: Foraminiferen-„Riffe" im karnischen Hallstätter Kalk des Feuerkogels (Steiermark, Österreich). Paläontol. Z. **43**, 177–193 (1969).

Wray, J.L.: Upper Devonian calcareous algae from the Canning Basin, Western Australia. Colorado School Mines Prof. Contr. 3, 76 pp. (1967).

Wray, J.L., Playford, P.E.: Some occurrences of Devonian reef-building algae in Alberta. Bull. Canadian Petrol. Geol. **18**, 544–555 (1970).

5 Environmental Significance of Foraminiferal-Algal Oncolites

T.M. PERYT, Warsaw, Poland

5.1 Introduction

Algal oncolites are regarded as indicators of shallow-water environment and the size of oncolitic forms is often regarded as an indication of turbulence: the bigger the forms, the stronger the turbulence assumed. Similar environmental significance is also usually ascribed to foraminiferal-algal oncolites. The well-known example of foraminiferal-algal oncolite is *Osagia*, which was ascertained by Johnson (1946, p. 1103) as being composed of intergrown blue-green algae (Porostromata) and encrusting foraminifera. *Osagia* is often regarded as environmental index fossil and clear, shallow (0–10 feet), warm, gently agitated water is assumed. There is no doubt that such interpretation is an overgeneralization. This is shown by two other examples of foraminiferal-algal oncolites: soft delicate "tender" oncolites from the Zechstein Limestone (Upper Permian) and *Sphaerocodium* from the Alpine Triassic.

5.2 Soft Delicate "Tender" Oncolites

In the basinal facies of the Zechstein Limestone of Germany (Füchtbauer, 1968), North Sea (Taylor and Colter, 1975), and Poland (Peryt and Peryt, 1975) a very specific type of algal oncolites/pisolites occurs. It is characterized by a very delicate wavy lamination and a great number of algal envelopes around the nucleus that as a rule is cryptocrystalline grain (Fig. 1A, B). Füchtbauer (1968) estimated the depth of origin of these delicate oncolites, called by him "tender" ones, as not less than 30 m. This supposition was accepted by Peryt and Peryt (1975), while Taylor and Colter (1975) supposed that if the oncolites are autochthonous, the depth limit was approximately 100–150 m. Foraminiferal-algal oncolites are associated with "tender" oncolites, so their depth of origin must be similar. Foraminiferal-algal oncolites are embedded in micritic matrix, often silty and sandy, with small amounts of skeletal material – mainly foraminifera (*Nodosaria-Agathammina* low-energy assemblage); in smaller amounts – ostracods and crinoids (Fig. 1A, B). Tests of sessile tubular foraminifera lie parallel to algal lamination, although in many cases they cover corroded algal structures (see Peryt and Peryt, 1975, Pl. 1, Fig. 1a, b). The latter case is often observed in the close neighborhood of areas of condensed profiles of the Zechstein Limestone. It should be stressed that foraminiferal-algal oncolites are occasionally found in the near-shore zone but their internal structure, as well as accompanying sediments, are quite different and are evidently of shallow-water origin.

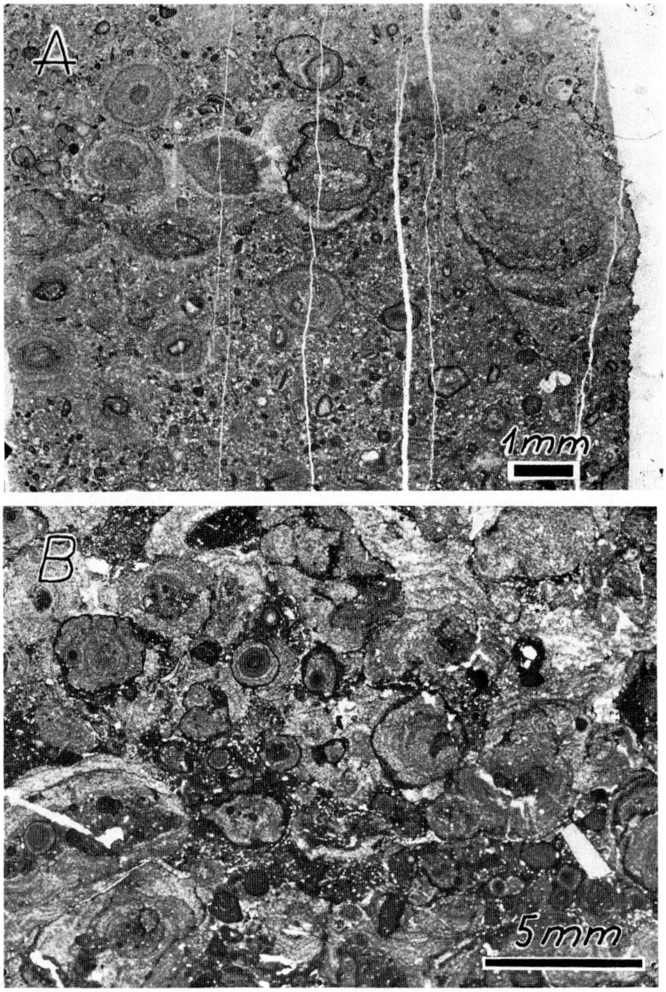

Fig. 1 A and B. Soft delicate "tender" oncolites from the Zechstein Limestone of Poland with sessile tubular foraminifera occurring on the outer algal laminae (*dark zigzag lines*). (A) example from the Banie 1 borehole, depth 3893.0 m. (B) example from the Gdańsk IG 1 borehole, depth 1640.7 m

5.3 Sphaerocodium

The genus *Sphaerocodium* was established by Rothpletz in 1890, and based on *Sphaerocodium bornemanni* from the Upper Triassic Raibl beds of the Eastern Alps. Wood (1948) restudied two other species of *Sphaerocodium* from the Silurian of Gotland and recognized the symbiotic intergrowth of the calcareous alga *Rothpletzella* and the encrusting foraminifera *Wetheredella*. In his description of *Sphaerocodium bornemanni*, Rothpletz (1891, p. 299) had already noted the intergrowth between algae and foraminiferal tests, growing in concentric zones. Wood (1948), Wray (1967), and Flügel and Wolf (1969) described several species of *Sphaerocodium* characterized by branching tubular filaments. On the other hand, ball-like forms with concentric laminae, and showing the main structural features of the original *Sphaerocodium bornemanni* Rothpletz, have been reported from numerous Upper Triassic localities in the German and Austrian Alps, and in Croatia. They have been determined as *Sphaerocodium*, quite often as *Sphaerocodium bornemanni*, and were interpreted as oncolites (Kraus, 1969, p. 106; Babić, 1970, p. 15). *Sphaerocodium*-bearing limestones have been described by many authors (e.g. Schuler, 1968; Jerz, 1966).

I have studied samples from one of the type-localities of *Sphaerocodium bornemanni*, the Zirler Klamm in Tyrol, Austria. The *Sphaerocodium*-bearing limestones are developed as thin beds intercalated in marly rocks (see Jerz, 1966, Figs. 2–4). Petrographically, the samples are pisosparrudites. Multilayered algal oncolites with numerous tests of sessile tubular foraminifera occur within a sandy matrix containing a significant amount of bituminous material, as well as skeletal fragments of echinoids, crinoids, foraminifera, dasyclads, and shells.

Sphaerocodium hitherto was reported as indicating shallow waters, whereas the rounded form was thought to record turbulent waters. It seems strange, however, that these algal balls do not occur in the northern, shallower part of the Raibl facies. This is explained by Jerz (1966, p. 78) as being due to higher salinity in that area. Regarding the well-known resistence of blue-green algae to limitating ecological conditions, such an explanation seems to be unconvincing. Schuler (1968, p. 34–35) also noted the association of *Sphaerocodium* with a condensation horizon; Fig. 2 in Angermeier et al. (1963) suggests a break in the sedimentation and also a condensation for the "Grenzoolith". In any case, there is no strong evidence for shallow water origin. On the contrary, both the microfacies of the *Sphaerocodium*-bearing limestones, as well as the lithological character of the section, suggests a deeper-water sedimentary environment.

5.4 Conclusions and Remarks

Both the Zechstein soft, delicate, "tender" oncolites and the Triassic *Sphaerocodium* have the following features in common:

1. They occur within silty and marly rock types.
2. They form layers insignificant in thickness.
3. Associated sediments and biotas testify to a low-energy sedimentary regime.
4. There exists indirect or direct evidence of a very slow rate of sedimentation and even condensation.

It seems that the above-listed features may serve as an additional criterion of non-shallow-water foraminiferal-algal oncolites (and also may be algal oncolites) in other formations.

Foraminiferal-algal oncolites in both described cases occur within rocks of non-shallow-water origin but their allochthonous origin cannot be excluded; therefore the oncolites may be derived from the neighbouring shoals. Recently, Buchanan et al. (1976) found algal-foraminiferal consortia similar to those described from the Late Paleozoic (for example *Osagia*) in modern carbonate sediments of the Berry Islands, Bahamas. They concluded that the preferred habitat of the algal nodules are the grass-filled channels of the oolite shoal. The nodules are not widespread, although some local transporta-

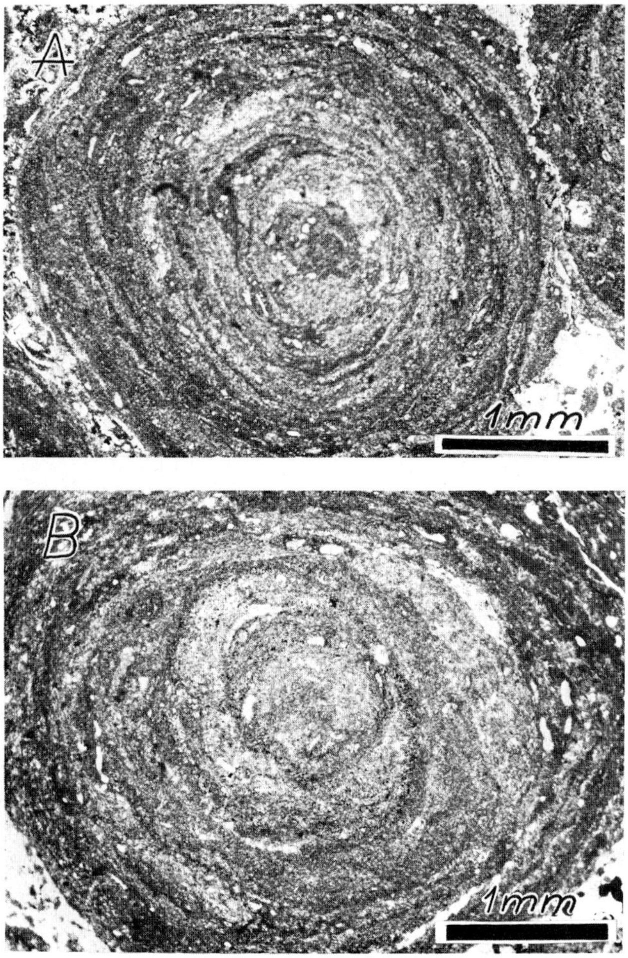

Fig. 2 A and B. *"Sphaerocodium"* from the Raibler beds of Zirler Klamm, Tyrol. Note the abundance of sessile tubular foraminifera inside the oncolites (A) as well as in the outer part of the oncolite (B)

tion may occur during heavy storms. In the Zechstein Limestone and the Raibler beds foraminiferal-algal oncolites are very common and seem mostly to be autochthonous. A detailed facies analysis seems to be necessary for all the different occurrences of foraminiferal-algal (and/or algal) oncolites.

Acknowledgments. I would like to express my sincere thanks to the following persons who helped me during my study of the problem: Hugh Buchanan, Peter Fay, Erik Flügel, G.E. Fogg, Z.J. Kotański, Danuta Peryt, T.S. Piatkowski, W.D.P. Stewart and J.L. Wray.

5.5 Bibliography

Angermeier, H.-O., Pöschl, A., Schneider, H.-J.: Die Gliederung der Raibler Schichten und die Ausbildung ihrer Liegendgrenze in der „Tirolischen Einheit" der östlichen Chiemgauer Alpen. Mitt. Bayer. Staatssamml. **3**, 83–105 (1963).

Babić, L.: Sphaerocodium or onkoids from the Upper Triassic dolomite of western Yugoslavia? Geol. vj. **23**, 11–20 (1970).

Buchanan, H., Streeter, S.S., Gebelein, C.D.: Recent algal-foraminiferal consortia from the Berry Islands, Bahamas. J. Sediment. Petrol., in press (1976).

Flügel, E., Wolf, K.H.: „Sphaerocodien" (Algen) aus dem Devon von Deutschland, Marokko und Australien. N. Jb. Geol. Paläontol. Mh. **1969**, 88–103 (1969).

Füchtbauer, H.: Carbonate sedimentation and subsidence in the Zechstein basin (northern Germany). In: Recent Developments in Carbonate Sedimentology in Central Europe. Müller, G., Friedman, G.M. (eds.). Berlin–Heidelberg–New York: Springer 1968, pp. 196–204.

Jerz, H.: Untersuchungen über Stoffbestand, Bildungsbedingungen und Paläogeographie der Raibler Schichten zwischen Lech und Inn (Nördliche Kalkalpen). Geol. Bavarica **56**, 3–102 (1966).

Johnson, J.H.: Lime-secreting algae from the Pennsylvanian and Permian of Kansas. Geol. Soc. Am. Bull. **57**, 1087–1120 (1946).

Kraus, O.: Die Raibler Schichten des Drauzuges Südliche Kalkalpen. Lithofazielle, sedimentpetrographische und paläogeographische Untersuchungen. Jb. Geol. B.A. **112**, 81–152 (1969).

Peryt, T.M., Peryt, D.: Association of sessile tubular foraminifera and schizophytic algae. Geol. Mag. **112**, 612–614 (1975).

Rothpletz, A.: Über *Sphaerocodium Bornemanni*, eine neue fossile Kalkalge aus den Raibler Schichten der Ostalpen. Bot. Zbl. **52**, 9 (1890).

Rothpletz, A.: Fossile Kalkalgen aus den Familien der Codiaceen und der Corallineen. Z. deut. Geol. Ges. **43**, 295–322 (1891).

Schuler, G.: Lithofazielle, sedimentologische und paläogeographische Untersuchungen in den Raibler Schichten zwischen Inn und Salzach (Nördliche Kalkalpen). Erlanger Geol. Abh. **71** (1968).

Taylor, J.C.M., Colter, V.S.: Zechstein of the English sector of the southern North Sea basin. In: Petroleum and the Continental Shelf of North-West Europe, Vol. 1 Geology. Woodland, A.W. (ed.). London: Applied Science Publishers 1975, pp. 249–263.

Wood, A.: *"Sphaerocodium"*, a misinterpreted fossil from the Wenlock Limestone. Proc. Geol. Assoc. **59**, 9–22 (1948).

Wray, J.L.: Upper Devonian calcareous algae from the Canning Basin, Western Australia. Prof. Contr. Colorado School Mines **3** (1967).

6 Stromatolites from the Precambrian Transvaal Dolomite of NE-Transvaal, South Africa

R. FÖRSTER, München, and H. WACHENDORF, Braunschweig, W. Germany

6.1 Introduction

Stromatolites of the South African Precambrian have been abundantly described. In 1935 Young compared the stromatolites of the Transvaal Dolomite with recent algal structures, at that time discovered by Black (1933) in the Bahamas. Truswell and Eriksson (1972, 1973, 1974) and Button (1971, 1973) gave a valuable classification of these Precambrian stromatolites and typified them as tidal-flat associations.

6.2 Geographic and Geologic Setting

The stromatolites were found during a visit to the Drakensberg escarpment in Eastern Transvaal. In the area between Pilgrims Rest, Ohrigstad and the Blyde River Canyon occur good exposures of the Transvaal Dolomite and of stromatolitic structures.

The Transvaal System rests unconformably on the basement of the Kaapvaal Craton, an epicontinental, flat-lying, sedimentary succession deposited in an early Proterozoic intracratonic basin. The Archaic granite is overlain by a sequence of clastic sediments with interlayered volcanics, which according to Button (1973), may reach a thickness up to 750 m. This Wolkberg Group passes transitionally upwards to the Black Reef Quartzite (max 15 m), which in turn grades into the Transvaal Dolomite. This mainly carbonatic sequence, with a thickness up to 2000 m and an extension of more than 100.000 km^2, is one of the oldest and most widespread early Proterozoic shelf-carbonate deposits of the world (Fig. 1). Radiometric datings of the lava just above the Transvaal Dolomite have yielded an Rb/Sr whole rock age of 2.224 ± 21 m.y. (Crampton, in Eriksson and Truswell, 1974), representing a minimum age for the dolomite. The uppermost part, the Pretoria Group, consists of approximately 7000 m of argillaceous and siliciclastic strata with interbedded volcanic horizons.

6.3 Lithographical Description

Several quartzitic beds are interlayered in the basal part of the Transvaal Dolomite, some of them showing an intensive red color. Recristallized grains show an internal zonal arrangement of hematite, relic of an original hematite crust and documenting a

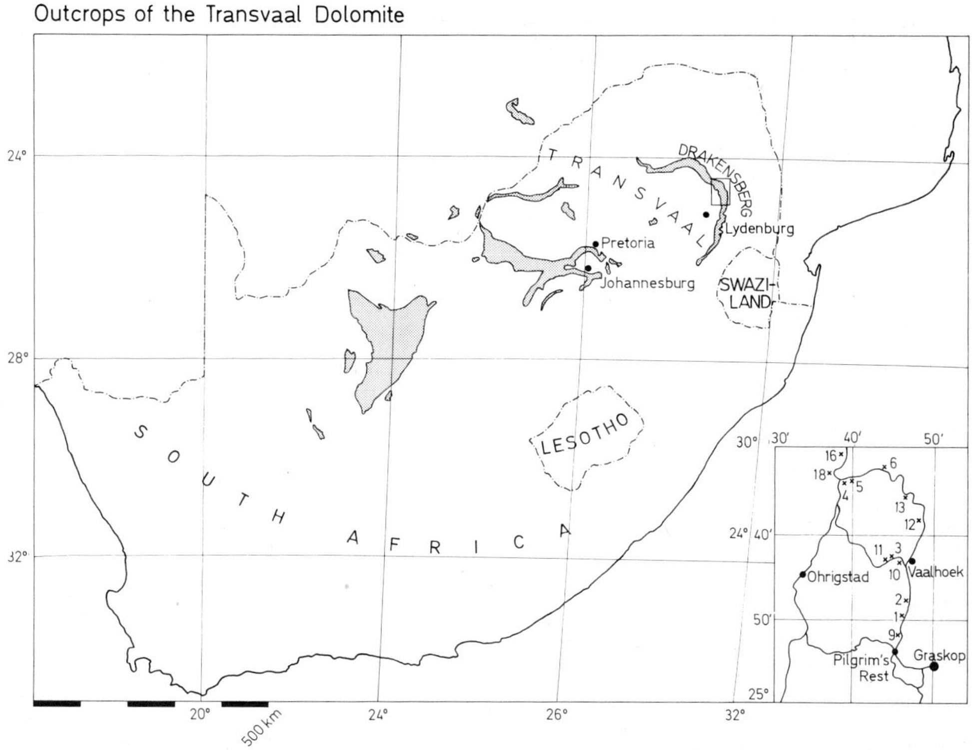

Fig. 1. Map showing the outcrops of the Transvaal Dolomite. Inset showing the area of study with positions of mentioned localities

Precambrian oxydative weathering. Black shales, occasionally concretionary and containing pyritic seams, attain a thickness of several meters and are confined to the lower parts of the sequence. Most parts of the Transvaal Dolomite are composed of sucessions of grayish to blueish, conspicuously banded and finely laminated dolomites or calcareous dolomites, with subordinate bands or lenses of cherts in its higher parts. The alternations are well documented by color differentiations and a delicate relief on weathered surfaces. With the exception of the more siliceous varieties, the total carbonate content lies above 90%, as shown by the representative sample 1b (Table 1). Microscopic studies reveal a fine-laminated mosaic of dolomicrite with intervening dolosparite, filling irregular lenses or layers. The original organosedimentary alternations are often partly obscured by recrystallization. Locally, the carbonates have been replaced by microcrystalline silica. The cherts form bands and nodules up to 25 cm thickness, generally in association with the domal structures. Relics of the presumably organic material are concentrated in opaque seams or as a dusty pigment in the carbonatic and siliceous matrix. Likewise, bands of oolitic dolomites are particularly associated with domal stromatolites. They may reach up to 30 cm and show occasionally a reversed grading. Even the intercolumnar spaces can be oolitic.

Table 1. Geochemical criteria of stromatolitic rocks. Ca and Mg have been determined by complexometry, total Fe-content by AAS; noncarbonate carbon was investigated by the combustion method

Sample	% Ca	% $CaCO_3$	% Mg	% $MgCO_3$	% Fe	Non-carbonate C %	Remarks (X-ray diffraction)
1 b	18.07	45.04	10.45	36.26	0.95	0.7	dolomite
2	22,04	55.05	12.40	43.0	0.48	non det.	dolomite
6 a	20.04	50.04	10.45	36.26	2.67	0.04	oolit. dolomite
12	17.64	44.04	9.48	32.88	1.36	0.10	calc. dolomite
13 a	13.23	33.03	4.86	16.86	0.90	non det.	calc. dolomite with chert
13 b	12.42	31.03	6.56	22.77	1.20	0.05	calc. dolomite with chert

6.4 Description of Stromatolites of the Drakensberg Area

Tabular thin sheets predominate among the stromatolitic structures. They are delicately banded with individual mm-sized laminae, and frequently they show an angular crinkling, constituting relics or imprints of ripple marks. Domal structures with diameters ranging from a few cm to several meters are the prevailing growth forms. Spheroidal forms and encapsulated or columnar structures are less common. Contrary to contemporaneous series in the Cape Province, branching columnar types have not been found here. Only passively branching stromatolites, rising from laterally linked structures, occur occasionally. A classification and a detailed description of contemporaneous stromatolites were given by Truswell and Eriksson (1972). Some examples from the Drakensberg area are demonstrated in Plates 1 and 2.

The lower part of the stratigraphic sequence studied shows a thick succession of most dark dolomites, with interlayered shales and quartzites. The middle part consists of thick, vast tabular bioherms with domal stromatolites of minor dimensions. The upper part is characterized by domal structures, increasing in number and size. The rate of chertification increases from the bottom to the top with a maximum in the so-called "Giant Chert".

The most impressive examples of large domal structures are exposed 6,8 km west of Vaalhoek, 150 m south of the road to Ohrigstad (Fig. 1, locality 11). Elongated hemispheroids with spaced lateral linkage (LLH-S) are distributed over a distance of several tenths of meters. The height of the single symmetrical structures ranges from 1.3 to more than 2 m, the laminae relief from 0.7 to 1.5 m and the shorter diameter of the axis from 2 to 4 m. The diameter of the long axis was not measurable in this exposure, but may reach 10 m at least. The spacing of the single mounds varies from 0.5 to more than 2 diameters. The inheritance is high. The minor internal structure is tabular laminated, in contrast to most of the smaller domes which show crinkled laminae or complex cauliflower structures.

The lithology is a dark grayish dolomite alternating with light bands of cherts up to 25 cm thickness. These chert bands are continuous across adjacent domes. The cherty bands disappear in the centers of the giant domes, and pass to seams of microcristalline silica. The laminae of the central part can be traced to a vertical diapir-like break leading upwards to the uninterrupted external chert bands.

6.5 Discussion

This impressive feature raises the question about the growth conditions of these giant domes. Concerning extant sediment-binding algal structures it has been generally presumed that the gross features of stromatolites are not dependent on specific algae groups, but are environmentally controlled. According to Logan et al. (1964) the algal laminites form in inner intertidal to supratidal settings, while larger domes develop in the outer intertidal zone of exposed headlands. In our case, features indicating a subaerial exposure are lacking. We suggest a shallow-water environment. The reported occurrence of a red quartzite may provide evidence of the existence of at least local oxygenic conditions *already at this time.* The biotic activities in and beneath the algal mats resulted in a production of gases. This accumulation of gases beneath the mats may be the most important factor permitting the effectiv "growth" of the giant domical structures. It seems reasonable to think that the exchange of gas between the bottom ooze and the open water was restricted. Locally these reducing conditions resulted in large accumulations of trapped gas bubbles (CH_4, H_2S, H_2), pushing upward the impervious algal carpet (see Kolbe in Remane and Schlieper, 1971). The pressure of gas beneath the unlithified and still growing algal mats caused the destruction and distortion of the central parts.

In all examples the original organosedimentary alternation has been replaced by dolomite and obscured by secondary recrystallization. An early diagenetic replacement by chert before the recrystallization of the carbonates is likely, but the mechanisms of silica enrichment is problematical. Precipitation of carbonates might have caused a relative enrichment in silica of the pore-solution. Siever (1962) suggested an influence of organic matter adsorbing silica which in turn could serve as nuclei for further enrichment. In the stromatolites under consideration chertification seems to start always in the laminae with a high organic content. During the disintegration of the organic material the CO_2 / $CaCO_3$ equilibrium might have been shifted, favoring carbonate solution and replacement by chert.

The rate of chertification changes with the facies. It increases, as already shown, from the bottom to the top of the Transvaal Dolomite, with a maximum in the "Giant Chert". Generally the chertification attains its maximum in the domical stromatolites. Contrary to what is found in the common domes of minor dimensions, the centers of the giant domes are devoid of chert; the replacement has especially influenced the external layers. The process of chertification seems to depend on the original carbonaceous/ organosedimentary facies and particularly on the grade of permeability. In general chertification follows the bedding. In the giant domes the thick cherty bands vanish, entering the distorted centers. But thin thread-like layers continue to the vertical break, which is mostly filled by chert. Often fragments of tabular sheets with a high silica content or entirely replaced by silica form breccias in the distorted centers, pointing to an early chertification.

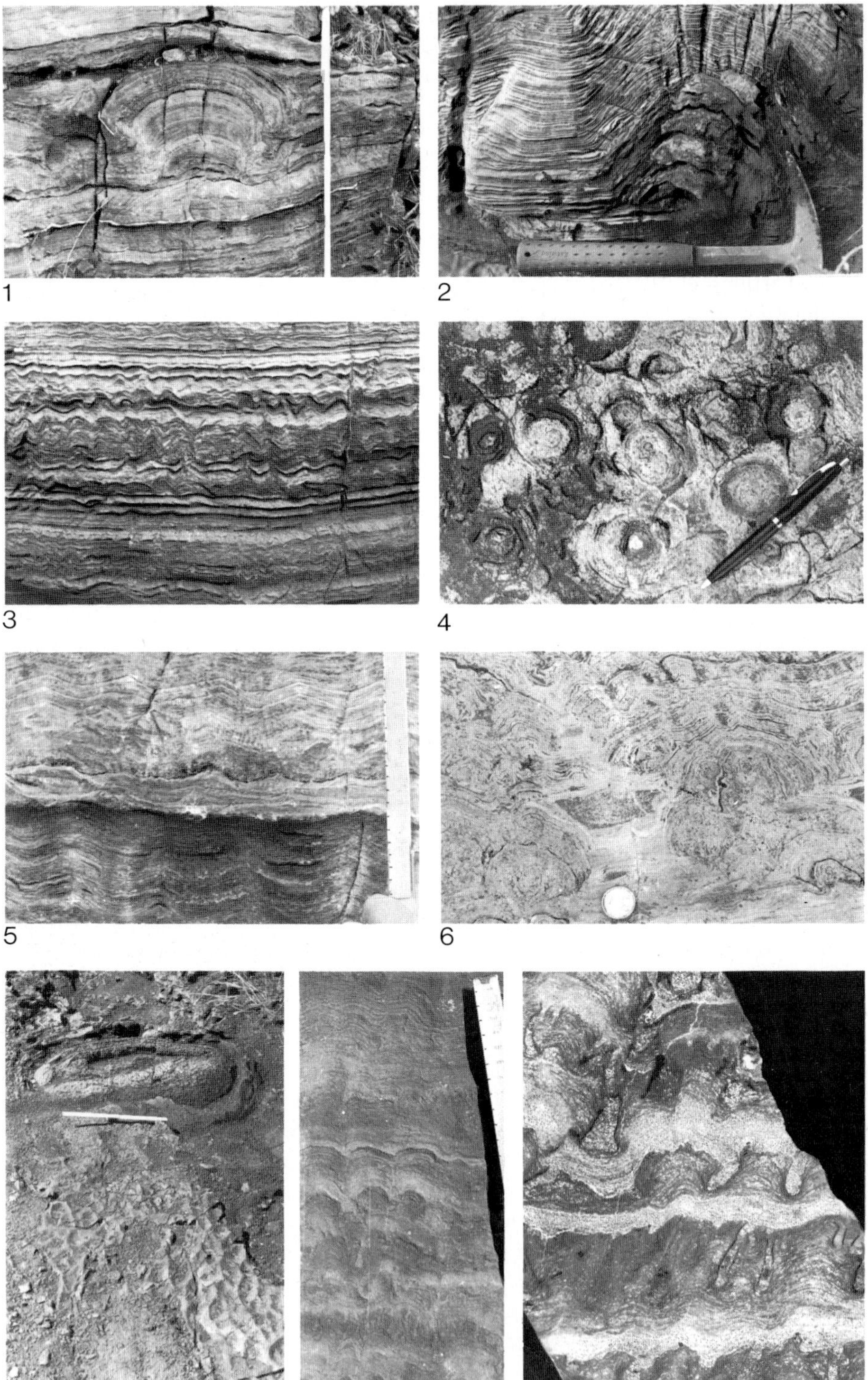

Plate 1. Legend see opposite page

◀ Plate 1/1. Small symmetrical domical structure with oversteepened flancs, embedded in a sequence of flat tabular stromatolites; several layers with angulate structures. (Length of scale 75 cm). Locality: 4.5 miles north of Pilgrims Rest (see text Fig. 1, locality Nr. 1)

Plate 1/2. Minor domical structure. Weathered surface shows a relief of delicately banded laminae structures. In the center the carbonates of crestal regions of single layers have been partly replaced by microcristalline silica. Locality: 4 miles south of "*the three rondavals*" branching (12).

Plate 1/3. Flat laminated tabular sheets with minor angulate crinkled structures. Diameter of angulate structures in the central part of the sequence 2–3 cm, relief 1–1,5 cm. (Height of total section 30 cm.) Locality: see Fig. 1 (1).

Plate 1/4. Contiguous to very closely spaced domical stromatolites. Plan view on the surface of a complete silicified layer of small closely spaced domical structures. Locality: branching of road to "The three rondavals", Blyde River Canyon (13).

Plate 1/5. Flat laminated tabular sheets with 3 cycles of different angulate crinkled structures, probably relics of superimposed ripple marks. Layers between cycle 2 and 3 (base of upper third) are corrugated. Locality: see Fig. 1 (1).

Plate 1/6. Irregular shaped spheroidal structures. In a sequence of flat, finely laminated tabular sheets spheroidal oncoid structures are developed. Oncoids, typical to the deeper parts of the intertidal environment with moderate water agitation have been encrusted by cauliflower-like domical structures which have been eventually covered by tabular flat laminated sheets again. Thin thread-like dark layers are composed of chert. Locality: see Fig. 1 (13).

Plate 1/7. Small elongate hemispherical stromatolite buried beneath dolomitic mudstones with interference ripple marks (length of scale = 40 cm). Locality: 5 km west of Vaalhoek, road to Ohrigstad (2)

Plate 1/8. Minor unlinked "columnar" structures. In a sequence of tabular sheets minor structures are developed. Structures rise from undulating mats, becoming increasingly a more convex to prolate shape, occasionally "columnar". The top of the single cycles, especially crestal regions of the prolate structures are often corrugated. The discrete variously shaped growth forms are separated by extremely narrow irregular channels, filled with coarse detritus. Depth of the channels up to 4 cm, thickness of the cycles up to 7 cm. Height of photographed section = 29 cm. Locality: road-cut 2.2 miles south of Abel Erasmus Pass.

Plate 1/9. Detail of cycles figured in 1/8. Cycles starting with undulating mats from which more convex to prolate growth forms rise. Crestal region partly corrugated. Narrow separating channels, in vertical sections resembling borings, but in horizontal sections forming irregular shaped channels. The lithology is a dark *recrystallized* dolomite. In the basal part of a cycle micritic dolomite with interlayered opaque organic materials (in photo: *light bands*); towards the top increasingly sparitic, especially in the surroundings of the channels (*dark seams*). Intercolumnar fill contains coarse detritus, often with grading. Scale 0.45 orig. size; direct print (*negative*) from peel

1

2

3

Plate 2. Legend see opposite page

◀ Plate 2. Giant structures 6.8 km west of Vaalhoek, 150 m south of the road to Ohrigstad (11)

Plate 2/1. Within a sequence of flat laminated tabular stromatolites a horizon of elongate hemispherical stromatolites with spaced lateral linkage is interlayered. Layers are continuous across adjacent domes.

Plate 2/2. Height of domes ranges from 1.3 m to more than 2 m, the short axis diameter from 2 to 4 m and the spacing of the single mounds varies from 0.5 to more than 2 diameters.

Plate 2/3. Layers of a dark dolomite alternate with light massive bands of chert up to 25 cm thickness. In the centers of the domes the cherty bands vanish, only continued by thread-like layers of microcristalline silica or fragments of tabular sheets with a high silica content. The laminae can be traced to a vertical diapir-like break, leading upwards to the uninterrupted external chert bands. Length of scale: 2 meters

---

Acknowledgments. Financial assistance from the Deutsche Forschungsgemeinschaft made this study possible. The work is a part of the „Afrika-Kartenwerk" (sheet Southeast Africa).

6.6 Bibliography

Button, A.: Early proterozoic algal stromatolites of the Pretoria Group, Transvaal Sequence. Trans. Geol. Soc. S. Africa **74**, 201–210 (1971).

Button, A.: The depositional history of the Wolkberg proto-basin, Transvaal. Trans. Geol. Soc. S. Africa **76**, 15–25 (1973).

Eriksson, K.A., Truswell, J.F.: Tidal flat associations from a Lower Proterozoic carbonate sequence in South Africa. Sedimentolgy **21**, 293–309 (1974).

Logan, B.W., Rezak, R., Ginsburg, R.N.: Classification and environmental significance of algal stromatolites. J. Geol. **72**, 68–83 (1964).

Remane, A., Schlieper, C.: Biology of brackish water. In: Die Binnengewässer **25**. Stuttgart: Schweizerbart 1971.

Siever, R.: Silica solubility, 0°–200° C, and the diagenesis of siliceous sediments. J. Geol. **70**, 127–150 (1962).

Truswell, J.F., Eriksson, K.A.: The morphology of stromatolites from the Transvaal Dolomite north-east of Johannesburg, South Africa. Trans. Geol. Soc. S. Africa **75**, 99–110 (1972).

Truswell, J.F., Eriksson, K.A.: Stromatolitic associations and their palaeoenvironmental significance: a reappraisal of a Lower Proterozoic locality from the Northern Cape Province, South Africa. Sediment. Geol. **10**, 1–23 (1973).

Young, R.B.: A comparison of certain stromatolitic rocks in the Dolomite series of South Africa with marine algal sediments in the Bahamas. Trans. Geol. Soc. S. Africa **37**, 153–162 (1935).

7 Environmental Significance of Upper Proterozoic Algal Stromatolites from Zambia

A. GUNATILAKA, Lusaka, Zambia

7.1 Introduction

Algal stromatolites in the Zambian Copperbelt from Mufulira were first reported by Malan (1964). They are confined to the Katangan Sequence, an 8000 m thick accumulation of Late Precambrian sediments deposited unconformably on an eroded basement surface and spanning a time interval of approximately 400–600 m.y. (Cahen, 1970).

The reported stromatolitic occurrences from Mufulira are confined to the Lower Roan successions (Fig. 1). The present study was carried out in six mine areas by examining underground exposures and over 40,000 meters of drill cores from the entire Katangan Sequence. Stromatolites from the uppermost Precambrian (Kundelungu) were not detected in the Copperbelt and a new occurrence is reported from the Kafue district (Fig. 1). Unlike the Roan sediments, the Kundelungu group is relatively unmetamorphosed and has provided excellent textural detail of the internal structures of its stromatolites. The distribution of the dolomitic bioherms is of primary significance in Copperbelt geology. They show a distinct relationship to the ore occurrences by being associated with lateral facies changes. The Cu-Co mineralization is always antipathetic to the stromatolite facies. Also, they can be used to trace the ancient shoreline of the Katangan sea, as the algal structures colonized lagoons, shallow embayments or tidal flats.

7.2 Roan Group Stromatolites

The first occurrence of the stromatolites in the Lower Roan is topographically controlled. As the old eroded valleys of the basement were invaded by the first marine transgressions, stromatolites colonized the summits of the paleoridges, forming reefal or biohermal accumulations. In the Copperbelt, the stromatolites and the associated sediments represent the whole spectrum of environments ranging from shallow subtidal → intertidal → supratidal, as in the Chambishi-Nkana basin (Fig. 1). At Nkana, the stromatolites are associated with large mud cracks and show crenulated laminations and "birds' eye" features. They colonized a large intertidal flat of predominant terrigenous deposition. The tidal flats in the Colorado River delta (Thompson, 1968) have many features in common with these Lower Roan occurrences. In many places, however, the algal structures appear to be comparable to some documented occurrences from Shark Bay, W. Australia (Clemmey, pers. comm. 1975).

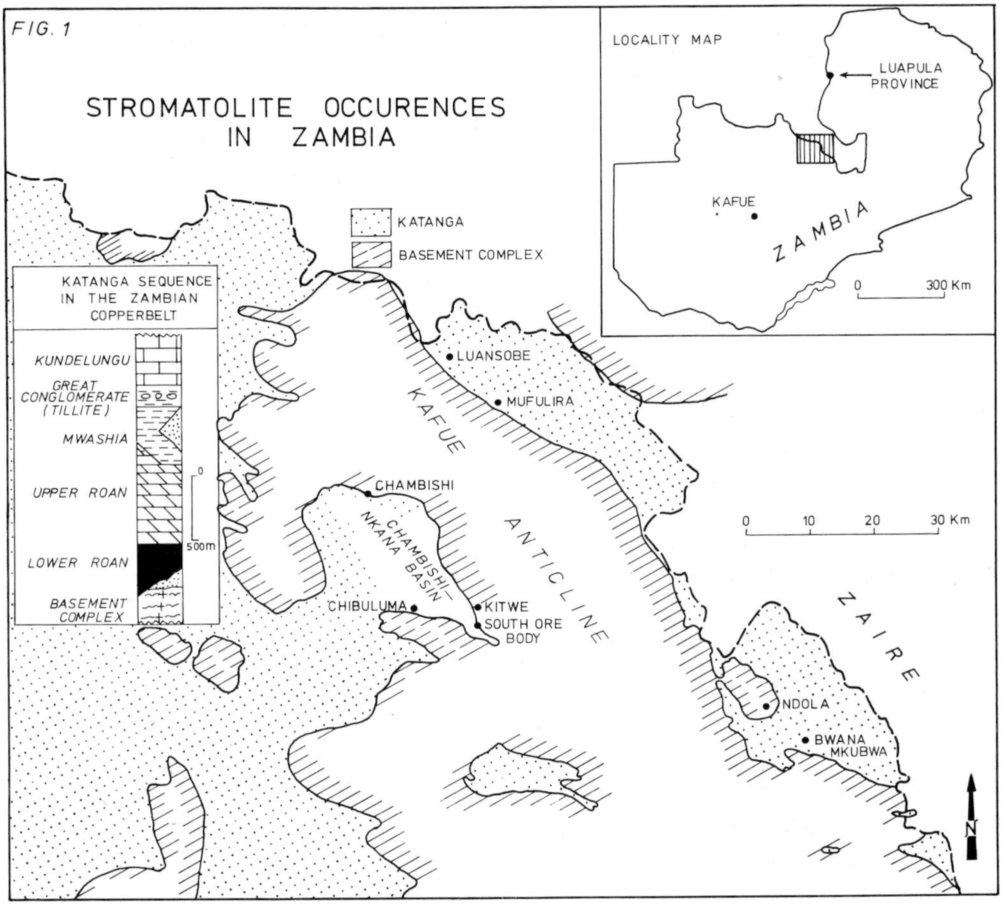

Fig. 1. Stromatolite occurrences in Zambia

In the Nkana North region, the following sequence has been observed (Fig. 2A): pebble conglomerate → cryptalgal laminites thickening upwards → stromatolitic domes (LLH or SH) → oncolites (SS)/algal biscuits → cryptalgal laminites. This is a typical shallow subtidal sequence of decreasing energy conditions, according to the model presented by Gebelein (1969, Fig. 19). In the same general area, a stromatolite bed is associated with a 2 m thick anhydrite band displaying characteristic "chicken-wire texture" (Fig. 2B). Such supratidal features are not laterally extensive and occur mainly as small pockets within the upper regions of the tidal flat.

In the predominantly clastic Lower Roan, the algal facies are the only carbonatic rocks. There is a lateral facies change away from the paleohighs into more siliciclastic and argillaceous lithologies. The copper grade is very low ($\ll 0.5\%$) or pyritic in the algal facies and increases to over 5% in the paleoslopes or basin margins, to decrease again in the paleovalleys (Mendelsohn, 1961; Binda and Mulgrew, 1973).

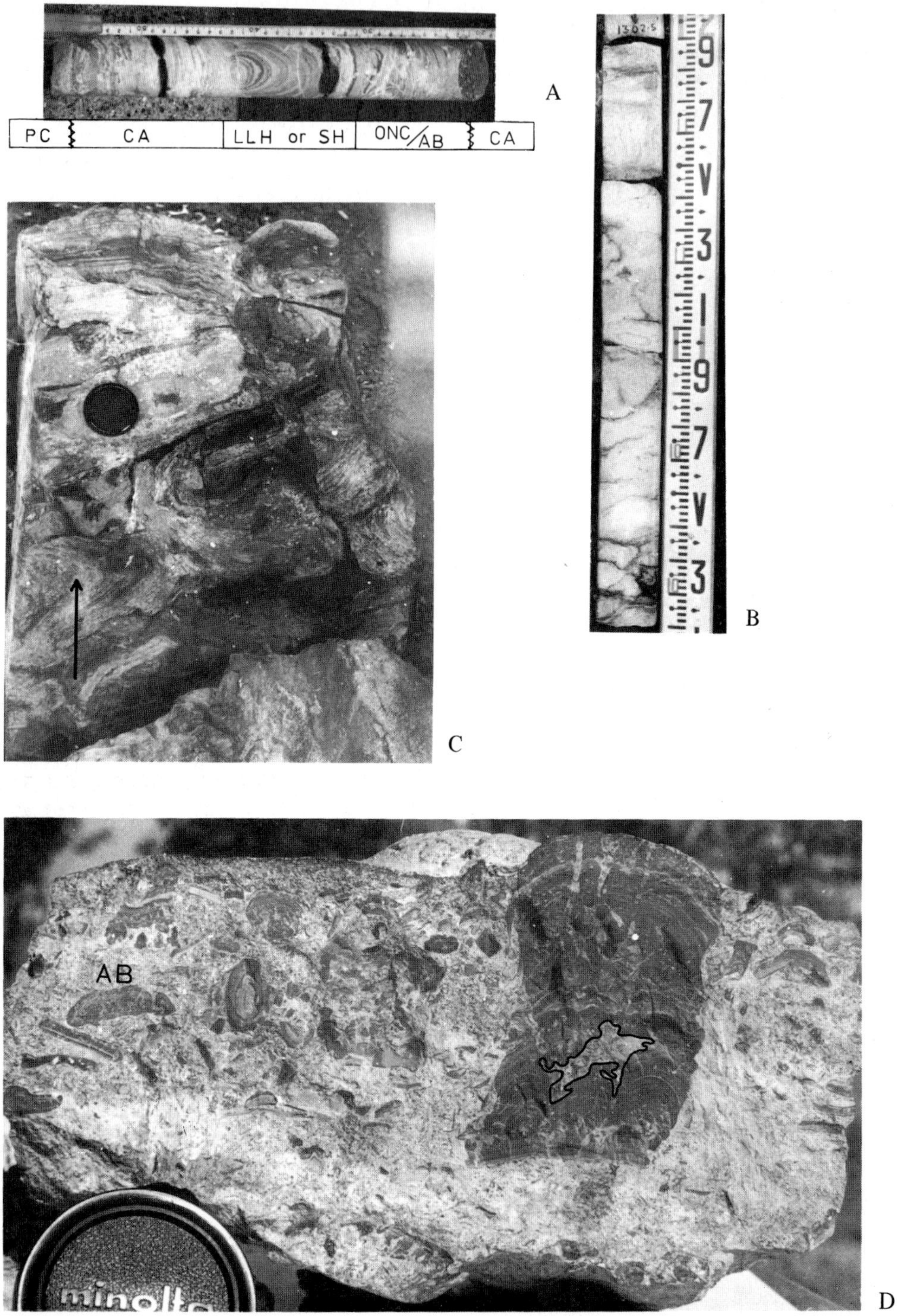

Fig. 2. Legend see opposite page

The lamination in the stromatolites were probably produced by the usual trapping-binding mechanism (Gebelein, 1969; Gunatilaka, 1975) as quartz, feldspars, and ferromagnesian minerals are present in significant amounts. Evidence for large scale carbonate precipitation within the laminar structure is lacking.

The Upper Roan sequence is predominantly dolomitic; the stromatolites show much the same environmental spectrum as in the Lower Roan. The algal structures are, however, more variable and show a higher proportion of algal biscuits in some areas; supratidal features are more widespread. The laminae show more evidence of carbonate precipitation. The sediments show only incipient Cu-Co mineralization – a factor which has handicapped detailed environmental studies.

7.3. Kundelungu Group Stromatolites

In the Kafue district (Fig. 1), over 350 m of a thick carbonate sequence have been reported by Pipping (in press). The lithologies comprize thick oolitic limestones, peloidal, stromatolitic, oncolitic, micritic, and intraclastic organo-detrital limestones. These probably represent a setting similar to the modern Bahamian-type carbonate environment and could be considered as a Precambrian analogue. The thick oolite sequences (representing a shelf margin carbonate environment) and the lack of typical supratidal evaporite deposits in the Kundelungu compares well with a Bahamian environment.

The stromatolites are mostly compound forms and occur in reef-like structures with evidence of having been formed in a highly agitated environment (Fig. 2D). The initial exposed surface was inhabited by colonies of blue-green algae. Due to the high energy, a continuous mat surface was not established. Rather, a mixed oncolitic and/or discontinuous mat surface (of small broken pieces) formed, in which the later higher relief stromatolites established themselves. These surfaces seem to have stabilized the bottom and appear to have fulfilled the same function as the Recent subtidal gelatinous mats reported from the Bahamas (Bathurst, 1967). The oncoids are very likely a morphological and ecological adaptation of the mat forming communities to specific energy conditions. Very often, intraclasts form the nuclei of the oncolites and they vary in size from a few millimetres to about 3 cms.

The most common sequence observed is oncolites → SH-Type stromatolites to → LLH-type structures, which appears to have formed on a erosional surface that was stabilized by a hard substrate (Fig. 2D). The passage from one form to another is very abrupt (often in less than 10 cms). The stromatolites themselves show evidence of a peri-

◀ Fig. 2 A–D. (A) Part of a subtidal stromatolite sequence from the Chambishi-Nkana basin. *PC*: pebble comglomerate; *CA*: cryptalgal laminite; *LLH*: laterally linked hemispheroids; *SH*: stacked hemispheroids; *ONC*: oncolites; *AB*: algal biscuits. (B) Association of stromatolites with anhydrite bed showing "chicken-wire texture", Nkana North region of the Copperbelt. (C) *Conophyton*-like stromatolite from the Lower Roan of Mufulira, in a dolomitic rock. Note the angular laminae. (D) Intertidal stromatolites from the Kundelungu Group in the Kafue district. The structure has grown on a oncolitic surface. Note the vertical cracks and scour features (enhanced in *black ink*). Organo-detrital fragments, large oncolites and algal biscuits (*AB*) surround the structure tidal or lower intertidal origin with periodic dessication due to alternate flooding and drying (vertical cracks, surface scours, crenulated lamination). The *in situ* algal structures are generally small (< 10 cms in height), rounded, and upright. Those formed earlier have a large surface diameter, whereas the later forms tend to be quite small due to overcrowding. The detrital fragments (dreived from intraclasts, oncolites, algal biscuits, algal chips etc.) show a very random orientation within the stromatolite layers, indicating that the reef-like structures were formed in areas of high agitation (Fig. 2D).

7.4 Discussion and Conclusions

The Lower Roan sediments in Zambia are thought to be younger than 1300 m.y. and older than 840 m.y., while the Kundelungu is probably not older than 720 m.y. (Cahen, 1970). Hence, a Riphean age for these deposits is implied. During this period, the predominantly clastic Lower Roan environment changed to one representing a somehow similar "Bahamian-type" carbonate environment in the middle Kundelungu times.

In the mine areas of the Copperbelt, algal stromatolites from the same stratigraphic horizon in the Lower Roan Group represents environments ranging from the subtidal → intertidal → supratidal, as in the Chambishi-Nkana basin. An expansive sabkha environment is not represented. Rather, the supratidal sabkha features are very localized in each area of deposition. The Kundelungu stromatolite reefs show mainly an intertidal distribution in a highly agitated environment.

Recent investigations in the very thick Upper Proterozoic carbonate sequences in the USSR seem to favor a morphogenetic evolution of these stromatolite forms, with subsequent modification by environmental factors (Raaben, 1969). In the Lower Roan of Zambia, the stromatolite beds are not more than a few meters thick, yet Riphean forms resembling *Conophyton* (Fig. 2C), *Collenia, Inzeria, Jurusania* and *Baicalia* may be *inferred* when photographs are compared with published material (Cloud and Semikhatov, 1969).

The data from the present study favor an environmental control of stromatolite distributions. In the Zambian context, the environmental control is suggested by the association of stromatolites with lithologies containing Cu-Co mineralization. The algal reefs grew in well-oxygenated environments having a good tidal exchange. Most current theories favor the initial localization of the stratiform Cu-Co ores in a shallow, low energy-reducing environment (now represented by the argillites and/or carbonaceous quartzites of a tidal flat). The algal facies are, without exception, either barren, or very low in ore. The ore emplacement adjacent to the reefs has been a factor favoring a non-epigenetic origin for these deposits, or rather a syngenetic-diagenetic origin.

Biostratigraphic zonation of the Upper Proterozoic using stromatolites is very much dependent on which point of view is accepted. In addition, stromatolite-building communities may have undergone significant ecological drifts in their habitats throughout their geologic history. Soviet techniques of analysis are only just beginning to be applied in the West and recent trends seem to indicate that the two schools of thought are arriving at a compromise (Hoffman, 1973). If successful, accurate interpretations of the causality of stromatolites and their geobiological significance may be hopefully in view, and interregional correlations *may* be attempted.

7.5 Bibliography

Bathurst, R.G.C.: Sub-tidal gelatinous mat, sand stabilizer and food, Great Bahama Bank. J. Geol. **75**, 736–738 (1967).

Binda, P.L., Mulgrew, J.R.: Stratigraphy of copper occurrences in the Zambian Copperbelt. Unpublished company report, Roan Consolidated Mines (1973).

Cahen, L.: Etat actuel de la geochronologie du Katangien. Annales Musee. Rech. Afr. Centr. **65**, 7–14 (1970).

Cloud, P.E., Semikhatov, M.A.: Proterozoic stromatolite zonation. Am. J. Sci. **267**, 1017–1061 (1969).

Gebelein, C.D.: Distribution, morphology and accretion rate of Recent subtidal algal stromatolites, Bermuda. J. Sediment. Petrol. **39**, 49–69 (1969).

Gunatilaka, H.A.: Some aspects of the biology and sedimentology of laminated algal mats from Mannar lagoon, N.-W. Ceylon. Sediment. Geol. **14** (1975), in press.

Hoffman, H.J.: Stromatolites: Characteristics and Utility. Earth Sci. Rev. **9**, 339–373 (1973).

Malan, S.P.: Stromatolites and other algal structures at Mufulira, Northern Rhodesia. Econ. Geol. **59**, 397–415 (1964).

Mendelsohn, F.: Geology of the Northern Rhodesian Copperbelt. London: MacDonald 1961.

Pipping, F.: Geology of the Mutapanda River area. Geological Survey of Zambia. Report **31** (in press).

Raaben, M.E.: Columnar stromatolites and Late Precambrian stratigraphy. Am. J. Sci. **267**, 1–18 (1969).

Thompson, W.: Tidal flat sedimentation, Colorado River Delta, north-western Gulf of California. Geol. Soc. Am. Memoir **107**, 133 p. (1968).

8 Stromatolites and Thrombolites in Precambrian/Cambrian Boundary Beds of the Anti-Atlas, Morocco: Preliminary Results

M. SCHMITT and W. MONNINGER, Würzburg, W. Germany

8.1 Locality

The studied section comprising the "Adoudounien" and the Lower Cambrian "Série schisto-calcaire" crops out on the northern slope of the Anti-Atlas near Tiout, 25 km South of Taroudannt. The Anti-Atlas provided, mainly by the work of Choubert (e.g. Choubert, 1963), many contributions to the stratigraphy of the Precambrian/Cambrian boundary.

8.2 The Section

The uppermost 200 m of the "Calcaire inférieur" consist of dark gray dolostones with a few interbeds of dolomitic marls.

The overlying Série lie de vin has a thickness of nearly 1000 m and is divided into three sequences:

1. Lower sequence: predominance of reddish dolomitic marls, occasionally limestones and dolostones,
2. Middle sequence: predominance of limestones and dolostones, the interbeds are dolomitic marls,
3. Upper sequence: similar distribution of limestones and dolostones as in the Middle sequence, the interbeds consist of marlstones and partly of shales.

The Calcaire supérieur, with a total thickness of 300 m, consists mainly of dark and black limestones; dolostones and marls are very rare, strata with chert are abundant. The overlying Lower Cambrian "Série schisto-calcaire" is an alternating sequence of marly shales and black limestones containing trilobites and archeocyathids.

Age: according to Choubert et al. (1973), the Adoudoudien – especially the Série lie de vin – is counterpart of the tillites of the Taoudenni-Basin in West Africa. A radiometric age of 650–620 m.y. is proposed.

Plate 1/1a,b. (a) Thrombolitic bioherm of conical shape, Upper sequence of the Série lie de vin. (b) Slab cut from the top of the same bioherm showing a stromatolitic columnar arrangement with a weak lamination (scale in mm) ▶

Plate 1/2, 3, 5. Relatively high-domed bioherms of stratiform thrombolites from the Middle sequence of the Série lie de vin (scale in cm)

Plate 1/4. Thrombolite from the same layer as shown in Plate 1/1a with thin incorporated tuffites (scale in cm)

Plate 1. Legend see opposite page

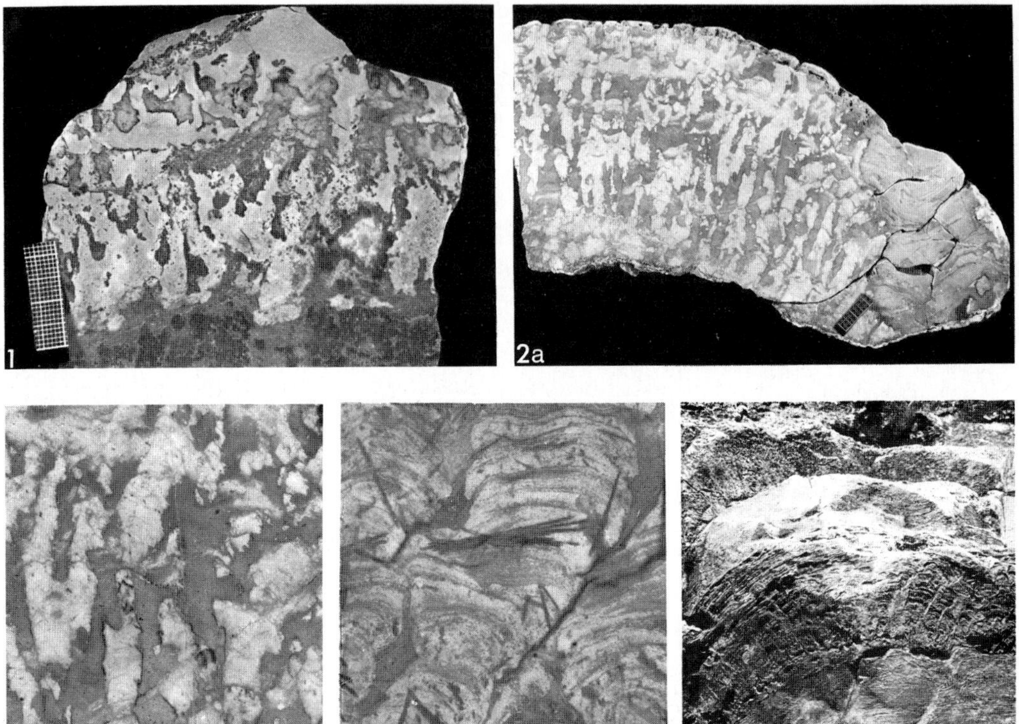

Plate 2. Legend see opposite page

8.3 Algal Structures

8.3.1 Distribution

The main development of stromatolitic and thrombolitic bioherms can be observed in the three sequences of the Série lie de vin.

The Lower sequence shows algal bioherms in the dolostones and limestones. Besides LLH stromatolites and columnar branching stromatolites tabular thrombolites occur, i.e. more or less weakly domed bioherms lying flat on the substrate. Some of the bioherms are strongly recrystallized or dolomitized. At the base of the Middle sequence, the so-called Barre de Tata, contains thrombolites in abundance. They form either low-lying bioherms or bioherms with a relatively high relief, some of them being conical. Several follow one above the other within one single bank. Their vertical arrangement is interrupted only by thin interbeds of marl. The Upper sequence shows a maximal development of algal bioherms. Thrombolites reach a thickness of almost 4 m and are up to 5 m large.

8.3.2 Internal Structures

8.3.2.1 Thrombolites

Thrombolitic bodies contrast strongly with the enclosing sediments, as weathering has outlined a wrinkled fabric in them. Polished slabs show clotty, rounded structures, partly elongated upward. Beside this clotted fabric (Aitken, 1967), the structures change within the thrombolites, especially in the Upper sequence of the Série lie de vin. The thick bioherms pass from a clotted structure at their base to a digitate and columnar structure towards their top. Occasionally a weak lamination is visible.

8.3.2.2 Stromatolites

In this paper only columnar stromatolites will be discussed. There is a considerable variation in the arrangement, linkage and branching of columns, as well as in the lamination and microstructure.

◄ Plate 2/1. Slab cut showing two different internal structures: thrombolitic clotted fabric at the bottom and columnar branching stromatoltic structure above (Lower sequence, Série lie de vin, scale in mm)

Plate 2/2a,b. (a) Slab cut of columnar stromatolite from the Lower sequence of the Série lie de vin. The form is similar to *Linella* and *Acaciella* (scale in mm). (b) Detail-photo of the slab cut shown in (a) (scale x 0.6)

Plate 2/3a,b. (a) Slab cut of columnar stromatolites with needle-shaped crystals, the crystals could have been gypsum and are replaced by calcite; Upper sequence, Série lie de vin (scale x 0.6). (b) Stromatolitic bioherm from which slab cut in Plate 2/3a is taken (scale in cm)

Plate 2/4, 5. Bioherms of columnar stromatolites from the Middle sequence of the Série lie de vin. The columns are well-outlined by the weathering (scale in cm)

From slab cuts parallel and perpendicular to the columns we have tentatively recognized forms that are very similar to those described by Walter (1972) from the Bitter Springs Formation of the Amadeus Basin in Australia. The stromatolitic material studied so far resembles strongly the groups named *Acaciella, Tungussia* and *Linella*, which are reported from the Late Riphean or Vendian. If these identifications are confirmed by further and more detailed investigations, then the Série lie de vin could be dated as Late Riphean.

Thrombolites so far known have no stratigraphic significance and they have not yet been reported from the Late Precambrian. The forms found in the Adoudounien are almost identical by their gross shape to those described by Aitken (1967) from the Cambrian and Ordovician of Alberta.

8.4. Environment

The sediments of the Série lie de vin seem to be deposited in warm and shallow water with changing salinity. Similar conditions might have existed in the Calcaire supérieur. Only the Lower Cambrian Série schisto-calcaire shows constant marine conditions.

The Série lie de vin, especially the Upper sequence consists of alternating sediments with a small scale vertical change in limestone and dolostone. The limestones are dark gray, finely crystalline, and contain thrombolites; the dolostones are light brown, finely to coarsely crystalline and contain stromatolites. It is remarkable that the large bioherms of columnar stromatolites are not, or only slightly, dolomitized, whereas the small LLH and SH algal structures consist mostly of dolostone. Accordingly, within a distinct facies, the lithological variation parallels variation in the algal structures.

Aitken (1967) locates the thrombolites in the lowest low tide zone, whereas the columnar and digitate stomatolites are situated in the intertidal zone. This relative gradation of water depth fits well with the observations made in our section. The action of the tide, however, has not yet been proved.

In many cases, the substrate of the thrombolites was a soft limy mud; indeed, the early consolidated thrombolites sank into the mud, which is recognizable in the downward bended lamination of the substrate.

There is no evidence of strong water energy during the growth of the thrombolites, as the enclosing layers are mostly even laminated mudstones with very few intraclasts. The layers which cover the thrombolites are generally marlstones. They often fill up the biorelief without any visible erosional discontinuity. The water energy in the environment where the columnar stromatolites grew is hard to decipher because of the recrystallization and dolomitization of the enclosing rocks.

Thrombolites and stromatolites were only found in the Adoudounien, the greater part of which is of late Precambrian age. The uppermost parts of the Calcaire supérieur are of Cambrian age, as dated by an abundance of trilobites and archeocyathids. The archeocyathids grew in thrombolitic bioherms similar to the ones described from the Upper sequence of the Série lie de vin. The rocks consist of dark gray and black limestones. Apparently, the archeocyathid-bioherms were build up in a biotope similar to that of the thrombolites, but this conclusion is still hypothetical.

8.5 Bibliography

Aitken, J.D.: Classification and environmental significance of cryptalgal limestones and dolomites, with illustrations from the Cambrian and Ordovician of south-western Alberta. J. Sedimentol. Petrol. **37**, 1163–1178 (1967).

Choubert, G.: Histoire géologique du Precambrian de l'Anti-Atlas. Notes Mém. Serv. Géol. Maroc **162** (1963).

Choubert, G., Boudda, A., Faure-Muret, A.: Essai de chronologie du Precambrian supérieur du Sud marocain. C.R. somm. Soc. Géol. France **1973**, 123–124 (1973).

Walter, M.R.: Stromatolites and the biostratigraphy of the Australian Precambrian and Cambrian Spec. Pa. Paleontol. **11**, 268 pp., 55 Figs., 34 Pls. (1972).

9 Riphean Stromatolites from Western Lower Himalaya, Himachal Pradesh, India

ANSHU K. SINHA, Delhi, India

9.1 Introduction

Stromatolites have been found to be useful tools in the correlation of the Late Precambrian or Riphean stratigraphic sequences, following rapid development of new methods of study and classifications in the USSR, Australia, and USA (Semikhatov, 1962; Krylov, 1963. 1967; Komar et al., 1965; Raaben, 1969b; Walter, 1972; Serevryakov and Semikhatov, 1974) and elsewhere. They are found all over the world, particularly profusely in Precambrian carbonate rocks. They are known to have survived moderate deformation and metamorphism (Valdiya, 1967; Walter, 1972). The columnar branching forms show definite evidence of evolutionary changes as emphasized by Glaessner (1971) and others. Stratigraphic units established on the basis of stromatolitic assemblages have been shown to be nearly isochronous, according to the absolute age-data (K-Ar on glauconite, Raaben, 1969b).

The systematic study of stromatolites from the Lower Himalaya is only a decade old. Misra and Valdiya (1961) reported an occurrence of *Collenia* from Gangolihat dolomitic limestone of Pithoragarh in Kumaun. The "pseudo-organic structure" in the carbonate rocks Himachal Pradesh reported by Oldham (1888) and Pilgrim and West (1928) were identified as a group of stromatolites (i.e. *Baicalica* by Valdiya, 1962, 1969). His study of stromatolites of the Gangolihat, Deoban, Tundapathar, and Shali rocks indicates the later Middle Riphean age of these formations. In Kumaun, similar forms under different names from the extension of Gangolihat formation have been described by Dixit (1966) from the Someshwar area, by Banerjee (1970) from the Pungar-Sarju Valley, by Misra and Kumar (1966) from the Badolisera area. In Jammu, Gupta and Dixit (1970), Gupta and Dixit (1971), Singh and Vimal (1972) and Raha and Sastry (1973) have found forms that give a Precambrian age to the Limestone.

The geologic setting of the studied area is shown on Fig. 1. Stromatolite localities are situated in the tectonic windows of the autochthonous or para-autochthonous units of the carbonate-terrigenous rocks of the Shali Formation and of the flyschoidal rocks of the Simla group (Srikantia and Sharma, 1971); these have been subsequently covered by three successive allochthonous units of Jaunsar-Blaini-Infra Krol-Krol Tal, Chail and Jutogh. More detail can be found in Pilgrim and West (1928), West (1939), Gansser (1964), Fuchs (1967), Srikantia and Sharma (1969, 1971), Valdiya (1970), Sinha (1972) and Srikantia (1973).

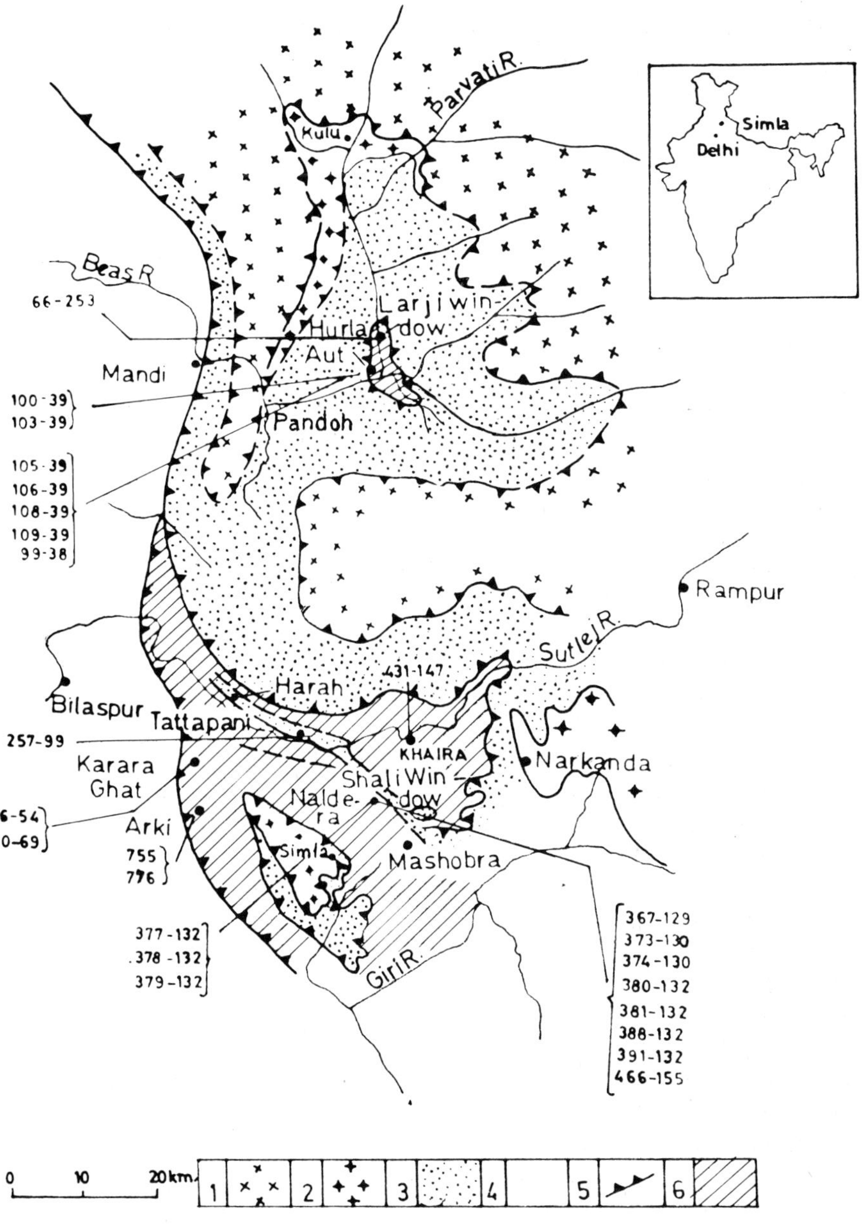

Fig. 1. Schematic tectonic map of Simla-Kulu region showing the location of stromatolite occurrences (after Ashgirei and Sinha). *1:* Upper Himalayan thrust-sheet; undifferentiated; *2:* Jutogh thrust sheet; *3:* Chail thrust-sheet; *4:* Thrust sheet of Jaunsar, Blaini, Infra Krol, Krol and Tal and other younger rocks; *5:* Thrust sheet surface; *6:* Autochthonous rocks (Riphean carbonate-terrigenous and flyschoidal group of Shali and Simla fms.)

9.2 Morphological Features and Classification

The morphological features of stromatolites were studied with special reference to the (1) mode of occurrence, (2) shape of individual columns, (3) size of columns, (4) ramification, and (5) form of stromatolitic microstrata (see Fig. 2).

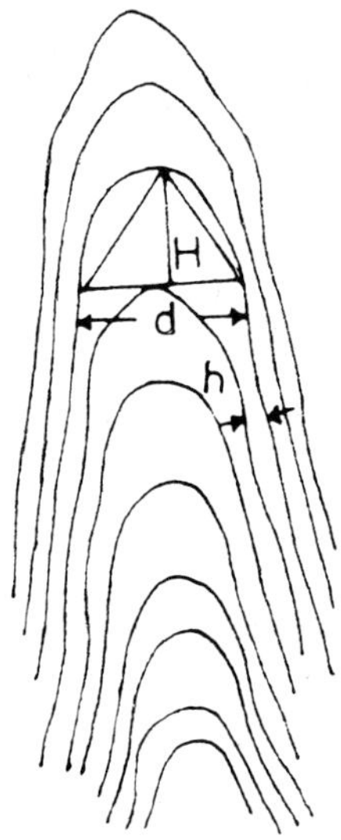

Fig. 2. The parameters in the description of crestal zones of *Conophyton* (*d*: width of crestal zone; *H*/*h*: coefficient of thickening: adapted after Komar et al., 1965; Walter, 1972)

On the basis of specific characteristics of microstrata, mode of ramification, and shape of the columns, the following classification according to Raaben (1969a) has been adopted:

1. Conophytonida: not ramified cylindroidal columns (*Colonella, Conophyton*),
2. Kussiellida: passively branching, cylindroidal columns (*Jurusania*),
3. Tungussida: actively branching, columns widening upward, cuplike, with divergent axes (*Tungussia, Baicalia*, and others).

9.3 Systematic Descriptions of Columnar Stromatolites

Group *Jurusania* Krylov (1963)

Jurusania himalayica f. nov. (Fig. 4, Plate 1/1, 2; Plate 2/1, 2, 3)

Type Specimen (S. No. 4535/775) Lodged at Geological Institute, Moscow, this specimen has subcylindrical columns with rounded or irregular, sometimes strongly elongated transverse sections. The diameter of columns varies from 1.0 to 4.0 cm, the height is

Fig. 3. Occurrences of stromatolites near Khaira, Shali-window Sutlej Valley

Fig. 4. Morphological reconstruction and column of *Jurusania himalayica* f. nov. S. No. 775, 776. Locality – Arki, Simla

6–8 cm or more. The lateral surface of the column is finely toothed, with separate and thin-hanging cornices and connecting small bridges. Enveloping of laminae is not characteristic. The distance between adjacent parallel and subvertical columns varies from 0.2–0.3 cm, or more. A few of the columns are branching. The broader column branches up into narrower ones, i.e., branching passively. Locally, the adjacent columns may coalesce.

The laminae, constituting the columns, simply arch upward without any overlapping margins, with some of the arcs usually well-rounded, sometimes a little edged. In rare cases this is complicated by additional buckling. The coefficient of convexity (h/d) varies in most cases from 0.1 to 0.4 and rarely goes up to 0.5. In the central part of

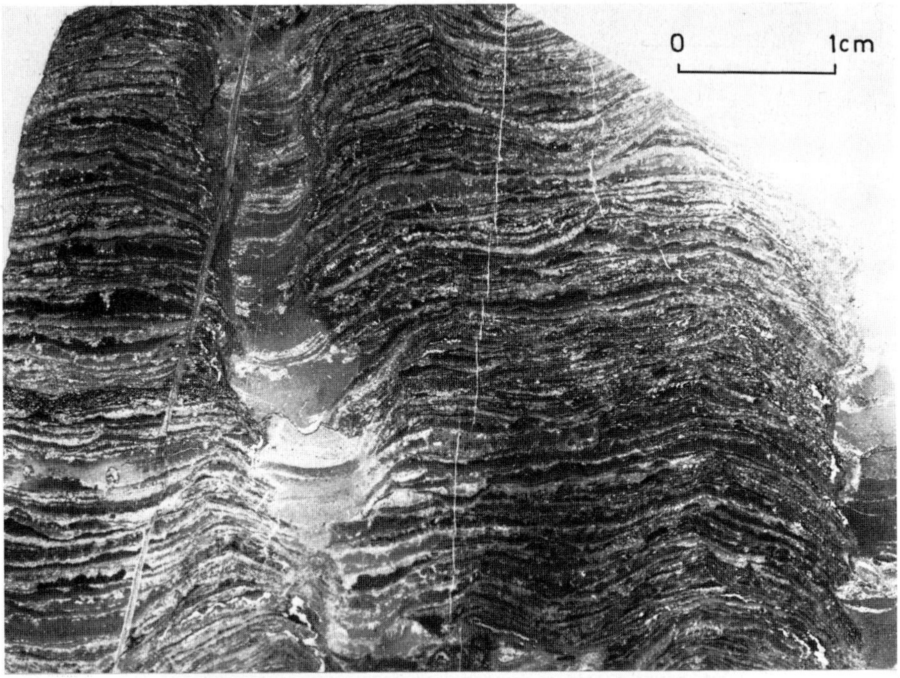

1

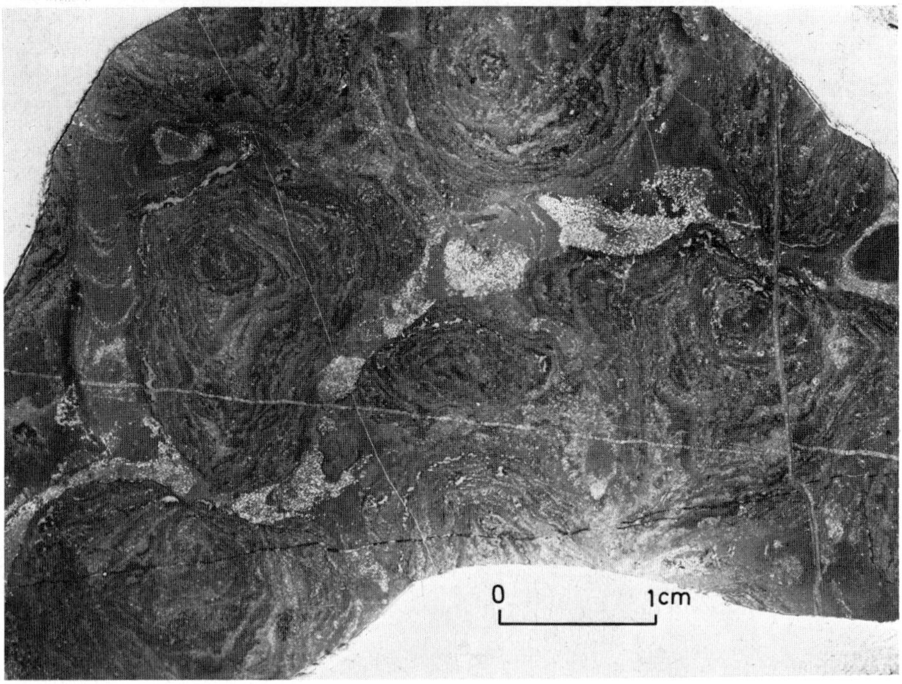

2

Plate 1/1, 2. *Jurusania himalayica* f. nov. Microstructure in thin section: longitudinal (Plate 1/1) and transverse (Plate 1/2) views. Photomicrograph. Locality – Arki, 20 km south-west of Simla. S. No. 775

1
2
3

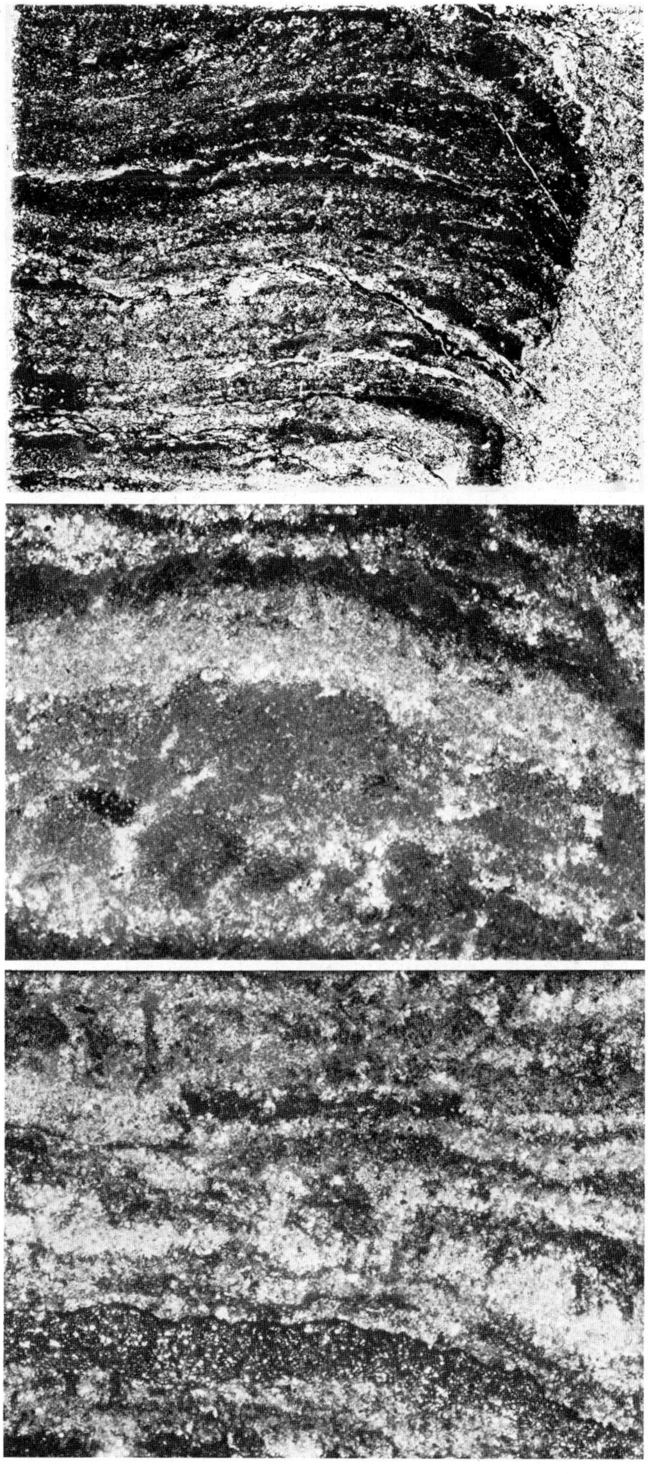

Plate 2/1, 2, 3. *Jurusania himalayica* f. nov. Microstructure in thin section showing spliting of thick laminae and dark micrite layers. Photomicrograph: Plate 1/1: x 8; Plate 1/2: x 20; Plate 1/3: x 20

column the laminae are comparatively thicker. Often the thick lamina splits up into two to four thinner interlayers. Lamination is distinct and fine. The particular textural features are represented by contiguous layers of micrites with thickness of 0.1 mm, but usually ranging from 0.2 to 0.4 mm. The micrite layers have dark, gray color, the color distribution being even or zonal. Dark dotted inclusions of about 0.01 mm diameter are observed either scattered or in patches. The lower boundary of the micrite laminae is even, while the upper part is usually uneven or sinuous. Micrite microstrata stand out in contrast to the uniformly crystallized carbonate (sparite), which constitutes the remaining part of the column.

Material. Two samples, S. No. 775 and 776 (lodged at Geological Institute, Moscow).

Locality. Arki (20 m WSW of Simla) Simla Slates (Kakarhatti, Lst.).

Age. Group *Jurusania* is widespread in the Upper Riphean and Vendian.

Remarks and Comparison. The texture of the present form of *Jurusania* from the Arki area differs from that of all the previously described forms of this group; this lead us to create a new form. The name is derived from Himalaya. In morphological construction of *J. himalayica* f. nov. all the basic characteristics of *Jurusania* Krylov group are present. *However, it has distinctive secondary features of its own, such as predominance of columns with irregular transverse sections and lesser average diameter.* Considering these features *J. himalayica* f. nov. is closer to *Jurusania nisvensis* Raaben and *J. sibirica* (Yakovlev). The present form differs from *J. nisvensis* by its thinner laminae, smaller size of cornices hanging at the margin of columns, and more regularity of the whole columnar surface. It strongly differs in texture from *J. sibirica* (Yakovlev).

Jurusania (?) sp. indet. (Fig. 5)

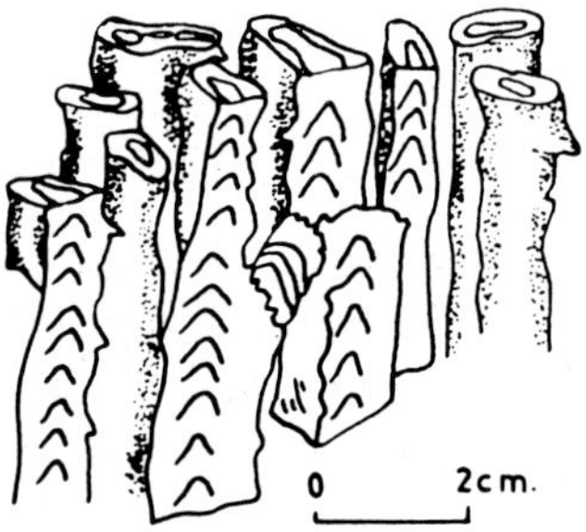

Fig. 5. Morphological reconstruction of fragments of *Jurusania* (?) sp..indet. S. No. 377–182. Locality – Naldera ridge Simla

This specimen has thin cylindrical columns, 6–7 cm high and 0.9–1,5 cm in diameter; transverse section of columns are irregular, rounded, or elongated; the lateral surface is uneven, finely indented, with discrete bigger cornices. Enveloping of laminae is not observed. Very often along the surface stylolitic sutures are present. All the structures have undergone deformation of recrystallization in higher degree, which makes definite identification difficult. Greater parts of the columns are mutually parallel, sometimes ramification of a broader column into two narrower parallel columns, without broadening of the branches. However, separate segments of columns are inclined with acute angles to the main axial direction, perhaps due to the secondary coalescense of these segments. In one of the samples, columnar structure passes into layered irregular wavy pillow structure from bottom to top. Inter-space between columns are usually 0.3–0.5 cm wide. In the intercolumnar space there are numerous furrows of stromatolitic laminae.

Lamination in the columns is quite distinct. Laminar arcs are high, often subconical. Coefficient of convexity is 0.5–0.8; sometimes up to 1.0. Microstrata of micrite represent the main element of microstructure. They extend without gaps, from one end to the other of the column. In this section they are uniformly grayish or spotted. The thickness of the lamina varies from 0.3–0.4 mm to 1.0 mm and more. At the periphery of the column the thickness of laminae somewhat diminishes. Micrite microstrata are divided in still thinner laminae of uniformly crystallized carbonate from 0.05 to 0.8–1.0 mm thick.

Remarks and Comparison. Affinity of this form with the group of *Jurusania* is most probable. The small average size of the columns distinguishes this form from all previously described forms. Amongst these only *J. nisvensis* Raaben shows smaller diameter of columns.

The above form of *Jurusania* differs from typical forms by the absence of tight edges in the periphery of the column. One texture of the lamina of the present significantly differs from that of *Jurusania himalayica* f. nov. recovered from the Arki area. In shape and texture it is closer to the *J. cylindrica* and *J. nisvensis*, but it differs from them by the dimension of the component columns. Unsatisfactory preservation of the material does not permit recognition of a new form. Perhaps this is the form that Valdiya (1969) called the *Collenia parva*, a Upper Beltian form.

Material: Three samples, S. No. 377–132, 378–132, 379–132.

Location: Naldera, Simla Slates (Naldera Lst.).

Age: Upper Riphean or Vendian.

Group *Tungussia* Semikhatov 1962

Tungussia (?) Semikh. sp. indet. (Fig. 6)

Fragmented sample of four incomplete columns diverging in different direction of the axis, the inclination of which with the surface of strata sometimes reaches up to 45°. Columns are irregular in form, are somewhat curved, vary in diameter from 1.5 cm at the base to 5 cm in the upper part. The lateral boundary is irregular with fringes, but in other parts completely or partially enveloping.

Lamination is distinct but changing in form, often deeply bulging or asymmetrical. At places enveloping boundary is conspicuously differentiated. Details of microstructure of laminae have been obliterated by recrystallization. Laminae are constituted of uniformly granular large crystals of calcite. Thickness of lamina varies from the fraction of 1–2.0 mm.

Material: S. No. 431–147.

Location: Khaira, Shali Formation, Sutlej Valley, NE of Simla.

Age: Middle Riphean to Vendian.

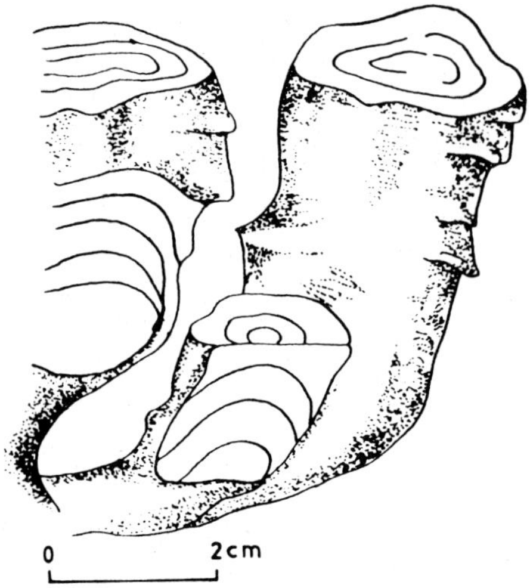

Fig. 6. Morphological reconstruction of *Tungussia* (?) Semikh sp. indet S. No. 481–147, Locality – Khaira, Shali tectonic window, Shali Fmn

Remarks: The characteristics of the lateral boundary of columns and the laminar form show that this form belongs to the group *Tungussia* of Semikhatov. But the absence of columnar structure with subhorizontal or low-angle axis makes identification doubtful, whereas recrystallization has made form-identification impossible.

Group *Conophyton* Maslov, 1937

Conophyton ex. gr. *cylindricus* (Grabau) (Fig. 7, Fig. 8)

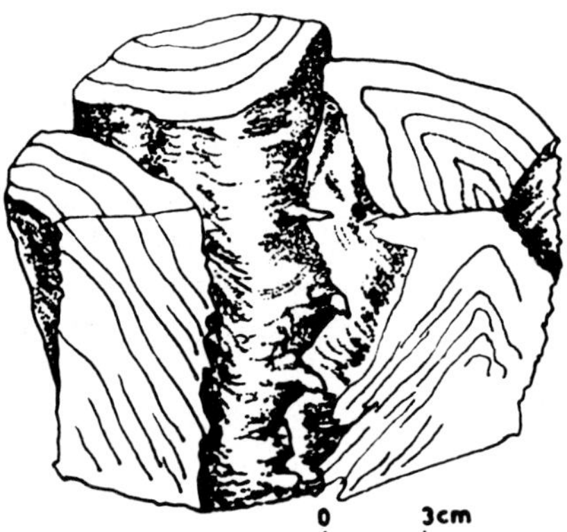

Fig. 7. Morphological reconstruction of fragments of *Conophyton* ex. gr. *cylindricus* (Grabau), S. No. 100–39. Locality – 1 km North of Aut, Shali Fmn

Columns are subcylindrical, with a diameter ranging from 6–8 cm, with rounded or oval form in transverse section. The disposition of columns is subparallel at the base. Ramification is not observed. The width of the interspace between columns does not exceed 3–4 cm. There is an irregular lateral columnar surface, with cornices and bulges. Laminae

Fig. 8. *Conophyton* ex. gr. *cylindricus* (Grabau). Microstructure in thin section (photomicrograph): thickening in the crestal part to be noted. Locality – 1.5 km north of Aut, Shali Fmn

are even, lenticular, thickening in the central part of column. They are composed of yellowish, transparent crystallised carbonates. Thickness of lamina is 1.0–1.2 mm but in the central part may reach up to 1.5 mm.

Material: S. No. 100–39, 103–39.

Location: Village Aut in the Larji Window, Shali Formation (Kulu).

Age: Lower and Middle Riphean.

Remarks: Recrystallization of carbonate of the laminae has obliterated the details of macrostructure, rendering determination of form impossible. The presence of conspicuous ribbon-like microstratum, allows us to place our specimen in the sub-group *Conophyton cylindricus* Maslov. *Conophyton* has been recorded in the Riphean of USSR, South Africa, Sahara, Manchuria, China, North America. In India, *Conophyton cylindricus* has been described by Valdiya (1969) from Semri series of Vindhyan system and Raha and Sastri (1973) from Jammu Limestone.

Group *Colonella* Komar, 1964

Colonella sp. indet. (Fig. 9)

The columns are unbranched and parallel to each other. Transverse section of columns is rounded, with diameter from 8 to 15 cm. The lateral boundary of columns are uneven, without enveloping. Laminae are gently bulging, with simple, fairly low arcs, their coefficient of convexity ranges between 0.3 and 0.4. Microscopic study shows uniformly thick (1.2 to 3.0 mm) microstrata which are composed of uniform yellowish brown crystallised carbonates.

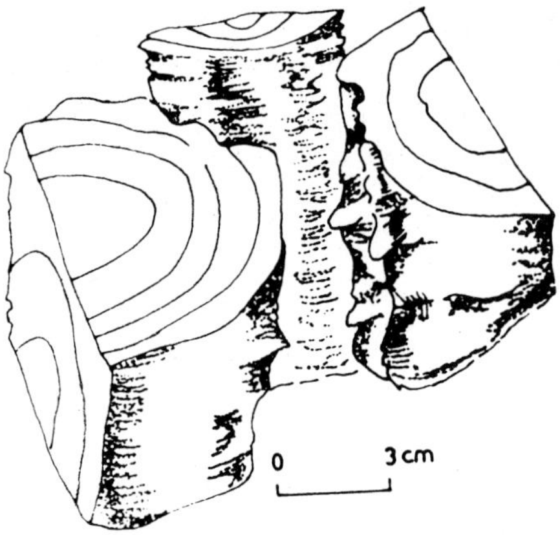

Fig. 9. Morphological reconstruction of *Colonella* sp. indet. S. No. 105–39. Locality – 1 km north of Aut, Shali Fmn

Material: 6 samples from two outcrops, S. No. 66–253, 105–39, 100–39, 108–39, 109–39, 99–38.

Locality: Village Aut, Larji Window Shali Formation (Kulu).

Age: Lower and Middle Riphean.

Remarks and Comparison: Simple ribbon-like texture of lamina is similar to that of *Colonella laminata* described by Komar (1964). Significantly, the reduced thickness of the ribbon-like microstrata (0.5–1.0 mm), and the elevated bell-shaped arc of the lamina, so characteristic of the latter, do not, however, permit definite identification of the form. Raha and Sastry (1973) described this form from Jammu Limestone. Valdiya's (1969) *Collenia columnaris* from Fawn Limestone (Vindhyan), and the Gangolihat Dolomites (Pithoragarh) is possibly *Colonella.*

9.4 Other Stromatolites

Mound-like Stromatolite (Fig. 10)

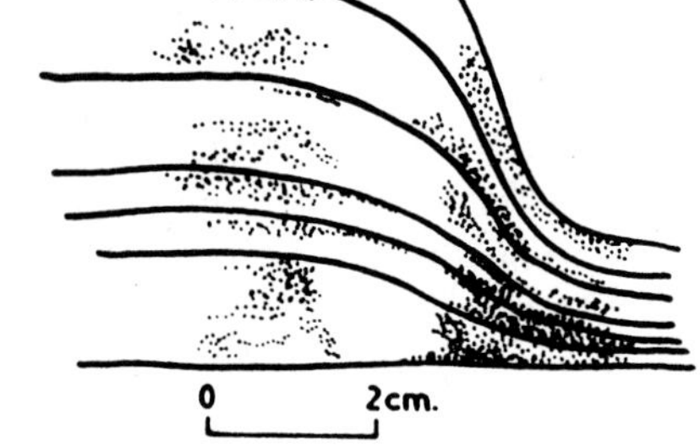

Abb. 10. Morphological reconstruction of fragment of mound-like stromatolite, S. No. 75–233, Locality – 500 m east of Hurla, Shali Fmn

The diameter is 11–12 cm; the thickness of laminae in the central part ranges from 1.5 to 2.5 mm and more, in the marginal parts sharply bending down and simultaneously pinching out. Lamination is distinct.

Remarks: It is difficult to identify the form, as the details have not been preserved. Valdiya describes it as *Collenia symmetrica* from Lower Shali Limestone (H.P.) and Thalkedar and Gangolihat Dolomite of Kumaun.

Material: S. No. 75–233.

Locality: Village Hurla, "Larji-Window", Shali Formation, Lower Himalaya (Kulu).

Age: Riphean.

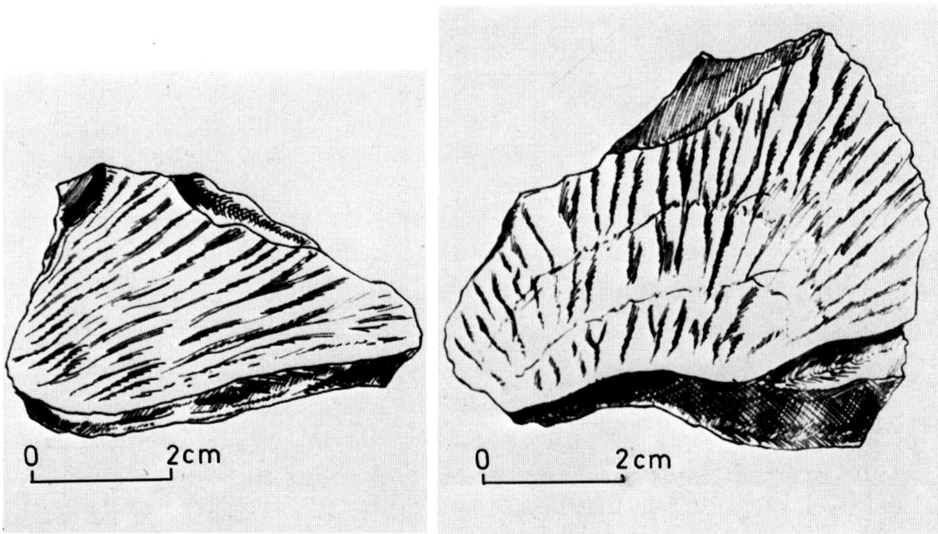

Fig. 11. Morphological reconstruction of *Newlandia* (like) structure. S. No. 267–99. Locality – Tattapani, Shali Fmn

Irregularia Korolyuk (?) (Fig. 12, Fig. 13)

Remarks: Laminar structure has thin wavy microstrata which is highly deformed and brecciated. This type of form is recorded from Jammu Limestone (Raha and Sastry, 1973).

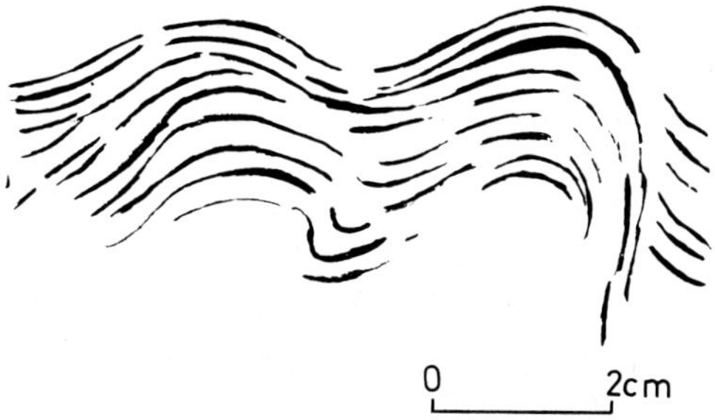

Fig. 12. Morphological reconstruction of *Irregularia*. S. No. 190–69. Locality – Kararaghat, Kakarhatti Limestone

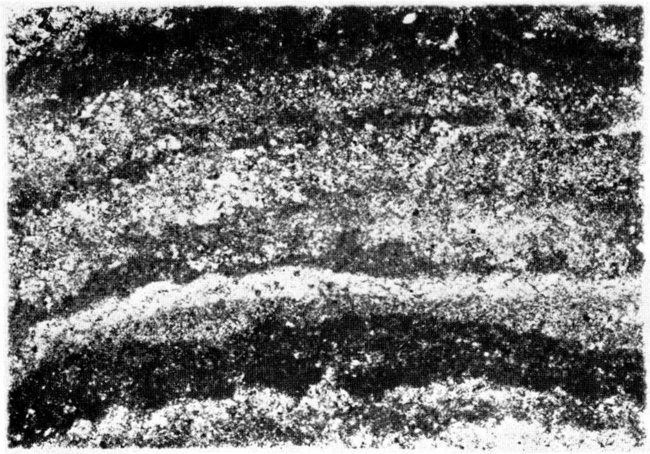

Fig. 13. *Irregularia* Korolyuk, Microstructure in thin section. Photomicrograph x 20. S. No. 388–132

Age: It indicates relatively younger age (Vendian). However, the definite age of this type of thinly laminar stromatolites is under further study.

Material: 156–54, 190–69, Locality – Kararaghat (Kakarhatti Lst), 367–129, 373–130, 374–130, 380–132, 381–132, 388–132, 391–132, 466–155.

Locality: Naldera-Mashobra, Simla Slates (Naldera Lst.).

Newlandia (Fig. 11)

Pre-Upper Riphean, however, in recent years this form has also been reported from younger sediments (i.e. Vendian).

Material: 257–99. Locality – Tattapani, Shali Formation.

9.5 Stratigraphical Importance of the Stromatolites

The studied stromatolites and their known stratigraphic range allow us to conclude that the rocks of Shali Formation, in the Beas valley, were deposited during the Lower Riphean (*Conophyton* and *Colonella* assemblages present), whereas in the Sutlej valley, the presence of younger forms, i.e. *Tungussia* and *Newlandia* would indicate a Middle Riphean age. Valdiya's (1969) proposal to fix age of Lower Shali Limestone, vis-à-vis the carbonate rocks of equivalent Himalayan formation in transition period between Middle and Upper Riphean, cannot be accepted in view of the presence of the *Conophyton-Colonella* assemblage. However, the record of similar forms (Raha and Sastry, 1973) in the Jammu Limestone demonstrates the correlation with the Larji window Shali formation.

In the light of the available data, it would be rational to give an approximate Middle Riphean age to the Shali rocks, although they may overlap the Lower Riphean and the Upper Riphean limits respectively.

On the other hand, the carbonate rocks of the flyschoidal Simla slates of Naldera and Kakarhatti can be dated with greater accuracy, thanks to the presence of the *Jurusania* group which is very characteristic of the Upper Riphean to Vendian. A new form has been recognised which has been named as *Jurusania himalayica*. Besides this, the other form of *Jurusania* group has been recognized in the carbonate rocks of the Naldera-Mashobra ridge. The presence of *Irregularia* at Kararaghat, Naldera, and Mashobra in the carbonate rocks of the Simla Slates extends its age up to the Vendian. It can be concluded with confidence that the Kakarhatti and Naldera limestones are definitely younger than the Shali Formation, where *Conophyton, Colonella* and *Newlandia* are found. *Jurusania* and *Irregularia* suggest a rather young Riphean to Vendian age for the Kakarhatti and Naldera limestones.

The carbonate-terrigenous rocks of the Pithoragarh-Tejam-Deoban-Shali-Jammu Belt, extending from the southeast in Kumaun to the northwest in Jammu (Misra and Valdiya, 1961; Dixit, 1966; Misra and Kumar, 1966; Valdiya, 1969; Banerjee, 1970; L.M. Gupta and Dixit, 1970; V.J. Gupta and Dixit, 1971; Raha, 1972; Singh and Vimal, 1972; Raha and Sastry, 1973), are thus characterized by Riphean stromatolites and supports the view of the existence of shallow marine basin during Riphean period (Ashgirei et al., 1975).

The abundance of *Jurusania* and other younger forms in Naldera and Kakarhatti limestones of the Simla Slates (Pilgrim and West, 1928) and Basantpur Formation (Srikantia and Sharma, 1971) leads to the conclusion that the flysch basin (Sinha, 1972, 1975) came into existence a little later than the Shali Basin.

Acknowledgments. The author expresses his deep sense of gratitude to Dr. (Mme) Maria Raaben for identification and to Prof. G.D. Ashgirei for guidance. Due acknowledgment and thanks is extended to Prof. I.C. Pande for supplying two samples and hosting me as a Visiting Research Scholar under the aegis of U.G.C. (India) scheme. Sincere thanks are due to Dr. K.S. Valdiya for going through the manuscript, with suggestions and comments, though the opinions expressed are entirely those of the author. Gratitude is expressed to Prof. A.G. Jhingran for encouragement and facilities. Dr. D.M. Banerjee's help for literature is gratefully acknowledged.

9.6 Bibliography

Ashgirei, G.D., Sinha, Anshu K., Raaben, M.E., Dmitrenko, O.B.: New findings on the geology of Lower Himalaya, Himachal Pradesh, India. Chayanica Geologica **1** (2), 143–151 (1975).

Banerjee, D.M.: A study of some stromatolites from the calc zone of Sarju – Pungar Valley Areas in Kumaun Himalaya. J. Palaeol. Soc. Ind. **14**, 66–76 (1970).

Cloud, P.E., Jr., Semikhatov, M.A.: Proterozoic stromatolite zonation. Am. J. Sci. **267**, 1017–1061 (1969).

Dixit, P.C.: A study of stromatolites from Girichhina area, District Almora, Kumaun Himalayas. Pub. Cent. of Advan. Study Geol., Chandigarh **5**, 83–92 (1966).

Fuchs, G.: Zum Bau des Himalaya. Wien 1967.

Gansser, A.: Geology of Himalayas. New York: Wiley Interscience 1964.

Glaessner, M.E.: Die Entwicklung des Lebens im Präkambrium und seine geologische Bedeutung. Geol. Rundschau **60** (4), 1323–1339 (1971).

Gupta, L.N., Dixit, P.C.: On the occurrence of Stromatolites from the "Great Limestone" north Katra (Jammu Province). Sci. Cult. **36** (3), 185–186 (1970).

Gupta, V.J., Dixit, P.C.: A note on algal stromatolites from the great Limestone of Riasi, J. and K. Sci. Cult. **37** (2), 253–354 (1971).
Komar, V.A.: Stromatolites of Upper Precambrian of the north of the Siberian Platform and their stratigraphic importance. Trudy Acad. Sci. USSR, Geol. Inst. 154 (1966).
Komar, V.A., Raaben, M.E., Semikhatov, M.A.: *Conophyton* in the Riphean of the USSR and their stratigraphic importance. Trudy Geol. Inst. Leningrad **131**, 77 (1965) (In Russian).
Krylov, I.N.: Columnar branching stromatolites of the Riphean deposits of the Southern Ural and their importance for the stratigraphy of the Upper Precambrian. Trudy Geol. Inst., Leningrad **69**, 133 (1963) (In Russian).
Krylov, I.N.: Riphean and Lower Cambrian stromatolites of Tien-Shan and Karatau. Trudy Geol. Inst., Leningrad **171**, 76 (1967) (In Russian).
Misra, R.C., Kumar, S.: A note on the occurrence of Stromatolities from the Thalkedar Limestones from Raintola, Distt. Pithoragarh, U.P. J. Pal. Soc. India 5.9, 31–33 (1966).
Misra, R.C., Valdiya, K.S.: The calc zone of Pithoragarh with special reference to the occurrence of stromatolites. J. Geol. Soc. India 278–90 (1961).
Oldham, R.D.: The sequence and correlation of the Pre-Territory sedimentary formation of the Simla region of the Lower Himalayas. Rec. Geol. Surv. Ind. **21**, 130–143 (1888).
Pilgrim, G.E., West, W.D.: The structure and correlation of the Simla rocks. Mem. Geol. Surv. Ind. **53** (1928).
Raaben, M.E.: Upper Riphean stromatolites (Gymnosolenida). Trudy Geol. Inst., Leningrad, 203–100 (1969a) (In Russian).
Raaben, M.E.: Columnar stromatolites and Late Precambrian Stratigraphy. Am. J. Sci. **267**, 1–18 (1969b).
Raha, P.K.: A note on the occurrence of algal stromatolites in the Jammu Limestone, Udhampur Dist., Jammu and Kashmir. Ind. Min. **26** (1), (Silver Jubilee Volume) (1972).
Raha, P.K., Sastry, M.V.A.: Stromatolites from the Jammu Limestone, District Udhampur, Jammu and their stratigraphic and palaeogeographic significance. Him. Geol. **3**, 135–147 (1973).
Semikhatov, M.A.: The Riphean and Lower Cambrian of the Yenisey Mountains. Trudy Geol. Inst., Leningrad **68**, 242 (1962) (In Russian).
Serebryakov, S.N., Semikhatov, M.A.: Riphean and Recent Stromatolites: A comparison. Am. J. Sci. **274**, 556–574 (1974).
Singh, P., Vimal, K.P.: Discovery of stromatolities from Sirban Limestone of Riasi. J. and K. State. J. Pal. Soc. India **15**, 6–9 (1972).
Sinha, Anshu K.: Tectonic position, Lithology and Age of Autochthonous Simla and Shali series and the Problem of Tectonic Unit Jaunsar, Blaini. Infrakrol, Krol and Tal in Himalaya (India). Unpublished Ph.D. thesis, Friendship Univ. Moscow 1972 (In Russian).
Sinha, Anshu K.: Calcareous Nannofossils from Simla Hills (Himalaya, India), with a discussion on their age in the tectono-stratigraphic column. J. Geo. Soc. India **16** (1), 69–77 (1975).
Srikantia, S.V., Sharma, R.P.: Shali Formation – A note on the stratigraphic sequence. Bull. Geol. Soc. Ind. **6** (3), 93–97 (1969).
Srikantia, S.V., Sharma, R.P.: Simla Group – A reclassification of the „Chail Series", "Jaunsar Series" and "Simla Slates" in the Simla Himalaya. J. Geol. Soc. Ind. **12** (3), 234–240 (1971).
Srikantia, S.V.: An outline Tectonic Picture of the Simla Himalaya with special reference to Shali and other structural Belts. Sem. Geodyn. Him. Region, Intra Union Comm. Geodyn., N.G.R.I. Hyderabad, India, 122–135 (1973).
Valdiya, K.S.: Note on the discovery of stromatolitic structure from the Lower Shali Limestone of Tattapani, near Simla, H.P. Curr. Sci. **31**, 64–65 (1962).
Valdiya, K.S.: Stromatolites of the Lesser Himalayan Carbonate formation and the Vindhyans. J. Geol. Soc. India **10**, 1–25 (1969).
Valdiya, K.S.: Simla Slates: The Precambrian Flysch of the Lesser Himalaya, its Turbidites, sedimentary structures and Palaeoccurrents. Geol. Soc. Am. Bull. **81**, 451–468 (1970).
Walter, M.R.: Stromatolites and the biostratigraphy of the Australian Precambrian and Cambrian. Spec. Paper in Palaeont. **11**, Palaeontol. Assoc. London 190 (1972).
West, W.D.: Structure of the Shali Window. Rec. Geol. Surv. Ind. **74**, 133–163 (1939).

10 Distribution of Stromatolites and Stromatolitic Reefs in the Precambrian of India

W.A. SCHNITZER, Würzburg, W. Germany

Contrary to extensive research in the Soviet Union, Australia, USA and Africa, the information on the distribution of Precambrian stromatolites in India is still not satisfactory. It is only in the last twenty years that stromatolites have been reported from the Vindhyans, the Cuddapah Basin and lately from the Himalayas (Srinivasa Rao, 1947; Mohan, 1968; Prakash, 1966; Lotze, 1967; Sarma and Nageswara Rao, 1969; Valdiya, 1969). The author has had the opportunity, from 1961 to 1969, to investigate almost all Precambrian sedimentary regions beginning from the Cuddapah, Kurnool, Bhima, Pakhal, Godavari, Bastar, Chhattisgarh, and extending up to the Vindhyans. It became evident that stromatolites are widely distributed in the Precambrian carbonate series (Fig. 1A). Accordingly, a biostratigraphical scale based on stromatolites and lithofacies, and checked by the few available radiometric ages can hopefully be in view for India too. Through the mapping of a large part of the Chhattisgarh Basin and detailed profile surveys in the Basins of Bastar, Kurnool, and Bhima, it was possible to establish a large cyclic division of the Late Precambrian (see Figs. 1A, B, C) which could be followed over a distance of more than 1000 km. The cycle starts with sandstone or dark limestones and ends up with thick red beds. On the basis of radiometric dating in the Cuddapah Basin, which gave an average age of 1400 m.y., the period of sedimentation considered here ranges from 1600 to 1160 m.y. (Aswathanarayana, 1962, 1964). Accordingly, the overlying Kurnool System and Chhattisgarh System should encompass a period of time between 1100 to 800 m.y. It is now possible to check how far the available stromatolites could be used for stratigraphical subdivision, as has been done in the Soviet Union and Australia (Salop, 1968; Glaessner et al., 1969, etc.).

In India the majority of stromatolites occur in formations younger than 1500 m.y. Up to now stromatolites have been reported from the following Systems in the Peninsular part of India.

10.1 Cuddapa System

Stromatolites are chiefly found in the Vempalle limestones and dolomites, and in the Tadpartri limestones of the Cheyair Series (Prasad and Verma, 1967; Viswanathiah and Aswathanarayana Rao, 1967; Sarma and Nageswara Rao, 1969; Schnitzer, 1969a, 1971). These are of *Collenia* and *Cryptozoon* type. The stromatolites are of considerable sizes, up to 6 m long, and build individual reef bodies.

According to the author's data, the stromatolites described by Lotze (1967) from Lower Cuddapah of the Godavari Zone do in fact belong to the younger Kurnool and Chhattisgarh System.

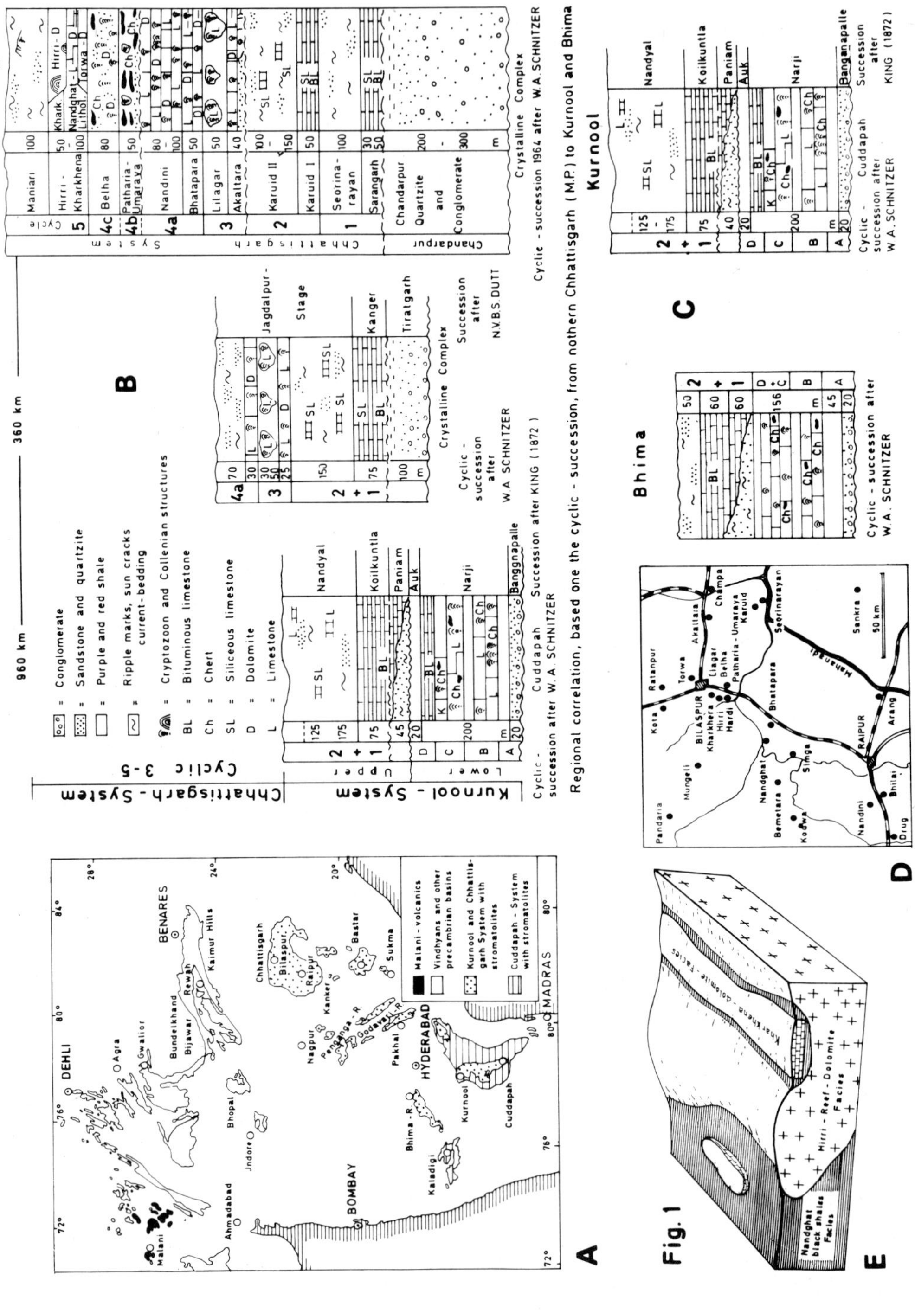

Fig. 1. Legend see opposite page

10.2 Kurnool System

A profusion of stromatolites occurs in the entire range of Narji stage (Cycles B and C) as large reef builders (Fig. 2G, H). Besides their occurrence in the Kurnool and dolomites of the Bhima Basin (Fig. 1B, C), as well as in the basins of Pakhal, Godavari, and Penganga (Lotze, 1967; Schnitzer, 1969, 1971).

10.3 Chhattisgarh System

The Chhattisgarh System, defined by the author (Schnitzer, 1969a, 1969b), contains a particularly great abundance of stromatolite-rich carbonate series. According to the classification of Logan et al. (1964), Cryptozoon and Collenia structures and their transitional varieties are the dominant types. Particularly SH and LLH types are present. The 3rd and 4th Cycles in the Bastar-Jagdalpur Basin are very rich in stromatolites and small reef structures in limestone marls (Fig. 2). In a completely developed cyclic sequence of the Chhattisgarh Basin almost all carbonate series were found to contain stromatolites, although they are mainly concentrated in Cycles 3 (Akaltara, Lilagar, Fig. 2D), 4 (Bhatapara, Nandini, Patharia-Umaraya, Belha, Fig. 2C, E), and 5 (Hirri-Kharkhena, Fig. 2B, F). They differ greatly from the stromatolites of the Cuddapahs and the Upper Vindhyans (see Fig. 2A, Bhander-limestone).

The 5th Cycle, nicely distributed between Raipur and Bilaspur, deserves particular attention. It contains large stromatolite reefs, which alternate with dark dolomitic shales and bedded dolomites. The respective position of each facies is shown in Fig. 1E. Whereas the Nandaghat-Facies and the fine-grained Kharkhena dolomite are deprived of stromatolites, the reef-structures (up to 100 m thick) of the Hirri dolomites display a large number of stromatolites, mainly on the reef edges (Fig. 2B, F). They seem to disappear in the inner parts of the reef bodies, but this may be due to diagenesis. Individual stromatolites attain a diameter of 1.5 m, which is the maximum size found in the Chhattisgarh System.

10.4 Vindhyans

Numerous stromatolites and small reefs are found in the Kajrahat limestone, Fawn limestone and Rohtas limestone of the Lower Vindhyans. The Bijaigarh-shale and, particu-

◂ Fig. 1 A–E. (A) Distribution of Cuddapah-Kurnool and Chhattisgarh Systems with special reference to stromatolites (according to the author's research in Peninsula India). (B) Regional correlation on the basis of cyclic sedimentation: northern Chhattisgarh (M.P.) – Kurnool Basin. (C) Correlation Kurnool Basin – Bhima Basin. (D) Type localities in the central part of the Chhattisgarh Basin. (E) Dolomite province of Hirri-Kharkhena: facies relationship between black shales of the Nandghat-facies, Hirri reefal dolomite with stromatolites, and the bedded Kharkhena dolomite facies

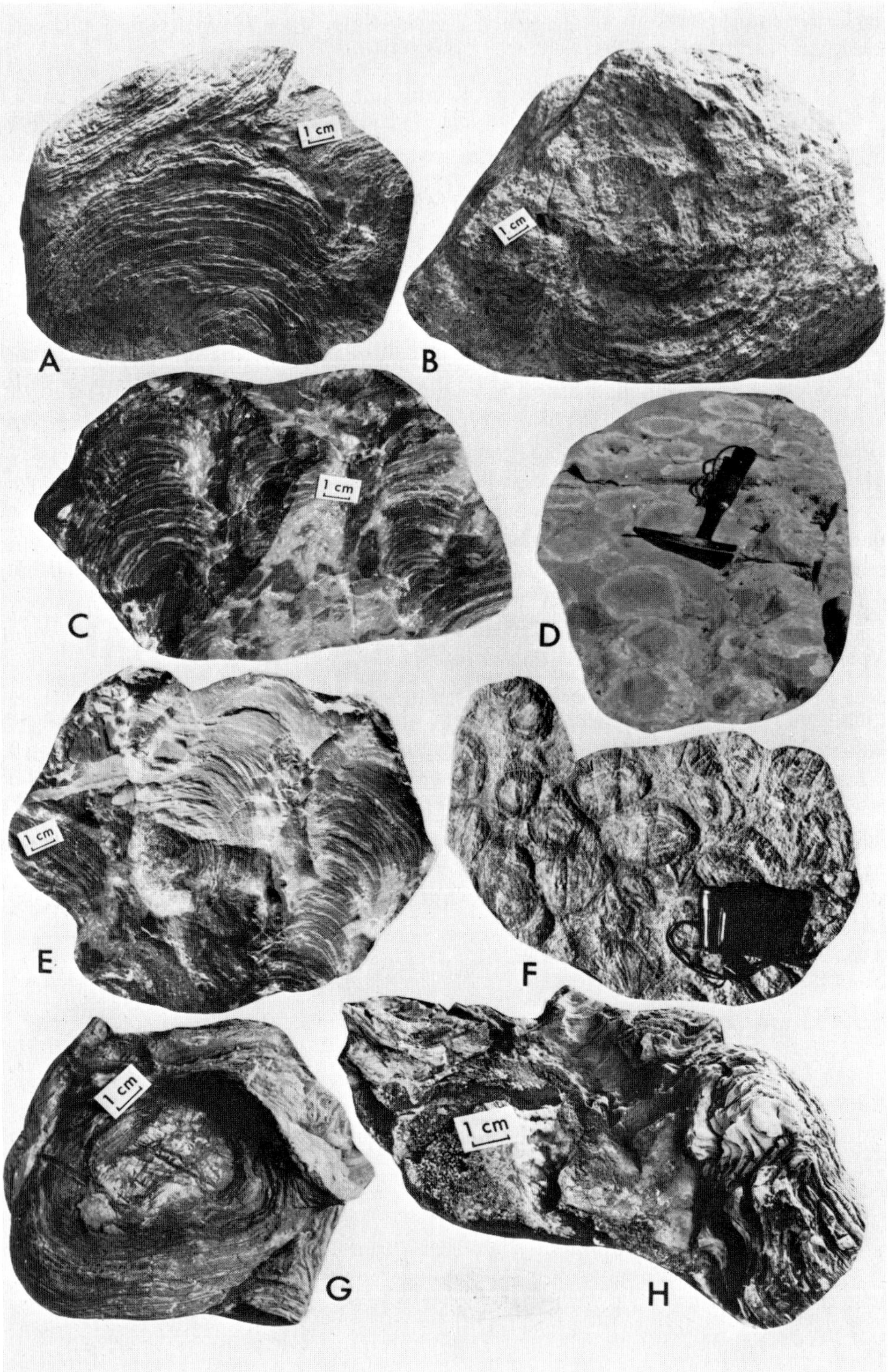

Fig. 2. Legend see opposite page

larly, the Bhander limestone in the Upper Vindhyans, also contain stromatolites (Prakash, 1968; Mohan, 1968; Valdiya, 1969; Schnitzer, 1969b, 1971; Fig. 2A).

On the basis of stromatolites, Valdiya correlated the Lower Vindhyans with the Lower Riphean (older than 1260 m.y.), the Upper Vindhyans with the Middle Riphean (1260–1000 m.y.), and the stromatolite-rich limestones of the Lower Himalayas with the Middle and part of the Upper Riphean.

In summary, the distribution of stromatolites in the Late Precambrian deposits of Peninsular and Extra-Peninsular India compares in abundance and diversity with what has been reported from other continents. They could, accordingly, be tentatively used in stratigraphic subdivisions in India. Very distinctive is the morphology of the stromatolites from the Cuddapah System to Kurnool and Chhattisgarh System and to the Upper Vindhyans. The stromatolites of the Lower Vindhyans show analogies with the forms from the Chhattisgarh System. At this stage, a detailed investigation of stromatolite-rich limestones and dolomite series, so widely developed on the Indian shield, is required for the definition of any biostratigraphical scheme. Radiometric data (which are so highly variable for example in the Vindhyans), geotectonic phases, from which some are recognizable world-wide, and particularly climatic factors (Tillites, Red-Beds, Kaolinite-crusts) and the fundamentals of lithological facial division, which are recognizable in many continents, must be included in the entire conception.

10.5 Bibliography

Aswathanarayana, U.: Age of the Cuddapahs, India. Nature **194**, 565–566 (1962).

Aswathanarayana, U.: Determination of Rocks and Geochronology of India. Intern. Geol. Congr. 22 Sess. India (1964).

Glaessner, M.F., Preiss, W.V., Walter, M.R.: Precambrian columnar stromatolites in Australia: Morphological and stratigraphic analysis. Science **164** (1969).

Logan, B.W., Rezak, R., Ginsburg, R.N.: Classification and environmental significance of algal Stromatolites. J. Geol. **72**, 1, 68–83 (1964).

Lotze, F.: Stromatolithen in jungpräkambrischen Kalken der Godavari-Zone. (Kurze Mitteilungen zur Geologie Indiens I.). N. Jb. Geol. Paläontol. Mh. 1967, 11, 677–683 (1967).

Mohan, K.: Stromatolitic structures from the lower Vindhyans, India, with additions from South Africa, Australia, and North Korea. N. Jb. Geol. Paläontol. Abh. **130**, 3, 335–353 (1968).

Prakash, R.: Shell-like forms in the Basal stage, Vindhyan system: communicated for publication in the Current Science 1967a. Cited in: Krishna Mohan, Stromatolitic structures from the Lower Vindhyans, India, with additions from South Africa, Australia, and North Korea. N. Jb. Geol. Paläontol. Abh. **130**, 3, 335–353 (1968).

◀ Fig. 2A–H. (A) Stromatolites from the Bhander Limestone, near Rewa (Upper Vindhyan System). (B) Stromatolites from the Hirri reef complex, Bilaspur district. The largest stromatolites may be up to 1.5 m in diameter. – Cycle 5. (C) Stromatolites, with SH- and LLH-types predominating, in limestones of Nandini near Raipur. – Cycle 4a. (D) Reefal stromatolites in the Lilagar Beds, Bilaspur district. – Cycle 3. (E) Stromatolites, mostly of SH- and LLH-types, from the Bhatapara and Nandini Limestones, near Raipur. – Cycle 4a. (F) Surface patterns of stromatolites bordering the Hirri reef-complex, Hirri Mines, Bilaspur district. – Cycle 5. (G and H) Silicified stromatolites from large reef-bodies, Narji stage (lower part of Kurnool system), near Kurnool. – Cycles B/C

Prasad, K.N., Verma, K.K.: Stromatolites from Vempalle formation of Cuddapah, Andra Pradesh. J. Indian Geosci. Assoc. **7** (1967).

Salop, L.: Pre-Cambrian of the U.S.S.R. XXIII. Intern. Geol. Congr. Prag, **4**, 61–73 (1968).

Sarma, S.R., Nageswara Rao, C.: Stromatolites from the Cuddapah basin. J. Indian Geosci. Assoc. **10**, 102–105 (1969).

Schnitzer, W.A.: Mineral deposits as a basis for India's economy. Nat. Res. Devel. **1**, 145–168 (1975).

Schnitzer, W.A.: Die jung-algonkischen Sedimentationsräume Peninsula-Indiens. N. Jb. Geol. Paläontol. Abh. **133**, 191–198 (1969a).

Schnitzer, W.A.: Zur Stratigraphie und Lithologie des nördlichen Chhattisgarh-Beckens (Zentral-Indien) unter besonderer Berücksichtigung von Algenriff-Komplexen. Z. deut. Geol. Ges. 1966, **118**, 290–295 (1969b).

Schnitzer, W.A.: Das Jungpräkambrium Indiens („Purana-System"). Erlanger Geol. Abh. **85**, 1–44 (1971).

Srinivasa Rao, M.R.: Algal structures from the Cuddapah limestone, Pre-Cambrian, South India. J. Mysore Univ. **5**, Sec. B. Sci., 67–72 (1947).

Valdiya, K.S.: Stromatolites of the Lesser Himalayan carbonate formations and the Vindhyans. J. Geol. Soc. India **10**, 1 (1969).

Viswanathiah, M.N., Aswathanarayana Rao, A.N.: Algal stromatolites from Cuddapah formations near Mutssukota, Anantapur District (A.P.) India. Indian Mineral. **8**, 62–65 (1967).

11 Sedimentpetrographical and Geochemical Studies of the Late Precambrian Stromatolites of India

N.K. SRIVASTAVA, Würzburg, W. Germany

11.1 Introduction

Precambrian stromatolites are widely distributed both in Peninsular and Extra-Peninsular India (Valdiya, 1969; Schnitzer, 1971). Although these bio-sedimentary structures have been used for stratigraphic correlation of various Precambrian sedimentary basins in India, their petrographic and chemical features have been hardly touched.

The main purpose of this report is to shed a light on the petrographic and geochemical nature of the stromatolites so profusely developed in the Late Precambrian sediments of Chattisgarh Basin in Central India (Fig. 1), the stratigraphic and geological aspects of which have been discussed by Dutt (1963) and Schnitzer (1971). Stromatolites of these deposits belong mainly to SH-V, LLH, and SS-C groups of Logan et al. (1964).

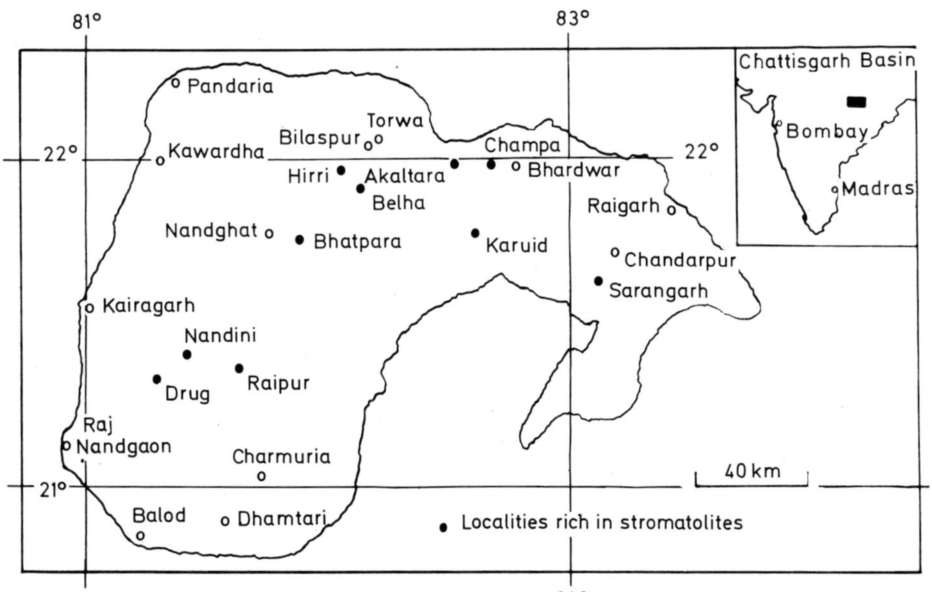

Fig. 1. Index map

11.2 Mineralogical Investigations

X-ray diffraction and staining methods revealed an overwhelmingly carbonate composition of the stromatolites; dolomite predominates in the samples from Bhatapara, Nandini, Belha and Hirri, where it reaches its maximum abundance (Fig. 2). Ferroan-calcite and ferroan dolomite are also present in vugs and pores, especially in samples from Nandini, Hirri, Akalatara, and Bhatapara. Silicification of stromatolites brought about noticeable changes in the mineralogical pattern along the profiles. Hirri stromatolites are particularly affected by the secondary silification.

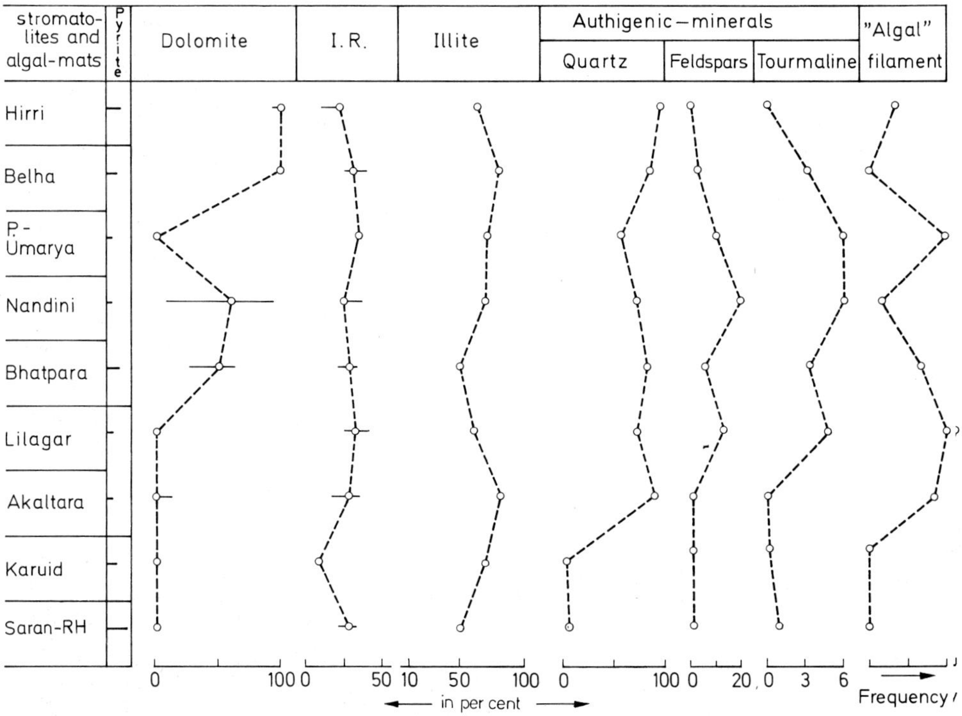

Fig. 2. Distribution of pyrite, dolomite, insoluble residues, authigenic minerals and "algal filaments" along the profile

The nature and quantity of insoluble residues vary considerably between samples from different sedimentary cycles (see Fig. 2), whereas samples from the same stratigraphic horizon show rather constant features. Dolomitization and micro-stylolitization have directly influenced the quality and quantity of the insoluble fractions; these are mainly composed of quartz (both authigenic and detrital), feldspars (generally fresh grains of albite, microcline, and orthoclase), heavy minerals (zircon, rutile, tourmaline, anatase, brookite, and glauconite), opaque minerals (pyrite, hematite, goethite, and magnetite) and clay minerals. The clay fractions revealed the predominance of hydromicas over the iron chlorites. Poorly crystallized kaolinite was present in a few samples only, although its content increased in both under- and overlying Red Beds (sandstones, siltstones, and mudstones). The highest concentration of hydromicas (up to 90%) was found in samples from Akaltara, Belha, and Hirri. A study of the authigenic minerals in the stromatolites and the associated rocks showed that stromatolites are richer in insoluble authigenic minerals (quartz, feldspars, pyrite, tourmaline, brookite, anatase, chlorite) than the equivalent carbonate and terrigenous rocks. This was explained by the particularities of stromatolitic buildup and of their environment of growth (Srivastava and Schnitzer, 1976).

11.3 Microtextural Investigations

In these sections, most of the bio-sedimentary structures found in the Chattisgarh basin show fine laminations with varying thickness (0.04–1 mm) formed by the interplay of carbonate minerals, terrigenous material, and "algal filaments". The laminations are mostly regular and the contacts with the wall rocks sharp (Fig. 3). Two main types of laminated structures are recognizable under the microscope. The first type shows flat, slightly crinkled, finely laminated (0.04–0.1 mm) microstructure composed of micritic calcite and displaying laminoid fenestral fabrics, similar to those described by Tyrrell (1969) and Davies (1970), from inter- and suptratidal areas. These structures are also characterized by the presence of microripples, and their highly bituminous and pyritiferous nature. The laminoid fenestral voids are partly filled with calcite, rarely ferroan dolomite, clayey material and cryptocrystalline silica. This type of structure is commonly developed in the Sarangarh and Karuid in the Chattisgarh basin, as well as in the Bastar basin further south.

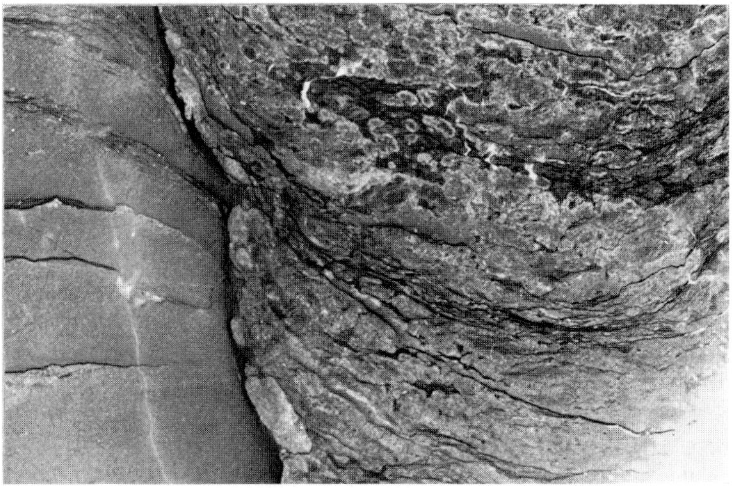

Fig. 3. Photomicrographs of Lilagar stromatolite. Note the complex internal structure and the sharp contact with the wall rock. x 50

The second type of laminated structure forms domes with a complex internal structure. Some samples show desiccation cracks suggesting occasional emergence (Bhatapara, Lilagar). One of the most important features of these algal structures is the presence of thread-like vermicular filaments (up to 10 micron in Ø) as shown in Fig. 4. These filaments resisted dolomitization, as completely dolomitized stromatolites may still show well-preserved filaments, as found in samples from Bhatapara, Nandini, and Hirri. The true nature of these filaments is under study. A further microstructure composed of clear calcite in a clustered pattern is widespread in highly silty Lilagar and Patharia-Umaraya stromatolites. Both primary and diagenetic vugs with radiaxial-fibrous and granular calcite, and dolomites of various generations are abundant in these stromatolites, which are profusely developed in the 3rd, 4th and 5th sedimentation cycles of the Chattisgarh System.

Fig. 4. Thread-like "algal filaments" in the dolomitic Hirri stromatolites. x 50

11.4 Geochemical Studies

Sr, Mg, Mn, Fe, Ni, Zn, Cu, Cr, Na, and K concentrations were determined by AAS from the HCl-soluble parts of the bio-sedimentary structures. Strontium displays its characteristic distribution: calcitic stromatolites show higher Sr contents (200–400 ppm) than those made of dolomite (30–150 ppm). Strontium content decreases with increasing dolomitization (Fig. 5). Richter (1970) observed similar trends in the Devonian reefs of the Eifel. The generally low Sr concentrations (lower than 400 ppm) point to a diagenetic alteration in an open sytem. Strontium released by diagenesis was flushed back to the sea-water, as no strontium minerals were found in the rocks of the Chattisgarh System. Higher Mg content was observed, as expected, in dolomitic samples from the upper part of the profile. Sarangarh, Karuid, Lilagar, and Patharia-Umaraya bio-sedimentary structures are poor in Mg (around 0.8%). The average in Mn content is highest in Lilagar samples (1030 ppm) and the lowest in Karuid (270 ppm) and appears to be controlled by the insoluble residue. Iron concentration varies in the same way and is controlled by the insoluble residue. Part of Fe substitutes for Mg and Ca in dolomites and calcites, as was shown by staining methods. Large scale postdiagenetic changes also contributed iron to the stromatolites, thereby eliminating the primary iron concentrations of bio-sedimentary structures from different stratigraphic horizons. The concentrations of sodium and potassium are partly controlled by the insoluble residues of the stromatolites. The average Na and K concentrations vary between 400–600 and 50–250 ppm respectively.

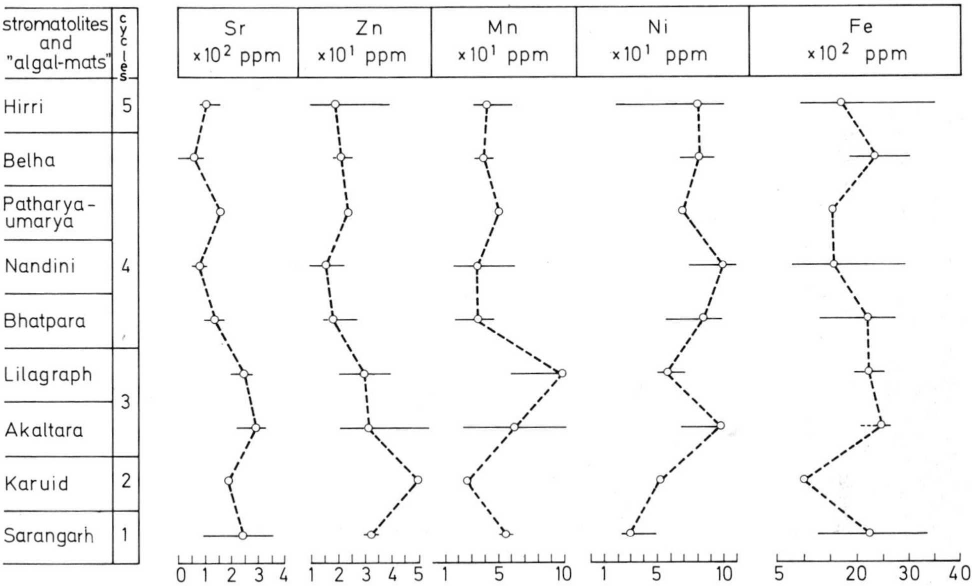

Fig. 5. Distribution of Sr, Zn, Mn, Ni, Fe in the stromatolites of the Chattisgarh System

11.5 Diagenesis

The diagenetic events undergone by the bio-sedimentary structures include dolomitization, silicification, cementation, and development of authigenic minerals. Dolomitization is nil in the Sarangarh and Karuid structures. In other samples, dolomite appears as thin layers, in cavities (both normal and ferroan dolomite), as intraclasts and as associated with the stylolites. On the basis of crystal size, alignment of dolomitic-front, and other textural features, both early and late diagenetic dolomites are found in the bio-sedimentary structures of the Chattisgarh basin. Late diagenetic dolomitization is evident when associated with stylolites, infilling of vugs and when forming coarse-crystalline discordant patches (Hirri, Belha, and Nandini). Early diagenetic dolomites are lamina-controlled and commonly have small crystals (less than 20 microns). Early diagenetic dolomitization is especially evident in Akaltara, Lilagar, and Patharia-Umaraya. Dedolomitization and calcitization processes were observed in some samples from Bhatapara and Hirri.

11.6 Conclusions

This study revealed considerable differences in the lithology, mineralogy, and geochemical aspects of bio-sedimentary structures collected from different stratigraphic horizons of the Chattisgarh System. The Sarangarh-Karuid type of bio-sedimentary structures probably formed in a low energy environment under reducing conditions. The digitate columnar structures forming domes from the upper part of the Chattisgarh System probably developed in a high-energy environment, for the stromatolites built large reefs (Nandini, Hirri). The presence of "algal filaments" in the bio-sedimentary structures of Bhatapara, Nandini, and Hirri is of great significance. In my opinion, qualitative and quantitative data on the mineralogy, fabrics, and chemistry of the stromatolites may help the stratigraphic correlation of other stromatolite-rich Precambrian sedimentary basins of India.

Acknowledgment. The author is deeply indebted to Prof. Dr. W.A. Schnitzer and to the Deutsche Forschungsgemeinschaft for help and assistance.

11.7 Bibliography

Davis, G.R.: Algal-laminated sediments, Gladstone Embayment, Shark Bay, Western Australia. In: Carbonate Sedimentation and Environments, Shark Bay, W. Australia. Tulsa 1970, pp. 169–205.

Dutt, N.V.B.S.: A suggested succession of the Purana Formations of the southern part of Chattisgarh, M.P. Rec. Geol. Surv. India **93**, Pt. 2 (1963).

Logan, B.W., Rezak, R., Ginsburg, R.N.: Classification and environmental significance of algal stromatolites. J. Geol. **72**, 68–83 (1964).

Richter, D.K.: Zur Petrographie und Genese der Dolomite in der Eifel. Diss. Bochum Univ., 1970, 133 pp.

Schnitzer, W.A.: Das Jungpräkambrium Indiens („Purana-System"). Erlanger Geol. Abh. **85**, 44 pp. (1971).

Srivastava, N.K., Schnitzer, W.A.: Authigenic minerals in the Precambrian sedimentary rocks of the Chattisgarh System (Central India). N. Jb. Mineral. Abh. **126** (2), 221–230 (1976).

Tyrrell, W.W., Jr.: Criteria useful in interpreting . . . . . carbonate units. In: Depositional Environments in Carbonate Rocks. Friedman, G.M. (ed.). Spec. Publ. 14, SEPM. Tulsa 1969, pp. 80–97.

Valdiya, K.S.: Stromatolites of Lesser Himalayan carbonate formations and the Vindhyans. J. Geol. Soc. India **10** (1), 1–25 (1969).

12 Algal Remains in the Lower Cambrian Bonavista Formation, Conception Bay, Southeastern Newfoundland

A.-St. EDHORN and M.M. ANDERSON, St. John's, Newfoundland

12.1 Introduction

Lower Cambrian rocks are present in a number of scattered outcrops in southeastern Newfoundland. Those considered here are exposed on the southwest side of Conception Bay (north side of the Avalon Peninsula), coastally at Bacon Cove and inland west of the southern end of Colliers Bay, and they belong to the Bonavista Formation, the lowermost of the three Lower Cambrian formations recognized in the region. The type section for this formation is on the northwest side of Smith Sound in the Bonavista map-area, where it has an estimated thickness of 195 m (Hutchinson, 1962). At the Conception Bay localities, however, the Bonavista Formation is represented by a much smaller thickness of sediments, about 15 m (difficult to estimate as partly concealed), which is probably stratigraphically equivalent to the upper part of the sequence at the type locality.

Fossil blue-green and red algae of the families Rivulariaceae and Epiphytaceae, respectively, and also blue-green algae of the genus *Girvanella*, were discovered in conglomeratic and arkosic beds, as well as in limestones, at both the localities mentioned above. These algae, initially, occupied pockets, hollows, and irregularities in the late-Precambrian erosion surface over which the sea advanced early in Cambrian time, i.e. at the time of formation of the conglomerate, and they became cemented into it by trapping and binding of silt and sand and by calcification, hence building up stromatolites between the pebbles and cobbles.

Within the small thickness of sediments that overlie the conglomerate, repeated, often abrupt, changes in the algal flora have been found to occur at successively higher levels. Thus, one or more species dominated the sea-floor for shorter or longer periods at different times. These changes in the flora appear to have been a response to slight variations in environmental conditions that took place from time to time during sedimentation. In the succeeding limestones (unrepresented or possibly concealed west of the southern end of Colliers Bay), only one type of alga is dominant, namely *Girvanella*, which suggests that environmental conditions were stable during the time that the limestones were being formed.

12.2 Geologic Setting

The Lower Cambrian sequence in southeastern Newfoundland has been described chiefly by Walcott (1888, 1900), Howell (1925), Hutchinson (1956, 1962), and Fletcher

(1972). The sequence is subdivided into three formations, redefined by Hutchinson (1962), in ascending order, the Bonavista, Smith Point, and Brigus Formations.

The Bonavista Formation, which is the only Lower Cambrian formation considered here, was named by van Ingen (1914), but he did not designate a type section. The latter was chosen later by Hutchinson (1962) as the sequence of Cambrian rocks on the north-west side of Smith Sound in the Bonavista map-area that he believed van Ingen had examined when he named the formation. Hutchinson (1962) estimated the thickness of the type section as 195 m, 135 m of exposed strata plus a possible 60 m of concealed basal beds. Some 15 m of the strata cropping out beyond the concealed beds that Hutchinson excluded from the type section, are, however, regarded by Jenness (1963) as representing the lowermost part of the Bonavista Formation.

The lower part of the Bonavista Formation exposed at Bacon Cove, southwestern Conception Bay, consists of a basal conglomerate overlain by a varying thickness of calcareous arkose that is followed above by pink to red partly stromatolitic limestones separated by thin layers of dark red siltstone. The conglomerate contains angular to rounded pebbles and cobbles of locally derived sedimentary and volcanic rocks in a calcareous arkosic matrix; in places it includes boulders of the underlying Conception Group (late-Precambrian) siltstones. The clastic section beneath the limestones varies rapidly in thickness, from being absent in places, to about a metre in depressions in the underlying steeply dipping Conception Group siltstones (McCartney, 1967).

Stromatolites are found in situ in the conglomerate, and they are also present in limey bands in the overlying arkose. The calcareous stromatolites in these beds vary in color from red to greenish; they are stratiform to columnar, with the latter up to a few centimetres high and about one centimetre in diameter. The columnar stromatolites are closely spaced or laterally linked, and occasionally exhibit branching.

The red to pink stromatolite-bearing limestones at the top of the basal section at Bacon Cove are overlain by a series of red mudstones. The thickness of the latter, some 5 m of which is exposed, cannot be determined as the mudstones are concealed above by Pleistocene glacial deposits.

The second locality at which fossil algae were found in a basal sequence of the Bonavista Formation is on the north side of the Conception Bay highway (lat. 47° 27′ 24″ N, long. 53° 14′ W) west of the southern end of Colliers Bay (part of Conception Bay), where the Lower Cambrian sediments rest unconformably on andesite of late-Precambrian age. The small, inconspicuous outcrop at this locality formerly occupied an area of several square metres but it has recently been largely destroyed during road-widening. The exposed sediments have a thickness of some 40 cm with individual layers measurable in millimetres or centimetres. The sequence begins with a basal conglomerate consisting of angular to well-rounded sand, granules and small pebbles (mainly fragments of quartz and andesite) in a calcareous silty matrix. This basal layer is overlain by dark-purple stromatolitic limestones containing thin intercalations of pink limestone or purple calcareous arkose. The stromatolites are essentially stratiform (up to a cm or two thick) with individual laminae planar to wavy.

12.3 Algal Remains and Stromatolites in the Bonavista Formation

Thin sections of specimens of stromatolite-bearing beds from Bacon Cove and west of the southern end of Colliers Bay have been examined, and they reveal the presence in these beds of fossil algae of the families Rivulariaceae and Epiphytaceae, and of the genus *Girvanella*; fossil algae have not previously been reported from the Bonavista Formation.

Three new species of rivulariacean algae have been found in the material studied; one belongs to the genus *Primorivularia* Edhorn 1973, and each of the others has been assigned to a new genus. Different growth patterns exhibited by the rivulariacean colonies, and the presence or absence of reproductive bodies, are the principal features that separate these genera. Descriptions of the new species and genera (Plate 1/1–4) are already in preparation and will be published elsewhere.

The rivulariacean algae are largely confined to stratiform stromatolites, and it is evident from the microstructure of the latter that they were formed by these blue-green algae.

The epiphytacean algae, probably representing at least two species, have not been classified, but they are reported here because of their participation in the formation of columnar stromatolites. Apart from the fact that several of the characters used for distinguishing one species of *Epiphyton* from another can only be observed in finely preserved material, many of the more than 80 species described to-date are very similar to one another. Consequently, it will be necessary to compare a great many individuals from beds of the same age before a decision can be reached regarding (1) the number of species present in beds of the Bonavista Formation, and (2) whether or not they belong to already-named species. Furthermore, it is necessary to be cautious in applying names to algae that are heavily iron-encrusted, like many of those from the Bonavista Formation, because it has been found from localities elsewhere, that where rivulariacean colonies have grown successively more or less one above the other, such individual sequences are, when iron-encrusted, confusingly like *Epiphyton* both in form and size. Rivulariacean colonies are generally from 50–100 μ in diameter; individual filaments are 1–1.5 μ wide.

Even though *Epiphyton* is not restricted to the Cambrian, the genus is now known to have been even more widely distributed during that period than when Harlan Johnson stated in 1966 that "*Epiphyton* is one of the most common and most widely distributed Cambrian genera." The finding of *Epiphyton* in the Bonavista Formation in southeastern Newfoundland adds another area to the many already recorded in which it flourished during the Lower Cambrian.

The first description of *Epiphyton* was given by Bornemann in 1886, and he tentatively referred it to the Siphonales. Since that time many different opinions have been expressed regarding its systematic position. Johnson (1966, p. 16) reviewed these opinions up to 1961: "Chapman (1918) and Priestly and David (1910) placed it close to *Solenopora*, Gordon (1921) considered it a green alga close to *Ortonella*, and Pia (1927) thought it belonged among the Cyanophycophyta, an idea shared by several Russian workers (Krasnopeeva, 1937). Maslov later (1956) tentatively put the genus among the green Codiaceae. Because of the microstructure she found, Korde (1955, and 1958/59) places *Epiphyton* among the red algae in the class Floridae and erects a family Epiphyta-

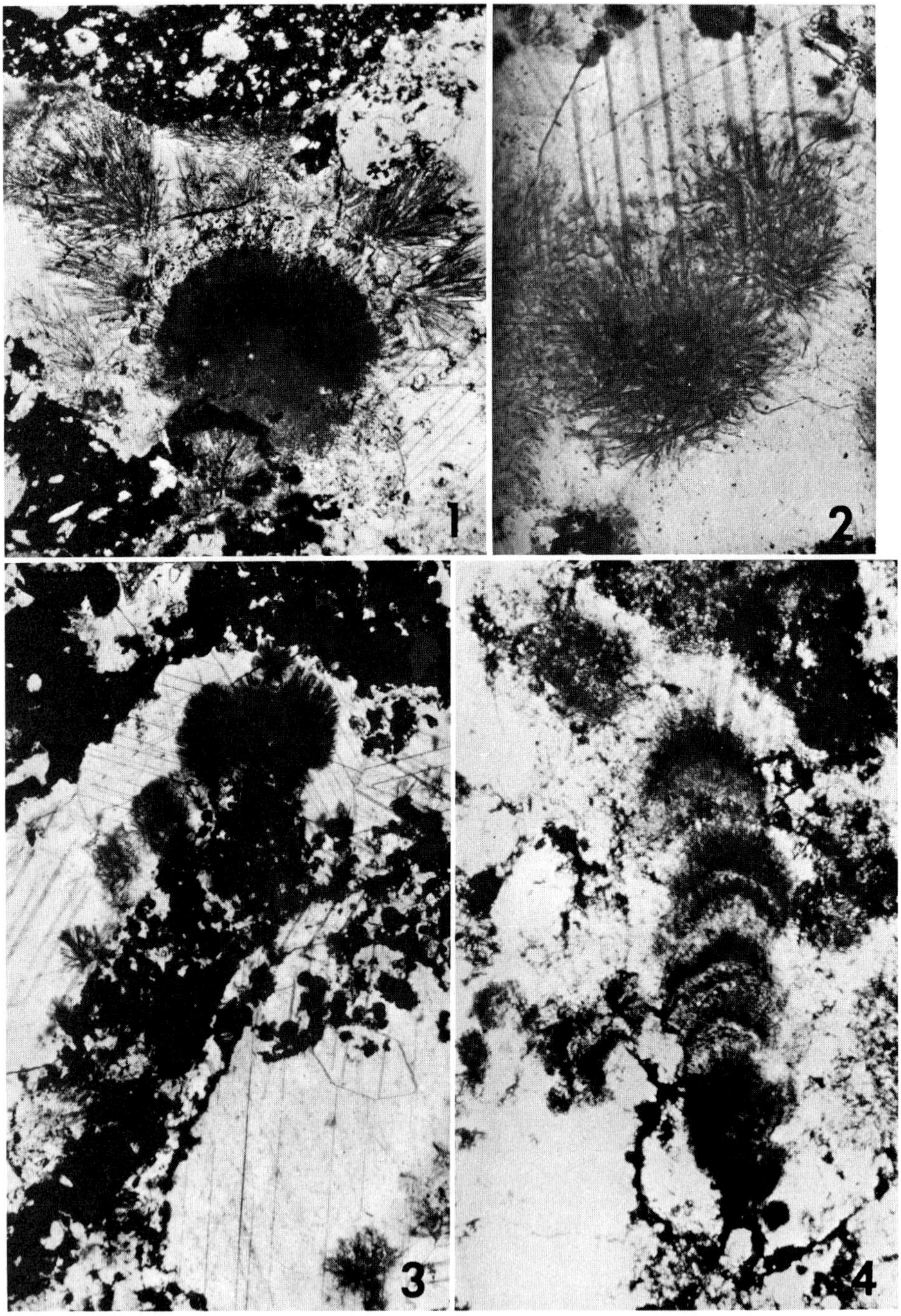

Plate 1. Legend see opposite page

ceae to contain the genera *Epiphyton* and *Chabakovia*". Korde actually considered *Epiphyton* to be a type of red alga two years earlier than noted by Johnson (Korde, 1959, 1961). Since 1961, Maslov (1963) has referred *Epiphyton* to algae of uncertain systematic position, Chuvashov (1965) has included it in the Porostromata of Pia (1927), and Riding and Toomey (1972) have suggested that it may be related to red algae included in the Corallinaceae, even though uncalcified genicular areas or joints characteristic of such forms have not been observed. The absence of joints was, in fact, one of Korde's (1969) main reasons for placing the genus in an independent family of red algae rather than in the Corallinaceae. Until such time as convincing evidence for an alternative assignment is forthcoming, the writers accept her placement of *Epiphyton* in the Epiphytaceae.

Girvanella is widespread at a number of levels in the Bonavista Formation southwest of Conception Bay, notably in limestones. This alga was first described by Nicholson and Etheridge (1878) who believed that it was a foraminifer; their original generic description is as follows: "microscopic tubuli with arenaceous or calcareous (?) walls, flexuous or contorted, circular in section, forming loosely compacted masses. The tubes apparently simple cylinders, without perforations in their sides, and destitute of internal partitions or other structures of a similar kind." This tubular microfossil, now known to be calcareous, has subsequently been considered a sponge, a green alga, and a blue-green alga (for details see Riding, 1975). Bornemann (1886) was the first to suggest a blue-green algal affinity for *Girvanella*, although he described it under the generic name *Siphonema* which is thus a synonym (Hinde, 1887). Bornemann (1886) and, considerably later, Frémy and Dangeard (1935) compared *Girvanella* to extant blue-green algae, while recently Lauritzen and Worsley (1974) have suggested that it represents members of the Oscillatoriaceae. However, only traces of filaments have so far been found within these tubes (Johnson, 1942; Wood, 1957), which are now generally regarded as calcified sheaths rather than the calcified walls of trichomes (Pollock, 1918; Seward, 1931; Klement and Toomey, 1967; Riding, 1972, 1975), and as calcification of sheaths is known to occur in the members of several families of blue-green algae (Riding, 1975), it is not possible at present to assign *Girvanella* to a particular family of the Cyanophyta. Furthermore, the range in diameter and thickness of the calcareous tubes, and the variety of growth habit exhibited by different species, suggest that *Girvanella* represents a number of different filamentous algae that may or may not be closely related. Nevertheless, in the absence of identifiable filaments, these forms can only be treated as species. Species are distinguished on the basis of the diameter of their tubes, wall thickness, type and frequency of branching when present, and general mode of growth (Johnson, 1966). At least two species of *Girvanella* are present in basal beds of the Bonavista Formation at Bacon Cove and west of the southern end of Colliers Bay.

◀ Plate 1/1. Rivulariacean colony of the genus *Primorivularia* with developing young colonies. x 85

Plate 1/2. Rivulariacean algae growing in tufts. x 125

Plate 1/3. Hemispherical colonies of rivulariacean algae with spore formation. x 58

Plate 1/4. Zoned colonies of rivulariacean algae forming "ministromatolites". x 160

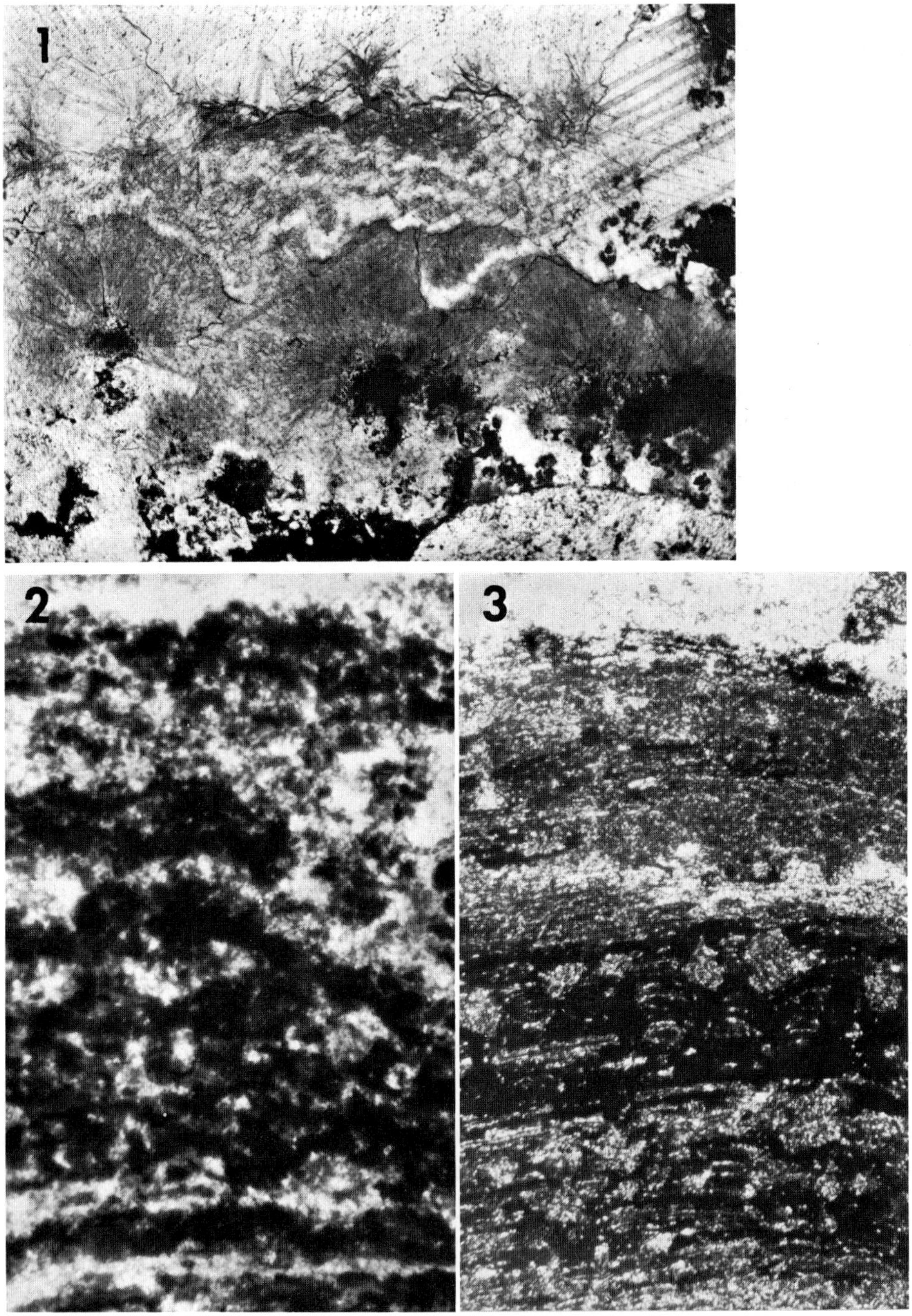

Plate 2. Legend see opposite page

Girvanella was reported by Ahr (1971) as having been primarily responsible for the formation of stromatolites in the Upper Cambrian Wilberns Formation of central Texas. No evidence has been found in the material studied from the Bonavista Formation that *Girvanella* forms stromatolites, although it does occur in thin layers within stratiform stromatolites west of the southern end of Colliers Bay. These layers are clearly intercalations, and their presence indicates that short-lived changes in environmental conditions enabled *Girvanella* to grow over the surface of the stromatolite-forming algae. At other times, however, the environmental changes were more prolonged, and on these occasions the layers formed by *Girvanella* were from one to several centimetres thick. *Girvanella* thus gave rise to thin limestones, characteristically pink in colour, and was evidently rock-forming; it also acted on occasion as a substrate for the stromatolite-forming algae.

Wherever *Girvanella* appears in the Bonavista sequence, stromatolite-forming and other algae are generally absent, and vice versa. Consequently, where they succeed one another, there is an abrupt change in the algal flora, which suggests that the environmental conditions that favoured *Girvanella* were inimical to the other species of algae, or that the latter, although tolerant of the changed conditions, were unable to compete with the growth of the former.

Thin *Girvanella* limestones are present in the Bonavista Formation throughout southeastern Newfoundland, and it is apparent that such limestones began to develop early in the Lower Cambrian rather than during the Middle Cambrian as stated by Johnson (1966). In his study of the Silurian of Gotland, Hadding (1959) distinguished between ball-forming *Girvanella* and mat-forming species that form dense limestones. Only the latter are associated with the Bonavista Formation.

12.4 Observed Relations Algal Growth/Stromatolitic Structures

Detailed studies of the rivulariacean colonies at different stages of growth and disintegration reveal a relationship between the growth pattern of the algae and the microstructure of the stromatolites that they form. There are colonies of rivulariacean algae (Plate 2/1, 2) showing beginning processes of forming stromatolites, compaction of colonies, zonation, calcification, disintegration, binding and trapping of debris and cementation into laminae. The laminae are dense, convex to planar, alternating light and dark bands of about equal width, 6–7 μ, sometimes disturbed by changes in sedimentation or disintegration of the algal layers. There are hemispherical patches or spaced microcolumns which can be traced vertically through the laminae because of their pronounced heavier outline (Plate 2/3), which probably represents the zonation in individual rivulariacean colonies. All three species of rivulariacean algae encountered in the rocks studied may have participated in the formation of the stromatolites but whether they were equally active builders has not yet been determined.

◀ Plate 2/1. Radiating, hemispherical colonies of rivulariacean algae. x 90

Plate 2/2. Stromatolitic laminae consisting of rivulariacean colonies. x 210

Plate 2/3. Compacted, dense laminae formed by rivulariacean colonies. Note faint outline of microcolumns. x 145

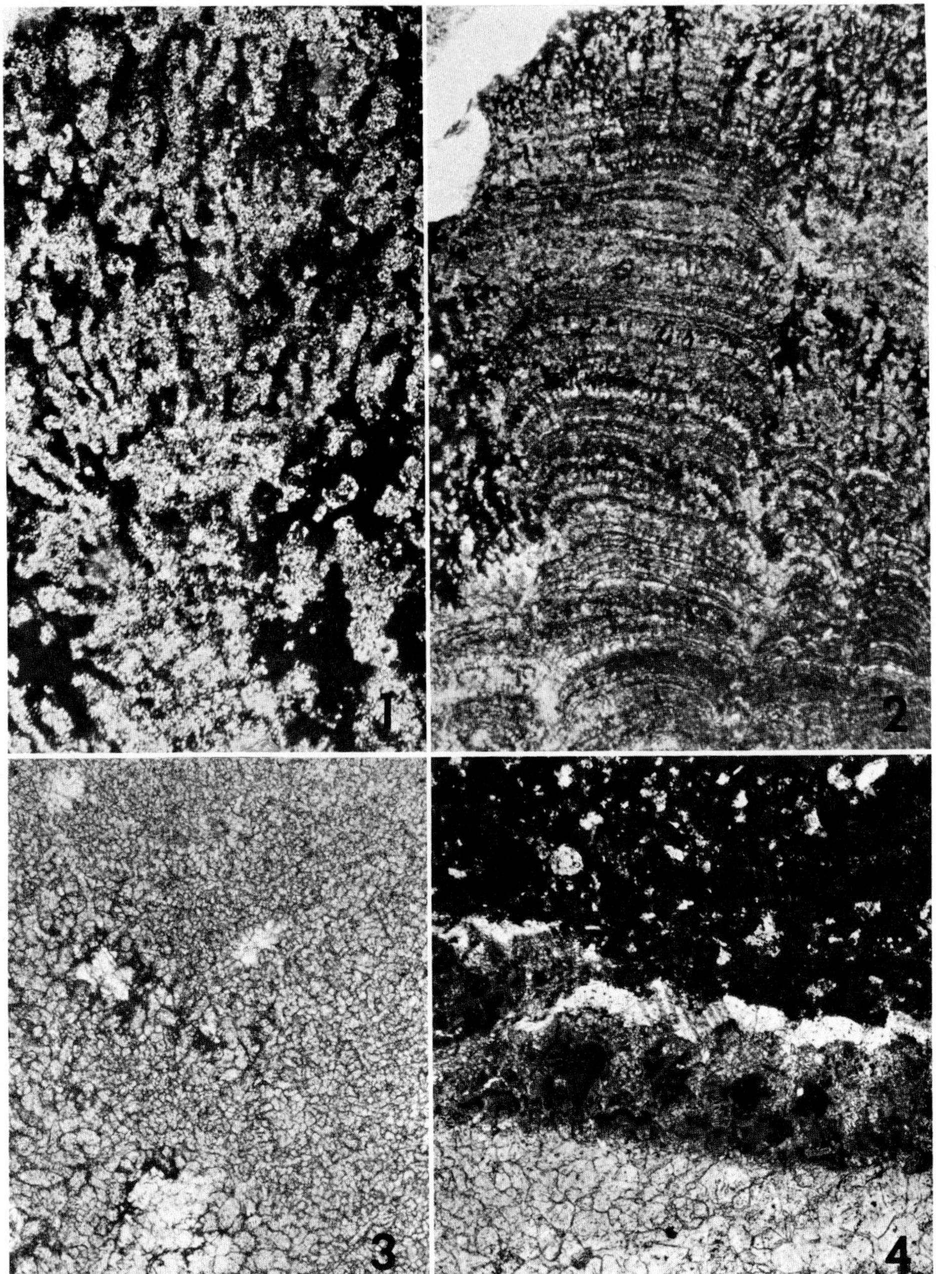

Plate 3. Legend see opposite page

From the observations made on these stromatolites, one can speculate on how the macrostructure of the developing stromatolites will be constructed, taking into account "normal" growth of the algae and undisturbed sedimentation. As mentioned earlier, the stromatolites formed by the rivulariacean algae are stratiform with the laminae planar to wavy and showing a high degree of inheritance. The upper surface of the stromatolites is wrinkled or mamillated.

Monty (1967) described recent algal mats formed by *Rivularia biasoletiana* and *Dicothrix bornetiana* in an intertidal environment bordering eastern Andros Island in the Bahamas; those of *D. bornetiana* are either flat mats up to 1 cm thick, or mamillated structures and small rounded heads 1–2 cm high. Both species build their mats on a rocky substrate or on previous mats formed by *Schizothrix calcicola*, which is similar to the behaviour of their fossil counterparts in the sense that they also colonized rocky surfaces or mats of *Girvanella* or other algae. The species of rivulariacean algae referred to from Andros island, unlike those found in the Bonavista Formation, are not stromatolite builders as the mats they form are nonlaminated.

A species of *Epiphyton*, which has a luxuriant growth as a result of repeated branching, forms stromatolites at several levels in the basal sequence of the Bonavista Formation at Bacon Cove. Small, well-preserved branching thalli of *Epiphyton*, individually 100 to 140 μ high and 20 to 40 μ wide, make up the laminae in linked columns, sometimes entering into a digitate pattern due to profuse branching (Plate 3/1). The spaced laminae have a crinkled, frilled pattern, and they are convex and vertically stacked and give the impression of being vertically striated because of the continuous growth of *Epiphyton*. The thickness of the laminae varies from 25 to 50 μ, depending on the extent to which the thalli were preserved by cementation and survived compaction. The thalli of *Epiphyton* can readily be distinguished in the columnar stromatolites illustrated in Plate 3/2, and also in the spaces between the columns.

The microstructure of the stromatolites studied was controlled by the algae that built them. Their gross structure also appears to have been biotically controlled to some extent, but this cannot be stated with more certainty until many more stromatolites of the same kind from other Lower Cambrian outcrops throughout southeastern Newfoundland have been studied in detail. It is clearly important, as pointed out by Larsson (1973), that the lamination morphology in stromatolites be carefully examined before any attempt is made to classify them entirely from their gross structure. If this is not done, the influence of biological processes in the shaping of stromatolites, which has been stressed by Walter et al (1973), Serbryakov and Semikhatov (1974), and other workers, notably in Russia, may be overlooked.

◀ Plate 3/1. Continuous growth of branching *Epiphyton*. x 140

Plate 3/2. Laminae in a stromatolite built by *Epiphyton*. x 33

Plate 3/3. *Girvanella* mat. x 33

Plate 3/4. *Girvanella* mat functioning as substratum for developing stromatolites. x 90

12.5 Conclusions

Species of Rivulariacean and Epiphytacean algae associated, respectively, with stratiform and columnar stromatolites in basal beds of the Bonavista Formation southwest of Conception Bay, southeastern Newfoundland, were responsible for their formation. The microstructure of the laminae in these stromatolites reflects the growth pattern of the algae that built them. The gross structure of the stromatolites also appears to have been shaped to some extent by the biological activities of the algae concerned, although this remains to be confirmed.

Girvanella was rock-forming early on in the Lower Cambrian, and gave rise to thin, dense limestones.

12.6 Bibliography

Ahr, W.M.: Paleoenvironment, algal structures, and fossil algae in the Upper Cambrian of Central Texas. J. Sed. Petrol. **41**, 205–216 (1971).

Bornemann, J.G.: Die Versteinerungen des Cambrischen Schichtensystems der Insel Sardinien nebst vergleichenden Untersuchungen über analoge Vorkommnisse aus andern Ländern. Erste Abt. Ksl. Leop.-Carol. Deut. Akad. Naturforsch. **51**, 1–147 (1886).

Chapman, F.: Report on the British Antarctic Expedition. Geology **2**, pt. 4, 81–83 (1918).

Chuvashov, B.I.: Foraminifera and algae from the Upper Devonian deposits of the western slope of the central and southern Urals (in Russian). Akad. Nauk SSSR, Uralsk. Fil., Sverdlovsk, Inst. Geol. Trud. **74**, 3–153 (1965).

Edhorn, Anna-Stina: Further investigations of fossils from the Animikie, Thunder Bay, Ontario. Proc. Geol. Assoc. Can. **25**, 37–66 (1973).

Fletcher, T.P.: Geology and Lower to Middle Cambrian Trilobite Faunas of the Southwest Avalon, Newfoundland. Ph. D thesis Univ. Cambridge 530 pp. (1972).

Frémy, P., Dangeard, L.: Sur la position systématique des Girvanelles. Bull. Soc. Linn. Normandie **8**, 101–111 (1935).

Gordon, W.T.: Scottish National Antarctic Expedition 1902–1904. Cambrian organic remains from a dredging in the Weddell Sea. Trans Roy. Soc. Edinburgh **52**, 681–714 (1921).

Hadding, A.: Silurian algal limestones of Gotland. Lunds Univ. Arsskr. N.F. Avd. 2, Bd. 56, No. 7, 26 pp. (1959).

Hinde, G.J.: Review of Dr. J.G. Bornemann (1886). – The fossils of the Cambrian strata of Sardinia. Geol. Mag. **34**, 226–229 (1887).

Howell, B.F.: The Faunas of the Cambrian Paradoxides beds at Manuels, Newfoundland. Bull. Am. Pal. **11**, No. 43, 1–140 (1925).

Hutchinson, R.D.: Cambrian stratigraphy, correlation, and paleogeography of eastern Canada. 20th Intern. Geol. Congr. Symp.: El sistema Cámbrico su paleogeografía y el problema de su base, Pt. II, 289–311 (1956).

Hutchinson, R.D.: Cambrian stratigraphy and trilobite faunas of southeastern Newfoundland. Geol. Surv. Can. Bull. **88**, 156 pp. (1962).

Jenness, S.E.: Terra Nova and Bonavista map-areas, Newfoundland. Geol. Surv. Can., Mem. **327**, 184 pp. (1963).

Johnson, J.H.: Permian Lime-secreting algae from the Guadalupe Mountains, New Mexico. Bull. Geol. Soc. Am. **53**, 195–226 (1942).

Johnson, J.H.: A review of the Cambrian algae. Colorado School Mines Quart. **61**, No. 1 162 pp. (1966).

Johnson, H.M.: Silurian *Girvanella* from the Welsh Borderland. Palaeontology **9**, 48–63 (1966).

Klement, K.W., Toomey, D.F.: Role of the blue-green alga Girvanella in skeletal grain destruction and lime-mud formation in the Lower Ordovician of west Texas. J. Sedimentol. Petrol. **37**, 1045–1051 (1967).
Korde, K.B.: Manual on collecting and study of fossil algae. M.-L. Publishing House Akad. Nauk SSSR 1–26 (1953).
Korde, K.B.: Algae from Cambrian deposits of the rivers Lena, Botoma and Amga (in Russian). Akad. Nauk SSSR, Inst. Paleontol. Trud. **56**, 79–91 (1955).
Korde, K.B.: New genera of the fossil Cyanophyta (in Russian). Paleontol. Inst. Acad. Sci. USSR, Pt. II, 113–117 (1958).
Korde, K.B.: The morphology and systematic position of representatives of the genus *Epiphyton* (in Russian). Akad. Nauk SSSR, Dokl. **126**, 1087–1089 (1959).
Korde, K.B.: Cambrian algae from the southeastern part of the Siberian Platform (in Russian). Akad. Nauk SSSR, Inst. Paleontol. Trud. **89**, 1–147 (1961).
Krasnopeeva, P.S.: Algae and Archaeocyathinae of the oldest formation in the Potkheen district in Khakassia (in Russian). West-Siberian Geol. Trust, Rec. Geol. Krasnojarsky Region **3**, 51 pp. (1937).
Larsson, K.: The lower Viruan in the autochthonous Ordovician sequence of Jämtland. Sveriges Geol. Unders. Ser. C., No. 683, 82 pp. (1973).
Lauritzen, Ø., Worsley, D.: Algae as depth indicators in the Silurian of the Oslo region. Lethaia **7**, 157–161 (1974).
Maslov, V.P.: Fossil algae of the USSR.(in Russian). Trudy Inst. Geol. Nauk. SSSR No. 160, 301 pp. (1956).
Maslov, V.P.: In: Fundamentals of Paleontology: Algae, Bryopisda, Psilopsida, Lycopsida, Arthropsida, Pteropsida (in Russian). Orlov, Yu A. (ed.). Akad. Nauk SSSR Moscow 1963, 261 pp.
McCarttney, W.D.: Whitbourne map-area, Newfoundland. Geol. Surv. Can. Mem. **341**, 135 pp. (1967).
Monty, C.L.V.: Distribution and structure of Recent stromatolitic algal mats, eastern Andros Island, Bahamas. Ann. Soc. Geol. Belg. **90**, 55–100 (1967).
Nicholson, H.A., Etheridge, R.: A Monograph of the Silurian Fossils of the Girvan District in Ayrshire. **I**, fasc. 1. Edinburgh–London: Blackwood 1878, 135 pp.
Pia, J.: Thallophyta. In: Handbuch der Palaeobotanik, Band 1: Thallophyta, Bryophyta, Pteridophyta. Hirmer, M. (ed.). München–Berlin: R. Oldenbourg 1927, p. 31–136.
Pollock, J.B.: Blue-green algae as agents in the deposition of marl in Michigan lakes. Ann. Rep. Michigan Acad. Sci. **20**, 247–260 (1918).
Priestley, R.E., David, T.W.E.: Geological notes of the British Antarctic Expedition 1907–1908. Congr. Geol. Intern. 11th, C. R. 767–811 (1910).
Riding, R.: Calcareous algae and some associated microfossils from Ancient Wall reef complex (Upper Devonian), Alberta. (Abs.). Am. Assoc. Petrol. Geol., Bull. **56**, 648 (1972).
Riding, R.: *Girvanella* and other algae as depth indicators. Lethaia **8**, 173–179 (1975).
Riding, R., Toomey, D.F.: The sedimentological role of Renalcis and Epiphyton in Lower Ordovician mounds, southern Oklahoma. J. Paleontol. **46**, 509–519 (1972).
Serebryakov, S.N., Semikhatov, M.A.: Riphean and Recent Stromatolites: a comparison. Am. J. Sci. **274**, 556–574 (1974).
Seward, A.C.: Plant life through the Ages, a Geological and Botanical Retrospect. Cambridge Univ. Press. 1931, 601 pp.
van Ingen, G.: Table of Geological Formations of the Cambrian and Ordovician Systems about Conception and Trinity Bays, Newfoundland. Princeton Univ. Contr. Geol. Newfoundland No. 4 (1914).
Walcott, C.D.: The stratigraphical succession of the Cambrian faunas in North America. Nature (Lond.) **38**, 551 (1888).
Walcott, C.D.: Lower Cambrian terrane in the Atlantic Province. Proc. Wash. Acad. Sci. **1**, 301–339 (1900).
Walter, M.R., Golubić, S., Preiss, W.V.: Recent stromatolites from hydromagnesite and aragonite depositing lakes near the Coorong Lagoon, S. Australia. J. Sedimentol. Petrol. **43**, No. 4, 1021–1030 (1973).
Wood, A.: The Type-Species of the Genus *Girvanella* (Calcareous Algae). Palaeontology **1**, 22–28 (1957).

13 Stromatolites from the Zechstein Limestone (Upper Permian) of Poland

T.M. PERYT and T.S. PIATKOWSKI, Warsaw, Poland

13.1 Introduction

Stromatolites were found on the Fore-Sudetic area and Łeba elevation area in the upper part of the Zechstein Limestone (Upper Permian), a carbonate member that underlies the evaporites of the Werra cycle (Fig. 1). According to the paleogeography and facies analysis stromatolites occur both in the nearshore and basinal areas, in the supratidal, intertidal, and subtidal zones as indicated by the fossil content and associated sedimentary structures. The aim of this paper is the comparison of stromatolites coming from these different environments.

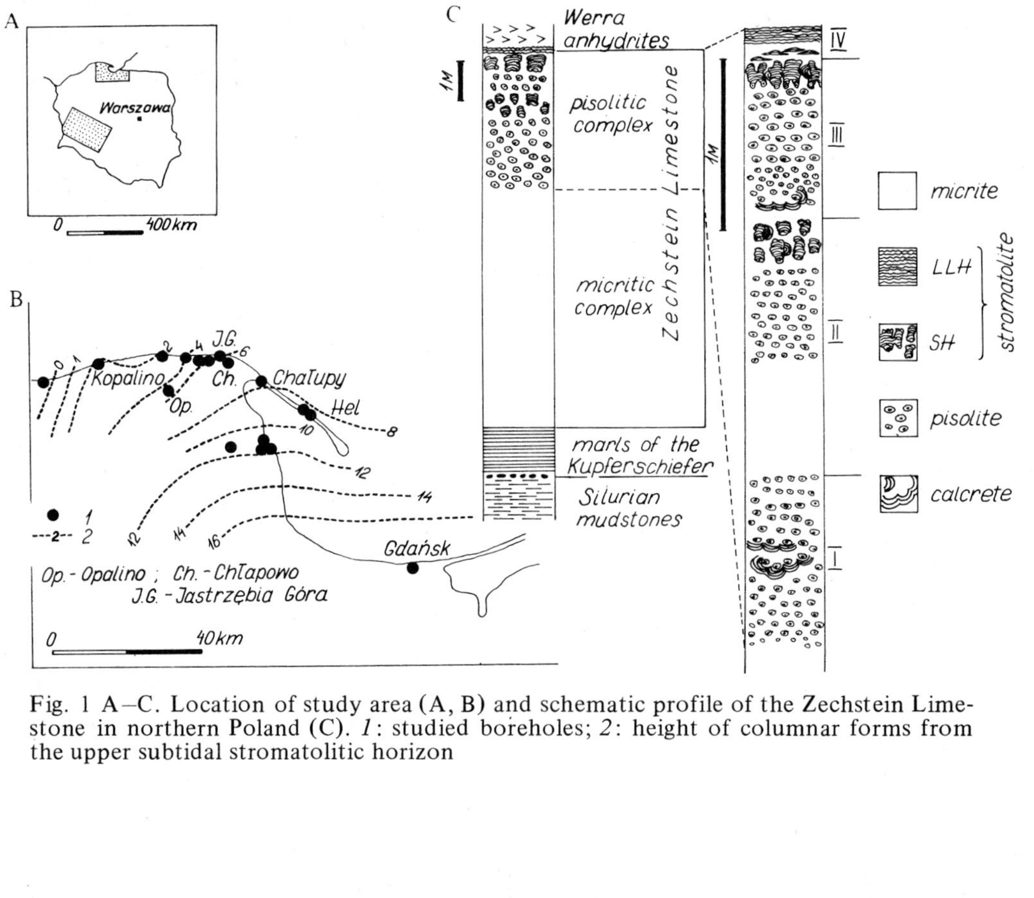

Fig. 1 A–C. Location of study area (A, B) and schematic profile of the Zechstein Limestone in northern Poland (C). *1*: studied boreholes; *2*: height of columnar forms from the upper subtidal stromatolitic horizon

13.2 Subtidal Stromatolites

13.2.1 Distribution

Widespread subtidal stromatolites were found in northern Poland where they form two horizons that can be traced on the area of more than 15,000 km^2 (Fig. 1). The thickness of the lower stromatolitic horizon ranges between 5 and 25 cm, while the thickness of the upper one is between 2 and 20 cm (Fig. 1B). In the lower stromatolitic horizon small irregular stromatolitic forms of height not exceeding 4 cm and width 3 cm occur

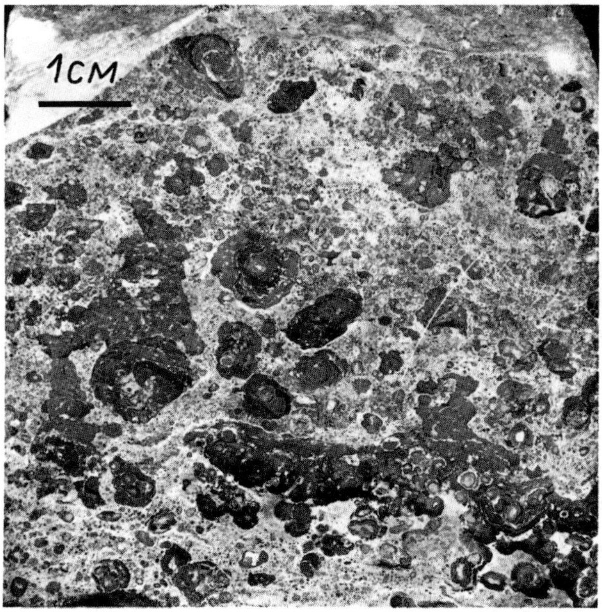

Fig. 2. Small irregular stromatolitic forms from the lower subtidal stromatolitic horizon. Hel IG 1 borehole, depth 1099.2 m

(Fig. 2), that are as a rule isolated and always grow on multilayered delicate oncoids whose diameter sometimes exceeds 1 cm. In the upper stromatolitic horizon large columnar forms, in height up to 17 cm and in width up to 15 cm, occur (Fig. 3, 4, 5), the dimensions decreasing in the north-western direction (Fig. 1B, 5). The columnar forms are very often distinctly asymmetrical in growth, both in vertical and horizontal section (Fig. 6). Because the directions of asymmetry vary during growth (Fig. 6) they may correspond with changes of current directions; unfortunately the studied cores were unoriented and therefore it is impossible to see any regional current pattern. Columnar forms are developed on oncoids (Fig. 5), more seldom on bigger skeletal grains (Fig. 4). At the beginning of growth they are small, later they merge, thus forming a massive body, and after this the partitioning of that body takes place (Fig. 5). It should be stressed that the described columnar forms eastwards from Chałupy occur also in the lower stromatolitic horizon, being associated with small irregular stromatolitic forms.

Fig. 3. Columnar forms from the upper subtidal stromatolitic horizon. Hel IG 1 borehole, depth 1098.2 m. Thin section

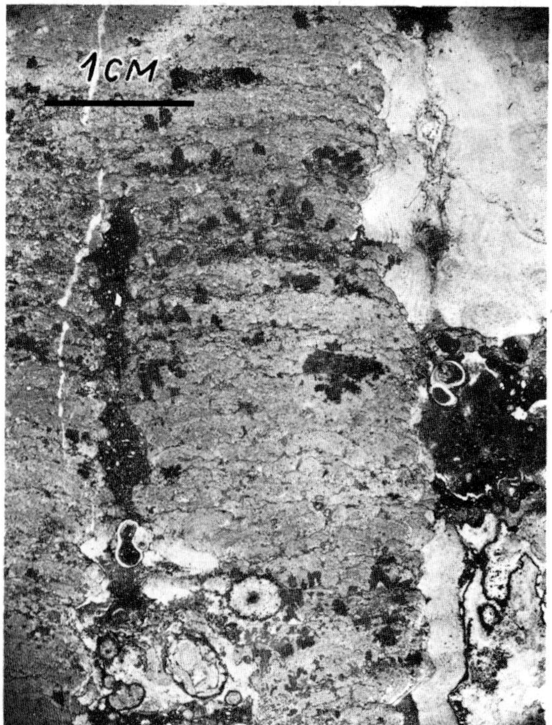

Fig. 4. Columnar forms from the upper subtidal stromatolitic horizon. Chłapowo IG 3 borehole, depth 951.2 m. Thin section

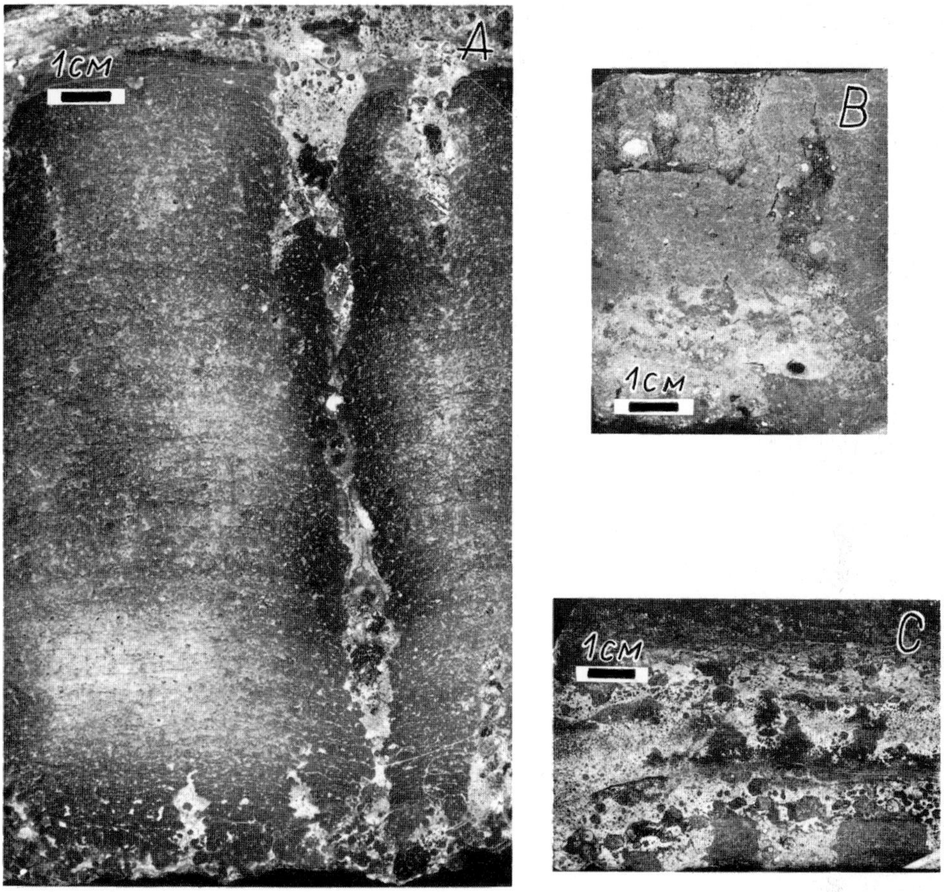

Fig. 5 A–C. Morphological differentiation of stromatolites from the upper subtidal stromatolitic horizon. (A) Gdańsk IG 1 borehole, depth 1640.0 m, (B) Kopalino IG 1A borehole, depth 763.6 m, (C) Opalino IG 1 borehole, depth 998.0 m

13.2.2 Microstructure

Stromatolites are composed of carbonate material, usually almost exclusively of limestone, with subordinate amounts of bituminous material. The subtile stromatolitic lamination is distinctly marked (Fig. 3, 4) and is shown by the alternation of light and dark laminae that form couplets about 0.15 mm in thickness. The average crystal size in light laminae is 0.035 mm, while in dark laminae it is 0.015 mm. Several to a dozen laminae build up small domes between 0.5 and 1.5 mm high (Fig. 7, 8). The sets of domes (high 0.5 to 5 mm) are separated by overgrowths of sessile tubular foraminifera and by laminae of terrigenous material (Fig. 7, 8). Within the sets, in their lower parts the domes grow independently, frequently overgrowing themselves; in the upper parts the domes unite, thus forming continuous algal laminae (Fig. 3; compare the lower and upper part of the stromatolitic body).

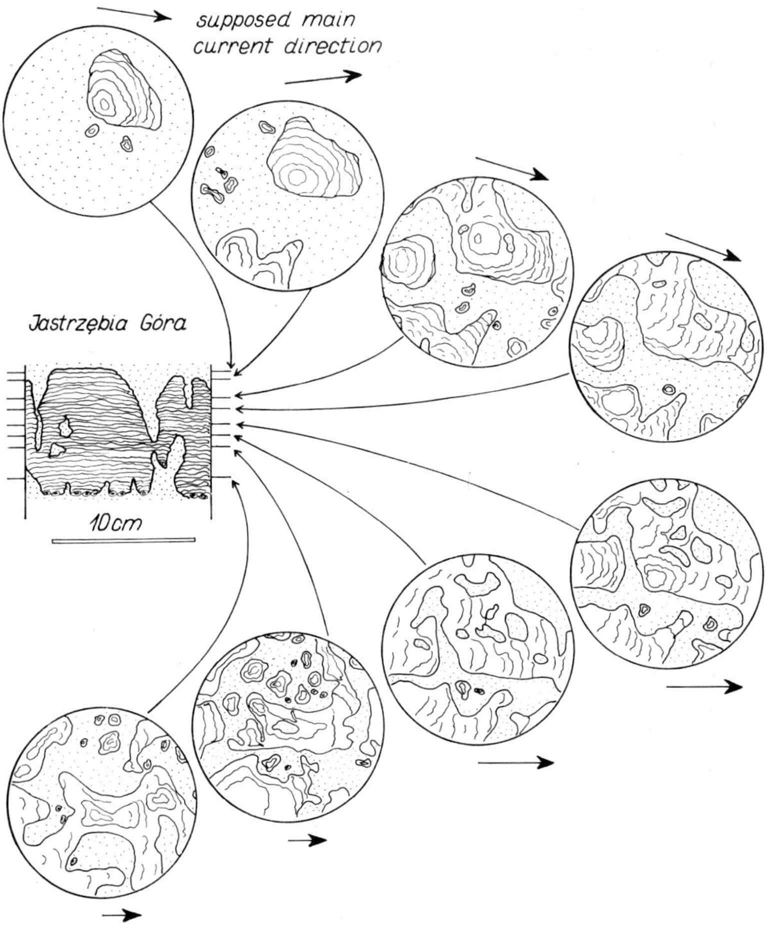

Fig. 6. Changes in the plan of stromatolitic body during growth. Example from the Jastrzebia Góra IG 1 borehole, depth 912.5 m

The h/l ratio within the observed domes ranges from 2 to 1/4, and sometimes the transition of domes into continuous laminae can be observed. The laminae roll up not before the end of the stromatolitic structure (Fig. 8). These continuous laminae are slightly, but distinctly undulated, and at the end of structure they are evenly drawn up while the outermost laminae frequently grow exceedingly (Fig. 8).

The traces of erosion can be recognized on the lateral sides as well as in the upper part of stromatolitic forms; the erosion is often marked by the overgrowths of sessile tubular foraminifera and by laminae rich in clay material (Fig. 8). Besides, the solution traces are found within the stromatolites; the originated cavities of variable shape and size are filled with dolomitic micrite cement and evidently discontinue original algal lamination.

Both the h/l ratio of domes and the amount of overgrowths of sessile tubular foraminifera show some evolutionary trends within the columnar stromatolites. In the lower

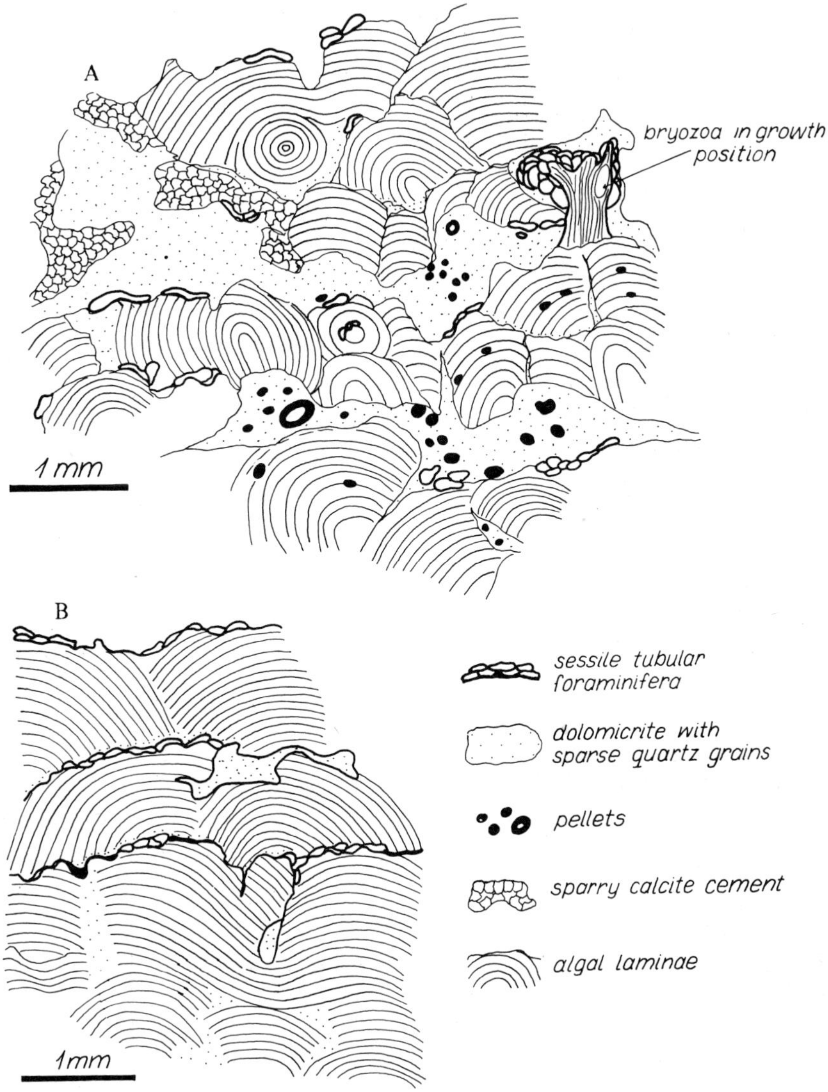

Fig. 7 A, B. Stromatolitic microstructures: (A) characteristic for the lower subtidal stromatolitic horizon and (B) characteristic for the upper subtidal stromatolitic horizon. (A) example taken from the Chałupy IG 2 borehole, depth 975.55 m, (B) example taken from the Hel IG 1 borehole, depth 1098.2 m

part of columnar forms the domes of h/l ratio = 2 resemble the algal biscuits and reach 5 mm in diameter. The sessile foraminifera initially occur in great number, but later they gradually, although very slowly, decrease in number. At the same time the h/l ratio decreases and in the middle part of the columnar forms the continuous laminae occur. In their upper part the columnar forms are composed mainly of continuous algal laminae and the content of clayey material separating the algal laminae gradually increases.

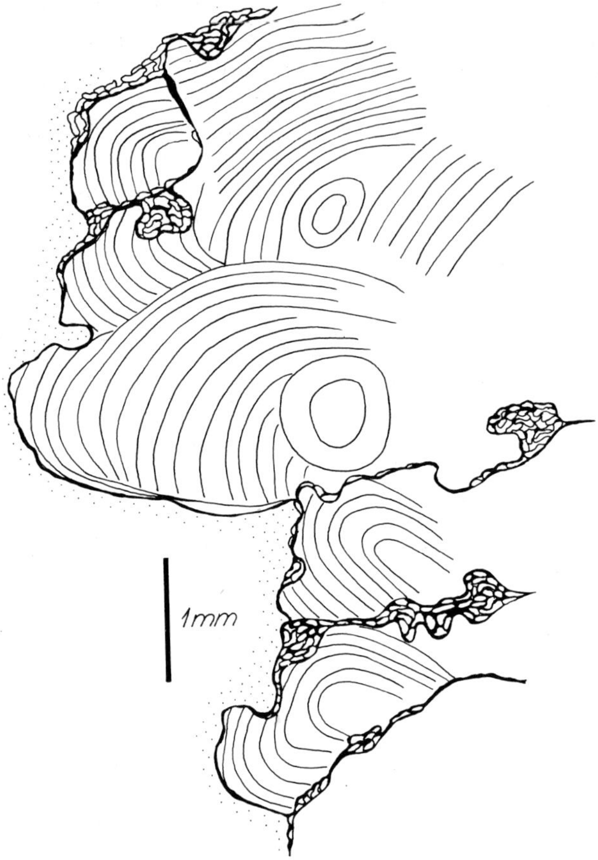

Fig. 8. Stromatolitic microstructure of the lateral part of columnar stromatolite. Example taken from the Hel IG 1 borehole, depth 1098.2 m

13.3 Shallow Subtidal–Intertidal Stromatolites

Shallow subtidal and intertidal stromatolites are of frequent occurrence in western Poland in the upper part of carbonate buildups, forming layers 1 mm to several tens of cm in thickness (Peryt, 1977). The stromatolitic beds are intercalated with bryozoan biolithities and/or bryozoan biomicrites. The bryozoa in growth position clearly indicate a subtidal environment. When bryozoa occur as fragmented grains within the stromatolitic bodies, an intertidal origin of stromatolites is equally as probable as a subtidal one. Bryozoan fragments are associated with minor amounts of foraminiferal, gastropod, and crinoid fragments, but usually the only organisms in thin micritic and/or biomicritic interbeddings within stromatolitic bodies are uniserial and sessile foraminifera.

Shallow subtidal-intertidal stromatolites generally exhibit a dense lamination (Fig. 9). Very characteristic feature is the abundance of sessile tubular foraminifera that occur both on algal laminae and within laminae (Fig. 9). The stromatolites form continuous (gently undulated) mats, which very seldom show sun-cracks and a more marked undulation (Fig. 9).

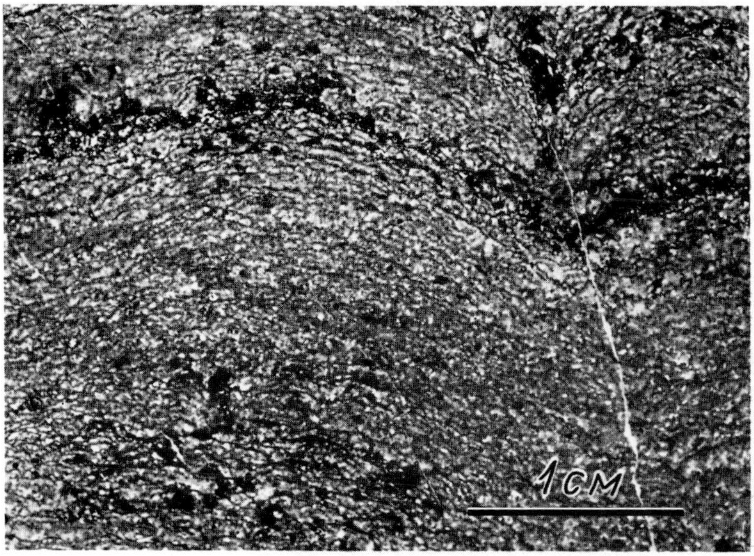

Fig. 9. Shallow subtidal-intertidal stromatolites of dense lamination, with sessile tubular foraminifera (*dark parts*). *Upper right corner*: sun-crack filled with algal growths is vissible. Zmigród 1 borehole (Fore-Sudetic area), depth 1408.6 m. Thin section

13.4 Peritidal Stromatolites

Widespread peritidal stromatolites comparable with those occurring in the recent sabkha environment of the Persian Gulf were found in western and northern Poland. They underlie the Werra anhydrites and are usually 10–20 cm thick. The peritidal stromatolites of western Poland are clearly bipartite: dolomitic mats with gas-formed structures occur in the bottom, in the upper part anhydritic-dolomitic mats with anhydritic laminae can be found (Peryt, 1977). Also in the intertidal zone smooths algal mats occur that are evidently stabilizers of pelletal deposits (Fig. 10). The characteristic feature of peritidal stromatolites is the alternation of two types of lamination: dense and grumous laminae. This is a diagenetic, not a primary, feature.

The bed-bearing peritidal stromatolites in northern Poland vary in thickness from 4 to 30 cm and are composed of several distinctly individualized layers; in one of them small sloping stromatolitic forms, in height up to 1 cm and in length up to 2 cm occur (Fig. 11). Stromatolites are calcitic and exhibit only a dense type of lamination. The gently domed morphology is strongly controlled by contemporaneous erosion caused by sediment-bearing waters (Fig. 12).

13.5 The Environmental Significance of the Stromatolitic Sequences

The sequential development of stromatolitic forms reflects the changing hydraulic conditions. Therefore the stromatolites can be used for paleoenvironmental reconstructions (Peryt, 1975; Gunatilaka, 1975).

Fig. 10. Smooth algal mats of intertidal origin stabilizing pelletal material. *Light parts of photo*: mainly anhydritic fillings. Radziadz 11 borehole (Fore-Sudetic area), depth 1600.5 m. Thin section

Fig. 11. Peritidal stromatolites and associated sediments (lower part of sabkha sequence) from the Hel IG 1 borehole, depth 1098.1 m. Thin section

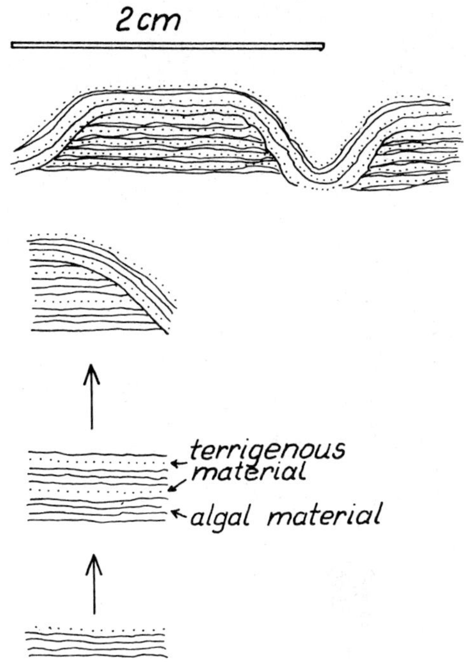

Fig. 12. Scheme illustrating the development of small sloping stromatolitic forms of peritidal origin, presented in Fig. 11

In the Zechstein Limestone of western Poland a distinct succession of particular stromatolitic types is observed, especially in the nearshore area (see Peryt, 1977). On the oncolites of shallow-water origin or on bryozoan biomicrites algal stromatolites grow, associated with bryozoans in growth position. Over these stromatolites dolomitic algal mats occur, that in turn pass upward in dolomitic-anhydritic algal mats of sabkha type. This succession of the major stromatolitic types seems to reflect the gradual shallowing of the sea (Peryt, 1977). The stromatolitic microstructures indicate several minor subcycles which reflect small fluctuations of the water-depth during the regression period.

In northern Poland columnar forms of SH type start growing on oncolites; the stromatolites again seem to reflect a gradual shallowing that ended with emergence and calichification (see Peryt and Piatkowski, 1976). It is striking that LLH structures never follow SH structures, although such a succession is common in the microstructure sequences. The topmost part of the Zechstein Limestone section is composed of intertidal stromatolites gradually passing in the supratidal mats of the sabkha type.

Studies on modern stromatolites have shown that gross morphology tends to reflect environmental conditions, while stromatolitic microstructures reflect changes in microbiotas (Logan et al., 1964; Gebelein, 1974). Despite the immense progress in our knowledge of recent stromatolites during the last two decades (Ginsburg, 1960; Logan, 1961; Logan et al., 1964; Monty, 1967, 1972; Gebelein, 1969; Davies, 1970; Gunatilaka, 1975 and others) the factors controlling the distinct forms of the stromatolites are not well known. It seems that, in the case of the Zechstein Limestone of northern Poland, the depth and velocity of currents, both related with each other, were the main controlling environmental factors. In the region of Gdansk and Hel stromatolites are very well developed; they reach 20 cm in height in the upper subtidal stromatolitic horizon and

10 cm in the lower subtidal stromatolitic horizon; they have narrow and fissure-like interareas. Contrary to this, in the WNW direction stromatolites diminish and occur in a diffuse manner; west of Kopalino they decline completely (cf. Fig. 1B). Due to greater current velocities in the western shallower part of the region only specific morphological types of oncolites were formed in environments with a strong clastic influx (mainly quartz). Relatively rapid currents caused the directional growth of stromatolitic forms and their asymmetry. In the eastern, deeper part of the region good conditions existed for the growth of stromatolites and stromatolites with well-developed axial symmetry were found.

The features discussed above are presented in Fig. 13, together with the recent models of the Shark Bay and the Bermuda areas. It is somewhat surprising that the model of subtidal stromatolites from the Zechstein Limestone of northern Poland is very similar to the recent model of intertidal stromatolites from Shark Bay, and quite different from recent Bermuda subtidal stromatolites, although microstructures of the Zechstein stromatolites are similar to macrostructures of Bermuda stromatolites. This finding indicates that paleoenvironmental conclusions based only on stromatolitic growth forms are very dangerous.

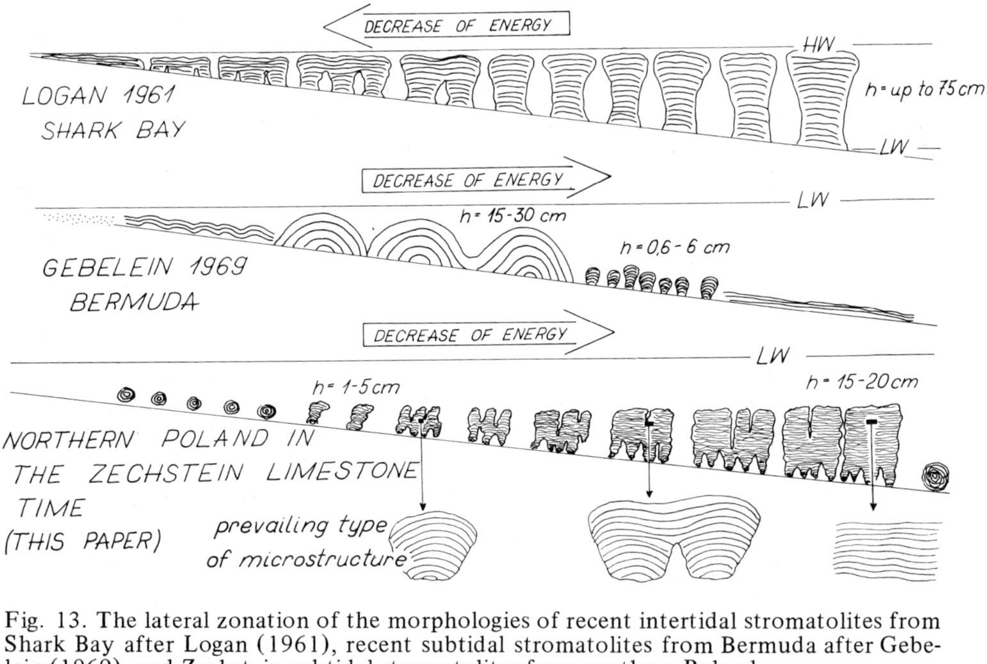

Fig. 13. The lateral zonation of the morphologies of recent intertidal stromatolites from Shark Bay after Logan (1961), recent subtidal stromatolites from Bermuda after Gebelein (1969), and Zechstein subtidal stromatolites from northern Poland

In northern Poland the stromatolitic microstructures change parallel with changes of macromorphology of stromatolites. This testifies that micromorphology is – at least to a certain degree – also environmentally controlled. Hubbard (1972) correlated the characteristic laminar fabrics with sedimentary rate and algal growth rate. However, in western Poland some types of lamination occur independently from sedimentary environment and sedimentary rate. Therefore, it seems likely that the biological composition of stromatolites was a reflection of environmental factors other than depth and sedimentary rate.

Acknowledgments. The authors would like to express their sincere thanks to Hans Füchtbauer, Claude Monty and Ryszard Wagner for encouragement and inspiring remarks and discussions. Discussions with Michael Gibbons, David Lemone, and Laszlo Rácz were also highly appreciated.

13.6 Bibliography

Davies, G.R.: Algal-laminated sediments, Gladstone embayment, Shark Bay, Western Australia. Am. Assoc. Petrol. Geol. Memoir **13**, 169–205 (1970).

Gebelein, C.D.: Distribution, morphology, and accretion rate of Recent subtidal algal stromatolites, Bermuda. Sediment. Petrol. **39**, 49–69 (1969).

Gebelein, C.D.: Biologic control of stromatolite microstructure: implications for Precambrian time stratigraphy. Am. J. Sci. **274**, 575–598 (1974).

Ginsburg, R.N.: Ancient analogues of recent stromatolites. Int. Geol. Congress, XXI Session, Norden, Part XXII, 26–35 (1960).

Gunatilaka, A.: Some aspects of the biology and sedimentology of laminated algal mats from Mannar Lagoon, northwest Ceylon. Sediment. Geol. **14**, 275–300 (1975).

Hubbard, J.A.E.B.: Stromatolitic fabric: a petrographic model. 24th Intern. Geol. Congr. Sect. **7**, 380–396 (1972).

Logan, B.W.: *Cryptozoon* and associated stromatolites from the Recent, Shark Bay, Western Australia. J. Geol. **69**, 517–533 (1961).

Logan, B.W., Rezak, R., Ginsburg, R.N.: Classification and environmental significance of algal stromatolites. J. Geol. **72**, 68–83 (1964).

Monty, C.L.V.: Distribution and structure of recent stromatolitic algal mats, Eastern Andros Island, Bahama. Ann. Soc. Géol. Belgique **90**, 55–100 (1967).

Monty, C.L.V.: Recent algal stromatolitic deposits, Andros Island, Bahamas. Preliminary report. Geol. Rundschau **61**, 742–783 (1972).

Peryt, T.M.: Significance of stromatolites for the environmental interpretation of the Buntsandstein (Lower Triassic) rocks. Geol. Rundschau **64**, 143–158 (1975).

Peryt, T.M.: Charakterystyka mikrofacjalna cechsztyńskich poziomów weglanowych cyklotemu pierwszego i drugiego na obszarze monokliny przedsudeckiej. Studia Geologica Pol. **54**, in press (1977).

Peryt, T.M., Piatkowski, T.S.: Osady caliche w wapieniu cechsztyńskim zachodniej cześci syneklizy perybaltyckiej. Kwart. Geol. **20**, 525–536 (1976).

14 A Possible Bacterial Origin for Famennian Micrites

A.F. MAURIN and D. NOËL, Paris, France

14.1 Introduction

Since Drew (1911), the role of bacterias for precipitation of calcium carbonates (marine) has been recognized. Their biochemical effects on organic sludge have been tentatively demonstrated by Trichet (1967–1972) and other workers. It should be stressed that blue-green algae play a passive role, as providers of organic substrate. Fossil bacterias (both filamentous or rod-like) have been mostly found in siliceous sediments, as organic remains trapped with inclusions in cherts (Schopf et al., 1965) or individually as silicified objects (Arbey et al., 1975).

The present paper deals with organic remains (diameter below the μ range) included within micritic limestone. The intimate association between some of our bacterias (?) and rhomboedric crystals is such that sedimentological relationship between them is suggested, thus improving our understanding of micrite formation.

To our knowledge, this is the first record of likely in situ organic nanno-objects within fossil carbonates, apart from the ubiquitous algal filaments of larger diameter, already noticed in optic-microscopy (Fry and McLaren, 1959).

14.2 General Consideration

The material investigated comes from a Famennian section (Palliser formation) in the Rocky Mountains of Alberta, Canada. There, the Palliser formation is made of two megacycles totaling a 300 meter section. The two cycles are respectively called Morro and Costigan members. Detailed thin section analysis on the upper megacycle (Costigan) indicates a sequential disposition of second-order cycles. While the megacycles range upward from emersive to open marine facies, these second-order cycles, whithin laminated lithology (base of each megacycle), exhibit a full array of marine to emersive facies indicative of prograding shebkha environments (Maurin, 1972). 15 second-order cycles have been counted over a 30 meter span of laminated sediments. Most of them are uncomplete, and three are larger than the average, owing to a greater thickness of open marine crinoidal facies (Fig. 1). Such a section is only partially dolomitized, dolomitization being restricted to the upper part of each cycle (supratidal) and within burrows of both inter and subtidal zones. This lack of diagenesis is thus a warranty for further analysis of original sediments.

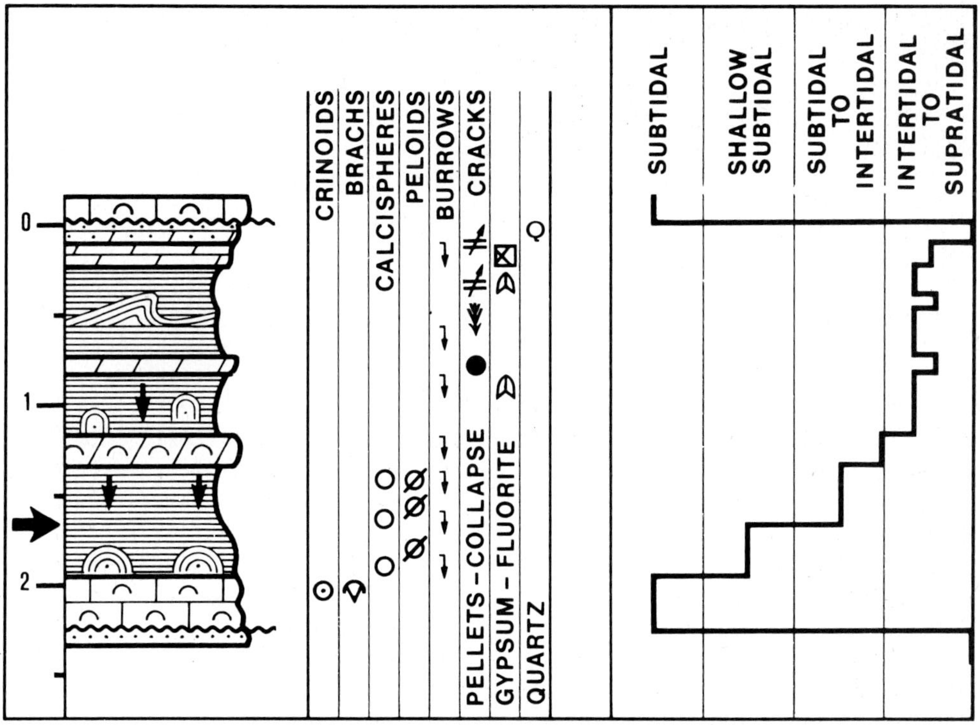

Fig. 1. Idealized cycle of Upper Palliser Section (Costigan). Laminites with stromatolites are *striped.* Location of studied sample is marked by *horizontal arrow*

14.3 Sedimentology of Micrites

Each second-order cycle begins with micritic carbonates, the subtidal origin of which is warranted by numerous criteria, including their sequential position (Fig. 1) and the presence of 2 mm-thick layers of calcispheres packstones (Fig. 2A). By using ultra-polished, then etched (NH_4Cl, 10%) sections, our SEM studies were originally aimed toward: 1. comparing these calcispheres (Fig. 2B) with cysts of the modern *Acetabularieae* (Rupp's, 1966, *Chalmasia antillana* revised by Valet, 1968), as illustrated by Marszalek (1975) in SEM microscopy. Famennian calcispheres range in diameter between 40 to 150 μ. Some are still present as hollow spheres with a very fine micritic wall, some are filled with micrite and exhibit a crystalline wall of possible diagenetic origin (Fig. 2B, lower right).

In the fossil record, these calcispheres packstones are considered as good indicators of shallow subtidal environments, compared to the location of our samples on Rodriguez Key and according to Marszalek (1975) who assigns a depth between 2 meters and mean sea level to such *reproductive layers.* 2. Our second purpose was the general study of micritic matrix surrounding calcispheres. Very fine micrite is present as peloidal-patches (Fig. 2C). Junctions between subspheric peloids produce a typical *tetraedric* pattern (Fig. 2D) of definite value for further studies.

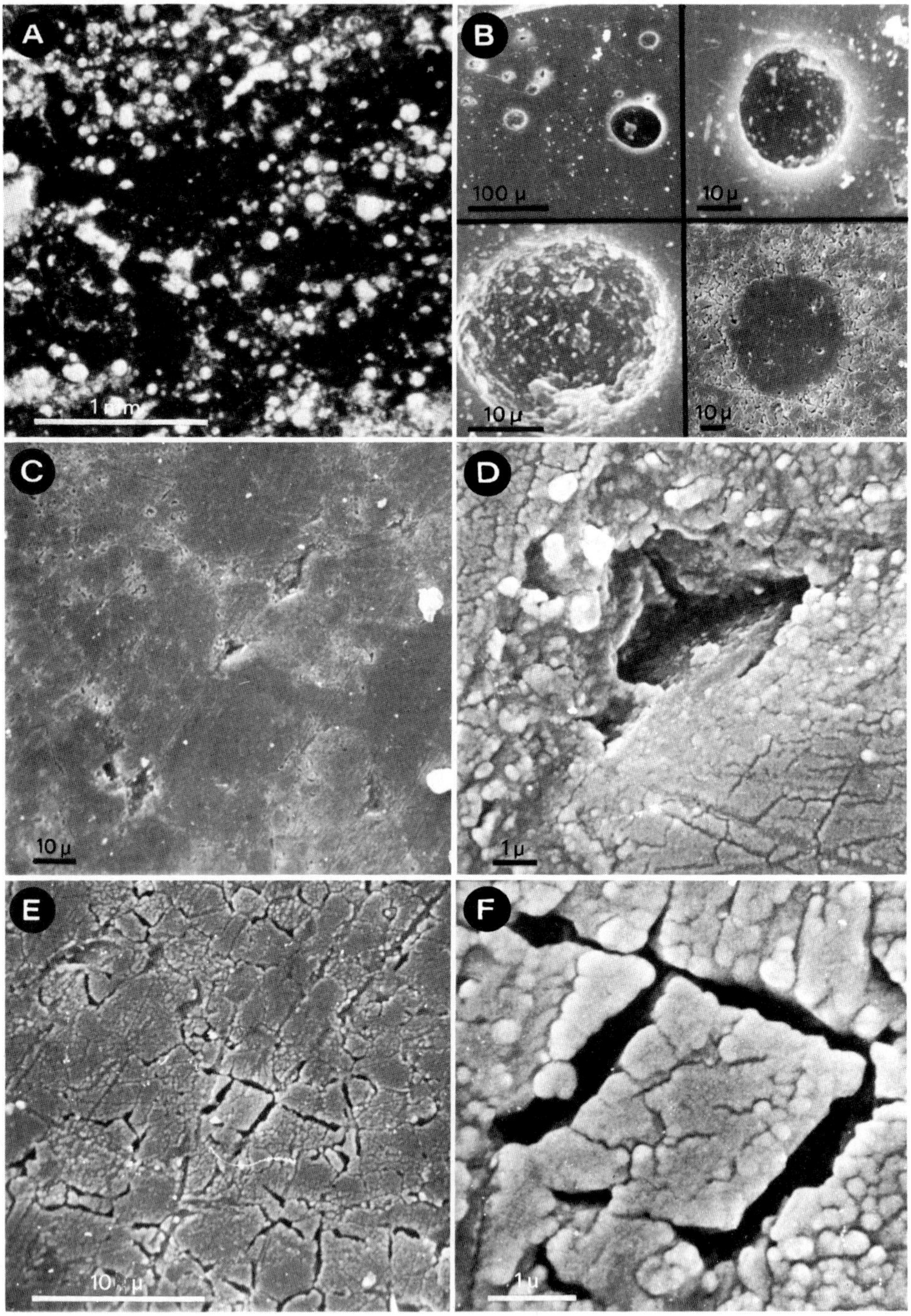

Fig. 2

Coarser micrite (1 to 4 μ) is also present as matrix for calcispheres or peloids (Fig. 2E, 3D). A mathematical program for classification of crystallographic patterns is under experimentation on a Texture Analyser (Leitz-École des Mines, Fontainebleau). Polished sections exhibit randomly some perfect sections of elementary rhombs in the 2 or 3 μ range (Fig. 2F). The following organic remains occur within either the coarser micrite or the patches (peloids) of finer material.

14.4 Organic Remains

Acid etching leaves smooth-shaped objects more or less protruding from the eroded calcitic surface. Assumption is made that these objects, resistant to acid attack, could be fossil bacterias still preserved as organic bodies. Four kinds have been recognized to-date: Dichotomous filaments (0.3 to 0.7 μ in diameter, over 14 μ in length (Fig. 3A), and shorter filaments, coiled (0.4–3.5 μ) (Fig. 3B). Both objects, although well exposed above, do penetrate below the calcitic surface, warranting their contemporaneity with micritic matrix (or anteriority). The third kind (Fig. 3C), in the form of a banana (0.4–2 μ), lies upon the surface, thus being of doubtful origin (it may have been introduced after etching and prior to metalization of the preparation). The last sort of objects, rod-like (Fig. 3D, E, F), deserves more attention: these are involved with a patch of fine micrite (crystals sizes 0.8 to 1.4 μ) and one of them is almost perfectly centered along a rhomboedron axis (Fig. 3E, F). Individual measures of our bacteria-like bodies are 0.2 μ in width and 1.8 to 2 μ in length. Circular sections (Fig. 3E) are likely to be normal sections of identical bacterias (?), but, as for the prominent sphere against the "banana" object of Fig. 3C, may represent sporular figures, although an artifact origin (dust, etc.) may also be involved.

More work is warranted in order to precise the flora (?) population, then to transfer the study toward biology.

14.5 Sedimentological Bearing

Trichet (1972, and from former work in his bibliography) gives a convincing biochemical explanation for carbonate nucleation within organic matter from decaying blue-green algae. Slight pH increment within weak amino-acids (high pH) initiates carbonates precipitation. Within *any kind of organic slush* this process could work, as already observed by L. and J. Morellet (1913, p. 31), on calcifying mucilage around *Acetabularieae* spores. Figs. 3E, F could represent a bacteria involved with the chemical process of carbonate

---

◀ Fig. 2 A–F. (A) Optical photograph, natural light, of calcispheres-peloids packstone. (B) SEM views of selected calcispheres. (C) SEM photograph of polished-etched section of micritic patches with tetraedric peloid junctions. (D) Detail of tetraedric junction. (E) Coarser micritic matrix. (F) Detail of a rhomboedron

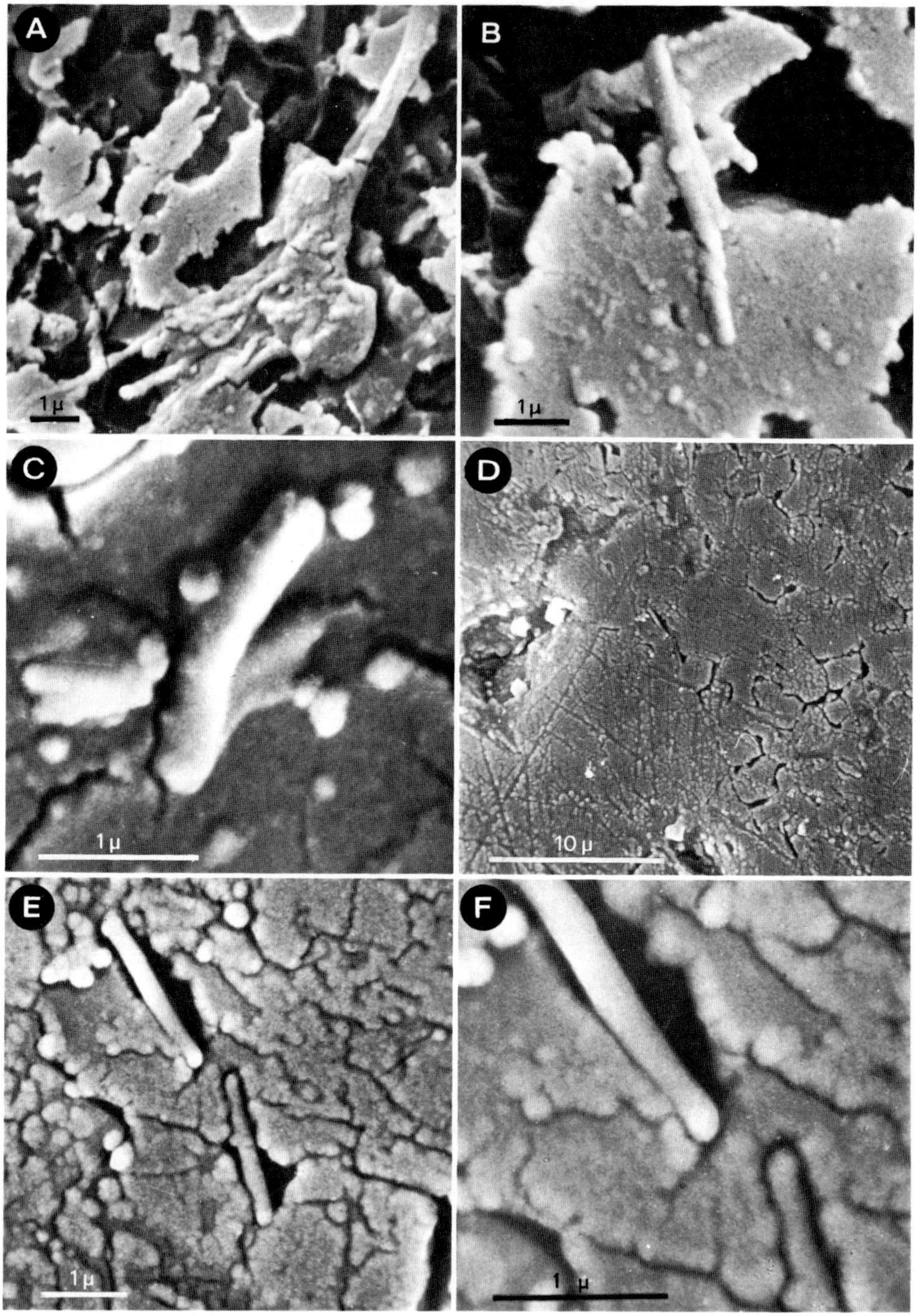

Fig. 3. Legend see opposite page

nucleation: the *axial* presence of such an object appears to be directly linked to the crystallographic pattern, thus, to the initiation of such a crystal. So, in our understanding, lower right bacteria of Fig. 3E and F could have been fossilized "*in the act*" of nucleating a rhomboedron (e.g. calcite or Mg-calcite). The 0.02 μ void between bacteria and the inner wall of crystal could represent shrinkage of the bacteria after death, the natural space formerly occupied by organic liquors or artificial solution created during etching. Contemporaneity between the two objects seems demonstrated. Chemical processes would involve bacterial destruction of amino-acids, CO_2 production, then carbonate nucleation within a *basic* micro-environment around the responsible bacteria.

Following Adolphe and Billy (1974), who observed the necessary *association* of several kinds of bacterials for calcite-nucleation, the four kinds observed here should give warranty for the biochemical phenomenon to occur.

14.6 Conclusions

It is not the contention of such remarks to account for *all micrites* observed in thin section. As observed by Valet (1968), Valet and Segonzac (1969), the whole of *Acetabularia* plants, except the cysts, disintegrates into aragonitic needles, thus creating sedimentary micrite. Only mucilage around spores, and decaying blue-green algae may transform into micrite through the process illustrated here. Such a process may help the understanding of formation of micritic envelopes, both aragonitic or calcitic.

At this point of investigation, availability of Mg in solution may account for the distinction between aragonite or Mg-calcite. However, we feel the major difference should be created by the varieties of bacterias. The involved biochemical reactions are a function of the specific *biochemical potential* of bacterias, thus leaving full possibility for development of any cristalline form or size.

14.7 Bibliography

Adolphe, J.P., Billy, C.: Biosynthèse de calcite par une association bactérienne aérobie. C.R. Acad. Sci. Paris **278**, Série D, 2873–2875 (1974).

Arbey, F., Bouyx, E., Devigne, J.P.: Une population de bactéries sporulales témoin d'un été glaciaire ou d'un interglaciaire éocambrien. C.R.IX Congr. Intern. Sédimentol., Thème 1, 51–54 (1975).

Drew, G.H.: The action of some marine denitrifying bacteria in tropical and temperate seas, and the bacterial precipitation of calcium carbonate in the sea. J. Marine Biol. Assoc. U.K., New Ser. **9**, 142–155 (1911).

Fry, W.L., McLaren, D.J.: Fungal filaments in a Devonian limestone. Geol. Surv. Can. Bull. **48**, 9 pp. (1959).

◄ Fig. 3 A–F. (A) Dichotomous filaments (length of upper long filament is 14 μ). (B) Coiled short filament. (C) "Banana"-like bacteria with possible sporular sphere. (D) View of micrite with rod-like bacterias in the center. (E) and (F) Close up of rod-like bacterias

Marszalek, D.S.: Calcisphere Ultrastructure and Skeletal Aragonite from the Alga *Acetabularia Antillana.* J. Sediment. Petrol. **45**, No. 1, 266–271 (1975).
Maurin, A.F.: Laminites et Stromatolites algaires du Famennien des Montagnes Rocheuses, Alberta, Canada (Abs.). C.R. Congr. Géol. Intern. Montréal, Sect. **7**, 586 (1972).
Morellet, L., Morellet, J.: Les Dasycladacées du Tertiaire parisien. Mém. Soc. Géol. France, Pal., XXI, **47**, 43 pp. (1913).
Rupp, A.: Origin, structure and environmental significance of Recent and Fossil Calcispheres. (Abstr.). Geol. Soc. Am. Spec. Papers **101**, 186 (1966).
Schopf, J.W., Baghoorn, E.S., Maser, M.D., Gordon, R.O.: Electron microscopy of fossil bacteria two billion years old. Science **149**, 1365–1367 (1965).
Trichet, J.: Etude du mécanisme de la nucléation du carbonate de calcium dans les dépôts algaires. Lien avec la biogéochimie comparée de quelques types de dépôts dus à des Cyanophycées (Polynésie Française). C.R. Congr. Geol. Intern. Montréal, Sect. 7, 631–638 (1972).
Valet, G.: Contribution à l'étude des Dasycladacées. Thesis Fac. Sci. Paris, 216 pp. (1968).
Valet, G., Segonzac, G.: Les genres Chalmasia et Halicoryne (Algues Acétabulariacées). Bull. Soc. Géol. France, **7**), XI, 124–127 (1969).

15 The Development of the Dasyclad Algae through the Ages

M. HERAK, V. KOCHANSKY-DEVIDÉ, and I. GUŠIĆ, Zagreb, Yugoslavia

15.1 Introduction

Evolution is a process in which the time factor is of crucial importance. Therefore its reconstruction requires, first of all, a chronologic dating of particular morphologic features appearing in different taxa, then the dating of various morphologic units in the form of taxa of different ranks, and, finally, the recent result of development in the past, expressed in a mosaic-like distribution pattern of taxa in particular ecological areas. Only in the individuals of the recent phase can the morphologic similarities be tested by genetic and partly physiologic investigations, as to in what degree they are real expressions of a true genetic relationship. In fossil individuals the relationship may be assumed only on the basis of morphologic features present in taxa linking in time.

The attempts to reconstruct the development of the dasyclad algae respected all the demands mentioned with the exception of the continuity of time. Therefore, it very often happened that morphologically close organisms were considered as parts of a taxon or of a development line despite the time gap between them. Temporal and morphologic gaps were mostly considered to be a consequence of incomplete exploration of fossils, with the conviction that new finds would fill up the existing gaps. But in many cases the gaps remained, requiring new approaches to explain them. Some new records, on the other hand, did not fit the assumptions. Thus the initial optimism was replaced by more criticism and almost a doubt as to whether true genetic relationship can ever be established. A classification resulted from this in which more attention has been devoted to differences than to similarities. Hence numerous new taxa and multiplication of names resulted from diagnoses and descriptions in which personal convictions outweighed the facts which could be recognized by every unbiased researcher interested in comparison of the taxa. Frequently, the sketched reconstructions of newly established species became more precise than the available fossil record could possibly allow. Thus, it was difficult to use the multiplying names in tracing the process of the evolution of the group as a whole.

In the beginning, due to the small number of known forms, genera and tribes were well defined, without overlapping, so that a conviction prevailed that considerable periods of time were necessary for some particular features to become predominant and to serve as a basis for different taxonomic categories. More attention was paid to recurrence of some features or taxa after a longer period of time, but without a satisfying explanation. Anyhow, the taxa were quite easily used as links of evolutionary lines, displaying clear progressive steps in the morphogenesis.

15.2 Brief Historical Review

J. Pia (1920, 1923), a passionate researcher of the dasyclad algae, considered the following features to be primitive: irregular thallus, aspondylity, dichotomous branching, endospore mode of reproduction, etc. Contrary to those forms, cylindrical thallus, euspondylity or metaspondylity, the presence of secondary, tertiary, etc. branches, the outward shifting of reproductive bodies, annulation, etc., were taken to be progressive. Irregular forms were considered to be confined to the Paleozoic, and the climax of the dasycladacean development was placed in the Middle Triassic, with a slow decline until recent times. It was through Pia that the interest in calcareous algae and so in Dasycladaceae grew steadily. The seemingly easy determinations on the basis of random sections were often misleading, without a critical distance as to the biologic variation range. Nevertheless, the increasing amount of data showed discrepancies in relation to the more or less idealized old schemes. The deviations were at first considered exceptions which confirmed the rule. But since they were constantly multiplying, they became a rather unpleasant obstacle which posed great difficulties.

It is interesting to note that Pia (1923) already noticed that the verticillate arrangement of branches occurred at least three times at different stages and in different lines of the evolution of the Dasycladaceae, i.e. in the genus *Teutloporella*, at the transition from *Macroporella* to *Stichoporella*, and in ancestral forms of *Oligoporella*. Kamptner (1958) pointed out that the verticills occurred independently no less than seven times in the course of the dasycladacean development, six times during the Paleozoic and only once in the Mesozoic (Dogger). Paleozoic examples are: at the transition from *Uraloporella* to *Teutloporella* and from *Rhabdoporella* to *Anatolipora*, and within the genera *Macroporella, Physoporella, Gyroporella,* and *Oligoporella.* In the Dogger, the verticills appeared at the transition from *Macroporella* to *Stichoporella*. It became obvious that the differentiation took place much earlier than was generally thought. Kamptner also pointed out that the shifting of the reproductive bodies towards the outside of the main tube had already started in the Paleozoic (*Koninckopora*).

Rezak (1959), at the XI International Botanical Congress, reported the verticillate arrangement in Paleozoic genera *Cyclocrinites, Coelosphaeridium, Primicorallina, Verticillopora*, and *Mizzia*. In *Primocorallina* he mentioned also the secondary and tertiary branches.

Kozlowski and Kazmierczak (1968) pointed out the existence of secondary and tertiary branches in Ordovician *Vermiporella*, although they doubted its dasycladacean nature due to the existence of transversal mebranes, oospores, etc.

Kochansky-Devidé and Gušić (1971) quoted several examples of repetition of identical or similar features in the dasycladacean development, e.g. the shape of the thallus (the string-of-pearls type, mushroom-shaped, spherical, etc.), the occurrence of secondary and higher branches, etc. In this connection, the authors recommend a more critical approach in distinguishing the genera, and in taxonomy in general. They also express some reserve concerning a former reconstruction of the tribe Diploporeae (Kochansky-Devidé, 1964), in which synchronous forms were supposed to belong to different genera (*Diplopora pusilla* and *Velebitella*), whereas forms with a large stratigraphical gap between them were considered to be members of one phylogenetic line (*D. pusilla* and *D. helvetica-hexaster*).

The accumulated new records, without clear delimitations, required a new attempt at explanation of evolutionary trends in the dasyclad algae. It was undertaken by Kamptner (1958). He retained Pia's subdivision into tribes (15 tribes, four of which are further subdivided into subtribes), and pointed out the existence of three parallel development lines in the Ordovician (Dasyporelleae, Primicorallineae and Cyclocorineae), without indicating a common ancestor. The Primicorallineae became extinct shortly after having appeared, i.e. already in the Ordovician. The Cyclocrineae were dominant in the Ordovician and kept occurring sporadically until the Permian. The Dasyporelleae are thought to be the real ancestor of all the younger forms (genera and tribes). However, within the Dasyporelleae two parallel groups may be distinguished arising from a common ancestor, probably in the Middle Cambrian. One group is represented by *Anthracoporella* and *Vermiporella* (including "*Pseudovermiporella*"). It became extinct in the Permian. Another group comprises *Dasyporella* and *Uraloporella*, lasting until the Upper Carboniferous. Before extinction, however, it gave rise to *Teutloporella* as a representative of a separate tribe. Finally, there remains the genus *Rhabdoporella*, standing alone in its tribe. In respect ot evolution of the younger Dasycladaceae, however, this genus has been of particularly great importance, because all of the later, better defined, genera and tribes, thought to be offsprings of *Rhabdoporella*. *Anatolipora* and *Dvinella* are considered to follow in a more or less straight line into *Diplopora*. Somewhere in the Devonian, i.e. before the appearance of *Anatolipora* (Lower Carboniferous), a group consisting of *Macroporella, Physoporella, Gyroporella, Oligoporella* and *Eogoniolina* branched off from the main evolutionary line, and most of Mesozoic, Cenozoic and recent forms are considered to represent a continuation of that group. An exception is made only by *Archaeocladus* (Permian), which should derive from *Dvinella* and give rise to *Palaeodasycladus* (Liassic) and *Conipora* (Dogger and Malmian). *Palaeodasycladus*, in turn, might continue into recent forms *Dasycladus, Batophora* and *Chlorocladus*.

Kamptner (1958) retained two ways of connecting the forms in the evolutionary lines. One is the linear way, in which two or more forms are connected with each other in a straight line, in spite of (sometimes) large gaps between them. The other way emphasizes the branching, the bifurcations being placed more frequently in the periods of gaps within the evolutionary lines.

Later on, many new forms were discovered, and the supposed life-span of some already known forms has been widened. A new attempt at solving phylogenetic relations was necessary. It was undertaken by Endo (1961). He introduced many new elements, based more on his own concept than on observed facts. According to Endo, the main elements for the explanation of relations are the following: sinuous bifurcating thallus; irregularly arranged branches; cylindrical thallus; perpendicular position of the branches; large spherical cavities, similar to sporangia, in cortical parts of the thallus; reproductive bodies; annulation; pear- or egg-shaped sporangia; etc. The criteria in using them have not been explained, therefore only his main original conclusions should be mentioned here. *Anthracoporella* and *Uraloporella* are thought to derive from a common ancestor which should be placed somewhere between *Amgaella* and *Vermiporella*, being similar to the latter. An important role is ascribed to *Sibiriella*, deriving, together with *Amgaella*, from an unknown ancestor. *Sibiriella*, in turn, gave rise to *Macroporella, Diplopora*, and *Gyroporella*, whereas an unknown ancestor similar to *Sibiriella* should have originated *Rhabodoporella, Kazachstanella, Urodella,* etc.

The steadily increasing number of new forms also required interventions in separate minor groups.

Ott (1965) presents the evolution of vestibules within the tribe Diploporeae, despite the gaps ranging, in extreme cases, from the Permian to the Cretaceous (*D. pusilla* – *D. johnsoni*).

Hurka (1968) is of the opinion that *Palaeoporella* unites the basic features of both Dasycladaceae and Codiaceae, probably representing the ancestor of both families.

Korde (1968, 1971) suggested the creation of a new family, Seletonellaceae, comprising the following tribes: Cambroporelleae, Amgaelleae, Seletonelleae, Rhabodporelleae, Cyclocrineae, Primicoralineae, and Macroporelleae. More attention is devoted to the features which are common to the group (family), than to genetic relations to the other groups of forms (family Dasycladaceae).

Recently, living Dasycladaceae (Dasycladales) also became an object of important studies. It is, first of all thanks to Valet (1968, 1969), that we owe a detailed analysis of the morphogenesis of several recent genera, which is important for an actualistic approach to the fossil forms often established on scarce remains. Valet made obvious that Dasycladales (here regarded as an order) are homogeneous in a cytological sense and, consequently, the differences within them may be the result of the variability expressed at lower taxonomic levels. It is interesting to note that, for example, morphologic differences between *Bornetella claviformis* and *B. subsphaerica* are greater than between the first mentioned species and the genus *Neomeris*. Further on, teratological cases (easily provoked in laboratory), concerning the position of the reproductive bodies, are also encountered in nature. The degree of calcification obviously depends on the intensity of light, but also on some other factors. This is proven by the following examples: *Batophora* has no calcification at all; *Dasycladus* is calcified only in the region of the main tube; *Neomeris* possesses a thin cortical layer enveloping only the secondary branches; and *Cymopolia* is completely calcified.

It became obvious from this that those fossil taxa which have been established on single finds may be highly problematic.

Segonzac (1968, 1971) pointed out the great variability in the vertical distance between the primary branches in *Cymopolia*. She is of the opinion that, generally, the variability in the shape of the primary branches may be used for specific diagnoses only. She considers the fertile bodies of the choristospore forms as secondary branches, discusses the relationship between the *Acetabularia*- and *Macroporella* group. *Sakkionella* is considered to represent the transition between the Neomerioideae and the Dasycladoideae, being tentatively placed within the Neomerioideae. The genera within those groups are distinguished by the number of the sterile secondary branches.

Elliott (1972) accepts the endospore, cladospore, and choristospore arrangements of the reproductive bodies as evolutive stages. However, the first evolutive tendency involved the elaboration of the thallus. The first stage was charcterized by an irregular stem-cell, sometimes recumbent, branching, or even anastomosing, with a dense and irregular arrangement of branches. *Mastopora* and related forms were the first which developed a large central cavity with a thin calcified cortex, similar to recent *Bornetella*. Trends of erection of the main stem could be noticed in *Vermiporella*. *Mizzia* displays a transition towards the formation of the verticills. Even some metaspondyle characteristics are recorded as early as in the Devonian.

A group of French authors (1975), after commenting on previous approaches to the classification of the dasyclad algae, recommends to distinguish the characteristics with decreasing values in order to establish a certain hierarchy among them, and not to request an imperative use of them. The disposition of the ramifications of the first order should be sufficient to define the family. The general shape of the thallus, the shape and order of the ramifications and the reproductive bodies would determine the genus or the subfamily, respectively. The dimensions (proportional relations), the nature and the degree of the calcification, etc., may define the species. The authors do not even exclude further modifications of the criteria.

15.3 Analytical Remarks

In all former attempts to establish the elements of evolutionary trends in the dasyclad algae, morphology was practically the only source of relevant data. Morphologic diversities were used for the creation of new taxa, while similarities were used as indications of genetic relations of different taxa, and, consequently, as a basis for the reconstruction of the evolutionary lineages. Taxonomists sometimes consciously follow some formalism, without pretension to create only genetic units. But in the evolutionary schemes the same units are used as more or less homogeneous genetic entities, hence consideration of direct connections of morphologically close taxa (genera), even if they are separated by a considerable time gap. The first appearance and probability of a continuous duration of a possible taxon (or feature) were treated as data of lesser importance.

In taxonomical interpretation in general, some 29 morphologic features may be encountered as elements of diagnoses of new forms. Adding the age, there are 30 data, different in rank, which are used more or less unanimously.

The main stem appears in 14 variations, position of branches in three, branching of higher orders in four, the form of the pores in five, and the position of reproductive bodies in three variations. Using the diagnoses only, we succeeded in selecting some 80 genera convenient for a correlation, in order to discover a common basis for a more or less unbiased approach to their evolution. The analysis has shown some unexpected results. There are two genera whose diagnoses lack all of the 29 mentioned features. The diagnoses of seven genera contain one feature only, 15 (possibly 17) genera two features. 34 genera three features, 20 genera four features, 13 genera five features, and seven genera six features. No less than 23 features are omitted. Taking into account that the features in question are not of the same rank, the diagnoses seem to be very poor.

It may also be interesting to mention that, on the other hand, the genus *Mastopora* alone, according to the diagnoses of its species, displays four types of main stem. Further on, the genera *Eodasycladus, Palaeodasycladus*, and *Petrascula* have three identical features, whereas the others are not given (*Eodasycladus* may have cladospore secondary branches?). Also *Linoporella, Acroporella, Furcoporella, Crinella, Euspondyloporella, Triploporella, Montiella, Meminella, Montenegrella*, and *Suppiluliumaella* have 3 main features identical. By the position of the reproductive bodies, they form two groups: endospore forms (*Linoporella, Acroporella*, and possibly *Furcoporella*), and cladospore forms (*Montiella, Meminella, Euspondyloporella, Triploporella, Montenegrella*, and

Suppiluliumaella), while *Crinella* does not show clear relations. *Lemoinella* and *Sakkionella* possess, according to their diagnoses, only two of the quoted three features, the others being omitted. That means that the creation of all mentioned genera has been based on some less obvious features which may be found in the descriptions, but not in the diagnoses.

As shown, there are not many genera with clear diagnoses, which may be used by all interested searchers, regardless of their personal convictions.

15.4 Suggestions

In such circumstances it seems very useful to reexamine the particularly important features in respect to their appearance and duration in the geologic past.

The cylindrical main stem continues in different taxons practically without interruption from the Cambrian to the recent time. It might at any time in the past show variations. All other types of the main stem appear more or less sporadically and therefore may be considered as deduced.

Apart from the most ancient known aspondyle forms (Cambrian, Ordovician) making more or less closed primitive group, the euspondility in the rest of the taxa of the geologic past was more continuous than either aspondylity or metaspondylity. That means that only the euspondylity can be considered as a linking element in an evolutionary continuation, while the other two may be reflections of the variability in different time intervals.

The prevalence of branches of the first of second or higher orders has not been registered in a longer continuation. Consequently, all of them might be the result of variability in shorter time intervals.

Among the branch forms, the acrophore type lasts continuously longer than other types. United with the trichophorous type, it would represent a consistent row of taxa, and, consequently, taken together they may be considered as an element of evolutionary persistence. Other types are mostly considered as undefined varieties, relevant to the placing of the reproductive bodies.

In all the cases mentioned, it is evident that the degree of continuity of a particular feature during the past is proportional to its regularity, and reversely proportional to the degree of its differentiation and specialization. That means that the regularity, non-differentiation and non-specialization may be considered as indications for the linking of different taxa throughout the time-space, and as a source for sporadic specializations and iterative revival of some features. These features might be repressed during shorter or longer periods of time due to the activation mechanism in the autoregulation system of the gene activity. In other words, it is possible that, in shorter or longer time intervals, some similar morphologic features reappear, without direct phylogenetic linking. This would mean that they are not necessarily to be taken together as signs of direct connections, for they may be only sporadic, independent and blind evolutionary offsets. If, in addition, the cases of convergencies are taken into consideration, it seems obvious that any conclusion based on a single feature may lead to an inadequate interpretation of the taxonomy.

should be categorized under one name, which, in addition to morphologic similarities, cover a more or less continuous time interval, to keep apart at least in the meantime forms which cannot be brought together in every respect.

The procedure we have followed and are proposing gives an opportunity for consequent emendation of all our statements and conclusions, on a basis which reduces unproven personal convictions to a minimum. However, in spite of the fact that in this way it may be impossible to find out the real interconnections, only the probable trend of evolution, it will be possible to prepare at least an "artificial" basis for a reliable application of fossil forms in the stratigraphy, without sometimes rather strange time gaps and homonymic interpretations.

15.6 Bibliography

Elliott, G.F.: Lower Palaeozoic Green Algae from Southern Scotland, and their Evolutionary Significance. Bull. Brit. Mus. Nat. Hist., Geol. **22**, No. 2, 355–376 (1972).

Endo, R.: Phylogenetic Relationships among the Calcareous Algae. Sci. Rep. Saitama Univ. (B), Comm. Vol. R. Endo, 1–52 (1961).

Groupe Français d'Étude des Algues Fossiles: Reflexions sur la systematique des Dasycladales fossiles. Étude critique de la terminologie et importance relative des critères de classification. Geobios **8** (4), 259–290 (1975).

Hurka, H.: Über den anatomischen Bau und die systematische Stellung des paläozoischen Algengenus *Palaeoporella* Stolley. Nova Hedwigia **15**, 571–582 (1968).

Kamptner, E.: Über das System und die Stammesgeschichte der Dasycladaceen (Siphoneae verticillatae). Ann. Naturhist. Mus. Wien **62**, 95–122 (1958).

Kochansky-Devidé, V.: *Velebitella*, eine neue jungpaläozoische Diploporeengattung und ihre phylogenetischen Verhältnisse. Geol. vjesnik **17**, 135–142 (1964).

Kochansky-Devidé, V., Gušić, I.: Evolutionstendenzen der Dasycladaceen mit besonderer Berücksichtigung neuer Funde in Jugoslawien. Paläontol. Z. **45**, 1/2, 82–91 (1971).

Korde, K.B.: Kembrijske vodorosli. Bjull. Moskov. obšč. ispit. prir. Otd. geolog. 5, **43** No. 5 (1968).

Korde, K.B.: K sistematike i evoljucii vodoroslej iz porjadka Dasycladales (Chlorophyta). Bjull. Moskov. obšč. ispit. prir. Otd. geolog. 2, **46**, No. 2, 134–135 (1971).

Kozlowski, R., Kazmierczak, J.: On two Ordovician Calcareous Algae. Acta palaeontol. pol. **13**, No. 3, 325–346 (1968).

Ott, E.: *Dissocladella cretica*, eine neue Kalkalge (Dasycladaceae) aus dem Mesozoikum der griechischen Inselwelt und ihre phylogenetischen Beziehungen. N. Jb. Geol. Paläontol. Mh, 683–693 (1965).

Pia, J.: Die Siphoneae verticillatae vom Karbon bis zur Kreide. Abh. zool.-botan. Ges. Wien **11** (2), 1–263 (1920).

Pia, J.: Einige Ergebnisse neuerer Untersuchungen über die Geschichte der Siphoneae verticillatae. Z. Indukt. Abstammungsl. **30**, 63––98 (1923).

Rezak, R.: Paleozoic Evolution of the Dasycladaceae. Proc. 9th Intern. Botan. Congr. 1959, Vols. II, IIA.

Segonzac, G.: Les *Cymopolia* (Dasycladacées) du Thanétien des Pyrénées. Bull. Soc. Hist. Nat. Toulouse **104** (3–4), 381–391 (1968).

Segonzac, G.: Une nouvelle Acétabulariacée tertiaire: *Rastroporella oviformis* n.g., n.sp. (Algue calcaire). Bull. Soc. Géol. France **13** (7), No. 1–2, 181–186 (1971a).

Segonzac, G.: Algues calcaires du Sparnacien de Lavelanet (Ariège). Dasycladales, Caulerpales et Cryptonémiales. Bull. B.R.G.M. (2), Sect. 4, No. 1, 5–19 (1971b).

Valet, G.: Contribution à l'étude des Dasycladales. 1. Morphogenèse. Nova Hedwigia **16**, 21–82 (1968).

Valet, G.: Contribution à l'étude des Dasycladales, 2 et 3. Nova Hedwigia **17**, 551–644 (1969).

16 Classification Criteria of Fossil Dasycladales

J.P. BASSOULLET, Paris, P. BERNIER, Villeurbanne, R. DELOFFRE, Pau, P. GENOT, Paris, M. JAFFREZO, Paris, A.F. POIGNANT, Paris, and G. SEGONZAC, Toulouse (French group studying fossil algae)

16.1 Introduction

From the Dasycladales, which are currently represented by forty species distributed amongst eight genera (Valet, 1969), a very large number of fossils (over 120 genera) were recovered, the first known of which date from the Cambrian.

Dasycladales are defined by a set of morphologic characteristics (siphon, "pleuridies" verticils, calcareous sleeve . . .) and strictly cytological and biological characteristics (uninuclear, plasts, reproduction, chemical nature of the shell constituents . . .). Obviously the latter are almost totally ignored in paleontological examination. The Dasycladales order includes, in particular, fossil species some of which have different characteristics from the present day species from which the order was defined (for example, the absence of verticils). It seems, therefore, that there is a contradiction here.

16.1.1 Connection between Present-Day Dasycladaceae and Dasyclad Fossils

There seem to be two possible ways to make this connection:

Firstly, from a pragmatic viewpoint, certain fossil organisms are attributed to the Dasycladaceae for a certain number of reasons: historically this is what happened at first. Secondly, once a certain number of fossils have been attributed to the Dasycladaceae, the connection with the present-day forms is considered from an evolutionary point of view, and then the fossils and the present-day species become linked gradually: this is phylogeny.

16.1.1.1 Evolution and Phylogeny. Thanks to phylogeny the connection between present-day and fossil Dasycladale species can be made. Valet (1968) recalls that the *Neomeris* and *Cymopolia* genera have been in existence since Aptian and Albian times and that the Acicular fossils are practically identical with present-day *Acetabularia*. But "for the other present-day Dasycladaceae, the fossil species are unknown". However, "as a result of studying the morphogenesis, it appears to be possible to establish certain relationships between the present-day and the fossil genera and species, but without giving any real value to such considerations". For example, the ancestor of *Dasycladus, Chlorocladus, Batophora* could be *Palaeodasycladus* and/or *Eodasycladus.*

It should be noted that the author, a botanist, does not go any further than the Liassic, and, moreover, that the problem of uniquely fossil Dasycladales has not been considered.

Some authors have put forward more complete phylogenetic trees (Pia, 1920; Kampter, 1958; Endo, 1961). These trees have obviously been established according to the study of evolutive tendencies.

According to Pia, the evolution shows itself at the level of:

1. the general shape of the thallus, e.g. from cylindrical to club-shaped and, from there, to spherical;
2. the "central stem" (its relative thickness decreased from the Triassic to the Cretaceous);
3. the shape and distribution of the verticil branchings;
4. the position of the reproductive organs.

Emberger (1968) wrote that, as far as he was concerned: "despite the many documents that we have on Dasycladaceae fossils, we cannot establish an accurate genealogy of these algae. We can only put forward the phylogenetic course which they followed". Although he appears to agree with Pia's opinion on the evolution of the branch distribution, he disagrees on the subject of the reproductive organs. He imagined, in the beginning "a primitive stock of fertile branches, leading in turn to free sporangia, corresponding to secondary axis, like present-day Dasycladaceae; in a parallel direction, another line, issue of the same stock, would have evolved by progressively transferring the sporangial function . . . to the ultimate telomes, second degree branches in known forms, to first order branches (*Clypeina, Triploporella*), then to the trunk". In this way the so-called "endosporate" taxa, like *"Diplopora"*, can be explained. This interpretation is probably based on the fact that "cladosporate" species have been identified ever since the Silurian age.

Kochansky-Devidé and Gušić (1971) and Elliott (1972) agree with Pia's general ideas on the Dasycladaceae evolutionary trends but they also bring them up to date. The characteristics which Pia thought had evolved gradually appeared much earlier than he supposed, for example: the euspondyl and metaspondyl genera have been known since the Carboniferous age. Kochansky-Devidé and Gušić conclude that "Dasycladacean evolution did not take place during the Mesozoic, but much before . . . the beginnings of the development of isolated characteristics should rather be looked for in the Paleozoic".

The conclusion must therefore be made, that Dasyclad evolution was much less linear than Pia wrote. For him, the Paleozoic age knew only the primitive genera; the Mesozoic was a transition period; the Cenozoic genera were developed and relatively close to present day genera. In fact the developed genera were known very early and genera with primitive characteristics remained in existence until quite late. Moreover, within a single genus, both developed characteristics and primitive ones coexist.

16.1.2 Conclusions

From this rapid review of the Dasycladales order, the following can be noted:

1. this order was established by botanists according to a few morphologic characteristics and many biological ones;
2. a large proportion of the latter are excluded from paleontological examination, but, by contrast, the order includes, in particular, fossil genera;
3. the connection between fossil and present-day genera, although obvious in some cases, is not at all possible in every case;

4. lastly, disregarding the evolutionary characteristics related to the branch distribution, the other evolutionary characteristics are either being brought into question (the position of the reproductive organs), or else cannot be adapted to fit in with the many discoveries made since 1920.

This article has been written bearing in mind these conditions and with the following aims:

1. to examine critically the terminology used in describing Dasycladale fossils,
2. to attempt to blend the language of the botanists with that of the paelontologists;
3. to examine the relative importance of the distinction criteria, and to look at the consequences of such an examination at a systematic level.

16.2 Paleontological Classification Criteria

16.2.1 History

According to Pia (1920) the perfect natural classification system would have the species gathered into sub-groups which coincidence better in all their characteristics, even the non-observed ones, than with any species of another sub-group of the same rank. In this way, if the relationships of a group of organisms have been established with certainty by comparative examination of quite a high, though limited, number of properties, we can look at its system without having to consider all the remaining characteristics separately.

Taking these principles as a base, Pia proposes a subdivision of the "Siphoneae verticillatae" into 11 tribes. The first three group together the paleozoic genera, the next four the mesozoic genera, afterwards three for the Triassic and the present day, and the last which is the most artificial and which has no name, was reserved for some genera which were little known at the time.

In his general definitions of these tribes, Pia uses very varied morphologic characteristics, such as the shape of the thallus, the absence or presence of a ramification of this thallus, the distribution and change of the branches (simple or with several orders), the position of the spores, calcification . . .

These are the same characteristics which helped him define the genus, while for the demarcation of the species he considers dimensional criteria, distribution details, branch morphology and also the shape of the calcification, all at the same time.

Thus, in the introduction to his work, Pia adopts the following scheme for describing new species:

1. General shape (cylindrical, club-shaped . . .)
2. Dimensions
3. Skeleton (articulation . . .)
4. Shape of the verticil branches (simple or branched, widening or thinning towards the outside, open or closed pores . . .)
5. Disposition of verticil branches (unsystematically, in verticils or in clusters)

6. Axial stem (cylindrical or choked . . .)
7. Reproductive organs

He adds that: "it is not always possible to answer all seven questions at once", but "most of the time that will not be necessary for the distinguishing process".

The classification schemes proposed by Emberger (1944, 1968) and by Kamptner (1958) repeat Pia's, with just a few variations in the list of characteristics.

According to Emberger (1944) "the natural classification of the Dasycladaceae is a difficult problem. At present only the guide lines can be recognized. The fossil shapes are infinitely more numerous than the living ones, and they are still not perfectly known". For the moment, "it is not possible to go beyond the tribe, and the most sensible way to arrange them is to classify them in the chronological order of their appearance. This will necessarily lead to a linear result, and cannot be translated into reality".

In 1961, Johnson also took up Pia's description scheme, and he puts the "dimensions" category in last place. This listing is perhaps suggestive of a particular kind of classification, but the author gives no further explanation.

It therefore appears to be possible to pose the problem of the hierarchy amongst the various classification criteria. In fact, it would be advisable to distinguish those which must be referred to in order to define the family, the tribe, the genus and the species.

Botanist's definitions which principally base their distinctions on the "verticillisation" and the specialization of the reproductive organs are difficult for paleontologists to apply without adjustments, in so far as not all fossil remains possess these characteristics.

The paleontological classification criteria will be examined here in the order listed by Pia (1920).

16.2.2 General Shape of the Calcification of the Thallus

16.2.2.1 The Division of the Principal Axis. The division of the main axis and the "dichotomy" of these divisions are the two principal characteristics which lead to the definition of the paleozoic genera.

Split thalli are found in present-day Dasycladale species: the *Cymopolia barbata* thallus is made up of a succession of calcified joints or segments between which secondary axes spring forth.

The presence of an eventual division cannot be observed in *Cymopolia* (known since the Cretaceous) its thallus appears as isolated segments within the sediments, owing to the absent, or else very weak, calcification at the joint level.

The split thallus does not appear to be a very important characteristic in present-day shapes: one genus may in fact display some split-thallus species (*Cymopolia barbata, Dasycladus ramosus*) and others with non-split thallus (*Cymopolia vanbossae, Dasycladus vermicularis*). This characteristic does, however, come into the definition of the Cymopoliae tribe (Valet, 1969).

16.2.2.2 Different Morphologies of the Thallus. The main types of thallus are, according to Pia:

1. "Stick-shaped" (or cylindrical) type, simple or branched; non-articulated shapes, e.g. *Macroporella, Gyroporella, Vermiporella;*

2. Articulated shapes, e.g. *Teutloporella nodosa*;
3. "Pearl necklace" (or moniliform) type, simple or branched: all species made up of articles which are moveable, spherical to cylindrical shape: e.g. *Cymopolia, Mizzia.*

The degree of mobility of the articles helps to make the distinction between the "stick-shaped" and the "pearl-necklace" type, according to Pia. For a paleontologist this characteristic seems difficult to appreciate.

1. "club-shaped" (or claviform) type: without a definite limit between the head and the neck: e.g. *Palaeodasycladus, Dactylopora, Petrascula;* made up of a cylindrical neck and a spherical or ovoid head: e.g. *Coniporella*;
2. "bowl" (or spherical) type: the last order branch extremities being on the surface of a sphere, e.g. *Goniolina, Cyclocrinus*;
3. "Umbel" type: e.g. *Acetabularia, Orioporella.*

16.2.2.3 Heteromorphosis Phenomena. According to Valet (1968), numerous heteromorphosis phenomena can be commonly observed on algae which have been cultivated in laboratories. These phenomena, some of which may occur in a natural environment, show the great morphologic variability of the Dasycladales and accentuate the analogies which exist between present-day species.

There are manifold morphologic modifications which affect different parts of the thallus: the main axis, the branches, the reproductive organs.

The results of experiments undertaken by botanists and observations from nature show that it is possible to see a certain variability in comparable shapes, from both the generic and the specific points of view. Of course, the paleontologist can only consider morphogenera and morphospecies, but such results must encourage a certain amount of care both in using determination criteria and at the time of the introduction of a new systematic datum from rare material/samples.

16.2.2.4 Using the Shape of the Thallus in Classification. This characteristic plays a role in Pia's definition of a tribe and a sub-tribe, but it is not the only distinguishing criterion as the same shape can be found in several tribes. Moreover, within the same genus, the species can have different shapes (e.g. *Diplopora clavaeformis*: club-shaped, and *Diplopora helvetica*: stick-shaped).

The situation is the same for Tertiary and present-day genera.

Although the morphology of the thallus has been used by some authors as a factor for the definition of the tribe through examination of present-day forms, it is in fact no longer thought advisable to lay too much stress on this characteristic in the systematic description of Dasycladales, beyond the definition of the species.

16.2.3 Dimensions

It seems that it was Pia who, in 1920, "standardized" the measuring of Dasycladaceae, in particular by creating a system of letters which stand for the different parameters, which helps to simplify the writing.

Modern writers frequently use this system. However, some parameters seem to have fallen more or less into disuse while others seem to have a narrower meaning. This system has, from time to time, been supplemented by the addition of new parameters.

If Pia's study (1920), devoted to Triassic Dasycladales, is examined, it is noticeable that the author only uses measurements to distinguish between the species in exceptional cases, when he does, it is always mixed with comments about the morphology. Numerous examples could be quoted.

The measuring may therefore, under certain conditions, be counted as a specific criterion. But again, it must not be considered in too exclusive a manner.

16.2.4 Other Characteristics of the Calcification of the Thallus

16.2.4.1 Nature of the Calcium Carbonate. The calcium carbonate which makes up the sleeve of the living, present-day Dasycladales is aragonite.

What was it in the past? Although several Tertiary Dasycladales have retained their aragonite texture, a large number of Dasycladale fossils are in a sparite calcite state. This sparite is probably not the original structure of these algae, but rather represents a recrystallization of the aragonite.

However, were all the fossil species originally aragonite? It appears that, further to the preservation method due to the surrounding environment, some particular species can be seen, at the moment, with micritic calcite, which gives a blackish test under the photonic microscope. Two examples are *Konickopora, Macroporella sellii.* In which case, is the calcite observed original?

Some species, like *Clypeina jurassica*, are made of fibrous calcite deposited perpendiculary to the surface of the different parts of the thallus. It seems, in this case too, to be difficult to understand, in the present-day, what mechanism led to the formation of such a structure.

16.2.4.2 Degree of Calcification. The degree of calcification intervenes at the level of systematic knowledge of Dasycladales. In fact, the state of preservation of the structure of the individual fossils is strictly subject to the degree of calcification reached at different parts of the alga.

According to the work of botanists, it appears that for the same present-day species, the calcification may vary with the environmental conditions: on artificially cultivated algae (principally: *Acetabularia mediterranea*), Puiseux-Dao (1963), and later Valet (1969) reached the same results: " . . . when the environment is favorable, this calcification hardly appears at all during the cultivation. It increases as soon as the environment becomes poorer, especially if the light is strong". The calcification also seems to be related to the biotope. However, it is sometimes used to distinguish certain present-day species: thus, among the Acetabulariae, *Acetabularia schenckii* possesses cysts united within a calcareous mass, characteristic of the species; the reproductive disc is completely calcified. Similarly, within the *Halycorine* group, the species *H. spicata* can, of course, be distinguished from *H. wrightii* by the smaller number of cysts, but more particularly by the presence of a large calcareous mass which joins them together.

In some fossil forms found in the same bed, variations in the calcification can be noted:

1. within an individual form, in the verticils (some verticil groups are little calcified compared to others), e.g.: *Larvaria reticulata*;
2. between the different individual forms of a single species (at the level of the shape of the calcification of the fertile ampulla, as in the case of *Neomeris arenularia* (Genot and Poignant, 1974).

L. and J. Morellet (1913) warned that paleontological study based entirely on the calcification "was subject to inevitable errors, the various parts of the same plant being able to present different calcifications, and, also, different (even generically different) plants being able to produce analogous calcifications". Distinction between tertiary species (*Neomeris*, for example) is, nevertheless, based on calcification characteristics, little of it being done at the reconstruction of the alga itself.

In the light of recent studies on present-day species, it now seems right that part of the difference in calcification between individual forms may be the result of different ecological conditions, and that the systematic distinction of species based essentially on the degree of calcification appears to be wrong.

16.2.4.3 Morphological Aspects of the Calcification: Joints. Under the heading "external joints", Pia groups together several characteristics:

1. splitting: presence between some verticils of traces of interruption of the calcification, e.g. *Physoporella pauciforata*;
2. annulation: presence of furrows between the calcified verticils. The alga is thus composed of a succession of links each comprising one or more verticils, e.g. *Neomeris annulata*;
3. perannulation: "the subdivision of the sleeve into moveable joints combined with the formation of verticil series" (Pia), e.g. *Teutloporella nodosa, Cymopolia.*

The "internal joints" or intusannulation correspond to a periodical extension of the calcareous deposit towards the interior axial cavity, e.g. *Palaeodasycladus mediterraneus.*

In fact, all these terms, with the exception of perannulation, do not really represent joints, and appear to describe the artificial characteristics which are dependant on the state of preservation and on the degree of calcification of the individual forms observed. They can therefore only be used to add details to a description and can have no specific value.

In conclusion, joints can be defined as a particular area of the thallus where the restricted or else non-existent calcification divides the sleeve into segments or joints, composed of one or more verticils, and which allows for a certain amount of flexibility, e.g. *Cymopolia, Mizzia, Teutloporella nodosa.*

16.2.4.4 The Inside of the Calcareous Sleeve. The term "axial cell", which is currently used in reference to the interior of the calcareous sleeve, wrongly interprets the biological reality of the alga since each individual form is unicellular and there is continuity between the principal axis and the ramifications.

We suggest, for the description of fossil forms, the term "main axis" and "the axial cavity":

1. the main axis (still called the axial siphon) is "the siphon primitively set up from the rhizoidal mass fixed in the substrate" (Valet, 1968). The calcification which limits the volume of the algal siphon will be found particularly well preserved in some fossil forms (e.g. *Neomeris fragilis*).
2. the term "axial cavity" will be used when the absence or the low degree of calcification, as in the conditions for fossilization, involve the disappearance of certain elements of the alga, like the principal axis and first order branches (e.g. *Neomeris scrobiculata* and *N. arenularia*).

16.2.5 Shape of the Ramifications

Pia (1920) used the terms "branch", "pore" and "assimilatory sterile hair". For the first, it seems more correct to substitute the term "ramification" which does not imply any relationship to superior plants and which is used for present-day Dasycladale species by Valet (1969). In the case of successive ramifications, the terms primary or first-order ramification, secondary or second-order ramification, are used.

The term "pore" refers to the empty space which corresponds to the passage of a ramification across the calcified sleeve surrounding the axial cavity.

On the other hand it seems difficult to use the term "assimilatory sterile hair" which Pia used as referring to "the parts of the branches or of their ramifications, the activity of which is assimilation of carbonic acid . . . ".

Botanists also use the term trichoblast meaning the "branched hyaline, caducous hairs" (Valet, 1968). These are not preserved in a fossil states. Sometimes their existence is presumed in old species (e.g. *Tersella*, Morellet, 1951, "protoplasmic hairs").

To distinguish the shape of the branchings, Pia used several classic terms:

1. the term "phloïophorous" is used for ramifications which increase in size towards the outside to form a cortical layer.
2. the "vesicular" type is applied to a form of ramification close to the phloïophorous type, with a distal part which increases very quickly to form a rounded vesicule.
3. the "trichophorous" type refers to ramifications which become thinner towards the outside and have a very tapering point, in the form of a hair, like in the *Teutloporella* and in the terminal ramifications of the *Neomeris.*
4. the "pear-shaped" type resembles quite closely the latter type, having a basal part very large in relation to the distal, which is narrow and elongated, e.g. *Physoporella.*
5. the term "akrophorous" is used by its author for ramifications which neither end in a tapering point nor in an enlarged part. This term can therefore be applied to both ramifications with short pointed ends and subcylindrical ones with rounded ends.

The term "rounded", which in other respects may not seem to fit etymologically, is not therefore very precise and does not indicate a geometrically defined form. Besides, it is not very often used.

It does not really seem necessary to introduce any new terms to designate other variations in the detail of the shape of the ramifications.

The ramifications may present one or more narrowings without any dichotomy arising at this level, e.g. *Palaeodasycladus*, or they may show a partial axial subdivision (as in the case of *Dinarella kochi*).

When describing the shape of the terminal ramifications of fossil species, it should be pointed out that if the distal part is preserved, or if it cannot be seen:

1. in the first case (closed pore), the shape of the ramification can be accurately reconstructed (as in the case of *Gyroporella* and *Palaeodasycladus*).
2. in the second case (open pore), when the ramification has no end, either because it has not been calcified, or because it has been destroyed by erosion, the real shape of the ramification can only be roughly interpreted.

The shape of the ramifications plays a changeable role in the classification. For example Pia sees it as a characteristic which, according to the sample, intervenes:

1. at the "tribe" level, e.g. Cyclocrinae, with "phloïphorous" ramifications which form a cortical layer.
2. at the generic level, e.g. *Gyroporella*, which can be distinguished by its vesicular type ramifications.
3. at the level of a species, e.g. the various *Diplopora* species have "trichophorous" type (*D. annulatissima*), "akrophorous" type (*D. hexaster*) or "phloïophorous" type ramifications (*D. praecursor*).

Sometimes two varieties which differ in the shape of their ramifications (*forma trichophora* and *forma vesuculifera*) are linked to the same species (e.g. *Diplopora uniserialis*).

16.2.6 Distribution of First-Order Ramifications

16.2.6.1 Classification. The ramifications can be placed anyhow along the axis, or even in verticils, according to the preferred planes.

A verticil is a group of ramifications whose insertion points are distributed symmetrically around the principal axis, in a plane perpendicular to the axis. This distribution is regularly repetitive.

To distinguish between the various distributions, Pia (1912 and 1920) created the terms:

Aspondyl: ramifications without any particular positioning, e.g. *Gyroporella*

Euspondyl: ramifications arranged in a preferred plane, in verticils, e.g. *Clypeina, Salpingoporella*

Metaspondyl: ramifications regularly spaced out, in clusters, e.g. *Diplopora*

It should be noted that the metaspondyls, already with verticils, are only a subdivision of the euspondyls. Moreover, this latter group could also be divided into:

simple verticils: close together: e.g. *Salpingoporella*, or far apart: e.g. *Clypeina jurassica*

multiple verticils: groups of simple verticils close together, these groups being regularly spaced along the axis, e.g. *Likanella.*

In the metaspondyl type the ramifications are distributed on several planes, but their insertion points on the principal axis are all situated on the same plane. This must not be confused with the euspondyl type, with multiple ramifications close together, distributed in groups along the axis (*Likanella*).

It is not always easy to recognize whether a form is of the aspondyl or the euspondyl types, when the ramifications are numerous and close together, as for example, in some *Macroporella* types. In addition, the term mesospondyl was introduced especially for this genus, referring to the distribution in longitudinal rows.

In some species, different "regions" are distinguished by the presence of verticils of different morphologies between the distal part and the proximal part of the alga, the Coniporae tribe (e.g. *Petrascula*).

16.2.6.2 Using this Criterion in Classification. In Pia's classification the distribution of the ramifications is an important characteristic. It may be noted, in fact, that in the definition of the three first tribes (Cyclocrinae, Primicorallinae, Dasyporellae) aspondyl disposition is an essential criterion; however, the author places aspondyl (*Macroporella*), euspondyl (*Actinoporella*) or metaspondyl forms (*Diplopora*) in the same Diploporae tribe.

The fact that a species possesses only first order or else higher order ramifications, is another characteristic which plays a part in defining several tribes (Teutloporellae, Diploporae); but forms with only primary ramifications and others with two ranks of ramifications can also be found within the same tribe.

Finally, some genera may possess both aspondyl and euspondyl type species at the same time: e.g. *Macroporella, Teutloporella.*

16.2.7 Reproductive Organs

In this field, at least as far as Dasycladales fossils are concerned, observations are fairly rare, except in the Tertiary where a large number of forms have reproductive organs, often comparable with those of present-day forms. For other forms, a great amount of importance is sometimes attached to interpretation (and even to the imagination): the relative size of the principal axis or of the ramifications is often the only "proof" available of endosporate or cladosporate in the studied alga (Elliott, 1972).

Finally, the obvious fact that paleontological observations are of course limited to the purely morphological appearance, and particularly to the position in the thallus of the elements which play a part in reproduction, should be repeated.

16.2.7.1 The Position and the Meaning of the Reproductive Organs. Pia (1920) suggested three terms of describing the different positions occupied by the "spores":

1. the "endosporate" type: the most primitive Dasycladaceae "spores" being formed in the axial cell.

Valet (1968) recalls that endosporates have been noted in present-day Acetabulariae, and that he had also observed them himself on artificially cultivated plants. According to him, this cannot be considered as an exceptional phenomenon.

2. the "cladosporate" type "groups together all the verticilled branches which play a sporangium part".

3. the "choristosporate" type: "the formation of the spores takes place in the individual sporangia which are formed laterally or at the distal end of the verticilled branches".

It would appear that these terms, if not their meaning, have been commonly agreed on by those describing new species, although the Morellet brothers, for example, never used them, in their important works on the Dasycladaceae of the Tertiary.

To summarize, from a strictly paleontological perspective, the question of Dasycladales reproductive organs can be put forward in the following manner.

There are two fundamental possibilities:

1. either the cysts (spores) are in the principal axis. This situation is known for relatively ancient fossil species. But, an interesting fact is that this phenomenon, although not systematic, is not exceptional among present-day Dasycladales.

2. or the fertile ampulla produce cysts, or not. These fertile ampulla always correspond to ramifications of one order or another. There are several points arising from this:
3. the term "choristosporate" should not be used, at least not in the sense advocated by Pia.
4. it seems that Pia's ideas on the terminal position of the sporangia should also be rejected; as these sporangia are ramifications, they are either in a lateral position or secondarily in a terminal position.

But if the fertile ampulla are still accepted as modified ramifications, it appears possible to establish, from a paleontological point of view, a difference between ramifications modified in function only, and those modified in both function and shape.

There are therefore different types to be considered:

cyst in the principal axis, e.g. *Diplopora*

- fertile ampulla
  - greatly modified ramifications
    - with cysts: e.g. *Acicularia, jodotella*
    - without cysts: e.g. *Cymopolia*
  - slightly modified ramifications
    - with cysts: e.g. *Triploporella fraasi*
    - without cysts: ? (paleontologically unknown)

16.2.7.2 Use of Reproductive Organs as a Distinguishing Criterion. According to Pia (1920) the reproductive organs intervene in the classification particularly at the tribe level.

In his scheme for the definition of Triassic genera, he hardly uses the reproductive organs at all. Otherwise, at the level of the species, for example for the *Diplopora* genus, Pia actually states that "the reproductive organs are either of the endosporate or the cladosporate type, both being able to coexist within the same species".

In the classification of tertiary forms adopted by Morellet (1913, 1922, 1939) great stress is laid on the reproductive organs. The majority of the Tertiary genera can be distinguished by:

the presence of *sterile ramifications and fertile ampulla* within the same verticils,

the position and shape of the reproductive organs.

The relative distribution of the second-order, sterile ramifications and of the reproductive organs may vary quite considerably from one genus to another, and represent a means of distinguishing between certain genera (e.g. *Neomeris* and *Larvaria*), according to the Morellet brothers.

Valet suggests (1969) some dichotomic keys in which the reproductive organ characteristics play an important role, at the level of both the genus and the species.

It would therefore seem that a certain amount of confusion exists as to the use of this criterion in classification.

16.3 Conclusions

That which fossilization has preserved is the only thing which should be considered in paleontology, and the importance of the fossilization is the principal cause of the disparities which exist between the languages of botanists and paleontologists. The present-day Dasycladales make up a relic group, and the morphological characteristics in their definition, which are directly applicable to exclusively fossil algae and infinitely more numerous, should be emphasized. It is useful to note that these morphological characteristics are used by Valet (1968–1969) in his descriptions of present-day Dasycladaceae.

16.3.1 Practical Use of Classification Criteria

Each characteristic can, it seems, be used at different levels; however if we take up the criteria defined in the preceding pages, three groups of decreasing importance appear to emerge. It is quite obvious that we are trying to establish a certain hierarchy among these criteria, and not to define an imperative, fixed use for them.

1. the disposition of the first order ramifications would come first; this is sufficiently important to define the family.
2. next, the general shape of the thallus (articulation, . . .), the shape and order of the ramifications, and the reproductive organs, which can determine according to the case considered, the taxons from the genus to the sub-family.
3. finally, the dimensions, the nature and degree of the calcification, and the internal distribution of the calcareous sleeve form a third group of criteria, which characterises, at most, the species.

We have considered the paleontological criteria of classification defined by Pia (1920) which, in his exhaustive work, has the great merit of clearly putting forward all the knowledge existing at the time about calcareous algae. We must pay tribute to his clearsightedness and his spirit of synthesis, even if in the course of the review we have undertaken, we cannot accept all his views. What is regrettable is that subsequently, successive authors have not thought it necessary to adopt a more critical position in reference to the terminology, the phylogeny and the classification which he set up. The evolution of the knowledge of fossil algae and the ideas on the whole series of problems concerning systematic, should have gradually modified the latter and more narrowly codified the determination criteria. Instead of which, we find ourselves today in the position of having an illogical, awkward and artificial classification, where, if one stone of the construction is moved, all the others became unsteady.

16.4 Bibliography

Bassoullet, J.P., Bernier, P., Deloffre, R., Genot, P., Jaffrezo, M., Poignant, A.F., Segonzac, G.: Réflexions sur la systématique des Dasycladales fossiles. Geobios **8** (4), 259–290 (1975).

Elliott, G.F.: Lower Paleozoic green Algae from southern Scotland and their evolutionary significance. Bull Brit. Mus. (Nat. Hist.) **22**, 4, 355–376 (1972).
Emberger, L.: Les plantes fossiles dans leurs rapports avec les végétaux vivants. Paris: Masson 1944, 492 pp.
Emberger, L.: Les plantes fossiles dans leurs rapports avec les végétaux vivants. Paris: Masson 1968, 758 pp.
Endo, R.: Phylogenetic relationships among the calcareous Algae. Endo comm. Vol., Sci. Rep. Saitama Univ., Ser. B, 1–52, 17 pl., 7 figs. (1961).
Genot, P., Poignant, A.: Contribution à l'étude, au microscope électronique à balayage, des Dasycladacées de l'Eocène du Bassin de Paris. Rev. Micropal. **17**, 2, 66–74 (1974).
Johnson, J.H.: Limestone building algae and algal limestones. Colorado School Mines, 297 pp. (1961).
Kamptner, E.: Über das System und die Stammesgeschichte der Dasycladaceen (Siphonae Verticillatae). Geol. Paläontol. Ann. Natur. Mus. Wien **62**, 95–122 (1958).
Kochansky-Devidé, V., Gušić, I.: Evolutions-Tendenzen der Dasycladaceen mit besonderer Berücksichtigung neuer Funde in Jugoslawien, Paläontol. Z. **45**, 1/2, 82–91 (1971).
Morellet, L., Morellet, J.: Les Dasycladacées du Tertiaire parisien. Mém. Soc. Géol. France **47**, 43 pp. (1913).
Morellet, L., Morellet, J.: Nouvelles contributions à l'étude des Dasycladacées tertiaires. Mém. Soc. Géol. France **25**, 2, 35 pp. (1922).
Morellet, L., Morellet, J.: Tertiary siphoneous algae in the W.K. Parker collection with description of some Eocene Siphoneae from England. Brit. Mus. (Nat. Hist.), 55 pp. (1939).
Pia, J.: Neue Studien über die triadischen Siphonae Verticillatae. Beitr. Paleontol. Geol. Öst. Ung. Orients **25**, 25–81 (1912).
Pia, J.: Die Siphonae verticillatae vom Karbon bis zur Kreide. Abhandl. Zool. Bot. Ges. **11** (2), 1–263 – Transl. M. Gubler, I.F.P., ed. Technip, 1961 (1920).
Puiseux-Dao, S.: Les Acétabulaires, matériel de laboratoire; les résultats obtenus avec les Chlorophycées. Ann. Biol. **2** (3–4), 99–154 (1963).
Valet, G.: Contribution à l'étude des Dasycladales. 1. Morphogenèse. Nova Hedwigia **16**, 21–82 (1968).
Valet, G.: Contribution à l'étude des Dasycladales. 2. Cytologie et reproduction. 3. Révision systématique. Nova Hedwigia **17**, 551–644 (1969).

17 Late Paleozoic Calcareous Red Algae

J.L. WRAY, Colorado, U.S.A.

17.1 Introduction

The first purpose here is to provide an inventory of important Late Paleozoic (Devonian-Permian) taxa that either belong to, or are believed to belong to, the marine calcareous Rhodophyta. This is followed by an examination of possible evolutionary patterns and their distribution in time and space. Secondly, this review discusses sedimentological and diagenetic aspects of Late Paleozoic calcareous algae in the interpretation of carbonate facies.

From a geological viewpoint, modern marine calcareous red algae belonging to the families Corallinaceae and Squamariaceae are important reef framebuilders and significant sediment producers over widespread geographical and bathymetrical realms. They are frequently preserved as fossils because they have thoroughly calcified skeletons resulting from calcium carbonate precipitation within cell walls. Unfortunately, the geological record of modern calcareous red algae can be traced backwards in time with certainty only to the middle part of the Mesozoic era. Thus, any systematic consideration of Paleozoic red algae is necessarily full of uncertainties because these fossils are extinct groups whose relationships to living taxa are unclear.

Despite problems of taxonomy and classification, fossil remains presumed to be calcareous rhodophytes are diverse and locally abundant constituents in the Late Paleozoic rock record, and thus are potentially important to many aspects of historical geology. This summary attempts to review current knowledge of the subject.

17.2 Classification and Description

The uncertain associations between living calcareous red algae and Paleozoic taxa presumed to be red algae necessitates a conservative approach to the classification of these fossils. Too often meager information on scattered occurrences of a few forms is organized into a classification, when in fact, the amount of data is wholly inadequate to even approximate their natural relationships. The prevailing arguments of the gross biological affinities of many ancient red algae, often whether they are plants or invertebrates, are reasons enough for caution.

Unless reasonably certain taxonomic affinities of fossil and living algae can be demonstrated, classifications serve no real purposes for understanding evolutionary patterns or for establishing the environmental ranges of fossil forms based on the habitats of analogous living species. In the absence of critical data, it seems more desirable to develop an empirical taxonomy useful for geological interpretations, rather than a scheme striving for logic, completeness, and similarity to classifications of extant taxa.

A generally accepted classification of living and extinct groups of calcareous red algae commonly preseved as fossils include:

Phylum Rhodophyta
  Class Rhodophyceae
    Subclass Florideae
      Order Cryptonemiales
        Family Solenoporaceae
          Gymnocodiaceae
          Squamariaceae
          Corallinaceae

Unquestioned representatives of the extinct families Solenoporaceae and Gymnocodiaceae are known in the Late Paleozoic. Forms believed to be ancestral to the modern Corallinaceae and Squamariaceae are relatively common, but their relationships to living taxa are not established. Other genera of questionable affinities are placed into informal categories for purposes of this discussion.

17.2.1 Solenoporaceae

Solenoporaceans seem to be the least controversial Late Paleozoic calcareous red algae. Species belonging to the genera *Solenopora* Dybowski, 1878, and *Parachaetetes* Denninger, 1906, are common and widespread representatives of the family. Both have been described often from Devonian, Early Carboniferous (Mississippian), and Permian rocks. *Parachaetetes* has only recently been reported from the Late Carboniferous (Pennsylvanian) in the United States (Heckel, 1975), where previously solenoporaceans were unknown. Of lesser importance, Carboniferous species of *Pseudochaetetes* Haug, 1883, have been described, and *Pycnoporidium* Yabe and Toyama, 1928, has been reported from the Permian (Endo, 1956).

Johnson (1960) reviewed aspects of the Paleozoic Solenoporaceae. More recently, Elliott (1965 and 1973) discussed the morphological features of the family, including reproductory bodies in Cenozoic forms; however, unequivocal reproductive organs have not been found in Paleozoic species. There is general agreement that the solenoporacean cellular structure is primitive and distinct from that of the modern Corallinaceae, although it is probable that solenoporaceans provided an ancestral stock to the true corallines.

17.2.2 Gymnocodiaceae

Two gymnocodiacean genera, *Gymnocodium* Pia, 1920, and *Permocalculus* Elliott, 1955, are known from the Permian. Representatives of this extinct family include forms that resemble the weakly calcified Recent *Galaxaura* belonging to the family Chaetangiaceae. Fossils occur as individual segments and locally are common in Permian carbonate facies.

17.2.3 Ancestral Squamariaceans

Fossil squamariaceans occur throughout the Cenozoic, and the genus *Ethelia* Weber van Bosse, 1913, has been reported from rocks as old as Early Cretaceous. There is now evidence that this family of calcareous rhodophytes occurs in the Late Paleozoic. Comparative studies of Holocene squamariaceans and Permian material suggest that some Late Paleozoic "phylloid algae" probably are related to the family Squamariaceae (Wray et al., 1975). This conclusion is based on comparable external morphologies, remnant internal structure in recrystallized thalli, and associated diagenetic fabrics characteristic of both Holocene and Permian material. Thus, modern calcareous squamariaceans are believed to be descendants of some Late Paleozoic phylloid algae, and the metastable aragonitic mineralogy of these algae can explain the pervasively altered condition of their ancient counterparts. Present data are insufficient to describe Paleozoic taxa.

17.2.4 Ancestral Corallines

One of the most diverse and common groups of Late Paleozoic red algae include forms that closely resemble members of the family Corallinaceae. Not only are these fossils potentially useful in understanding the origin and evolution of modern calcareous rhodophytes, but some are widespread and abundant constituents which were influential in the deposition of carbonate facies.

The genera *Archaeolithophyllum* Johnson, 1956; *Cuneiphycus* Johnson, 1960; *Katavella* Chuvashov, 1965; *Tharama* Wray, 1967; and *Lysvaella* Chuvashov, 1971, are important taxa within this artificial group. Morphologically all of these genera are characterized by well developed cellular tissue. *Archaeolithophyllum* (Early Carboniferous-Permian) and *Lysvaella* (Permian) have tissue differentiated into hypothallium and perithallium, in addition to reproductive bodies embedded in the perithallic tissue. *Cuneiphycus* (Late Carboniferous) has undifferentiated tissue composed of large cells and no known reproductive structures. Originally considered to have a segmented thallus, *Cuneiphycus* is now known to have a rigid branched habit. The Devonian genus *Katavella* has an unsegmented cylindrical thallus made up of differentiated tissue composed of prismatic cells and external reproductory organs. The apparently undifferentiated tissue in *Tharama* (Devonian) consists of very small rectangular cells, resembling coaxial hypothallic tissue. According to Riding and Wray (1972), *Tharama* resembles *Epiphyton* and other Early Paleozoic genera belonging to the family Epiphytaceae Korde, 1959, which have been assigned to the red algae. The relationship between the Epiphytaceae and "ancestral corallines" is not clear.

Archaeolithophyllum appears morphologically to be the most advanced genus within this group of ancestral coralline algae. Also, it is the most abundant, and geographically and stratigraphically the most widespread. Careful morphological studies of well preserved specimens may yet positively determine the true systematic position of this important Late Paleozoic alga.

17.2.5 Other Late Paleozoic Red Algae

The taxa considered here include *Ungdarella* Maslov, 1950; *Komia* Korde, 1951; *Foliophycus* Johnson, 1960; *Stacheia* Brady, 1876; *Stacheoides* Cummings, 1955; and *Aoujgalia* Termier and Termier, 1950. These genera are usually treated as a related and homogeneous group, although the botanical affinity of these fossils has been seriously questioned. All are characterized by a fibrous wall structure, lack distinctly cellular tissue, and have no preserved reproductory organs. They have at one time or another been assigned to the red algae in the family Ungdarellaceae Maslov, 1956.

These fossils seem to be divisible into two groups. One group includes *Stacheia, Stacheoides* and *Aoujgalia*; and the other, *Ungdarella, Komia* and *Foliophycus*. The *Stacheia* group is restricted to the Early Carboniferous and was succeeded by the *Ungdarella-Komia* group in Middle Carboniferous time. Stacheiinids have been considered to be foraminifera, although the wall structure is not typical of Paleozoic foraminifers; thus they are often classified as red algae (Petryk and Mamet, 1972). On the other hand, Wilson et al. (1963) and Wilson (1969) concluded that *Ungdarella* and *Komia* are stromatoporoids, but this interpretation has not been generally accepted. Most recently, these Late Paleozoic microfossils, along with others, have been assigned to the calcareous sponges (Termier et al., this volume).

Despite the large amount of disagreement concerning the classification of these fossils, some genera are common constituents in Late Paleozoic carbonate sediments and may occur in bank-like accumulations. Many forms have restricted stratigraphic ranges, and therefore are useful in stratigraphic correlations and age determinations.

17.2.6 Reinterpreted Red Algae

The Late Devonian genus *Keega* Wray, 1967, from Western Australia was originally described as a red alga of uncertain affinities, but probably related to the crustose Corallinaceae. This fossil has distinct tissue differentiation and reproductive organs similar to the modern genus *Lithophyllum*. However, subsequent study of specimens from Alberta (Wray and Playford, 1970) cast doubt on the algal affinity of *Keega*.

Riding (1974) made a thorough study of type material of *Keega*, in addition to specimens from Canada. As a result, he reinterpreted this fossil as a stromatoporoid, specifically the basal layer of a laminar form, as opposed to the more common dendroid form, of the genus *Stachyodes*. Riding demonstrated that the internal structure, or "cellular tissue", of the fossil is a combination of primary megastructure and diagenetically altered microstructure, and that the "conceptacles" in *Keega* are fortuitous sections of tubes which form branching networks within the stromatoporoid.

This presumably correct reinterpretation of *Keega* is significant because previously it was assumed that this algal was among the earliest "ancestral corallines", and also one of the most advanced structurally. Now it seems that *Archaeolithophyllum* from the Carboniferous and Permian occupies this position.

17.2.7 Problematical Red Algae

Several Late Paleozoic taxa have been considered problematical red algae, but only three genera are reviewed here.

The encrusting laminar habit of *Archaeolithoporella* Endo, 1959 (Permian), is well known and was originally described as having a morphology similar to the modern genus *Lithoporella*. Based on reexamination of the holotype and study of numerous new Late Permian specimens, Babcock (1974) shows that *Archaeolithoporella* is a very simple laminar structure, without cells. He concludes that there is no morphological evidence to justify classification of this fossil among the coralline red algae; however, it is considered to be of organic origin, and probably algal. This is an important reinterpretation, because this generic name has become entrenched in the literature as a fossil crustose coralline red alga.

Litostroma Mamet, 1959 (Carboniferous) and *Stenophycus* Fenton 1943 (Devonian) have both been classified as red algae of uncertain affinities. Neither genus has been reported to occur abundantly or widespread; consequently, their significance as a part of the presumed red algal flora in Late Paleozoic time is unknown.

17.2.8 Evolutionary Patterns

Lemoine (1911) suggested that the two main morphological stocks of the modern crustose Corallinaceae, namely the Lithothamnieae and Lithophylleae, evolved from the Solenoporaceae. Evolutionary patterns proposed later by Johnson (1960, Maslov (1962), and Wray (1970) modified Lemoine's original scheme to include Late Paleozoic "ancestral coralline" genera. Although differing in detail, each scheme derives the Lithothamnieae from *Solenopora* during the early Mesozoic, and the Lithophylleae and articulated corallines from *Parachaetetes* in the middle Paleozoic by way of Late Paleozoic "ancestral corallines".

Chuvashov (1971) has suggested an alternate evolutionary scheme for Paleozoic rhodophytes. The family Epiphytaceae initiates the series in Cambrian time. Representatives of the Epiphytaceae gave rise to the Devonian family Katavellaceae Korde, 1966, and this group in turn evolved the true Corallinaceae in the Carboniferous and Permian, represented by *Archaeolithophyllum* and *Lysvaella*. The Late Paleozoic Ungdarellaceae are presumed to have developed in parallel to this series because of completely different morphology. Similarly, the Solenoporaceae are considered to be a separate and distinct evolutionary pathway.

Korde (1973) has outlined a very elaborate classification of Cambrian red algae. The scheme proposes two new classes of red algae which are subdivided into six new orders and seventeen new families. Two previously recognized families, Solenoporaceae and Epiphytaceae, are incorporated into the new classification, but it is unclear how other Late Paleozoic red algae may be related to these earlier Paleozoic groups.

The recognition here of probable "ancestral squamariaceans" in the Late Carboniferous and Permian adds another phylogenetic lineage to possible evolutionary patterns of fossils considered to be red algae. However, the morphological characteristics of both the "ancestral corallines" and "ancestral squamariaceans" are not sufficiently definitive to classify these fossils in the Corallinaceae and Squamariaceae. The apparent lack of remains of these two groups of calcareous algae in the Triassic and early Jurassic is also a problem in establishing the continuity of lineages from the Recent back to the Late Paleozoic. Moreover, the limited number of recorded occurrences of some ancestral coralline genera are insufficient to understand even the relationships of taxa within this group.

In summary, more comprehensive research and new discoveries are needed to establish better the affinities of Late Paleozoic calcareous red algae to living groups. Because of the uncertain taxonomy of these fossil algae, it is difficult to outline meaningful evolutionary patterns. At best, only suggestions and hypothetical phylogenetic relationships based on available distributional data can be offered (Fig. 1).

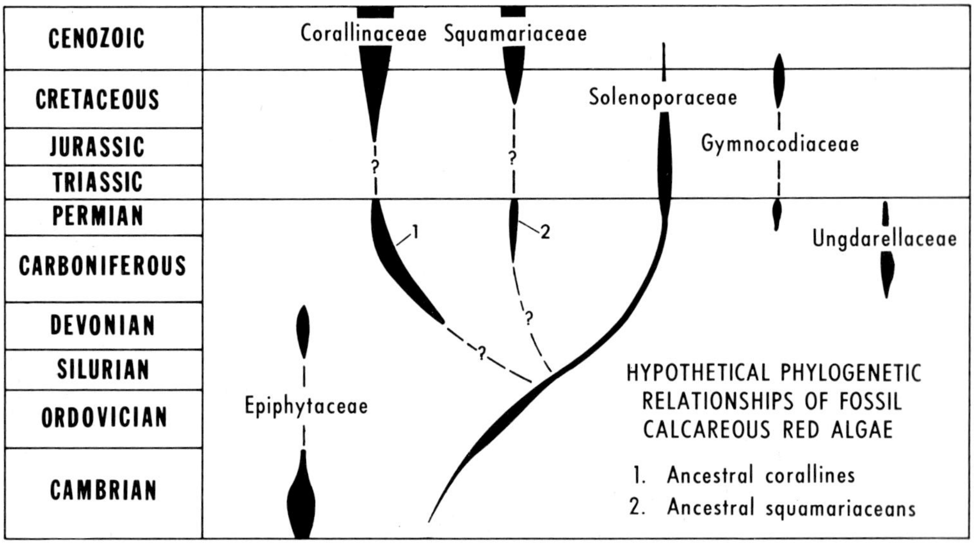

Fig. 1. Suggested hypothetical phylogenetic relationships of groups of fossils presumed to be calcareous red algae

17.3 Distribution in Time and Space

The distribution of Late Paleozoic red algae can be treated much more objectively than their assumed phylogenies. In general, there is a wide variation in both the chronologic and spatial distribution of these fossils. Some genera have been reported from numerous localities on several continents, and thus are known to be geographically widespread. Other forms seem to be much more restricted regionally, but this conclusion may be based on insufficient records of occurrence.

The known chronologic ranges within the Late Paleozoic of genera discussed here are shown in Fig. 2. Most genera are considered long-ranging taxa during the Late Paleo-

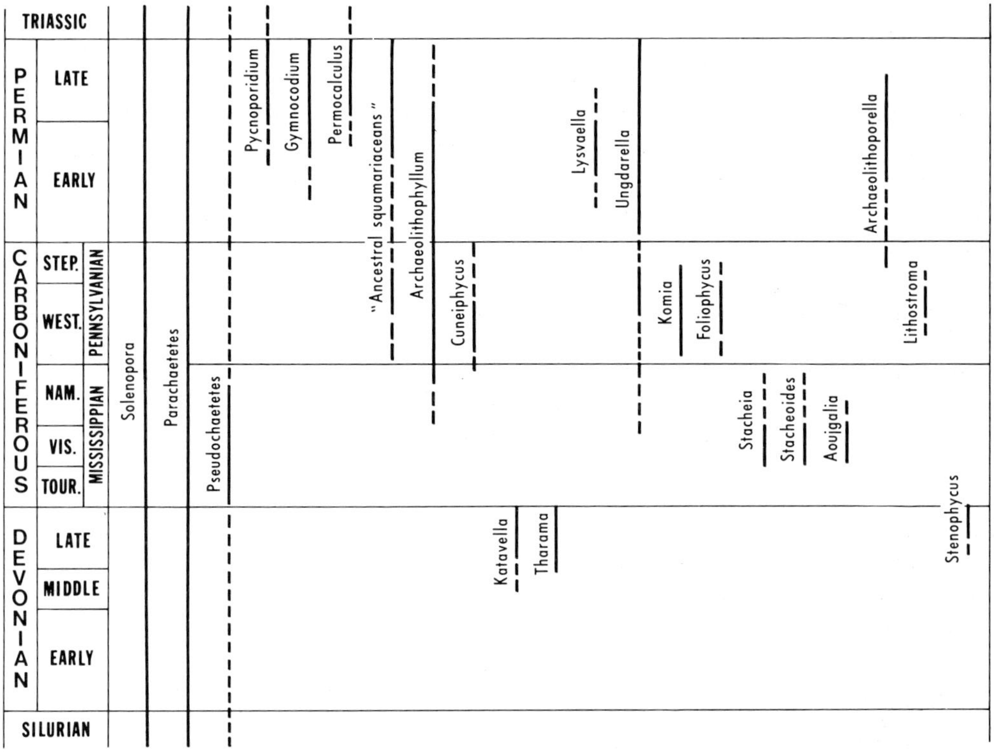

Fig. 2. Chronologic ranges of Late Paleozoic taxa considered to be calcareous red algae. Ranges are approximate and based on several sources of data

zoic, with the exception of some ungdarellids which have time-stratigraphic value. Other genera with short age ranges are ones known from only a few localities, and thus their geologic ranges are assumed to be poorly documented.

17.4 Paleoenvironmental Distribution

Late Paleozoic red algae are potentially important fossils for interpreting sedimentary environments, because they occur in abundance and appear to be restricted to specific depositional facies. No doubt individual taxa of Paleozoic benthonic red algae had very definite ecological tolerances, but much empirical distributional data are required to develop these patterns.

Fossil calcareous rhodophytes seem to present a difficult problem in paleoenvironmental studies because high-order taxa of living red algae are known to have broad ecological ranges; only lower-level taxa show ecological restrictions. Consequently, valid generic and specific concepts of Late Paleozoic taxa must be developed to have value in paleoenvironmental interpretations.

The majority of Late Paleozoic calcareous red algae grew most commonly in open-marine, carbonate shelf environments. As a group, these algae tolerated a variety of current velocities ranging from quiet water, in which dominantly mud-rich facies were deposited, to agitated environments characterized by grain-supported fabrics with little mud-sized carbonate and in which skeletal grains are generally fragmented and abraded.

Some genera, notably *Archeolithophyllum*, were predominantly encrusting organisms and served as framebuilders in the development of reefs and other carbonate accumulations (Wray, 1964). This genus has been observed to occur in both mud-rich and mud-free facies, but it is not known whether various species of this genus inhabited specific kinds of sea bottom conditions.

Heckel (1975) noted that Pennsylvanian solenoporids occur preferentially in mud-free grainstones deposited in agitated environments. On the other hand, delicately branching forms, such as *Ungdarella* and *Cuneiphycus*, have been observed most often in mud-rich rocks, suggesting quiet water deposition.

Fragments of gymnocodiaceans occur in mud-rich carbonate facies associated with foraminifers. Elliott (1958) reported these kinds of "algal debris facies" in the Cretaceous of the Middle East and interpreted them as having formed in very shallow waters.

The seemingly wide variation of facies occurrences of Late Paleozoic calcareous rhodophytes as a group, suggest that individual taxa can provide important clues to ancient environments. To accomplish this goal there is a need for quantitative distributional data of taxa obtained from numerous carefully studied stratigraphic sections within a given region. This information should be interrelated with studies of other skeletal constituents, in addition to depositional structures and fabrics in the same rocks. The resulting observed associations of specific kinds of algae and specific facies can decipher paleoecological factors, such as water depth, current energy, and turbidity, as a part of an overall understanding of carbonate sedimentological regimes. This knowledge also can be used in predictions of the extent and patterns of algal facies. This empirical approach is believed to be more promising than one relying on comparative studies of living red algae and their presumed ancient counterparts because of their uncertain biological relationships.

17.5 Sedimentology and Diagenesis

Late Paleozoic calcareous red algae provide a variety of sizes and shapes of carbonate grain constituents. These range from rather uniform cylindrical shapes less than 1 mm (e.g., *Permocalculus*) to phylloid shapes several centimeters in length (e.g., *Archaeolithophyllum*). These skeletal algae sometimes were deposited in sufficient abundance to constitute the dominant biotic constituent in the rock. For example, *Archaeolithophyllum* and *Komia* may make up 50% of the volume of Late Carboniferous (Pennsylvanian) facies in central and western United States.

Aside from the quantitative contribution of Late Paleozoic red algae to the formation of carbonate facies, some types controlled the distribution of sedimentary particles at the time of deposition. The two most important influences in the depositional environment were: (1) localization of carbonate sediments through mechanisms of trapping and binding fine carbonate sediment, and (2) formation of initial porosity beneath or between leaflike fragments because of a sheltering effect.

The "trapping and binding" influence is illustrated by the rigid encrusting thalli of "ancestral corallines" and "ancestral squamariaceans" that produced "boundstones" comparable to modern reef frameworks. These kinds of algal fabrics were presumed to have had potential wave-resistant qualities. "Trapping" of fine-grained carbonate sediment was also due to the baffling action on currents, mainly reduction of velocities, about large phylloid or branching thalli.

Phylloid algae were capable of providing a self-supporting skelatal framework at the time of deposition. This kind of depositional fabric preferentially excluded sedimentary particles because of an "umbrella" or sheltering effect, which resulted in the formation of interconnected void space. A combination of various kinds of sedimentological influences of Late Paleozoic red algae has resulted in numerous initially porous carbonate accumulations, including reefs and banks, some of which are oil and gas reservoirs.

The original mineralogy of "ancestral corallines" and "ancestral squamariaceans" is believed to have been an important factor affecting the early diagenesis of algal facies, but much of the evidence for the initial composition of these algae is necessarily based on supposition because of their uncertain relationships to modern taxa. The preservation of cellular tissue in *Archaeolithophyllum* ranges from good to completely obliterated, suggesting that this alga was originally composed of calcites with variable magnesium contents. Calcites with high-magnesium contents are most soluble during postdepositional modifications, whereas low-magnesium calcites are least soluble. By analogy with Recent crustose Corallinaceae, the kind and extent of diagenetic alterations in Late Paleozoic algae may be related to their initial mineralogy.

The pervasive alteration to sparry calcite of "ancestral squamariacean" thalli is believed to be related to their original metastable aragonitic composition. Studies of Holocene squamariaceans in the Caribbean region (Wray et al., 1975) demonstrate very early diagenetic alteration which eradiacates internal cellular tissue. This process also produced associated radially-arranged fans of aragonite crystals precipitated on the surface of thalli, commonly in voids beneath crusts. Examples of recrystallized Late Paleozoic squamariacean-like algae and associated diagenetic crystalline fabrics are remarkably similar to Holocene specimens. These kinds of interparticle aragonite cements have essentially filled all of the available void space in some Paleozoic rocks.

In addition to the original mineralogical composition of Late Paleozoic calcareous algae, the nature of the rock matrix, chemistry of formation waters, and other factors undoubtedly were important in promoting or retarding diagenesis. Because these ancient calcareous red algae may occur in large sedimentary accumulations of economic significance, an understanding of their initial composition and mode of diagenesis can offer clues to interpreting the nature and distribution of porosity in carbonate sediments.

17.6 Summary

1. Late Paleozoic calcareous red algae are a diverse assemblage of fossil plants and their skeletal remains are common sedimentary constituents.

2. The taxonomy, classification and phylogenetic relationships of Late Paleozoic calcareous fossils presumed to belong to the Rhodophyta are largely conjectural because of the uncertain affinities to living taxa.

3. The flora belongs to the families Solenoporaceae, Gymnocodiaceae, and two informal groups, "ancestral corallines" and "ancestral aquamariaceans". Other taxa are considered to be red algae of uncertain affinities and problematical red algae, and some of these most probably belong to invertebrate groups.

4. Late Paleozoic calcareous red algae are useful in reconstructing paleoenvironments through empirical associations, although new data are required to utilize their full potential.
5. Late Paleozoic calcareous red algae may be quantitatively important and widespread skeletal constituents in marine shelf carbonates and some forms have exerted physical controls on sedimentation. "Ancestral corallines", "ancestral squamariaceans", and other forms developed self-supporting skeletal frameworks and contributed to the formation of reefs and banks. The original mineralogy of "ancestral corallines" and "ancestral squamariaceans" was an influential factor in early diagenesis.

17.7 Bibliography

Babcock, J.A.: Role of algae in Capitan Limestone (Permian, Guadalupian), Guadalupe Mountains, West Texas and New Mexico. Ph. D. thesis., Univ. Wisconsin (1974).

Chuvashov, B.I.: A new genus of Late Paleozoic red algae. Paleontol. Zhur. **2**, 85–89 (1971).

Elliott, G.F.: Algal debris facies in the Cretaceous of the Middle East. Paleontology **1**, 254–259 (1958).

Elliott, G.F.: Tertiary solenoporacean algae and the reproductive structures of the Solenoporaceae. Paleontology **7**, 695–702 (1965).

Elliott, G.F.: A Miocene solenoporid alga showing reproductive structures. Paleontology **16**, 223–230 (1973).

Endo, R.: Stratigraphical and paleontological studies of the later Paleozoic calcareous algae in Japan, XI. Saitama Univ. Sci. Rept., Ser. B, **2** (3), 279–305 (1956).

Heckel, P.H.: Solenoporid red algae (*Parachaetetes*) from upper Pennsylvanian rocks in Kansas. J. Paleontol. **49**, 662–673 (1975).

Johnson, J.H.: Paleozoic Solenoporaceae and related red algae. Colorado School of Mines Quart. **55** (3), 1–77 (1960).

Korde, K.B.: Cambrian algae. Acad. Sci. USSR., Trans. Paleontol. Inst. **139**, 1–349 (1973).

Lemoine, M.: Structure anatomique des Mélobésiées. Ann. Inst. Océanogr. Monaco **2** (2), 1–213 (1911).

Maslov, V.P.: Fossil red algae of the U.S.S.R. and their facies relationships. Acad. Sci. USSR., Geol. Inst., Trudy **53**, 1–222 (1962).

Petryk, A.A., Mamet, B.L.: Lower Carboniferous algal microflora, southwestern Alberta. Can. J. Earth Sci. **9**, 767–802 (1972).

Riding, R.: The Devonian genus *Keega* (Algae) reinterpreted as stromatoporoid basal layer. Paleontology **17**, 565–577 (1974).

Riding, R., Wray, J.L.: Note on the algal genera *Epiphyton, Paraepiphyton, Tharama,* and *Chabakovia*. J. Paleontology **46**, 918–919 (1972).

Termier, H., Termier, G., Vachard, D.: Sur la systématique et la phylogénie des Moraramminida et des Aoujgaliia. (This volume).

Wilson, E.C.: No new *Ungdarella* (Rhodophycophyta) in New Mexico. J. Paleontol. **43**, 1245–1247 (1969).

Wilson, E.C., Waines, R.H., Coogan, A.H.: A new species of *Komia* Korde and the systematic position of the genus. Paleontology **6**, 246–253 (1963).

Wray, J.L.: *Archeolithophyllum,* an abundant calcareous alga in limestones of the Lansing Group (Pennsylvanian), southeastern Kansas. Kansas Geol. Survey Bill. **170** (1), 1–13 (1964).

Wray, J.L.: Algae in reefs through time. Proc. North Am. Paleont. Conv. Chicago 1969, 1359–1373 (1970).

Wray, J.L., James, N.P., Ginsburg, R.N.: The puzzling Paleozoic phylloid algae – Holocene answer in squamariacean calcareous red algae (abst.). Ann. Meet., Am. Assoc. Petrol. Geol. Soc. Econ. Paleontol. and Mineral. Abstr. **2**, 82–83 (1975).

18 The Mesozoic Red Algae: A General Survey

A.-F. POIGNANT, Paris, France

The principles of classification are considered before briefly describing the Corallinaceae, Solenoporaceae, Gymnocodiaceae and Squamariaceae. Phylogenesis is then dealt with, first in general way and afterwards applied to red algae. Finally a new phylogenetic schema is proposed.

18.1 General Remarks

The current principles of classification of algae are based on the color of the pigments, the chemical nature of the reserves (starch, fats), the type of reproduction cells, and the morphological characteristics.

It would seem obvious that only the morphological characteristics (nature, general aspect, and size of the thallus, dimensions of cells, size and architecture of the reproductive bodies, etc. . .) could be of use for the classification of fossil forms.

I shall deal essentially with the Mesozoic representatives of the families Solenoporaceae, Corallinaceae, Gymnocodiaceae and Squamariaceae. Naturally, the Palaeozoic and Cenozoic red algae must also be considered as well as the algae of uncertain affinities.

18.2 Classification

From amongst the algae, the calcareous forms as they fossilized, are the most likely to have survived to this day. Consequently, we have only a very incomplete inventory of the fossilized forms of algae living at a given time.

The classification of fossil algae always presents a problem. Most of the fossil algae found from the Tertiary onwards show types and families which still exist, and for which the classification of present-day algae may be used. However, the palaeoalgologist has been obliged to create an independent and original classification for the types and families existing only in fossil form, using only the criteria which are identifiable on the fossil. In this way, the Rhodophytaceae appear as a structurally very developed group, differing from the others by the type of reproduction and by the fact that nearly all its representatives are calcareous.

We may adopt certain principles and consider that each taxonomy must be applied to the subject in question, starting from the apparently logical principle that there is no general classification.

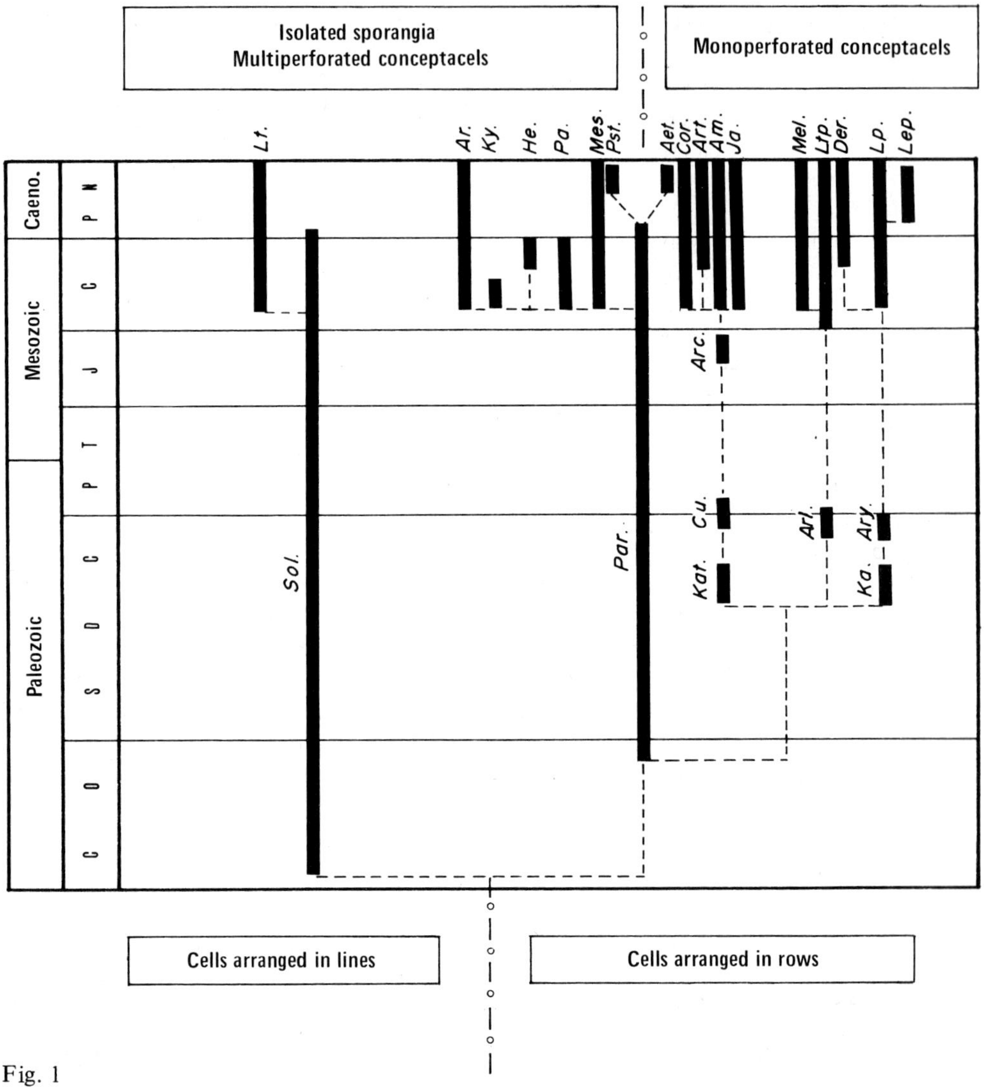

Fig. 1

The Aristotelian or creationist classification (also known as the traditional classification) proceeds by analogy and bases its groupings on the nonhazardous distribution of characteristics. In Linne's time this classification was a system of identification. In this way, a group is defined by the presence in its members of series of correlative characteristics which are not encountered together at the same time outside this group. An individual is thus attributed to a group by comparison of the easily observed characteristics. As may easily be imagined, this will lead to the break-down of the notion of species and to confusion between species and individuals.

Thus, the typological species may be defined as a collection of identical or very similar individuals, compared to a determined type.

The "statistic" classification. The organisms are classified according to their differences or their degrees of likeness. All the characteristics must be used, and for this it is necessary to have an objective definition of the unity of the characteristics and to translate it into numbers. All the members of a group must, therefore, possess the same number of characteristics; most often, only a few representative traits are chosen. We may therefore envisage here the biological species, by stating that the basic unit is no longer the individual but a population.

In such case, the fundamental criterion is the possibility of inter-breeding. However, the reproduction of algae is not the same as the reproduction of other organisms. We may wonder what happens if the temperature varies, and whether the type of algal reproduction varies also. We may wonder if a modification of the environment corresponds to a modification of the species and, in this case, if it is still the same species.

The "phylogenetic" classification considers that a natural group shows a common "ancestor-species", but the number of generations which occur and the evolution are uncertain. If it is true that classification cannot be based directly on phylogeny, it is nevertheless certain that an identical structure supposes a descent from a common ancestor. Thus, a group tracing its origin to a particular ancestral species will always present correlative characteristics in its members.

Local complications must be envisaged: populations descending from the same, but geographically distant species, will not be slow in revealing differences of varying importance, which will lead to new or geographical species being envisaged.

These difficulties may easily be foreseen in proposing a classification of living forms based on morphology, anatomy, cytology, cytochemistry, and ontogeny, whereas the paleontologist can only use morphology and a few other elements. In his case, the classification of species cannot be based on the evaluation of differences between individuals but may succeed by a quantitative study of morphological variations. According to Tintant (1966), we must therefore compare the type with the population, and not the population with the type, for the determination of the species.

In the case of algae, a statistical examination of a population would seem to be rather difficult to carry out, due to the incomplete fossilization.

The reproductive systems of the red algae are complex and the same plant in its sexual (male or female) or asexual form can give morphologically similar individuals, but, in addition, the alternation of the generations is not obligatory.

We must, therefore, consider which criteria are to be selected for the determination of red algae. This choice will play a part in the phylogenetic research.

The external morphology would not appear to be entirely satisfactory as a criterion, because it is not always available and can be sensitive to environmental conditions. On the other hand, the major criterion would indeed seem to be the study of tissue, of the internal morphology, of the arrangements of cells, or their size . . . Other arguments would support such a choice, in particular the fact that a case must be foreseen that algae whose reproductive organs are external and (or) non-fossilizable, or simply unrecognized, would be eliminated from our classifications.

We may remember that the same specific name is given to a sexual (gametophyte) or asexual (sporophyte) form, which means that they have similar isomorphic morphologies and we should then avoid becoming entangled in the intricacies of the reproduction of fossilized forms. This does not imply that we must ignore the examination of the reproductive systems, which, in certain quite frequent cases, will involve uncertainty between different species, or even between different families.

In conclusion, it must be admitted that the species, such as it may be envisaged amongst the red algae, is a rather doubtful notion, and that the family, which we consider here as an essential category, does not always appear to have very solid foundations.

18.2.1 The Corallinaceae

The Melobesioideae and the Corallinoideae are calcareous and consequently easily fossilizable; the Schmitzielloideae, which are non-calcareous, are practically unknown in the fossil state.

The Corallinaceae have closely packed, sectional rectangular cells, arranged in rows or threads, forming a hypothallus, a perithallus, or even a mesothallus.

The asexual reproductive organs are sporangia:

contained in conceptacles which are closed and limited by a wall: the roof is pierced by one or several orifices (*Lt, Meso, Lp* . . .)

developed in or partially immerged in the perithallic tissue, each sporangium emits its spores by an orifice situated in its upper part (*Ar* . . .).

Only *Agardhiellopsis*, which does not belong to the Corallinaceae, has spores shooting within a sorus which is not limited by a wall.

The sexual reproductive organs (male conceptacles with antheridia and female conceptacles containing first the oogonia, and later the spores resulting from the germination of the egg) are of no use for the determination of algae.

This large family quite certainly originated before the Carboniferous, but literally exploded at the Lower Cretaceous.

Two sub-families may be distinguished:

1. Melobesioideae. Nodular or ramified, non-articulated algae in thick or thin crusts. The tissue is composed either of a basal hypothallus whose cells are parallel to the substratum or else of a medullary hypothallus and a perithallus with cells perpendicular to the substratum (*Ar, Lp, Lt, Melo, Lith, Meso, Kym, Para, Hem* . . .);
2. Corallinoideae. Branching, articulated algae with reproductive organs in conceptales situated on the external parts of the joints. The cells are generally in concentric rows (*Cor, Am, Ja* . . .).

18.2.2 The Solenoporaceae

The Solenoporaceae represent an extinct group of marine organisms (Cambrian to Tertiary), which had its climax during the Jurassic.

The organisms are nodular or encrusting, formed of closely packed threads of vertically divergent cells, with polygonal sections (solen = Tube) and reproductive organs which have only very rarely been recognized and then are still uncertain. The cells are much bigger than those of the Corallinaceae. It should be noted that the nodular forms seems to be the oldest and would have been replaced by the ramified forms. Moreover, the arrangement of the concave walls is curious in that filament protrudes from divided cells.

The Solenoporaceae lived in association with the Tetracorals and Tabulata, which are very different from the later Hexacorals from the ecological point of view. The Solenoporaceae were ill-adapted to their environment and were gradually replaced by the Corallinaceae during the Cretaceous.

It should be noted that in the Corallinaceae the conceptacles are in well-defined cavities, which does not seem to be the case for the Solenoporaceae which have a fusion of sporangia and did not form conceptacles. The size of the reproductive organs would seem to depend on the number of combined sporangia. Thus we have a case of asexual reproduction with the Solenoporaceae. This is of the greatest interest for, as Lemoine remarks, "in the types with asexual reproduction without the formation of conceptacles, only the presence of sexual conceptacles may be attributed to the Corallinaceae family" (1971).

It has been suggested that the Solenoporaceae had sufficiently large cells (*Lithoporella* also had large cells, but nevertheless belong to the Corallinaceae) for the reproduction of spores to take place without the need of favorable, noncalcified external location. It is only occasionally that these algae had an original reproductive system without this being at the same time an evolutionary advantage.

Finally, we may note that it is very difficult to differentiate between hypothallus and perithallus. This observation, together with the type of reproduction, the absence of branching, and the stratigraphic place are arguments in favor of their being the probable ancestors of the Corallinaceae.

The classification is as follows:

Solenopora, alga with thin horizontal walls, widely spaced or even absent;

Pseudochaetetes, alga with thick horizontal walls, frequently and irregularly spaced;

Parachaetetes, alga with thick horizontal walls, numerous, regularly spaced and giving the tissue the appearance of a meshwork.

18.2.3 The Gymnocodiaceae

The term Gymnocodiaceae recalls the first attribution, that of the Codiaceae. In fact, the algae for which this name has been retained have nothing in common with the real Codiaceae.

All the forms included in the Gymnocodiaceae have now disappeared. Fossilized specimens are quite frequent, with thalli either segmented or not. The degree of calcification is very variable. The sporangia were interna, within the terminal segments. Some Gymnocodiaceae have the same characteristics as the present day *Galaxaura*, which are algae of the tropical seas.

18.2.4 The Squamariaceae

In 1969, Lemoine summarized our knowledge on this group as follows: "In this family, the reproductive organs are situated on the surface and are not protected by conceptacle walls; their conservation in the fossil state is rather unlikely. The structure generally appears more variable within one thallus than within that of the *Melobesia*; for these reasons, this family does not seem to be of any great use for stratigraphy".

The thallus is formed of a basal hypothallus, overlaid by a perithallus. It is fixed to the substratum by rhizoids.

Very numerous Palaeozoic, Mesozoic, and Cenozoic red algae exist, having uncertain affinities. They are certainly of stratigraphic and phylogenetic interest.

18.3 Phylogenesis

Before touching the complex and delicate problem of the phylogeny of red algae (Poignant, 1974), a few general problems should be considered first.

It now appears evident that, during the course of time, the morphology of organism is progressively modified. This correct, but perhaps simplified, conception of evolution leads us to believe that all living creatures derive from one single primitive species. For the geologist, appearances are different: due to unexplained phenomena, i.e. gaps of all types existing in every geological cross-section, the impression is one of discontinuity, perhaps artificial, which leads to an impression of irregularity in evolution. It may be expected that the evolutionary movement is essentially irregular and that it advances at varying speeds. We may finally remember that the biochemical methods used in phylogenetic research on living organisms cannot be applied to fossils.

It has often been said that megaevolution (appearance, development, and extinction of new groups) or phylogeny is the prerogative of organisms with sexual reproduction and that if these living creatures abandon this means of reproduction, the effect would be as if they had excluded themselves from it definitively (Thaller, 1972). Algae therefore pose a particular problem.

It may be considered that after a rapid phase, when a type establishes itself (Typogenesis of diversification), it then becomes stable (Typostasy or "acme"), and later disappears (Typolysis or "decline"). However other possibilities may exist (Lehman, 1962):

Anagenesis: appearance and transformation of the characteristics. The rate of anagenetic evolution is that of the morphological transformation;

Cladogenesis: multiplication of the number of lines of issue. The rate of cladogenetic evolution is then that of the appearance of sub-groups;

Orthogenesis: directed evolution. Successive changes are made in the same direction.

It would seem quite obvious that in the natural state, species are adapted to their environment, which, logically, brings us to the conclusion that if numerous species have become extinct, they were not sufficiently adapted. It may also be noted that never single characteristic is modified, but always several.

Briefly, it is difficult to understand the means because of the duration of the phenomenon and, moreover, other complications intervene, i.e.:

Speciation, which is the creation of distinct species, due to ecological and (or) biological isolation. There may exist at the same time a high rate of speciation and a low rate of cladogenesis, i.e. the species which appear compensate for the species which become extinct. It may be emphasized that if certain biotypes are devoid of representatives, an adaptation of the species may take place; if the biotypes are represented, then the conquest will only take place if the adaptation is sufficiently developed.

Parallelism, which corresponds to the acquisition by different related evolutionary lines, of a characteristic which the common ancestor did not possess.

To this panoramic view of the problem may be added the law of recapitulation which states that ontogeny is the rapid and shortened recapitulation of phylogeny and seems to tend towards a goal; that of the creation of a viable individual. We may immediately cite Cabioch (1966), who notes that the first stages of spore development of the Corallinaceae are curiously identical for the *Lt, Lp, Meso, Cor,* and *Jan*, which apparently do not appear on any genealogical tree.

The law of internation states that the organs situated on the exterior penetrate gradually deeper into the organisms during evolution and are covered by other formations. But as with the preceeding law, its application to algae poses a problem of principle, since the examples are chosen from amongst the higher organisms. A contrary hypothesis may be formulated, whereby the internal reproductive elements of the body of the green algae migrate towards lateral sporangia.

Finally, as an additional complication, plants often produce hybrids and these, when sexually isolated, may be the starting point of a new species.

The problem will become even more complex in the case of the study of red algae, if only because of the mediocrity of the anatomical criteria, which are not sufficiently marked and which are not always enough for the definition of the species. We may state immediately that we are obliged to confine ourselves to the typological species, since easy access to the population is not possible. Thus, the species is defined on a narrow basis and where there are numerous species or numerous individuals it is very difficult, if not impossible, to detect the direct connection between the species. It is evident that in any attempt to create a genealogical tree, the branching will not be clear, as sufficiently adequate fossils are lacking.

In defining the groups and placing them on the phylogenetic tree, we are led to make value judgements such as "primitive character, advanced, simple, complicated character, etc. . .". However a more critical examination will abolish lapidary formulations: "the simple is older than the complex, evolution has a goal, ontogenesis recapitulates phylogeny, evolution takes place in succeeding jumps, etc. . . ".

We might also say that classic phylogeny looks towards the common ancestor, while modern phylogeny must ask why the descendants of one common ancestor are different. In a simple, but apparently logical way, the characteristics of the ancestor must be compared with those of the descendants.

18.4 Phylogeny of Certain Red Algae

It was in 1894, that Brown, for the first time, suggested a phylogenic connection between the Solenoporaceae and the Corallinaceae. It is most probable that essentially two lines of descent evolved from one common, unknown, and probably Precambrian ancestor:

One line would correspond to the Solenoporaceae; there are numerous types and species in the Palaeozoic, but they became rarer after the ecological changes of the beginning of the Mesozoic, not having been able to adapt. These algae, which appeared in Antarctica and Siberia (present-day position) separated into *Solenopora* and *Parachaetetes* in the Silurian. They disappeared on the Mesogean coasts.

The other line would correspond to the Corallinaceae; they were curiously in reserve throughout the Palaeozoic waiting for the decline of the Solenoporaceae to be well-marked before invading their ecological biotope, beginning on the Mesogean coasts. We may note that a "test" took place in the Permo-Carboniferous with highly evolved forms, but it lasted only a very short time.

We may already conclude that the evolutionary lines of descent must be studied by the comparison of tissue, of by that of the reproductive organs, or else by that of the external morphology of different algae.

18.4.1 Comparison of External Morphologies

If we first examine external morphologies, we may note, as a certain number of authors did, that the first Solenoporaceae were tuberculated. These authors identified the tubercules as the equivalent of ramifications. The same is true for the Corallinaceae, whose first representatives are ramified and which later evolve towards crustose forms.

For example, Maslov (1956) believed that the crustose forms were later and more specialized than the ramified forms. However, the same author notes that the ramified forms are also highly specialized, and that they occupy a higher level than the tuberculated forms.

These remarks, which are correct but difficult to put in order, should be remembered and, moreover, we may note that Maslov (1956) considered that *Lithothamnium* and *Archaeolithothamnium* became separated from the stock of "*Solenopora*" at the Triassic-Jurassic boundary, with *Mesolithon* being taken as the direct descendant. This opinion does not respect the preceding point of view, since there existed at the same time both the ramified and crustose *Lithothamnium* and *Archaeolithothamnium*. Finally, the very ancient genus *Archaeolithophyllum* exists both in branches and in crusts, and this completely changes the problem and puts it back into the Palaeozoic.

Thus, it is impossible to come to any logical conclusions on the basis of the examination of the crustose and (or) ramified forms only.

The relation of the articulated forms with the other forms has held the interest of scientists. Thus, the crustose forms can be considered as the forerunners of the articulated forms. From the stratigraphic point of view, the reasoning holds good because in effect the articulated forms appear *after* the other types they are therefore perhaps the end of the evolution.

On the other hand, if we believe *Archamphiroa* to be the eldest known form of the Corallinoideae (Callovian), the reasoning is no longer valid, since the massive appearance of the crustose forms appeared in the Lower Cretaceous. We should, therefore, note the existence of *Archaeolithoporella* (Permo-Carboniferous), the first crustose form, which may be attributed to the monostromatic Corallinaceae. Here again the problem must be solved during the Palaeozoic since it cannot be solved in the Mesozoic.

Thus, the hypothesis of complicated forms evolving towards simple forms cannot be wholly accepted anymore, strangely enough, than can that of an evolution from the simple to the complicated.

18.4.2 Comparison of Tissue

From the examination of tissue it is generally admitted that *Lithothamnium* descends from *Solenopora* (aspect of cells in vertical lines) and that *Lithophyllum, Mesophyllum,* and *Archaeolithothamnium* descend from *Parachaetetes* (cells in rows).

This extremely simple representation has the great advantage of being logical. It is certainly insufficient and does not settle all the questions. Some authors have made different proposals and submit that *Archaeolithothamnium* descends from *Parachaetetes* and produces *Lithothamnium* and *Mesophyllum*. Other scientists believe that *Archaeolithothamnium* comes from *Solenopora* and produces *Lithothamnium*.

Nevertheless, some of these speculations should be immediately discarded and in particular that which would have *Archaeolithothamnium* as the ancestor of the Corallinaceae. This hypothesis is false and unrealistic.

Certain phylogenetic hypotheses have been suggested. Two of these are worthy of consideration:

Pfender (1926) made a detailed study of the *Archaeolithothamnium* of Provence and suggests two lines of descent, or more accurately, a hypothesis on the filiation of the species of Provence:

by *Ar. turonicum, Ar. dehornae,* . . . we obtained *Lithothamnium,*

by *Ar. hippuritorum, Ar. Penicillum,* . . . we obtain *Lithophyllum.*

This hypothesis cannot be accepted because *Lithophyllum* and *Lithothamnium* could only appear at the end of the evolution, i.e. at the Upper Senonian, which is false.

Maslov (1956) proposed a hypothetical relationship between certain species of *Archaeolithothamnium*, starting from *Ar. rude* (which is stratigraphically correct). This evolution would give:

a group whose perithallic filaments become wider;

a group whose perithallic filaments retain a constant width.

We may note that the increase in width in the first group is not very noticeable and that the constancy of width in the second group is not evident.

Finally, we may note that Pfender's and Maslov's concepts cannot be accepted, due to the fact that the criteria chosen are too imprecise and not very significant.

Maslov also proposes a hypothetical relationship between certain species of *Lithophyllum*;

forms whose perithallic filaments become wider;

forms whose perithallic filaments become narrower.

The arguments in favor of this hypothesis do not seem any more convincing than the preceding ones.

Once more, Maslov proposes a hypothetical relationship between certain species of *Lithothamnium*:

a strain of forms with perithallic filaments of constant width;

a strain whose perithallic filaments become wider;

a strain whose perithallic filaments become narrower.

In the last case, Maslov's hypothesis seems to be confirmed by observation, except for a few details.

We have then an original attempt to find a hypothetical relationship between certain species of *Lithothamnium, Archaeolithothamnium*, and *Lithophyllum* by the examination of the width of the perithallic cells. An objective and accurate study cannot confirm this point of view; the differences in measurements are too great or, on the contrary, too insignificant. Weak points of the hypothesis are the proposal of filiation for 33 species of *Lithothamnium*, which represents no more than a fifth of the total, and the choice of certain measurements, e.g. a difference between 6–13 $\mu$and 7–12 μ would appear illusory. Finally, the stratigraphic attributions are far from certain and the palaeoecology has not been taken into consideration, although certain factors are important in the determination of the size of cell.

These hypothesis cannot, therefore, be accepted, but at least they have the merit of existing since only a few authors have had the courage to make this kind of proposal.

18.4.3 Comparison of the Reproductive Organs

The phylogeny of certain red algae may also be tackled by the study of their reproductive systems.

Without coming back to the question of the reproductive systems of the Solenoporaceae, we must immediately note that for these algae there are still many unknown, badly defined, or imprecise facts.

In contrast, there is more accurate and abundant information for the fossilized Corallinaceae.

There are two types of asexual reproduction:

by the presence of conceptacles: *Mesophyllum, Lithophyllum;*

by the absence of conceptacles: *Archaeolithothamnium, Kymalithon, Hemiphyllum, Paraphyllum, Agardhiellopsis.*

Sexual reproduction takes place by the conceptacles, and, according to Lemoine "the only form of Melobesiae where sexual reproduction is unknown is *Lithoporella*". The isolated sporangia embedded in the tissue probably reunite as they penetrate the tissue and this merging leads to sori and, better still, to conceptacles with several pores. The number of pores diminishes and the conceptacles take on a curved form, finally becoming monoperforated. It is difficult to imagine the contrary and this is probably the origin of the notion that *Archaeolithothamnium* is the ancestral form.

Stratigraphically, the monoperforated conceptacles seem to be the first to appear. Also, ecological modifications have probably influenced the morphology or the type of reproductive organs, although the means remains unknown.

It should be remembered that Pfender believed that there are two groups of *Archaeolithothamnium*, one evolving towards *Lithothamnium* and the other towards *Lithophyllum*. But these two lines of descent have gone through a stage in which the conceptacles had numerous exiting pores. It should also be emphasized that conceptacles have been recognized with certainty as early as the Permo-Carboniferous (*Archaeolithophyllum, Archaeolithoporella*).

These different considerations would seem to be in opposition; the lines of descent envisaged by the study of tissue are different from those suggested by the study of reproduction. . . Again, according to Lemoine, if the development of conceptacles and reproductive organs is the important criterion, it would be logical to state that *Lithothamnium* descends from *Archaeolithothamnium* and *Lithophyllum* from *Lithothamnium,* but if we consider tissue development, *Archaeolithothamnium, Mesophyllum, Lithophyllum* descend from *Parachaetetes* and *Lithothanium* from *Solenopora,* but it is still true that *Lithophyllum* could descend from a common ancestor, independently of *Archaeolithothamnium*, " . . . by their structure, *Paraphyllum* and *Hemiphyllum* could be connected with *Lithophyllum*; whilst their asexual reproductive system separates them from it completely" (Lemoine, 1969).

This, then, is our dilemma: is, from the phylogenetic point of view, the tissue more important than reproduction? Foslie believed that reproduction was the essential criterion, whilst Lemoine preferred tissue. The main disadvantage is that the lines of evolution obtained by either means give contradictory results, and on the very rare occasions when there seems to be a degree of agreement, then this is generally opposed by the stratigraphy. This, then, is the problem of the phylogeny of the red algae.

18.4.4 Ontogenesis and Morphogenesis of Present-Day Algae

Taking into account the preceeding remarks, it would seem imperative to consider the ontogeny and the morphogenesis of present-day algae, even if the criteria used by algologists have very little in common with those of the paleontologists.

Cabioch (1972) has reconstitued and classified the stages of the realization of the structures:

A certain number of combinations between the morphogenetic and ontogenetic characteristics may be distinguished. Several tribes can then be defined from these combinations. Within these tribes, Cabioch has noted that the adult structure of certain present-day forms corresponds to the juvenile stages of forms with a more complex adult structure. She, therefore, proposes that these tribes should be interpreted as parallel evolutionary lines, whose form would be stages, but without the most simple structure being necessarily the most primitive.

Cabioch believes that evolution is seen in the middle types, either as a regression (producing forms with persistent juvenile characteristics) or as an advance (leading, by stages, from simple crustose thalli to ramified or even articulated thalli).

Cabioch also notes that, although *Sporolithon (Archaeolithothamnium)* is not the oldest form known in the fossilized state, and although its structure is relatively evolved, we may suppose that it belongs to an archaic type because of its cytological particularities.

18.5 Proposed Phylogenetic Schema (Fig. 1)

It would seem interesting, or even necessary, to propose a schema, while recognizing that any such attempt will require criticism or modification.

This hypothetical schema is based on time (stratigraphy), the nature of tissue (cells in lines or in rows), and the structure of the reproductive organs (isolated sporangia, sori, and multiperforated/monoperforated conceptacles). These are the only arguments available to the paleontologist.

We may recall that there are only two forms which have cells arranged in lines (*Solenopora* and *Lithothamnium*), all the other forms having cells in rows. In the case of the reproductive organs, naturally only asexual reproduction can be envisaged here and in no case sexual reproduction.

18.6 Conclusions

The most interesting feature of the red algae is their phylogeny. We are obliged to admit that with the present state of our knowledge the phylogenetic factors are not fully understood. It is difficult to explain the extraordinary proliferation of forms and individuals during the Lower Cretaceous, as apparently they were "not prepared". It is interesting to note that the processes of evolution seem to be quite comparable, at least on a large number of points, to those of the Angiospermes and Dasycladaceae during the same period.

The many remaining questions have their solutions perhaps in the Palaeozoic, although the inventory of the fossil forms is still very incomplete. We must hope, that one day this description of fossilized red algae may be completed.

18.7 Bibliography

Brown, A.: The structure and affinities of the genus *Solenopora* together with descriptions of new species. Geol. Mag., Ser. 4, **1**, 154–161; 195–283 (1894).

Cabioch, J.: Sur le mode de développement des spores chez les Corallinacées. C.R. Acad. Sci. (Paris) **262**, 2025–2028 (1966).

Cabioch, J.: Etude sur les Corallinacées. Croissance et mode de réalisation du thalle; conséquences systématiques et phylogénetiques. Cah. Biol. Mar. **12**, 121–186; **13**, 137–287 (1972).

Lehman, J.P.: Paléontologie et théories modernes de l'évolution. Année Biol. **1**, fasc. 7–8 (1962).

Lemoine, M.: Les Algues floridées calcaires du Crétacé du Sud de la France. Arch. Mus. Hist. Nat. (Paris), (7) **10**, 130–240 (1969).
Lemoine, M.: Remarques sur la reproduction des Algues calcaires fossiles: Melobésiées; la systématique, la phylogénie. Rev. Algol. **10** (2), 152–160 (1971).
Maslov, V.: Algues calcaires fossiles de l'U.R.S.S. Acad. Sci. U.R.S.S. **160**, 296 p. (trad. BRGM, France) (1956).
Pfender, J.: Les Mélobesiées dans les calcaires crétacés de la Basse-Provence. Mém. Soc. Géol. France, N.S. **3** (2), no. 6, 32 p., 10 pl. (1926).
Poignant, A.F.: A propos de la phylogénie des Algues fossiles. C.R.Acad. Sci. (Paris) 1711–1714 (1974).
Thaller, L.: Phylogenèse. Encycl. Univ. (1972).
Tintant, H.: La notion d'espèce en paléontologie. Mises a Jour (Paris) **1**, no. 3, 274–294 (1966).

19 A Comparison of the Anatomical Structures of a Recent and a Fossil Species of the Corallinaceae

M. MASSIEUX, Montpellier, France

19.1 Introduction

In 1970 Elliott gave a detailed description of a calcareous alga (*Rhodophyta*, of the family Corallinaceae), forming protuberance-covered nodules, from early and middle Miocene of South-West Iran. With an anatomy similar to a previously described genus, *Aethesolithon* Johnson, he gave it the name *Pseudaethesolithon iranicum*. This form is characterized by a monostromatic hypothallus with prismatic, oblique cells on the support, followed by a perithallus made of elongated, polyhedral cells disposed in superposed and juxtapositioned lenses which constitute the protuberances of the algal nodule.

Oriented sections made from fragments of non-decalcified protuberances from a specimen of *Hydrolithon reinboldii* identified by Foslie (1901) from the Rikitea, Tuamotou Archipelago (French Polynesia), showed that this corallinaceae possessed the same anatomical structure as the alga from the Miocene. The material, from the Seurat legacy, deposited in the Cryptogamy Laboratory of the Paris National Natural History Museum, was kindly lent by Madame M. Lemoine, to whom I owe particular thanks.

This form was mentioned for the first time in 1901, by Weber van Bosse and Foslie, under the name *Lithophyllum reinboldi*, with a short description without figuration. In 1904 Foslie (in Weber van Bosse and Foslie) designated it *Goniolithon reinboldi* Weber van Bosse and Foslie. The author enlarged upon the preceding description and gave illustrations. In 1905 he divided the genus *Goniolithon* to include a subgenus, *Hydrolithon* Foslie, for the species such as *reinboldii* which are provided with a hypothallus composed of a single bed of cells. Then in 1909, he elevated the subgenus *Hydrolithon* to the rank of genus with four species, including *H. reinboldii* Weber van Bosse and Foslie, and completed then the spelling of the specific name. In 1929 the external aspect of this alga figured under the name of *Goniolithon reinboldii* Weber van Bosse and Foslie., in a posthumous memoire written by H. Printz.

In 1911 Lemoine described and figured it under the name of *Poroliton reinboldi* Weber van Bosse and Foslie, whereas in 1953 Mason chose *H. reinboldii* (Weber van Bosse and Foslie) as type species of the genus *Hydrolithon*, a choice accepted by Adey in 1965, and in 1970, under the denomination of *Hydrolithon reinboldii* (Weber van Bosse and Foslie, 1901; Foslie, 1909).

19.2 Comparison between *H. reinboldii* and *P. iranicum*

19.2.1 External Aspect

Same external aspect of the specimens of *H. reinboldii* figured by the authors and of *P. iranicum* figured by Elliott (Figs. 1 and 2): nodule more or less round, only presenting the rounded extremities of the protuberances.

19.2.2 Internal Structure

A. Hypothallus. Similar to *Pseudaethesolithon* (Elliott, Pl. 2, Figs. 1 and 2), *H. reinboldii* adheres to a support (Figs. 2–5) by an unistratified hypothallus with monoclinic prismatic cells, rectangular base (Fig. 3), vertical longitudinal walls perpendicular to the support and to the perithallus (Fig. 4), and oblique transversal walls (Fig. 5); a disposition already remarked upon by the preceding authors.

Dimensions of the hypothallus cells of the living alga (in mm): height: 0.022 (0.026 –0.036) 0.048; width: 0.015 (0.017) 0.022; thickness: 0.008–0.013;

Dimensions given by Elliott: length: 0.036–0.045; width: 0.014–0.018

B. Perithallus. Above the hypothallus, the latter constitutes practically all of the alga in *H. reinboldii* (Figs. 1 and 2), as in *P. iranicum* (Pl. 2, Fig. 3), and in particular nearly all of the tissue of the protuberances. It shows no differentiation into medullary and peripheral tissues. It is made up of a juxtaposition and a superposition of lenticular zones composed of three thicknesses of cells in *H. reinboldii* (Figs. 1, 2, 4, 7, 8, 10–13), of two thicknesses of cells in *P. iranicum* (Elliott, Pl. 1, Figs. 1, 3, and Pl. 2, Figs. 1, 3).

The thickness of the lenses in *H. reinboldii* averages from 0.070 to 0.085 mm with an extreme value of 0.098. The length is very variable, according to the situation of the section on the lens. It was not possible to observe, with the limited material available, marked interruptions and resumptions of growth which could suggest phases of rest and periodic and rhythmic renewal of activity by the meristem. Annual growth seems to be indicated by several thicknesses of lenses, since generally, the conceptacles are enrobed by at least 3 to 4 thicknesses of lenses (Fig. 8).

The perithallial cells have lateral vertical walls that are intermeshing (Figs. 1, 2, 7) in *H. reinboldii*, but with very irregular polyhedral transverse sections (Fig. 6), a disposition closely comparable to that given by Elliott for *Pseudaethesolithon* (Pl. 1/1–3).

The perithallic lenses of the living alga are constituted in depth by 3 beds of large cells. On the surface, they have a layer of small cubic, cortical cells, also calcified from 0.010 (0.011) 0.017 mm in size, the development of which during growth was not possible to follow. They are covered by epithallial cells of 0.011 mm in width and 0.004 mm in height, and whose lateral and upper walls are not calcified (Fig. 12). These characteristics of the surface of the perithallus are clearly indicated by Weber van Bosse and Foslie in 1904, Fig. 21, and also this drawing does not indicate, in depth, the presence of cubic, cortical cells.

The dimensions of the big perithallus cells of *H. reinboldii*, in longitudinal section, are the following:

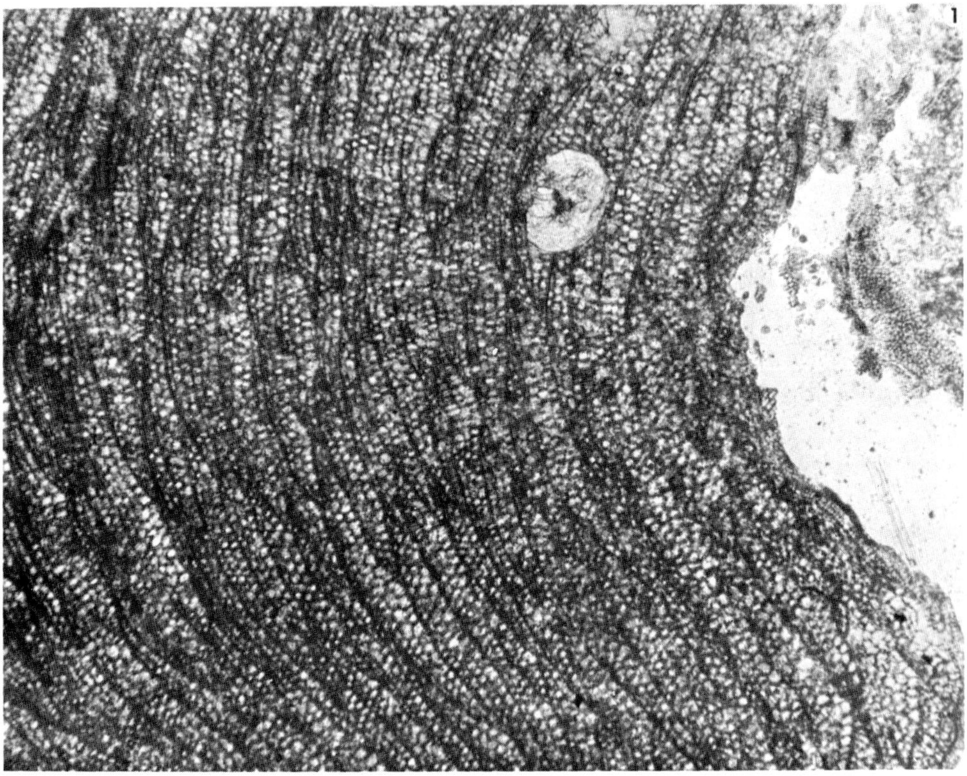

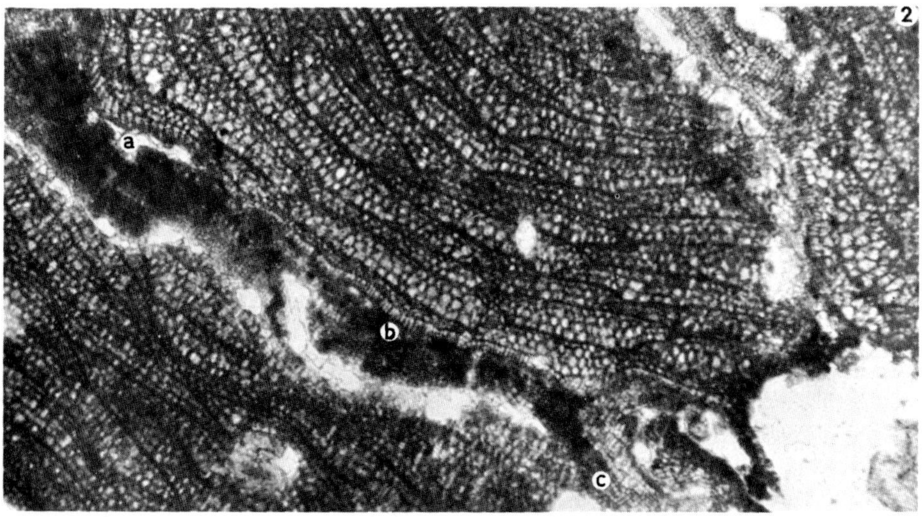

Figs. 1 and 2. Legends see opposite page

First layer (base of the perithallial lense) (in mm): height: 0.022 (0.031) 0.063; width: 0.013 (0.022) 0.063

Second layer: height: 0.020 (0.022) 0.044; width: 0.013 (0.022) 0.031

Third layer: height: 0.017 (0.022) 0.039; width: 0.013 (0.017) 0.023

These measurements were taken in the thickest part of the lens. The size of these cells diminishes towards the periphery. It seems that there was a decrease in size of the cells from the base to the summit. Finally, one must take into account cell fusion. The great differences observed in the measurements of cells in the same layer could be a result of the more or less marked importance of the lateral cellular fusion.

In transverse section, these cells show a very polymorphous polyhedral contour (Fig. 6), mor or less angular, whose maximum length from one angle to another can be as much as 0.040 mm. Between these polyhedral cells occur beaches of little rounded cells, of 0.009 mm diameter, which could represent either the small cells which constitute the lateral part of the lenses, or the small cells constituting the roof of the conceptacles.

Elliott gives as maximum dimensions for the large cells of the fossil alga a size range extending from 0.045 to 0.063 mm, and for the transverse sections, sizes as large as 0.063 mm. He also notes small cells of 0.009 mm in diameter.

At the surface of the perithallus sections of *H. reinboldii* support-cells of trichocytes or megacells (Fig. 13) occur, here and there, which do not occur in depth (dimensions of the support-cells, Fig. 13: height: 0.035 mm; width: 0.017 mm). These support-cells are inserted between the perithallial cells and cubic, cortical cells. Weber van Bosse and Foslie show them in 1904, text-fig. 21, under the name of heterocysts. Not observed in depth, they might be trichocytes with ephemeral megacells (Cabioch, 1971, pp. 174–181). They could not, in consequence, be seen in the fossil.

C. Conceptacles. Among all the conceptacles sectioned, only a single aperture has been observed for those possessing them. According to Adey (1965, 1970), Adey and MacIntyre (1973), and the previous authors who have studied *H. reinboldii*, the asexual conceptacles have only a single aperture. Whereas according to Elliott the asexual conceptacles of *P. iranicum* would be of the *Lithothamnium* type, with numerous openings.

Dimensions of the conceptacles (measurements taken on conceptacles with an aperture (in mm): height: from 0.125 to 0.254; width: from 0.159 to 0.275; diameter of the conceptacles in transverse section: from 0.212 to 0.233

◄ Fig. 1. Longitudinal section of a protuberance showing, at the base, the monostromatic hypothallus, then the perithallus formed by superposed and juxtapositioned lenses of perithallial cells, constituting by it self the protuberance, and an included conceptacle. Preparation 7, x 52. Compare with Elliott, Pl. 1, Fig. 3

Fig. 2. This preparation shows the hypothallus fixed on a fragment of *Porolithon onkodes*, and showing the three principle sections of its cells: (a) longitudinal section with the transverse walls inclined on the support; (b) transverse section with the longitudinal walls perpendicular to the support and to the perithallus; (c) horizontal section showing their rectangular base. Figs. 3, 4, and 5, show the details of these three sections. This hypothallus, monostromatic, is followed by the perithallus forming lenticular beds. Preparation 2, x 52

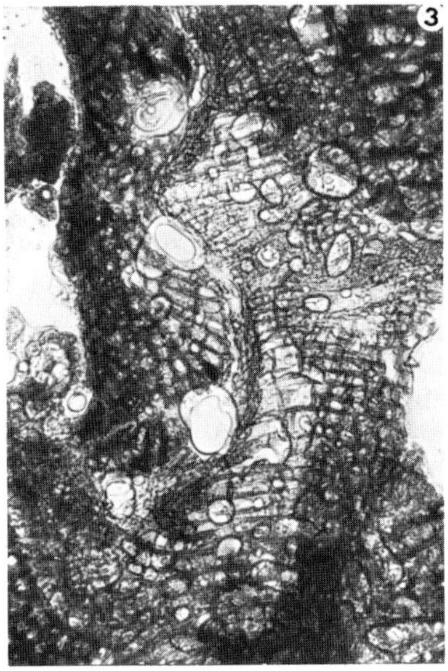

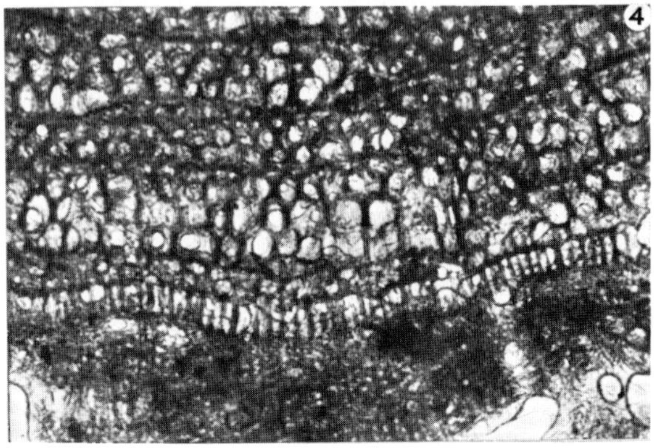

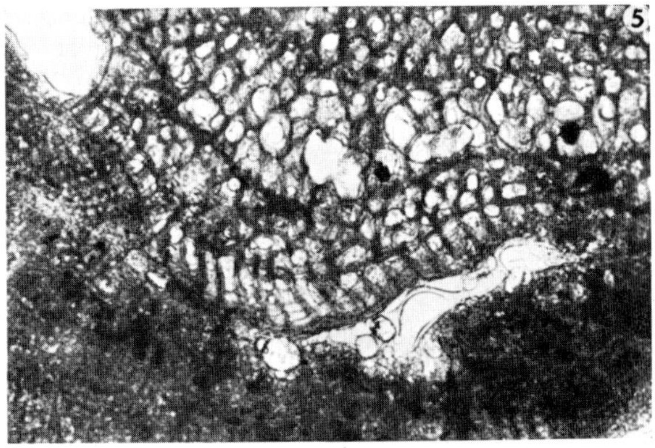

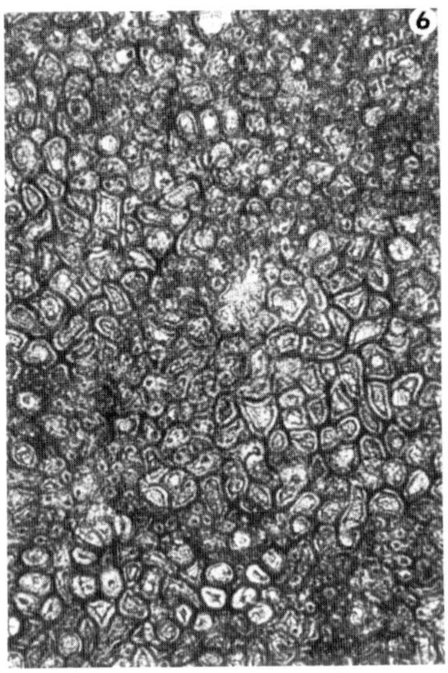

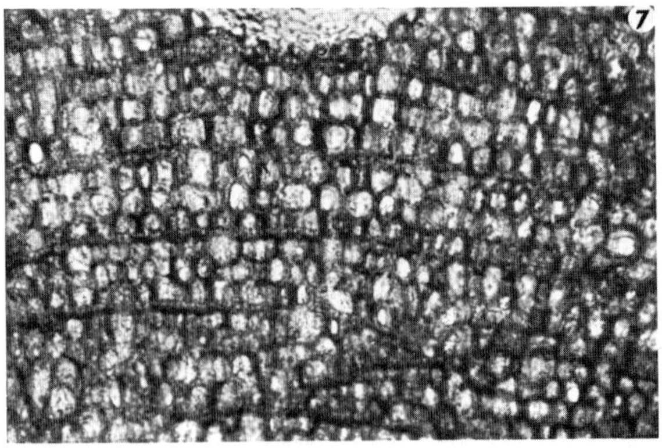

Figs. 3–7. Legends see opposite page

Dimensions of the spores emission tubes: length: from 0.032 to 0.063; width of the tube taken near the middle, the latter narrowing from the base to the summit: 0.021 (0.032) 0.041

Dimensions of the conceptacles of Pseudaethesolithon: height: 0.127; diameter: 0.236

On the preparation made, in certain cases, the conceptacles are disposed in concentric rows. Elliott notes similarly that several conceptacles can be disposed in rows in the fossil.

19.3 Conclusions

By this comparative study an attempt has been made to demonstrate the similarity existing between these two calcareous algae, which seem to belong to the same genus: same anatomical structure of the tissues, same degree of size of the different elements which compose them: hypothallial and perithallial cells and conceptacles.

It seems to be adequate, moreover, to regroup the genera *Hydrolithon* (type-species *H. reinboldii*), represented by its fossil form *H. iranicum,* and *Aethesolithon* Johnson 1964, in the tribe of the *Hydrolithoneae*, family Corallinaceae, subfamily Melobesioideae, characterized by: (1) an unistratified hypothallus, (2) a very irregular perithallus with large cells presenting cellular fusion, (3) asexual conceptacles with a single aperture, (4) trichocytes with ephemeral megacells.

This tribe is made up of 2 genera: Genus *Hydrolithon* Foslie 1909, type species *reinboldii,* including living and fossil species: *H. iranicum* (Elliott, 1970) Massieux 1975, from the early and middle Miocene of Iran.

Genus *Aethesolithon* Johnson 1964, characterized by protuberances presenting a medullary perithallus and a peripheral perithallus, with two fossil species, but which could include living species not yet inventoried: *A. problematicum* Johnson 1964, type species of the genus, from the early Miocene of Guam, and *A. grandis* Johnson 1964, from late Miocene-Pliocene of Guam.

◀ Figs. 3–5. Preparation 2, details of the hypothallus and of the beginning of the perithallus of the Fig. 2 section

Fig. 3. Horizontal section showing the hypothallus as an unistratified bed, with rectangular based cells as well as a marginal lobe.

Fig. 4. Transverse section of the hypothallus, showing the longitudinal walls of the cells, perpendicular to the support and to the perithallus, as well as the first perithallial cells

Fig. 5. Longitudinal section of the hypothallus, showing the transverse walls of the cells, inclined on the support, as well as the first perithallial cells. Compare with Elliott, Pl. 2, Figs. 1 and 2

Fig. 6. Transverse section of the perithallus showing the very irregular polyhedral contour of the large perithallial cells, and beaches of little cells with very reduced light. Preparation 4. Compare with Elliott, Pl. 1, Fig. 2

Fig. 7. Vertical section of the perithallus, showing the perithallial lenses made up of three beds of cells. Preparation 7. Compare with Elliott, Pl. 1, Fig. 1.

All sections x 145

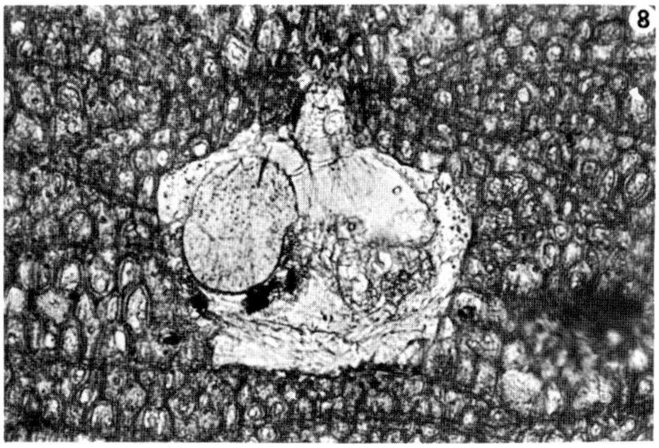

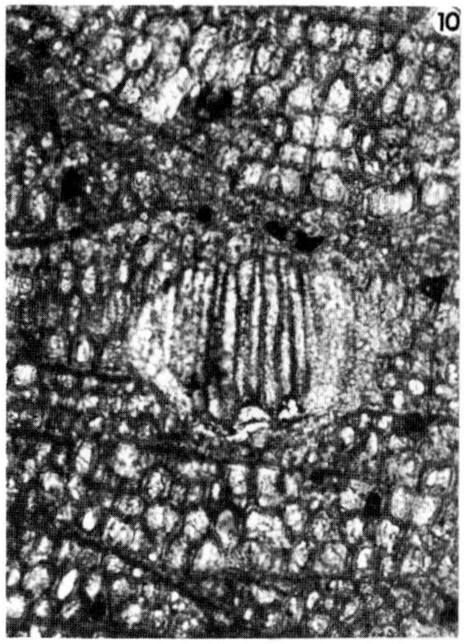

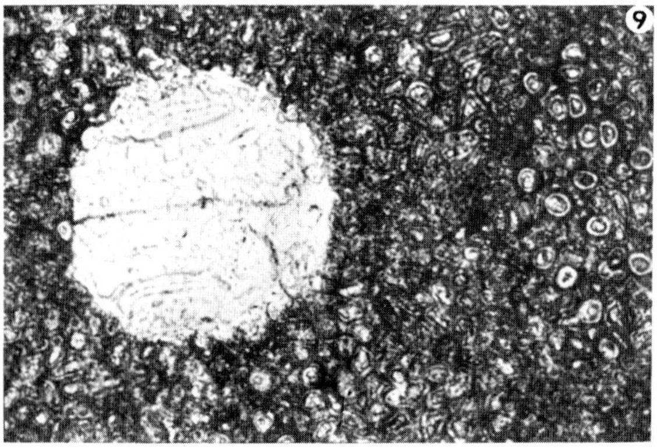

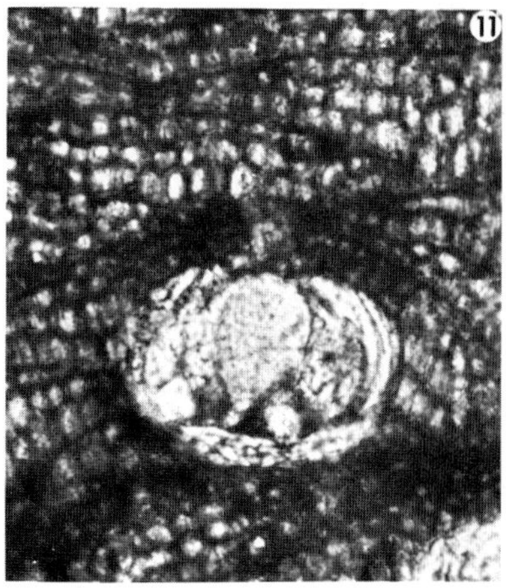

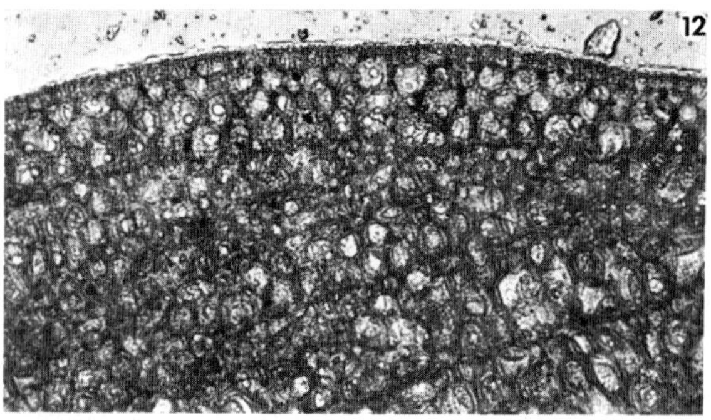

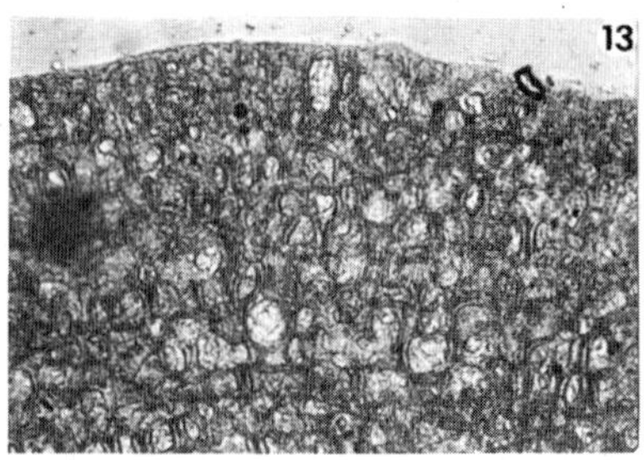

Figs. 8–13. Legends see opposite page

Two other of Johnson's species seem to belong to the genus *Hydrolithon*: these are: *Aethesolithon guatemalaensum* Johnson 1965, from the early Miocene of Guatemala, a species which shows only fructified crusts, without protuberances differentiated into a medullary and peripheral perithallus, as well as *Lithothamnium heterothallum* Johnson and Ferris 1950, from the Miocene of Lau, Fiji Islands, which Johnson in 1964 referred to the genus *Aethesolithon*, and which equally does not show protuberances with a perithallus differentiated into medullary and peripheral parts.

It should be noted that Johnson knew *H. reinboldii*, which he cited and figured in some recent material from Guam, in 1964, under the name of *Goniolithon reinboldi* Weber van Bosse and Foslie.

The presence of *Hydrolithon*, as early as the Miocene, and of *Aethesolithon*, demonstrates the antiquity of the group. It follows that one cannot derive *Hydrolithon* from *Aethesolithon*, as was suggested, hypothetically, by Adey and MacIntyre in 1973.

Numerous points remain to be elucidated by the detailed study of the living forms in order to better understand the fossil species, in particular: (1) Hypothallus–perithallus relations, genesis of the perithallus and formation process of the perithallial lenses, evolution of the perithallial cells and the process of cellular fusion. (2) Development of the cubic, cortical cells during the growth of the alga. (3) Study of the reproduction, principally to remove the contradiction of multipore asexual conceptacles in *H. iranicum* and unipore asexual conceptacles in all the other species of *Hydrolithon* and in *Aethesolithon*; cyclic and periodic fructification, or extending throughout the year. (4) Study of other living species of *Hydrolithon* and search for living forms which might have a protuberance structure of the *Aethesolithon* type.

◄ Fig. 8. Longitudinal section of a conceptacle passing through the spores emission aperture. The conceptacle occupies three thicknesses of perithallus lenses. The opening of the aperture is centered in the middle of a lense of perithallial tissue. Preparation 1. Compare with Elliott, Pl. 1, Fig. 3, and Pl. 4, Fig. 3–6

Fig. 9. Transverse section of a conceptacle showing its globular aspect. Preparation 5. Compare with Elliott, Pl. 4, Fig. 1

Fig. 10. Longitudinal section of a conceptacle showing very probably some sterile cells whose walls seem calcified. Preparation 6. Compare with Elliott, Pl. 3, Fig. 2

Fig. 11. Longitudinal section of a conceptacle, showing very probably, in the center, a sporangia, and at the side, elongated sterile cells whose cellular content has been replaced by a clear calcite which brings out the dark walls. Measurements of the sporangia section: height: 0.095 mm; width: 0.074 mm; height of the pedicel: 0.021 mm. Preparation 6

Fig. 12. Longitudinal section of the surface of a protuberance showing the three layers of perithallial cells, and a layer of cubic, cortical cells with calcified walls, and very flattened epithallial cells with non calcified lateral and external walls. Preparation 2

Fig. 13. Longitudinal section of the surface of the thallus showing the base of a trichocyte or megacell, inserted between the perithallial cells and the cortical cells. Preparation 1

All sections: x 145

19.4 Bibliography

Adey, W.H.: The Algal Tribe Lithophylleae and its included genera. In: Studies of *Lithophyllum* and related genera. Part II, Quart. Colorado School Mines **60** (2), 66–102, 12 pl. (1965).

Adey, W.H.: A revision of the Foslie crustose coralline Herbarium. Kong. Norske Vidensk. Selsk. Skrift. **1**, 46 p. (1970).

Adey, W.H., MacIntyre, I.G.: Crustose coralline Algae: a Re-evaluation in the Geological Sciences. Geol. Soc. Am. Bull. **84**, 883–904 (1973).

Cabioch, J.: Études sur les Corallinacées. Croissance et mode de réalisation du thalle; conséquences systématiques et phylogénétiques. Thèse Doct. es-Sci. Nat.: Robin et Mareuge, 288 p., 12 pl., 40 figs. Cahiers de Biologie marine, Paris, I – caractères généraux de la cytologie, **12**, 121–186 (1971); II – Morphogenèse; conséquences systematiques et phylogénétiques **13**, 137–287 (1972).

Elliott, G.F.: *Pseudaethesolithon*, a calcareous Alga from the Fars (Persian Miocene). Geol. Romana **9**, 31–42 (1970).

Foslie, M.: Three New Lithothamnia. Kong. Norske Vidensk Selsk. Skrift 1901–1, 1–5 (1901).

Foslie, M.: New Lithothamnia and systematical remarks. Kong. Norske Vidensk. Selsk. Skrift. 1905–5, 1–8 (1905).

Foslie, M.: Algologiske Notiser VI. Kong. Norske Vidensk. Selsk. Skrift. 1909–2, 1–63 (1909).

Foslie, M., Printz, H.: Contributions to a Monograph of the Lithothamnia. Kong. Norsk. Vidensk. Selsk. Museet, 60 p., 75 pl. (1929).

Johnson, J.H.: Limestone-Building Algae and Algal Limestone. Colorado School Mines, 297 p., 139 p. (1961).

Johnson, J.H.: Fossil and recent calcareous Algae from Guam. Geol. Survey USA, Professional Paper, Wash. **403**–G, 40 p., 15 pl. (1964).

Johnson, J.H., Ferris, B.J.: Tertiary and Pleistocene Coralline Algae from Lau, Fiji. Bernice P. Bishop Museum, Honolulu, Hawai Bull. **201**, 1–27 (1950).

Johnson, J.H., Kaska, H.V.: Fossil Algae from Guatemala. Prof. Contrib. Colorado School Mines **1**, 152 p., 47 pl. (1965).

Lemoine, M.: Structure anatomique des Melobesiees. Application à la classification. Ann. Inst. Ocean., Paris **2**, 213 (1911).

Mason, L.R.: The crustaceous coralline algae of the Pacific coast of the United States, Canada and Alaska. University of California, Publ. Botany, Berkeley **26** (4), 313–390 (1953).

Weber van Bosse, A., Foslie, M.: The Corallinaceae of the Siboga Expedition. Siboga Expeditie LXI. Leiden: Brill 1904, 110 p.

20 Reproduction Cycle of an Oncolithic Red Alga (Nemalionales) of the Lacustrine Miocene of Euböa (Greece)

J.-P. ADOLPHE, Paris, France

The nodules which have been investigated were found by Claude Guernet during geological studies on the Greek island of Euböa. Due to the wide extension of a Miocene lake, these nodules are distributed over a large area intercalated in marls as free or cemented forms.

Macroscopically light and dark zones can be seen. After the removal of the carbonate a red coloured organic tissue is found which consists of highly twisted tubes of red algae. The diameter of these tubes reaches 20 μ, mostly only 2–3 μ. The structure consists of an axial filament with four cross-shaped arranged cells (pleuridia) and V-shaped elements within the cortex. Cup-shaped cells indicate a differentiation of the cortex. Very probably the algae belong to the Nemalionales (see Bourrelly, 1970, and Magne, 1961) resp. to the Lemaneaceae.

Several reproduction elements are developed:

1. Female elements: the cells of the archegonium are covered with a trichogyne often showing spermatia.
2. Male elements: the antheridium has a shape like a pine-cone, it consists of a pedicellarium and large cells with spermatocysts releasing the spermatia.
3. Bush-shaped gonioblasts, formed by the fertilization of the oospheres together with the production of cystocarpes.
4. The cystocarpes produce cystospores which are released during the mature phase by an ostiole. Each cystospore developes a thallus identical with that described above.
5. Chromatic reductions may lead to the formation of typical tetraspores connected with the pedicellarium like group 2.

The tetraspores will form a female or male thallus. The reproduction cycle consists of three phases – gametophytic, carpophytic, and tetrasporophytic, see Fig. 1.

Until now no fossil representatives of the red algal group Nemalionales have been known. As has been shown by Adolphe and Rofes (1973) the Nemalionales entered the lacustrine environment during the Miocene forming nodules similar to those in recent rivers. It is striking that there are no associations of these red algae with blue-green or green algae.

The material described is of great importance because of the possibility of recognizing a complete reproduction cycle and studying alternations during the fossilization.

Pectine and cellulose have been destroyed only partially as also amidone, as determined by chromatography. According to Adolphe and Rofes (1973) fundamental differences exist in the precipitation of $CaCO_3$ by fungi or encrusted algae and the precipitation of $CaCO_3$ by bacteria. Bacteria contribute to the degradation of cellulose materials of water plants.

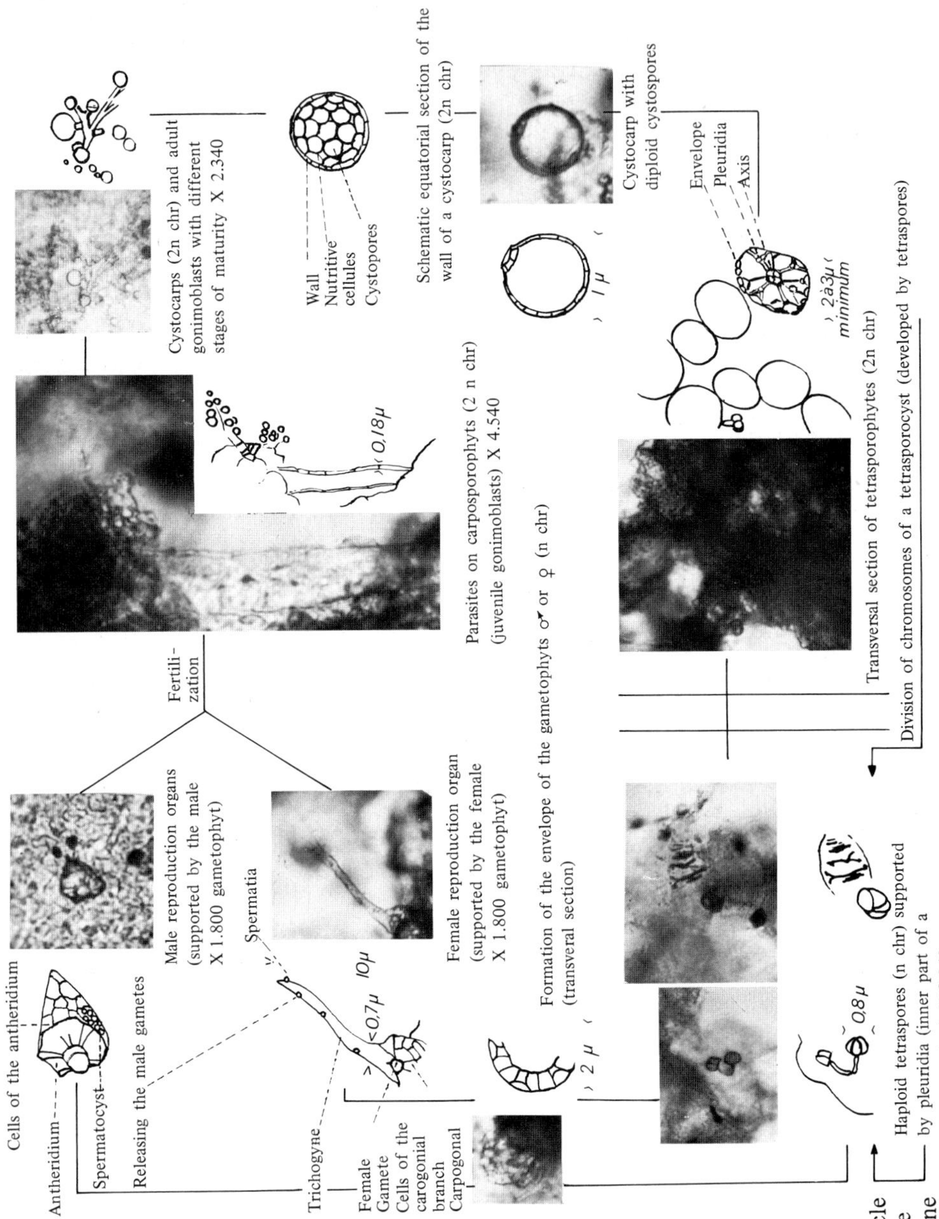

Fig. 1. Reproduction cycle of an oncolithic red algae (Nemalionales). Lacustrine Miocene, Euböa (Greece)

The ecological factors important for these oncolithic red algae can be summarized as follows: a relatively constant warm climate, seasonally fluctuating water levels with a general desiccation during spring to autumn and short-lived precipitations during the rest of the year. Similar facies are known from Miocene deposits of Algeria, the southwestern Aquitaine, Libanon, and Sicily; further investigations will show if there were time-equivalent ecological conditions in the Mediterranean area during that part of the Tertiary.

20.1 Bibliography

Adolphe, J.P., Rofes, G.: Concrétionnements calcaires de la Lévrière (cours d'eau expérimental; affluent de l'Epte, sous-affluent de la Seine, département de l'Eure). Bull. A.F.E.Q. **2**, 79–87 (1973).

Adolphe, J.P., Billy, C.: Biosynthèse de calcite par une association bactérienne aérobie. C.R. Acad. Sci. (Paris) **278**, 2873–2875 (1974).

Bourrelly, P.: Les Algues d'eau douce. Collection Faunes et flores actuelles. Paris VIe: Boubee 3, 1970.

Guernet, C.: Etudes géologiques en Eubée et dans les région voisines (Grèce). Thèse de Doctorat d'Etat, Univ. P. et M. Curie, Paris, 1971.

Magne, F.: Sur le cycle cytologique de Nemalion helminthoides (Velley) Batters. C.R. Acad. Sci. (Paris) **2**, 157–159 (1961).

Magne, F.: Sur la caryologie de 2 Rhodophycées considérées jusqu'ici comme cycle cytologique entierement haplophysique. C.R. Acad. Sci. (Paris) **252**, 4023–4024 (1961).

21 Problems of Affinity in Palaeozoic Calcareous Algae

R. RIDING, Cardiff, Great Britain

21.1 Introduction

It is proposed to reappraise current concepts of affinity in Palaeozoic calcareous algae. The need to do this arises from the questionable affinities often ascribed to Palaeozoic fossils conventionally regarded as calcareous algae. Commonly fossil algae are wrongly placed, e.g. a blue-green may be ascribed to a green algal group, but also non-algal fossils, particularly foraminifers and sponges (including stromatoporoids), are regarded as algae. These mistakes have usually been made and perpetuated by workers specializing in fossil algae. Consequently they have been accepted uncritically by others concerned with sedimentological and palaeoenvironmental problems with the result that errors have inadvertently been made. The outcome is that our ability to evaluate realistically the diversity, evolution, sedimentological importance and environmental significance of Palaeozoic calcareous algae has been gretly limited.

The extent of these difficulties necessitates restriction of the discussion here to major groups and a few particular examples; numerous genera and small groups are not mentioned at all. Also, since the prime purpose at this point is just to *document* problems of affinity, the wider significance of points made here is only briefly mentioned. The taxonomy of extent algae follows Fritsch (1935, 1945).

21.2 Palaeozoic Calcareous Algae

21.2.1 Recognition

Criteria for recognition must derive from algal morphology in general and the morphology of Recent calcareous algae in particular. That such an obvious statement should have to be made reflects on the present state of systematic work on these fossils.

Recent calcified algae include members of the blue-green, green, and red algae together with the yellow-green coccolithophores, one genus of brown algae (*Padina*), and some dinoflagellate cysts. No coccoliths, calcified browns, or calcified dinoflagellates are known from the Palaeozoic. The principal extant algal groups which have been compared with Palaeozoic fossils are the red Chaetangiaceae, Squamariaceae, and Corallinaceae; green Dasycladaceae, Codiaceae, and Characeae; and blue-greens. The morphology, calcification, ecology and other details of most of these Recent calcified algae are to be found in reviews such as those by Johnson (1961) and Milliman (1974, pp. 51–75).

21.2.2 Major Groups

A general classification of Palaeozoic calcareous algae is shown in Table 1. This indicates the main groups together with their conventionally accepted positions. In this section the problems of affinity and content of each major group are briefly reviewed.

Table 1. Currently accepted major groups of Palaeozoic calcareous algae

Blue-greens	coccoid forms
	Porostromata
Greens	Dasycladaceae
	Codiaceae
	Charophyta
	phylloid algae
Reds	Solenoporaceae
	Gymnocodiaceae
	ancestral corallines

21.2.2.1 Blue-Green Algae. Blue-greens include both filamentous and coccoid forms. Filamentous calcified blue-greens have tubiform skeletons and are referred to the Porostromata (see below, 21.2.2.2). It is not certain whether coccoid blue-greens calcify during life. Such a process would seem to be fundamentally disadvantageous since the cells would become entombed in $CaCO_3$, but it could take place in environments of rapid precipitation. Coccoid forms are known to give rise to calcareous spherulites (Buchbinder et al., 1974, and references) probably due to post-mortem bacterial attack (Krumbein and Cohen, 1974), but it is debatable whether these should be regarded as skeletal fossils since the degree of morphological detail retained seems slight. At present such spherulites are the only calcified remains of coccoid blue-greens known from Palaeozoic rocks.

However, non-filamentous calcareous Palaeozoic fossils have been referred to the blue-greens, notably *Renalcis* Vologdin, *Izhella* Antropov, and their synonyms. But these have relatively large, complex chambered skeletons and are now recognized to be Parathuramminacean foraminifers (Riding, 1972; Riding and Brasier, 1975).

21.2.2.2 Porostromata. This group of tubiform microfossils was erected by Pia (1927, p. 37) to contain fossils which he believed to be calcareous blue-greens but which he was unable to place in extant groups. Johnson (1961, p. 194) decimated the Porostromata by removing five genera to the Codiaceae (see below, 4). However, no calcified codiaceans are known to resemble these genera and Johnson's action was unjustified (Riding, 1975, p. 174). There is good evidence from Recent analogues that *Ortonella* Garwood and *Girvanella* Nicholson and Etheridge represent the calcified sheaths of filamentous blue-greens (Monty, 1967; Riding, 1977). It is likely that several other porostromate genera have a similar affinity, although this has still to be established, but some of the larger complexly-branched forms may represent other types of filamentous algae. It is convenient to retain these morphologically similar forms in one group, the Porostromata, until their affinities are better known. Consequently *Hedströmia* Rothpletz, *Ortonella* Garwood, *Bevocastria* Garwood, *Cayeuxia* Frollo, and *Garwoodia* Wood are here returned to the Porostromata.

21.2.2.3 Dasycladaceae. Work by Pia half a century ago has helped to make this one of the best known groups of fossil calcareous algae. Nevertheless, it presents a number of problems.

Pia (1920) recognized three tribes of Palaeozoic dasyclads (Dasyporelleae, Cyclocrineae, Primicorallineae) distinguished primarily by the form of the central stem and the number and size of branches. (It should be noted that the genus *Dasyporella* Stolley has been removed from the Dasycladaceae and tentatively placed in the red algae by Johnson and Konishi, 1959). These three groups differ from most post-Palaeozoic tribes by having their branches arranged irregularly, instead of in whorls, and by lacking signs of sporangia. However, Rezak (1959) proposed an additional Palaeozoic tribe, the Verticilloporeae, having branches in regular whorls, but these fossils have since been recognized to be crinoids.

A significant group of Palaeozoic fossils whose relationship to the algae in general, and dasyclads in particular, is still uncertain is the Receptaculitaceae. These are macrofossils, globular, ovoid or conical in form, which resemble sponges. But Hinde's (1884) suggestion that they are sponges was soon followed by Rauff's (1892) comparison of them with algae, and a dasycladacean affinity has especially been favoured in more recent years (Kesling and Graham, 1962; Nitecki, 1967, 1972, and references; Byrnes, 1968). Yet doubts continue to be raised concerning their affinity. Rietschel (1969), after careful consideration, decided that the receptaculitids are algae, but that they represent a distinct group of green algae separate from the dasycladaceans. Foster (1973) agreed that receptaculitids cannot be assigned to any extant group of sponges or algae, but was of the opinion that the evidence slightly favours a sponge affinity. The debate hinges on interpretation of the external morphology and internal structure together with the mode of growth and life-orientation of the organisms. The external surface of receptaculitids shows a neat arrangement of plates, variously termed facets or meroms, arranged in circlets centred upon poles at each end of the skeleton. One pole (the nucleus or enclosed end) is surrounded by circlets of few (initially eight) plates. The other pole (the apex or apertural end) bears circlets with many plates. The apex usually includes an aperture or is poorly preserved. The sponge hypothesis regards the aperture as an osculum, which would be orientated upward. Rietschel (1969) also regards the apex as being upward but believes it was closed during life. He regards it as the growing end and thus a site of weak calcification. In contrast Nitecki regards the apex as being downward and interprets the enclosed (nuclear) end as the generative pole. Campbell et al. (1974), in an important contribution, support and differ from aspects of both views. They agree with Nitecki that the apex was downward, but believe (with Rietschel) that this end was also the site of growth. Study of the mode of intercalation of the plates (Gould and Katz, 1975) supports the latter view that the plates are generated at the apex and are oldest at the nucleus. The apex-down life orientation of receptaculitids results in an enclosed skeleton which is difficult to equate with sponge morphology, but their downward generative poles also provide a contrast with Recent dasycladaceans. The mode of growth of receptaculitids is plant-like, algae are the only plants known to calcify significantly, and receptaculitids do bear general structural resemblances to dasycladaceans. Thus, Rietschel's view that receptaculitids are green algae similar to, but also somewhat different from, dasyclads provides the most plausible current explanation of their affinity.

A related problem concerns the taxonomic position of the Cyclocrineae. Most workers have regarded this group as being separate from the receptaculitids, but Nitecki (1972, p. 13) considers them a receptaculititd group. The debate concerning receptaculitid affinity must therefore also involve the cyclocrinitids which have been widely accepted as dasycladaceans.

Subsequent to Pia's work there has been a tendency to regard smaller and morphologically rather simple Palaeozoic fossils as dasycladaceans. These include a whole range of tubiform microfossils such as *Dvinella* Khvorova, *Uraloporella* Korde, *Anthracoporellopsis* Maslov, and *Kamaena* Antropov which are of Upper Palaeozoic age and which form a distinctive, if broad, morphological group (Riding and Jansa, 1974). A similar genus, *Donezella* Maslov, has been referred to the red algae by Maslov (Johnson, 1963) and to the Codiaceae by Rácz (1965). These genera do not exhibit well-defined pores, although a fibrous wall structure tends to be interpreted in this sense, and they are commonly septate. It is difficult to see how they have come to be regarded as dasyclads; they do not resemble dasycladaceans in general morphology, there are no signs of the relatively large pores shown by even small irregular Palaeozoic forms such as *Vermiporella* Stolley and *Epimastopora* Pia, which are often regarded as dasyclads, and their septate nature specifically excludes a dasycladacean affinity for them since dasyclads are multinucleate and siphoneous in their organization. The size and morphology of these small fossils are more consistent with a foraminiferal affinity (Riding and Jansa, 1974) but attributions of dasycladacean affinity persist. Some difficulties in resolving this problem stem from disagreement concerning the septate nature of one of the genera, *Uraloporella* (Riding and Jansa, 1974; Mamet and Roux, 1975). Even so, clearly septate forms have been referred to the green algae (Mamet and Roux, 1975) without any mention of the nonseptate nature of extant calcified green algae. This debate has been expanded by the suggestion that a number of these genera belong to the sponges (Termier et al., 1975). These rather neglected fossils are locally important sediment producers and mound-builders in the Upper Palaeozoic; resolution of their affinities is a matter of importance.

21.2.2.4 Codiaceae. Pia (1927, p. 59) recognized only two Palaeozoic genera as members of the Codiaceae: *Palaeoporella* Stolley and *Gymnocodium* Pia. He later removed *Gymnocodium* to the red algae (see below, 8). Subsequent work has added a number of Palaeozoic genera to the codiaceans, e.g. *Lancicula* Maslov and *Litanaia* Maslov from the Devonian, *Palaeocodium* Chiarugi and *Orthriosiphon* Johnson and Konishi from the Carboniferous, and *Succodium* Konishi from the Permian. But there have also been two other significant developments: first, the general referral of phylloid algae to the Codiaceae (see below, 2.2.6); second, the transferral by Johnson (1961, pp. 96–99) of the genera *Hedströmia, Ortonella, Bevocastria, Cayeuxia*, and *Garwoodia* to the Codiaceae from the Porostromata. Johnson appears to have assumed that a non-septate, branched, tubiform morphology indicates a codiacean affinity. But this overlooks the profound differences in size and shape between these fossils and extant calcified codiaceans which are both large and irregular, as well as the close similarity between forms such as *Ortonella* and calcified blue-greens. These genera are returned to the Porostromata (see above, 2.2.2).

21.2.2.5 Charophyta. Charophyte oogonia are known from the late Silurian onward. The morphologically similar group known as umbellids has long been regarded as foraminiferal but has more recently been compared with the charophytes, especially by Russian workers. Charophytes and umbellids are similar in both having two polar pores. Furthermore, Mamet (1970) has used the presence of an apical cover lid in umbellids to argue against a foraminiferal affinity for them. However, charophytes also lack this feature and differ from umbellids in their wall structure; the umbellid wall is fibrous while that of the charophytes is laminated. Hence, umbellids appear to differ from both foraminifers and charophytes. These points have been made by Peck (1974) who recommended that umbellids should for the present be regarded as microproblematica.

Langer (1975) has drawn attention to morphological differences between *Sycidium* Sandberger and other charophytes. He has suggested that *Sycidium* represents a new group of green algae similar to, but distinct from, the Charophyta.

21.2.2.6 Phylloid Algae. Leaflike Upper Palaeozoic fossils termed phylloid algae by Pray and Wray (1963, p. 209) form a homogeneous group in terms of their external morphology, but are heterogeneous in their probable affinities. The principal genera are *Calcifolium* Shvetzov and Birina, *Anchicodium* Johnson, *Ivanovia* Khvorova, *Archaeolithophyllum* Johnson, and *Eugonophyllum* Konishi and Wray. *Archaeolithophyllum* has a differentiated cellular internal structure which suggests a red algal affinity (see below, 9). The other genera have usually been regarded as codiaceans (see Johnson, 1963, p. 18), but Perret and Vachard (1975) have suggested that *Calcifolium* is a calcareous sponge while Wray et al. (1975) have noted their similarity to Recent squamariacean red algae.

Eugonophyllum and *Ivanovia*, especially, are very similar in external appearance to squamariaceans. It is difficult to pursue the comparison, however, due to the poor internal preservation of phylloid algae, although it is interesting that this preservational feature too is consistent with a squamariacean affinity since the latter have an aragonitic mineralogy. But *Eugonophyllum* has internal, or partially internal, reproductive-like structures which suggest that it is not a squamariacean and *Ivanovia* is the phylloid genus which at present compares best with the Squamariaceae (Wray, pers. comm.).

21.2.2.7 Solenoporaceae. Since the end of the last century the view that this important group belongs to the red algae and is closely related to the Corallinaceae, has been widely accepted (Johnson, 1960, pp. 1–2). Nevertheless, significant differences exist between corallines and solenopores (Johnson, 1961, p. 72). Solenopores generally have larger cells which are commonly polygonal, rather than rectangular, in transverse section; cross-partitions of the cells are less regularly developed than in corallines, and tissue differentiation and reproductive structures are rare. These features should discourage too close a comparison being made with the Corallinaceae. Indeed, even the algal nature of the group, while plausible, is not certain, although this is seldom admitted. Of particular concern is the homeomorphy of solenopores with some tabulate corals. There are, of course, also important differences in skeletal detail, notably septa and mural pores, but all tabulates do not show these features and the similarities in internal structure between solenopores and some tabulates is striking. The only obvious difference in some cases is size, most tabulates being 5–10 times larger than solenopores. It is curious that the Solenoporaceae is such a neglected group since it comprises sedimentologically important, long-ranging macrofossils. Further work to elucidate their affinity, including their relationships to the Corallinaceae, is overdue.

21.2.2.8 Gymnocodiaceae. Pia at first referred *Gymnocodium*-material to the dasycladacean genus *Macroporella* Pia, but subsequently came to regard it as probable codiacean and erected the genus *Gymnocodium* (Pia, 1920). Later he again revised his view, compared *Gymnocodium* with *Galaxaura*, and consequently placed it in the red algae (Pia, 1937). Elliott (1955) supported this interpretation, erected another genus, *Permocalculus*, and placed both genera in a new family, the Gymnocodiaceae.

However, some doubt remains concerning both the degree of similarity of *Gymnocodium* and *Permocalculus* and also their affinities. *Gymnocodium* does resemble codiaceans while *Permocalculus* is perhaps more similar to some dasycladaceans. Praturlon (1963) has found *Atractyliopsis*, a dasycladacean stem-filling, in a few specimens of *Permocalculus*. These difficulties in distinguishing between some Codiaceae, Dasycladaceae, and Chaetangiaceae may in this case be due to morphological convergence within these groups. They emphasize the problems of confidently interpreting algal morphology, even in relatively well studied genera.

21.2.2.9 Ancestral Corallines and Other Red Algae. This heterogenous and difficult group of Upper Palaeozoic fossils is broadly divisible into cellular and fibrous forms. Genera with cellular structure include *Archaeolithophyllum, Cuneiphycus* Johnson, and *Keega* Wray and have been termed 'ancestral corallines' (Wray, 1971). The genetic precision implied by this term is somewhat misleading and one of these genera, *Keega*, has been reinterpreted as a laminar form of the stromatoporoid *Stachyodes* Bargatzky (Riding, 1974). Nevertheless, *Archaeolithophyllum* does show a distinctly cellular structure together with tissue differentiation which suggests a relationship to crustose coralline algae. (Its leaflike form permits it to be placed in the phylloid algae – see above, 6 – on grounds of purely external appearance). To the coralline-like cellular subgroup should be added *Archaeolithoporella* Endo which shows superimposed crusts and conceptacle-like structures reminiscent of the extant crustose coralline *Lithoporella* Foslie (Johnson, 1963, p. 98).

Fibrous forms include *Ungdarella* Maslov, *Komia* Korde, and *Foliophycus* Johnson. The tufted, strand-like internal appearance of these cylindrical branching fossils is probably due in part to diagenetic alteration which may obscure a finer cellular structure. The affinity of these forms is obscure, although a relationship with red algae is conventionally implied, and *Komia* has also been regarded as a stromatoporoid (Wilson et al., 1963). The degree of separation between these genera also requires clarification. *Foliophycus* closely resembles *Ungdarella* and well preserved material from the south-western U.S.A., referred to *Ungdarella* by Toomey and Johnson (1968) has been claimed to be *Komia* by Wilson (1969).

21.3 Discussion

A radical revision of the classification of Palaeozoic calcareous algae is not intended here. Instead it will be shown how the problems outlined above modify our views of currently accepted major groups. Two types of modification have to be considered. The first concerns the *validity* of conventional major categories of Palaeozoic calcareous algae, the second concerns the *content* of these categories.

21.3.1 Major Groups

These problems of affinity necessitate few changes in the status of the major groups themselves. The phylloid algae and ancestral corallines etc. are heterogeneous groups, and the status of the Gymnocodiaceae has been tentatively questioned, but the only group dealt with here which appears to lack constituent genera is that of the calcified non-filamentous blue-green algae, which previously contained genera such as *Renalcis*. It is quite likely that representatives of this group, apart from diagenetic spherulites, will be recognized in the Palaeozoic, but since it is not possible at present to confidently refer any genera to this group it is not included in the revised scheme shown in Table 2. The phylloid algae is a somewhat different case. Our improved understanding of the probable affinities of several of the genera is best reflected by their removal to other groups, although the term retains morphological value particularly with regard to sedimentological studies. Since the essential concern here is for classification according to affinity, the category of phylloid algae ought to be omitted. However, several phylloid genera are still of quite uncertain affinity and so it is convenient to retain the group for them.

Table 2. Revised scheme of major groups of Palaeozoic calcareous algae indicating misplacements and misidentifications

(Coccoid blue-greens)	includes foraminifers
Porostromata	
Dasycladaceae	includes crinoids and (?) foraminifers
Receptaculitaceae	separate from dasyclads
Codiaceae	includes porostromates
Charophyta	umbellids and *Sycidium* separate
Phylloid algae	includes ancestral coralline, squamariaceans, and (?) sponge
Solenoporaceae	
Gymnocodiaceae	may include green algae
Squamariaceae	from phylloid algae
Ancestral corallines etc.	includes stromatoporoids

21.3.2 Content of Major Groups

The cases of mistaken identity which can be recognized necessitate either the real location of misidentified algae to other algal groups, or the removal of non-algal fossils to their correct phylum. These changes are summarized in Table 2. They affect the content of every group except the Solenoporaceae, and only the Solenoporaceae and Porostromata stand to lose no genera.

Misplaced algae include porostromate members of the Codiaceae, ancestral coralline and probable squamariacean members of the phylloid algae, and possibly the gymnocodiacean genera *Gymnocodium* and *Permocalculus*.

Non-algal and probable non-algal fossils include the Renalcidae (foraminifers), Verticilloporeae (crinoids), *Donezella* and *Uraloporella* etc. (foraminifers?), *Calcifolium* (sponge?), and stromatoporoids included in the ancestral corallines, etc. The affinity of the umbellids is uncertain.

21.3.3 Implications for Palaeozoic Calcareous Algae

The major groups listed in Table 1 thus fall into four larger categories: *bona fide* algae, probable algae, uncertain affinity, and non-algae.

Fossils referred to the green algal groups (Dasycladaceae, Codiaceae, charophytes) appear mainly to be algae. Fossils referred to the Solenoporaceae, Porostromata, Squamariaceae, and Receptaculitaceae are, together with some 'ancestral corallines', probably algae and we can note here that the better understood porostromate genera appear to be calcified filamentous blue-green sheaths. Fossils of uncertain affinity include some phylloid algae (*Calcifolium* may be a sponge), some ancestral corallines etc. (such as *Cuneiphycus* and *Ungdarella*), small septate tubiform fossils (*Donezella, Uraloporella*, etc.) which have usually been called green algae but which could well be foraminifers, and umbellids. Non-algal fossils, previously classed with the algae, include the Renalcidae, crinoids constituting the Verticilloporeae, and the stromatoporoid genus *Stachyodes* which was named *Keega.*

These observations place constraints on our views of the diversity and sedimentological importance of calcareous algae during the Palaeozoic. Apparent low diversity of calcareous algae in parts of the Palaeozoic has been attributed in part to lack of searching. Johnson and Konishi (1958, p. 3) made this point with respect to the Devonian. It is unfortunate that subsequent studies of Devonian algae made up for this deficiency by recognizing non-algal fossils as algae. Our knowledge is doubtless still very incomplete but it does now seem more likely that low diversity is a real feature of calcareous algae in the Palaeozoic generally, This view is strengthened by our ability to recognize all the major groups of *bona fide* and probable algae, with the exception of the charophytes (late Silurian onward) and squamariaceans (Carboniferous onward), virtually throughout the era. If the present is any guide this low diversity should not be regarded as especially unusual; in the Recent there is an enormous variety of algae but rather few of them calcify.

It is now clear that several of the sedimentologically more important Palaeozoic calcareous 'algae' are not *bona fide* algae. These include *Renalcis* etc., *Keega* (*Stachyodes*), and *Donazella* and *Uraloporella* etc. Paleoalgology's loss is another discipline's gain and these clarifications of affinity have significance, which extends beyond the purely sedimentological, for at least two non-algal groups. Recognition of the stromatoporoidal nature of *Keega* is of interest because it has drawn attention to structural modification of the basal part of the coenosteum in some laminar stromatoporoids (Riding, 1974). Realization of the foraminiferal nature of renalcids provides insight into the early history of calcareous foraminifers because renalcids are known from the oldest Cambrian strata and possibly even from the very youngest Precambrian (Riding and Brasier, 1975). Demonstration of the foraminiferal affinity of *Donezella* and *Uraloporella* etc. would be important for the study of Upper Palaeozoic foraminifers because it would modify current rather conservative views of the morphology of Palaeozoic foraminifers and would add a significant number of genera, including several Devonian forms.

Acknowledgments. I am grateful to several colleagues for discussion and advice which helped to clarify a number of the points made here, especially Graham Elliott, Wolfgang Krumbein, Ernst Ott, Siegfried Rietschel, and John Wray. My participation in the Erlangen Symposium, which facilitated most of this discussion, was made possible by grants from the Royal Society and University College, Cardiff.

21.4 Bibliography

Buchbinder, B., Begin, Z.B., Friedman, G.M.: Pleistocene algal tufa of Lake Lisan, Dead Sea area, Israel. Israel J. Earth Sci. **23**, 131–138 (1974).

Byrnes, J.G.: Notes on the nature and environmental significance of the Receptaculitaceae. Lethaia **1**, 368–381 (1968).

Campbell, K.S.W., Holloway, O.J., Smith, W.D.: A new receptaculitid genus *Hexabactron*, and the relationships of the Receptaculitaceae. Palaeontographica **146**, 52–77 (1974).

Elliott, G.F.: The Permian calcareous alga *Gymnocodium*. Micropaleontology **1**, 83–91 (1955).

Foster, M.: Ordovician receptaculitids from California and their significance. Lethaia **5**, 35–65 (1973).

Fritsch, F.E.: The Structure and Reproduction of the Algae. Vol. I. Cambridge: Univ. Press 1935.

Fritsch, F.E.: The Structure and Reproduction of the Algae, Vol. II. Cambridge: Univ. Press 1945.

Gould, S.K., Katz, M.: Disruption of ideal geometry in the growth of receptaculitids: a natural experiment in theoretical morphology. Paleobiology **1**, 1–20 (1975).

Hinde, G.J.: On the structure and affinities of the family of the Receptaculitidae, including therein the genera *Ischadites*, Murch. (*Tetragonis*, Eichw.); *Sphaerospongia*, Pengelly; *Acanthoconia*, gen. nov.; and *Receptaculites*, DeFrance. Quart. J. Geol. Soc. London **40**, 795–849 (1884).

Johnson, J.H.: Paleozoic Solenoporaceae and related red algae. Colorado School of Mines Quart. **55**, 3, 1–77 (1960).

Johnson, J.H.: Limestone-Building Algae and Algal Limestones. Boulder: Colorado School of Mines, 1961.

Johnson, J.H.: Pennsylvanian and Permian algae. Colorado School of Mines Quart. **58**, 3, 1–211 (1963).

Johnson, J.H., Konishi, K.: A review of Devonian algae. Colorado School of Mines Quart. **53**, 2, 1–84 (1958).

Johnson, J.H., Konishi, K.: Studies of Silurian (Gotlandian) algae. Colorado School of Mines Quart. **54**, 1, 3, 131–158 (1959).

Kesling, R.V., Graham, A.: *Ischadites* is a dasycladacean alga. J. Paleontol. **36**, 943–952 (1962).

Krumbein, W.E., Cohen, Y.: Biogene, klastische und evaporitische Sedimentation in einem mesothermen monomiktischen ufernahen See (Golf von Aqaba). Geol. Rundschau **63**, 1035–1065 (1974).

Langer, W.: Ein neuer Fund von *Sycidium* G. Sandberger im westlichen Devon. First Intern. Symp. Fossil Algae, absts., 18 (1975).

Mamet, B.L.: Sur les Umbellaceae. Can. J. Earth Sci. **7**, 1164–1171 (1970).

Mamet, B., Roux, A.: *Jansaella ridingi*, nouveau genre d'Algue? dans le Dévonien de l'Alberta. Can. J. Earth Sci. **12** (1975).

Milliman, J.D.: Marine carbonates. Berlin–Heidelberg–New York: Springer 1974.

Monty, C.L.V.: Distribution and structure of Recent stromatolitic algal mats, eastern Andros Island, Bahamas. Ann. Soc. Géol. Belg. **90**, 3, B55–B100 (1967).
Nitecki, M.H.: Systematic position of receptaculitids. Abst. Geol. Soc. Amer. Ann. Meet. 165–166 (1967).
Nitecki, M.H.: North American Silurian receptaculitid algae. Fieldiana, Geology **28**, 1–108 (1972).
Peck, R.E.: On the systematic position of the umbellids. J. Paleontol. **48**, 409–412 (1974).
Perret, M.-F., Vachard, D.: Sur l'appartenance du genre *Calcifolium* (Algae auct.) aux Pharétrones (Porifera). C. R. Acad. Sci. Paris **280**, D, 2649–2652 (1975).
Pia, J.: Die Siphoneae verticillatae vom Karbon bis zur Kreide. Abh. zool.-botan. Ges. **11**, 2 (1920).
Pia, J.: Thallophyta. In: Handbuch der Paläobotanik. M. Hirmer (ed.), Pt. 1. München–Berlin: Oldenbourg 1927, pp. 31–136.
Pia, J.: Die wichtigsten Kalkalgen des Jungpaläozoikums und ihre geologische Bedeutung. Proc. 2nd Congr. Avanc. Et. strat. Carb. **2**, 765–856 (1937).
Pray, L.C., Wray, J.L.: Porous algal facies (Pennsylvanian), Honaker Trail, San Juan Canyon, Utah. In: Shelf Carbonates of the Paradox Basin, Four Corners Geol. Soc. Symp., 4th Field Conf. 1963, pp. 204–234.
Praturlon, A.: Dasycladaceae from Upper Permian of the Dolomites. Geol. Romana **2**, 119–150 (1963).
Rácz, L.: Carboniferous calcareous algae and their associations in the San Emiliano and Lois-Ciguera Formations (Prov. León, NW Spain). Leidse Geol. Med. **31**, 1–112 (1965).
Rauff, H.: Untersuchungen über die Organisation und systematische Stellung der Receptaculitiden. Abh. Math.-Phys. Classe K. König. bay. Akad. Wiss. **17**, 3, 645–722 (1892).
Rezak, R.: New Silurian Dasycladaceae from southwestern United States. Colorado School of Mines Quart. **54**, 1, 115–129 (1959).
Riding, R.: Calcareous algae and some associated microfossils from Ancient Wall reef complex (Upper Devonian), Alberta. Abst. Bull. Am. Ass. Petrol. Geol. **56**, 648 (1972).
Riding, R.: The Devonian genus *Keega* (Algae) reinterpreted as a stromatoporoid basal layer. Palaeontology **17**, 565–577 (1974).
Riding, R.: *Girvanella* and other algae as depth indicators. Lethaia **8**, 173–179 (1975).
Riding, R.: Calcified *Plectonema* (blue-green algae), a Recent example of *Girvanella* from Aldabra atoll. Palaeontology **20**, 33–46 (1977).
Riding, R., Brasier, M.: Earliest calcareous foraminifera. Nature **257**, 208–210 (1975).
Riding, R., Jansa, L.F.: *Uraloporella* Korde in the Devonian of Alberta. Can. J. Earth Sci. **11**, 1414–1426 (1974).
Rietschel, S.: Die Receptaculiten. Senckenbergiana Lethaea **50**, 465–517 (1969).
Termier, H., Termier, G., Vachard, D.: Sur la systématique et la phylogénie des Moravamminida et des Aoujgaliia. First Intern. Symp. Fossil Algae, abst., 40 (1975).
Toomey, D.F., Johnson, J.H.: *Ungdarella americana*, a new red alga from the Pennsylvanian of southeastern New Mexico. J. Paleontol. **42**, 556–560 (1968).
Wilson, E.C.: No new *Ungdarella* (Rhodophycophyta) in New Mexico. J. Paleontol. **43**, 1245–1247 (1969).
Wilson, E.C., Waines, R.H., Coogan, A.H.: A new species of *Komia* Korde and the systematic position of the genus. Palaeontology **6**, 246–253 (1963).
Wray, J.L.: Algae in reefs through time. Proc. 1969 N. Am. Paleontol. Conv. **J**, 1358–1373 (1971).
Wray, J.L., James, N.P., Ginsburg, R.N.: The puzzling Paleozoic phylloid algae: a Holocene answer in squamariacean calcareous red algae. Am. Assoc. Petrol. Geol., Dallas Conv. Abstr. 82–83 (1975).

22 Receptaculitids are Calcareous Algae but no Dasyclads

S. RIETSCHEL, Frankfurt/M., W. Germany

During the International Symposium on Fossil Algae at Erlangen, the receptaculitids were mentioned by several speakers. Riding, in his contribution on "Problems of affinity in palaeozoic calcareous algae", dealt with their nature, assuming that they might be sponges. Discussions during the symposium convinced me of the usefulness of such a statement, which should help to resolve some of the former questions about the nature of receptaculitids. I thank, among others, especially Robert Riding for his persistent questions in discussions.

Receptaculitids were regarded either as animal remains (mostly sponges) or as calcareous algae. Their assignment to calcareous algae and comparison with dasycladaceen algae dates back to Rauff (1892). Some authors followed his interpretation, many did not. During the recent years the algal hypothesis has again been favored, mainly by Byrnes (1968), Nitecki (1967, 1970) and Rietschel (1969, 1970). Different lines of argument and two different or actually opposite interpretations of the life position of the receptaculitid body have led to two different hypotheses of the systematic position of receptaculitids amongst fossil and living algae. If we follow Nitecki's line of argument, we regard, roughly speaking, the cyclocrinitids – undisputed dasycladaceen algae for about 100 years – as receptaculitids and then the receptaculitids as dasyclads. I tried (1969) to give an analysis of the anatomic features of receptaculitids in consideration of biologic evidence, a method, which was recently supported by some results of Gould and Katz (1975). After comparing the receptaculitids with different groups of animals and plants, I proposed a systematic position in the Thallophyta but saw some evident differences to dasyclads.

So I shall firstly summarize the main arguments for the plant nature of the receptaculitids:

1. Receptaculitids have a "skeleton" of single elements (meroms) arranged in a peculiar manner, which is only explicable by a polar growth (substantiated by Gould and Katz, 1975).

2. The merom elements of a single specimen (organism) are of the same shape, but vary in size and proportions with regard to their position to the growing pole.

3. Besides the growing of the whole body by rise of new meroms at the vegetating pole, there was a distinct individual growth of each merom. This growth was apparently accompanied by increasing calcification, and finally ceased.

4. The weak calcification in the early stages of the meroms is responsible for their poor preservation. Therefore most, but not all specimens, now show an opening instead of the growing pole (often misunderstood as, e.g., the osculum of sponges).

5. The primary calcification, controlled by the organism, only took place in the cell wall of the meroms. Calcification often decreases to the inner part of the body, especially near the growing pole.
6. The receptaculitid body did not stop growing, even in very advanced stages, and never reached a fixed number of meroms. The growth of mature individuals decelerated, but reinforcement of growth was possible.

These essential features of the mode of growth are more typical of plants than of animals. They conform well with the unlimited manner of plant growth, during which new organs – similar in shape but with varying proportions – are produced throughout the life of the organism.

The interpretation of Nitecki and Byrnes, which postulates a life position upside down, is not supported to date by considerations of functional morphology. In my paper of 1969 I was reticent in my course of reasoning, but since then further investigations have so far endorsed my opinion, that I now feel absolutely sure about the plant nature of receptaculitids. I may also mention:

7. In 1969 I published (Fig. 12) a series showing ontogenetic stages of receptaculitids, which was based on hypothetical considerations, this has now been confirmed by a series of complete specimens showing different growth stages.
8. There are now more complete specimens, which show both poles of the body well preserved. There should no longer be any doubt, that receptaculitids had a totally closed apex (upper pole), and that the generally found apical opening is due to incomplete preservation.
9. By abundant material it is now possible to show that what are regarded as the main features of the receptaculitids change in a certain way from ordovician to permian time. This confirms that they are really of importance, and makes it possible to construct a phylogenetic scheme for the whole group.

When we compare the receptaculitids (in a strict sense) with other groups of lower plants, we might assign them to green algae, according to some morphologic features and their main life habitat in shallow waters. But in any case, we should distinguish and separate them from dasycladaceen algae. I again present three main arguments:

10. Receptaculitids had a calcified cell wall, while in dasyclads the calcification takes place in a mucus between the branches. This physiologic difference is generally manifested by different modes of preservation.
11. Up to date we have no definite reproductive organs of receptaculitids, and so we cannot be sure whether or not they had a reproduction mode similar to that of the dasyclads.
12. There are some significant morphologic differences between receptaculitids and dasyclads.

I therefore assume: receptaculitids are calcareous algae, but no dasyclads!

22.1 Bibliography

Byrnes, J.G.: Notes on the nature and environmental significance of the Receptalulitaceae. Lethaia **1** (4), 368–381 (1968).

Gould, S.J., Katz, M.: Disruption of ideal geometry in the growth of receptaculitids: a natural experiment in theoretical morphology. Paleobiology **1** (1), 1–20 (1975).

Nitecki, M.H.: Systematic position of receptaculitids. Geol. Soc. Amer. Ann. Mtg. 165–166 (1967).

Nitecki, M.H.: North american cyclocrinitid algae. Fieldiana, Geol. **21**, 182 pp. (1970).

Rauff, H.: Kalkalgen und Receptaculiten. Sitz.-Ber. Niederrhein. Ges. Natur-Heilkunde 74–90 (1892).

Rietschel, S.: Die Receptaculiten. Senckenbergiana lethaea **50** (5/6), 465–517 (1969).

Rietschel, S.: Rekonstruktionen als Hilfsmittel bei der Untersuchung von Receptaculiten (Receptaculitales, Thallophyta). Senckenbergiana lethaea **51** (5/6), 429–447 (1970).

23 On Moravamminida and Aoujgaliida (Porifera, Ischyrospongia): Upper Paleozoic "Pseudo Algae"

H. TERMIER, G. TERMIER, and D. VACHARD, Paris, France

Ischyrospongia were established by Termier and Termier (1973) as a class of sponges including the recent order Sclerospongia (Hartman and Goreau, 1970) together with the two fossil orders, Stromatoporoidea and Pharetronida. These fossil groups both have a hypercalcified, fibrous and porous skeleton preserving the canaliferous meshwork of the water system. Spicules are rare or absent. Many of them display a yellowish, hyaline, calcitic test.

In 1975 we included in Ischyrospongia two new orders of palaeozoic microorganisms: Aoujgaliida and Moravamminida. The aim of this paper is to determine their systematic contents.

23.1 Order Moravamminida (Termier et al., 1975)

23.1.1 Diagnosis

Often sessile organisms, sometimes coiled in the initial stadium. Adult stage cylindrical, tubular, erected and nearly always branching. Central cavity undivided or irregularly septated. Septa perforated by a central gullet. Tubes usually solitary, sometimes clustered. Wall yellowish, hyaline, perforated, mono- or oligocrystalline. Simple, bi- or trifurcated perforations.

23.1.2 Order Composition

Family 1: Kettneramminidae (Termier et al., 1975). Non septated skeleton. Simple perforations. Genera: *Kettnerammina* (Pokorny, 1951; Middle Devonian), *Vasicekia* (Pokorny, 1951; Middle Devonian), *Pseudoissinella* (Mamet and Rudloff, 1972: *Saccorhina?* Bykova, 1955, Middle Devonian – Lower Carboniferous), *Issinella?* (Reitlinger, 1954; Upper Devonian – Lower Carboniferous).

Family 2: Moravamminidae (Pokorny, 1951, *emend.*) . Septated, often ramose tubes and simple, bi- or trifurcated perforations.

Subfamily 1: Moravammininae (Pokorny, 1951, *emend.*). Initial stage fastened or coiled around a holder. More or less spread out. Septae irregularly set. Simple or bifurcated perforations. Genera: *Moravammina* (Pokorny, 1951: *Litaya* Bykova, 1955; Middle Devonian – Lower Carboniferous), *Evlania* (Bykova, 1952).

Subfamily 2: Palaeoberesellinae (Mamet and Roux, 1974, *emend.*). Initial stage not coiled but fixing collars sometimes present. Septation nearly always regular, completed

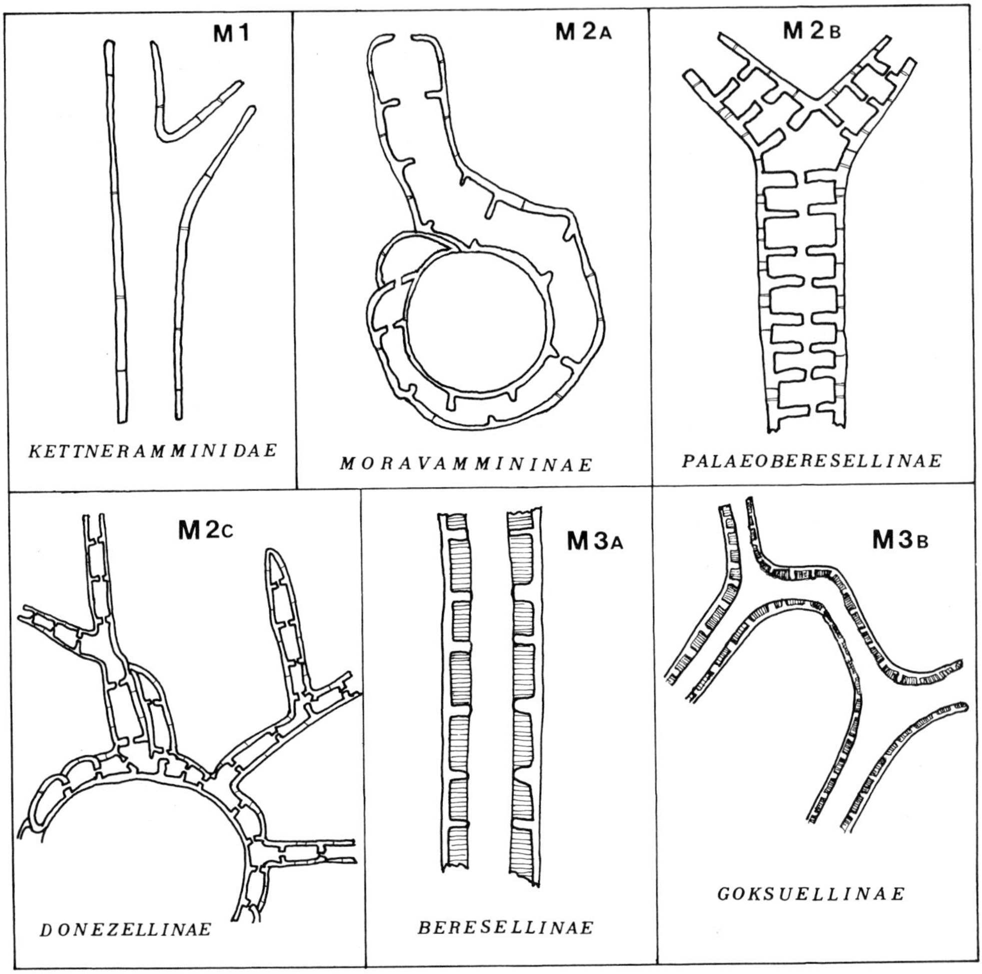

Fig. 1. Moravamminids. (M 1–3) Diagram of each family and subfamily displaying morphologic salient features. (M 1) Kettneramminidae: undivided tubes. (M 2) Moravamminidae: septated tubes. (M 2a) Moravammininae. (M 2b) Palaeoberesellinae. (M 2c) Donezellinae. (M 3) Beresellidae: distinctive wall (M 3a) Beresellinae. (M 3b) Goksuellinae

by aligned gullets. Simple, bi- or trifurcated perforations. Genera: *Palaeoberesella* (Mamet and Roux, 1974; Lower Carboniferous), *Exvotarisella* (Elliott, 1970; Lower Carboniferous), *Kamaena* (Antropov, 1967; Middle Devonian – Middle Carboniferous), *Pseudokamaena* (Mamet, 1972; Middle Devonian – Lower Carboniferous), *Parakamaena* (Mamet and Roux, 1974; Middle Devonian – Lower Carboniferous), *Alanyana?* (Güvenç, 1967; Upper Permian), *Proninella* (Menner and Reitlinger, 1971; Middle – Upper Devonian), *Anthracoporellopsis* (Maslov, 1956; Middle Carboniferous).

Subfamily 3: Donezellinae (Termier et al., 1975, completed). Fastened initial stage, partly coiled, from which arise numerous isolated or tied tubes. Regular and complete septation. Simple perforations. Genera: *Kamaenella* (Mamet and Roux, 1974; Lower

Carboniferous), *Praedonezella* (Koulik, 1973; Lower – Middle Carboniferous), *Donezella* (Maslov, 1929, non *auct.*; Middle Carboniferous), *Bevocastria?* (Garwood, 1931; Lower – Middle Carboniferous).

Family 3: *Beresellidae* (Maslov and Koulik, 1956, *emend.*). Isolated or tied cylindrical tubes with undivided internal cavity. Two-layered wall: yellowish external layer forming periodic re-entering at regular intervals, internal layer crossed by thin and very close, simple, bi- or trifurcated perforations.

Subfamily 1: Beresellinae (Maslov and Koulik, 1956). Isolated and nonbranched tubes. *Beresella* (Maslov and Koulik, 1956; Middle – Upper Carboniferous). *Uraloporella* (Kordé, 1950; poss. *Samarella*, Maslov and Koulik, 1956; Lower – Middle Carboniferous), *Dvinella* (*Dvinella*; Khvorova, 1949; Middle Carboniferous), *Dvinella* (*Trinodella*; Maslov and Koulik, 1956; Middle Carboniferous).

Subfamily 2: Goksuellinae subfam. nov. Branching tubes. Simple perforations. Genus *Goksuella* (Güvenç, 1966: *Donezella auct. pars*).

23.1.3 Systematic Position

Moravamminids differ from Foraminifera by lack of a proloculus; by the branching mode of perforations and frequent forking (almost unknown in hyaline Foraminifera). They are distinguished from all fossil algae by an important transverse septation united to a persistence and regularity of an axial gullet passing through the septa, to the fastening model and to the wall aspect, both in ordinary light and through crossed nicols (Echinodermic extinction). Moravamminids seems to be in the neighbourhood of Sphinctozoa, from which they differ by smaller size, usual forking and simpler structure. The internal cavity of Moravamminida is to be interpreted as a sponcocoel, perforations as inhalant pores, gullets and openings as oscula. The whole is asconoid-shaped.

23.2 Order Aoujgaliida (Termier et al., 1975)

23.2.1 Diagnosis

Erected or creeping microorganisms. Irregular shape, either conic, cylindrical, dendroid, or spreading and incrusting. External surface often embossed by knobs. Skeleton shaped by superposed laminae and/or radial pillars, both united in a more or less set network. Yellowish, hyaline, granular, porous wall. Simple, cylindrical pores, often choked by recrystallization.

23.2.2 Order Composition

Family 1: Aoujgaliidae (Termier et al., 1975). Complete overlaying of the structural elements (laminae, pillars) in successive layers around the holders (enveloping growth). Genera: *Aoujgalia* (Termier and Termier, 1950; Lower – Middle Carboniferous), *Stacheoides* (Cummings, 1955; Lower – Middle Carboniferous), *Epistacheoides* (Petryk and

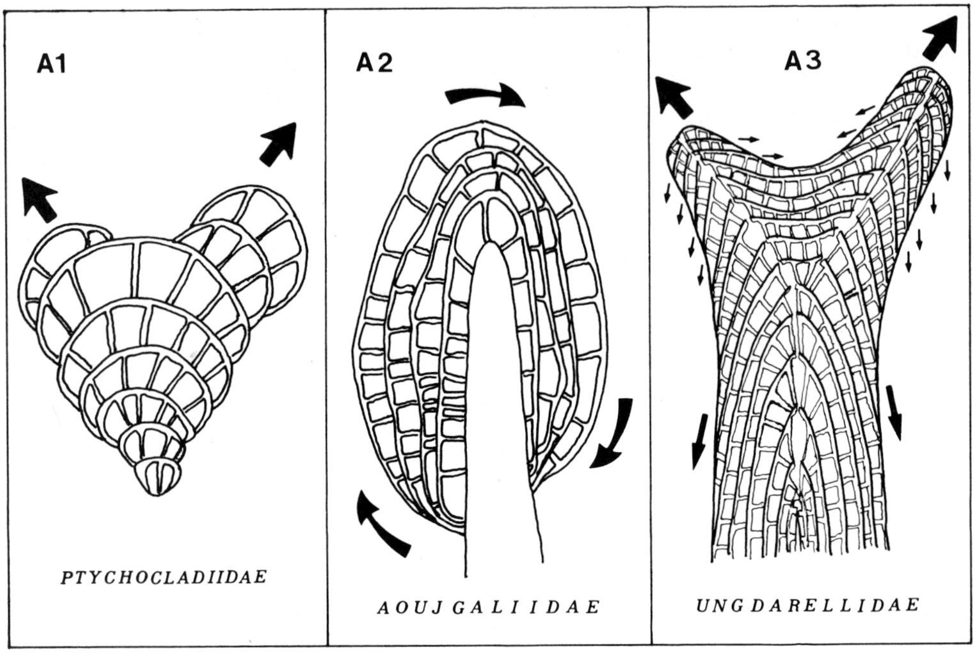

Fig. 2. Aoujgaliids. (A 1–3) Diagram of each family displaying the different growth modes. (A 1) Ptychocladiidae: directional growth. (A 2) Aoujgaliidae: enveloping growth. (A 3) Ungdarellidae: mixed growth

and Mamet, 1972; Lower Carboniferous), *Pseudokomia?* (Racz, 1964; Middle Carboniferous), *Pseudostacheoides?* (Petryk and Mamet, 1972; Lower Carboniferous).

Family 2: Ptychocladiidae (Elias, 1950, *emend.*). Uniseriate spread of successive layers partly overlaying, (directional growth). Genera: *Ptychocladia* (Ulrich and Bassler, 1904; Middle – Upper Carboniferous), *Stacheia* (Brady, 1876; Lower – Middle? Carboniferous), *Fourstonella* (Cummings, 1955; Lower Carboniferous – Permian?), *Cuneiphycus* (Johnson, 1960; Middle – Upper Carboniferous), *Amorfia?* (Racz, 1964; Middle Carboniferous).

Family 3: Ungdarellidae (Maslov, 1956, *emend.*). Generally branching shape with both directional extension and enveloping laminae (mixed growth). Genera: *Ungdarella* (Maslov *in* Kordé, 1951; Lower Carboniferous – Permian: *Foliophycus*, Johnson, 1960), *Komia* (Kordé, 1951; Lower Carboniferous – Permian), *Fasciella?* (Ivanova, 1973; *Shartymophycus*, Koulik, 1973; Lower – Middle Carboniferous).

23.2.3 Systematic Position

Aoujgaliids differ from attached foraminifera by lack of a proloculus and initial planispiral stage, by a different mode of septation and perforation. They are distinct from Rhodophycophyta by their rather coarse reticulate arrangement, and the wall character. Aoujgaliids seem to be very similar to some Stromatoporoids such as Idiostromatids, from which they are separated by smaller size and the absence of astrorhizal features and internal septated tubes.

23.3 Bibliography

Hartman, W.D., Goreau, T.E.: Jamaican coralline sponges: their morphology, ecology and fossil relatives. Symp. Zool. Soc. London **25**, 205–243 (1970).

Termier, H., Termier, G.: Stromatopores, Sclérosponges et Pharétrones: les Ischyrospongia. Ann. Mines et Géol. Tunis, Livre Jubil. Marcel Solignac **26**, 285–297 (1973).

Termier, H., Termier, G., Vachard, D.: Recherches micropaléontologiques dans le Paléozoïque supérieur du Maroc central. Cahiers Micropal., in press (1975).

24 Statistical Analysis of Fossil Algae

H. ZORN, Frankfurt/M., W. Germany

24.1 Introduction

Taxonomic differentiation of fossil algae at species level is often problematic, because of the high variability. Moreover there are some simply formed groups (e.g., blue-green algae), which have few attributes for differentiation. Therefore it is necessary to gather all the information available about morphologic features. Statistical analysis is an effective way to do this because:

1. Variation can be obtained quantitatively in order to recognize changes in space and time.
2. Relations between morphologic characters (variables) can be described exactly in order to differentiate between growth-dependent and independent variation and to compare various samples.

If we look at taxonomic studies on fossil algae, we mostly find some data, but statistical methods are rarely used. Are paleonotologists afraid of mathematics? The intention of this paper is to show by some examples, that many taxonomic problems can be solved by simple statistical methods.

24.2 Examples of Statistical Analysis of Fossil Algae

24.2.1 Variation of One Variable

24.2.1.1 Range. The difference between the largest and smallest items in a sample is called range. This measurement is easy to compute and therefore it is widely used in paleonotologic studies. But as can easily be seen, the range is clearly effected by even a single outlying value. One can expect large samples to have much greater ranges than small ones.

In his study on Permian algae of the Carnic Alps Homann (1972) described some species of *Girvanella*, a filamentous alga. He mentioned not only the total range of the filament diameter, as most authors do, but also the range of each sample. These ranges vary greatly within one species and sometimes do not overlap at all. The total range of a species therefore is much larger than those of a single sample. Is it only due to different sample sizes or are the differences caused by environmental or phylogenetic changes? Do two or more species that have large ranges, occur in those samples? None of these questions can be answered, if we do not know much more about the data.

24.2.1.2 Frequency Distribution. Much more information can be gained from the frequency distribution of the data than from simple ranges. Wood (1957) dealt with *Girvanella problematica* Nicholson and Etheridge, the type species from type locality. He wanted to remove the discrepances between various descriptions by different authors. Therefore he measured the inner and outer diameter of tubes of about 600 specimens and presented frequency distributions, which show two distinct maxima. Wood concluded, that at least two species occur in his sample.

If the distribution has only one peak and is nearly symmetrical, it can be well described by two statistical parameters, the arithmetic mean and the standard deviation. Such bell-shaped distributions (normal distributions) are found in many cases.

24.2.2 Relations between Variables

24.2.2.1 Recognition of Relations. Correlation between morphologic features in most cases is evident, because they are controlled by growth. But one cannot presume this in all cases, as will be demonstrated by the next example.

Several authors have studied dasycladacean *Clypeina jurassica* Favre, which is very common in the back-reef facies of the uppermost Jurassic and basal Cretaceous. One important feature is the number (w) of rays per whorl. Data given by various authors differ greatly. Some quote very small ranges for w (11–13), others quote very large ranges (11–20). In the first case one would like to assume that w remains constant during growth, whereas high ranges may be explained by an increase of rays during growth. Remane (1969), who has measured a great number of individuals, shows in a scatter diagram (his Fig. 48), that there is an increase of rays, when the diameter of the skeleton grows. The variability of w is therefore mainly the result of growth.

24.2.2.2 Ratios. Most paleontologists use the ratios of two parameters, if they are growth dependent. However, as stated by Zorn (1972), ratios are size independent only if a linear relation exists between the parameters and if this straight line of correlation goes through the zero point. Growth lines are normally parabolic (allometric growth). In many cases, especially in fossil algae, we do not know the youngest growth stages and therefore we can approximate the growth line by a straight line. But this line does not run through the zero point (see Fig. 1).

Ranges and standard deviations of ratios are strongly affected, if we exchange the denominator and the numerator (Zorn, 1972: 369ff.).

24.2.2.3 Growth Line and Dispersion around Growth Line. A statistical comparison of growth-dependent features is possible only if a growth line is constructed. An appropriate procedure is to calculate the reduced major axis, described in detail by Imbrie (1956). He also mentioned statistical test for the discrimination of two reduced major axes.

This method was used by the author in 1974 to discriminate between two very similar and highly variable species of Triassic dasycladaceans, *Diplora annulata* and *philosophi.* He concluded, that the type material of *D. philosophi* belongs to *D. annulata* and the samples later described as *D. philosophi* var. *exuberans* and var. *philosophi* do not belong to one species (Fig. 1). The slopes of the reduced major axes differ significantly.

Variation, which is not dependent on growth (shape variability), is described by the dispersion of measured points around the reduced major axis. Imbrie (1956) used the standard deviation of diagnonal distances measured from the reduced major axis as a parameter of shape variability. Zorn (1975) stated that the vertical distances are much more appropriate than diagonal ones. The values (s_k) of the samples mentioned above are shown in Fig. 1.

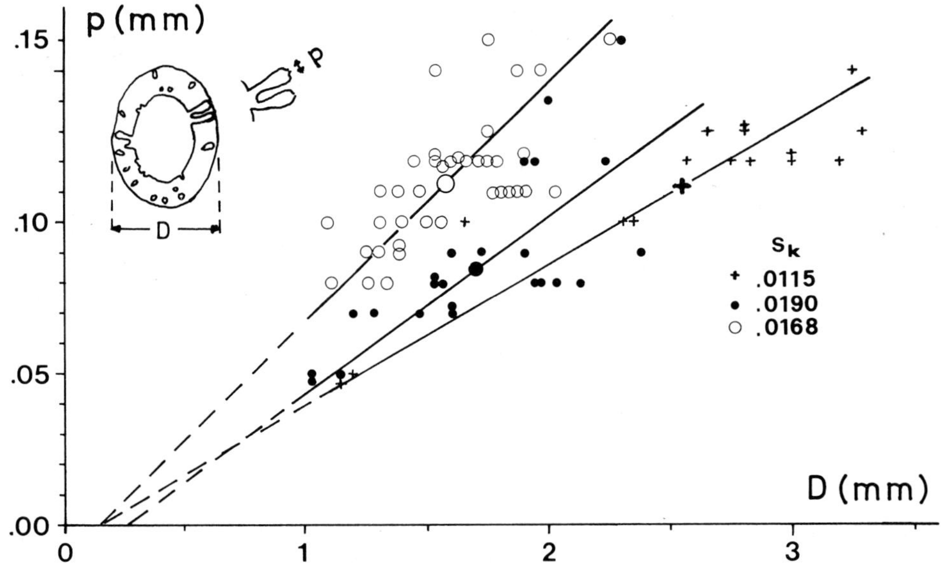

Fig. 1. Scatter diagram and gowth lines of outer diameter *D* and pore diameter *p* from *D. philosophi*. s_k: standard deviation of vertical distances. +: type material of *D. philosophi*; ○: *D. philosophi* var. *exuberans*; ●: var. *philosophi*

The measurement of shape variability is not only important for discrimination of taxonomic groups but also for ecologic and phylogenetic investigations (see sect. 3).

24.2.2.4 Discriminant Functions. A powerful statistical tool for the discrimination of taxonomic groups is the discriminant function. It enables us to consider more than two variables in one procedure and to reduce it to an univariate problem. Davis (1973) considers it at length and presents a computer program.

An application of this method is shown in Fig. 2. The same samples as those of Fig. 1 are projected onto discriminant function lines. Beside D and p a third variable (inner diameter of the skeleton) was added. There is a clear discrimination between type material of *D. philosophi* (triangles) and var. *exuberans* (circles). The other comparisons show overlapping of discriminant scores (dots belong to var. *philosophi*). The distance measure D^2 (called Mahalanobis' distance) enables us to test whether the distance differs significantly from zero. All values in Fig. 2 are significantly different from zero.

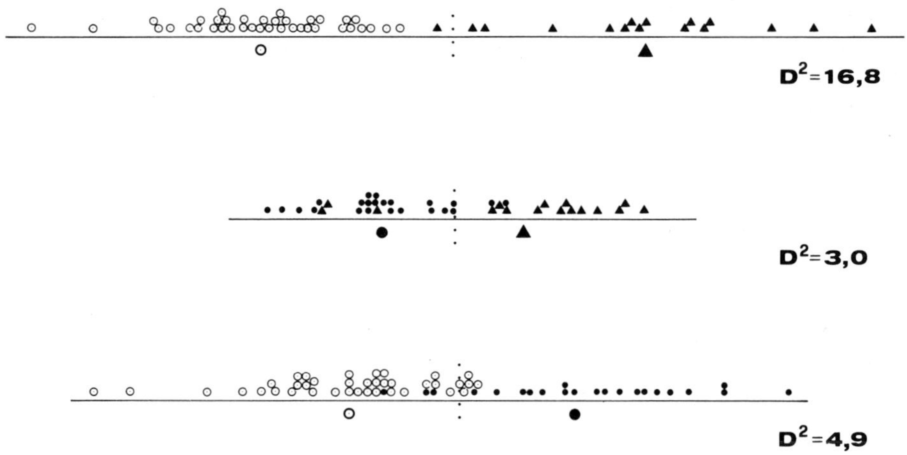

Fig. 2. Discriminant Scores of three variables (outer and inner diameter of skeleton, pore diameter) from *D. philosophi* (same samples as in Fig. 1; ▲ = + in Fig. 1)

24.3 Species Concept and Statistical Analysis

Paleontologic species are mostly based on single specimens from different localities of different ages. The variability of such a collection consists of three components:

1. Variation within a population,
2. Changes in space (environmental influence, geographic isolation),
3. Changes in time (phylogenetic trends).

Such a method of sampling and description does not allow the distinction of the above components to be made qualitatively or quantitatively. Moreover, one cannot assume that the variation of these specimens is a good estimate of the real variation of a species. Definition and discrimination of taxonomic units becomes problematic and subjective.

Therefore the taxonomist has to start with a sample (in a statistical sense), which is collected from one locality and one bed. Fossil algae are often so abundant, that one can obtain such a sample without problems. The next step should be the comparison of samples from different localities of the same age, but from different facies, in order to recognize ecologic and geographic influences on shape and size of skeletons. After the ecologic component of variation is known, samples of different ages can be compared to determine phylogenetic trends.

This procedure certainly takes up a lot of time and is not always feasible, but such steps as are possible should be taken. The high variability mostly found in algal remains should not be regarded as a drawback, which creates a lot of difficulties in taxonomy, but as an opportunity to discover clues about biology (functional morphology), environment and phylogenetic trends.

24.4 Bibliography

Davis, J.C.: Statistics and data analysis in geology. New York–London–Sydney–Toronto: J. Wiley and Sons, 1973.

Homann, W.: Unter- und tief-mittelpermische Kalkalgen aus den Rattendorfer Schichten, dem Trogkofel-Kalk und dem Treßdorfer Kalk der Karnischen Alpen (Österreich). Senckenbergiana lethaea **53**, 135–313 (1972).

Imbrie, J.: Biometrical methods in the study of invertebrate fossils. Bull. Amer. Mus. Nat. Hist. **108**/2, 216–252 (1956).

Remane, J.: Morphological and statistical analysis of *Clypeina jurassica* (Favre) (Dasycladaceae) by means of serial sections. Geol. Romana **8**, 85–116 (1969).

Wood, A.: The type-species of the genus *Girvanella* (calcareous algae). Palaeontology **1**/1, 22–28 (1957).

Zorn, H.: Sind morphologische Merkmale von Organismen durch Verhältniszahlen quantitativ zu erfassen? N. Jb. Geol. Paläontol. Abh. **140**, 354–377 (1972).

Zorn, H.: Variationsstatistische Untersuchungen an *Diplopora annulata* (Schafh.) und *D. philosophi* (Pia), Kalkalgen der alpinen Mitteltrias. Eclogae geol. Helv. **67**/1, 1–30 (1974).

Zorn, H.: Statistische Maßzahlen für die Variabilität wachstumsabhängiger, morphologischer Merkmale. N. Jb. Geol. Paläontol. Abh. **148**/1, 1–11 (1975).

25 Skeletal Structures of Some Calcifying Algae

G. FLAJS, Bonn, W. Germany

25.1 Introduction

The examination of ultrastructures of calcified hard parts, especially in molluscs, coelenterates, brachiopodes, and bryozoans during recent years has revealed a considerable amount of new information on physiologic, systematic, phylogenetic etc. questions. Although calcifying algae are of high paleontologic, ecologic, sedimentologic etc. interest, it is striking that hitherto – except in the coccolites – only sporadic information has existed on the morphologic features of skeletal ultrastructures, as well as on the mode of precipitation of calcium carbonate (e.g., Baas-Becking, 1931; Pobeguin, 1954; Lowenstam, 1955; Wilbur et al., 1969; Arnott and Pautard, 1970; Bayley and Bisalputra, 1970; Marszalek, 1971).

This preliminary paper is based on the investigations of numerous recent and fossil calcifying algae, especially green and red algae. Surfaces, broken as well as polished and etched sections of the calcified skeletons, have been examined by means of the SEM (Cambridge S 4) and the energy-dispersive X-ray spectrometer (ORTEC). I am very indebted to Dr. H. Riedel and Dr. U. Passauer (Museum of Natural History, Vienna, Botanical Dept.) for permission to use the algae herbarium, and to Dr. H. Kollmann (Paleontological Dept. of the Museum) for a large amount of material from J. Pia's collection of fossil algae. The Deutsche Forschungsgemeinschaft supported the investigations within the promotion of the Forschergruppe Biomineralisation.

25.2 Ultrastructures

25.2.1 *Chlorophyta*

Material: about 70 species of 34 genera (recent Codiales, Dasycladales. Caulerpales; fossil representatives of nearly all tribes of Dasycladaceans and Codiaceans).

The morphology of the needle-shaped aragonite crystals on calcified species (Pobeguin, 1954) hitherto examined is very similar in all species. There are no major differences between those species that precipitate aragonite in spaces separated from direct contact with seawater, and those species with encrusting surfaces.

The classical and best known example is *Halimeda* (Pl. 1/1). As pointed out by Wilbur et al. (1969), the mineralization starts on the surfaces of utricles and medullary filaments. The authors could not detect any traces of organic matrix influencing the calcification.

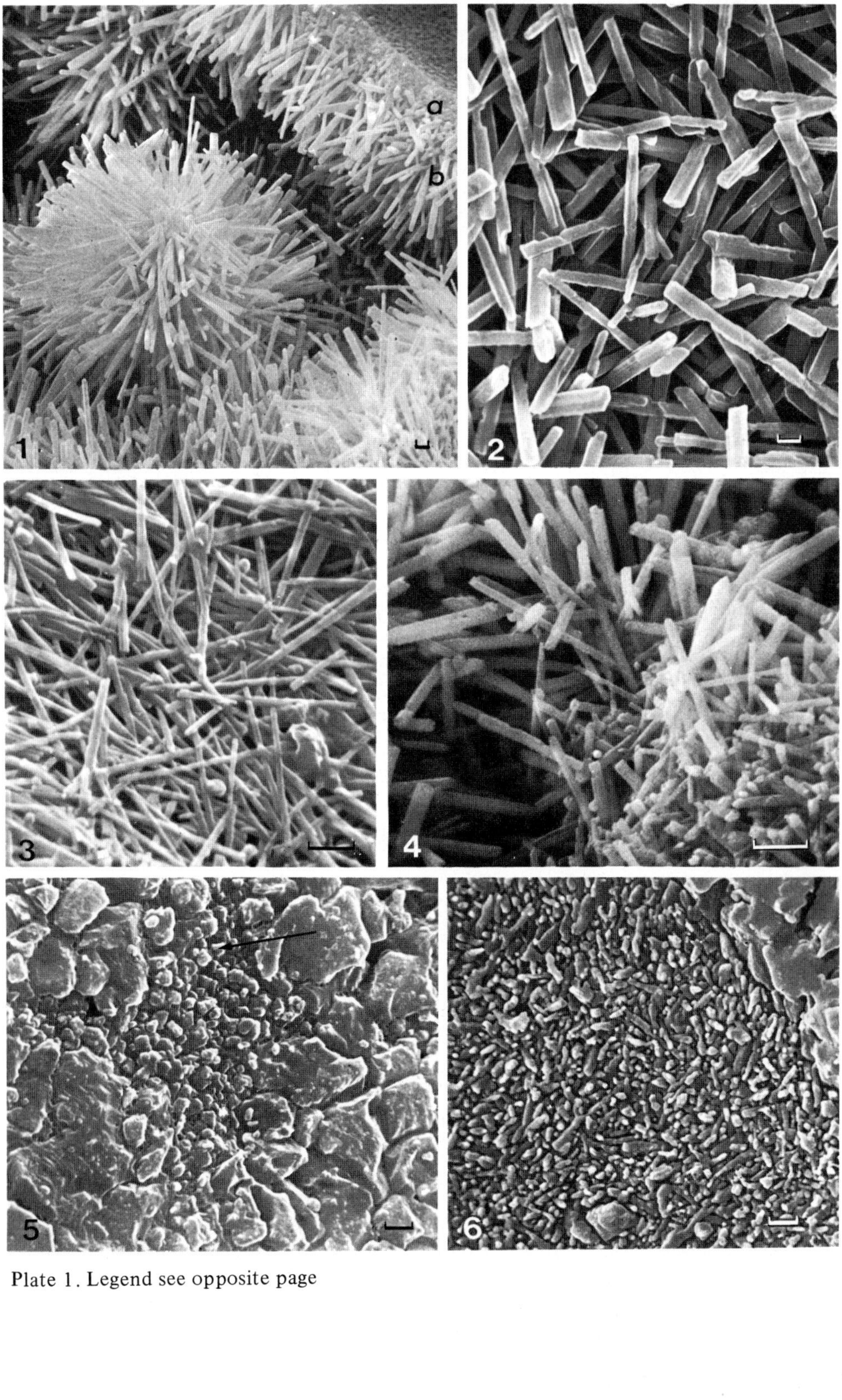

Plate 1. Legend see opposite page

In the typical case (Pl. 1/1), three stages of calcification can be distinguished in *Chlorophyta* (cf. Wilbur et al., 1969):

1. An impregnation of filament walls with ± rounded, very fine-grained aragonite, followed by the formation of small (less than 1 μ) ± granular crystals with well-defined crystal faces.
2. The growth of long aragonite neeedles normally to the surface of filaments.
3. A filling-out of the remaining spaces with irregularly arranged aragonite needles.

Stage 2 may sometimes be lacking. In such cases, the first layers of needle-shaped aragonite crystals are arranged tangentially to the filament walls.

Except for *Penicillus* (Marszalek, 1971), the distribution and arrangement of needles does not show any systematic value. Length and shape of the needles (stage 3) may exhibit, in some special cases, certain differences between single species. Wilbur et al. reported significant differences in crystal size and shape between laboratory grown *Halimeda monile* and *H. incrassata*, Marszalek for *Penicillus capitatus* and *P. dumentosus.* In my material, *Cymopolia mexicana* (from three localities) showed in all cases long needlcs (Pl. 1/3), differing from all other recent and fossil *Cymopolia*. On the other hand, great varieties of needle length, even within the same thallus, are recognizable (Pl. 1/4). Furthermore, the influences of ecologic factors (esp. light intensity, temperature, salinity, etc.) on the crystal morphology are widely unknown. Perhaps statistical work can reveal some relations between crystal morphology, ecologic conditions and systematic position.

The type of calcification described above is common in aragonitic *Chlorophyta* and also in calcifying *Phaeophyta*. It is a simple case of calcification and not a skeleton formation in the true sense, with the possible exception of stage 1, which leads to a stiffening of the filaments.

In the fossil forms, species of Dasycladaceans and Codiaceans with ± preserved primary structures show comparable skeletal ultrastructures (Pl. 1/2, 4 and 6). The value of the aragonite needle structure for classification of fossil forms with doubtful systematic position is restricted, because, e.g., most worms and some sponges exhibit very similar structures.

All figures are SEM-micrographs. Scale in all figures is 1 μ, except Pl. 2/1 and 5 (scale = 10 μ

◀ Pl. 1/1–4. Chlorophyta (broken sections)

Pl. 1/1. *Halimeda tuna* Lam., (Bahamas); the three stages of calcification (1, 2, 3; see sect. 2.1) clearly visible; x 2100

Pl. 1/2. *Neomeris arenularia* Mun.-Ch. (Eocene, Le Fayel/France); typical aragonite needles of fossil Dasycladacean; x 3300

Pl. 1/3. *Cymopolia mexicana* AG. (Capo Blanco/Mex.); long aragonite needles; x 6900

Pl. 1/4. *Neomeris scrobiculata* Gümbel (Eocene, France); varying crystal size in stage 3 (see sect. 2.1); purely aragonitic preservation, diagenetically not altered; x 3300

Pl. 1/5. Rhodophyta – *Lithothamnium* sp. (Upper Tertiary, Vienna Basin; strong diagenetic alteration of primary structure (cf. Pl. 2/2–4); *arrow* indicates skeleton; x 4150

Pl. 1/6. *Tubiphytes* sp. (Upper Carboniferous, Carnic Alps/Austria); possible relicts of aragonite neddle structure; x 3900

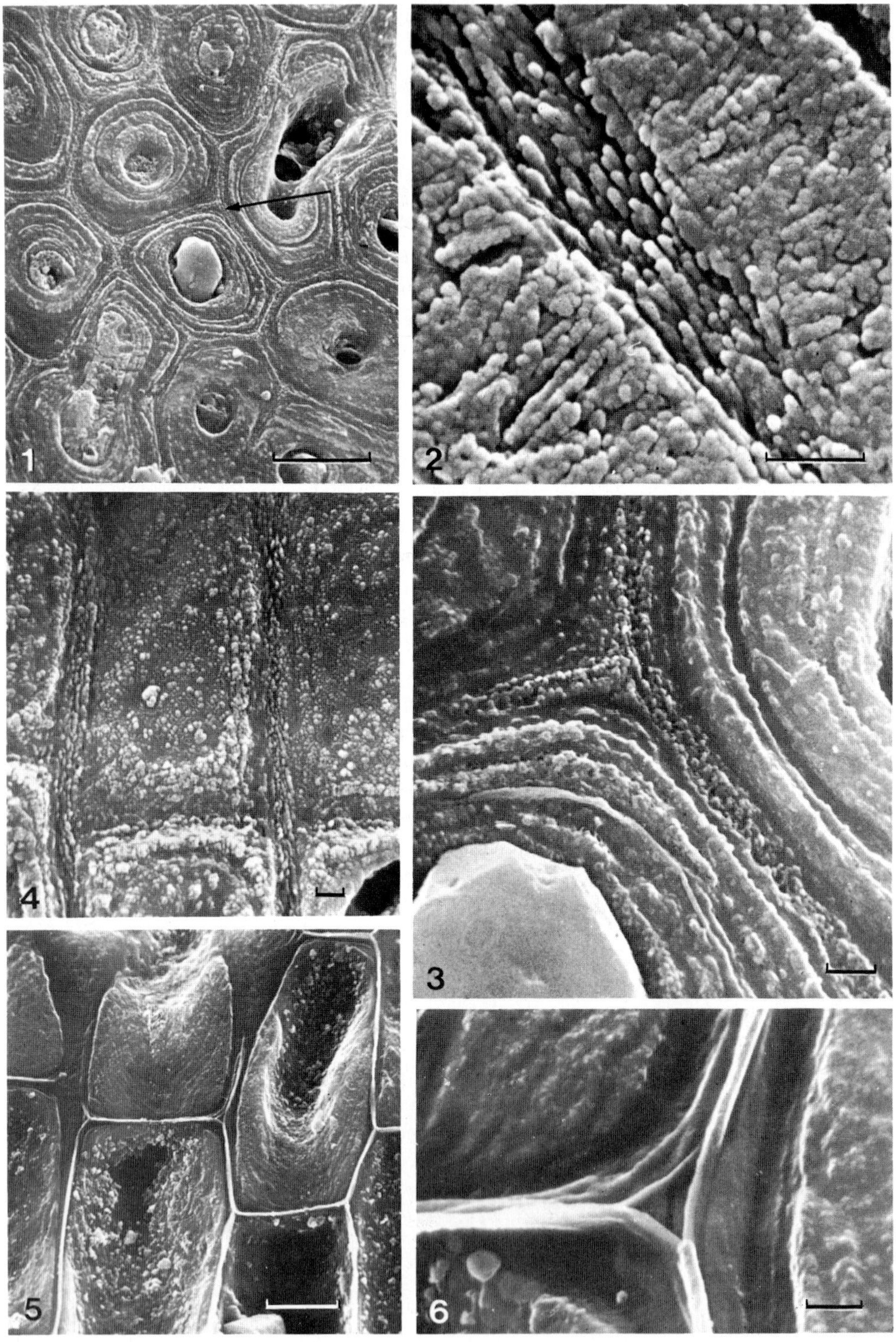

Plate 2. Legend see opposite page

25.2.2 *Rhodophyta*

Material: about 30 species of 15 genera. For ecologic purposes, some common genera of Corallinaceae from climatically very different regions (Red Sea, Mediterranean Sea, North Sea, Arctic Sea) have been studied.

In contast to the green algae, the cell walls themselves become calcified. Following Bayley and Bisalputra (for some Corallinaceae), the cell layers near the surface are not yet calcified. Calcification occurs in deeper layers and is related to degeneration of the cells and to the formation of numerous vesicles. The cell walls can be divided into layers. "Each cell is surrounded by a layer of the amorphous material. . . This amorphous material appears almost as a continuum between adjacent cells on the apex, and calcification is laid down within it" (Bayley and Bisalputra, 1970; p. 89; see also Arnott and Pautard, 1970. Figs. 9 and 43).

25.2.2.1 Aragonite Precipitating Rhodophyta (Squamariaceae). Since only three species of one genus (*Peyssonelia*) have been examined hitherto, these first results have to be confirmed by further investigations. In etched sections, the polygonal outline of the single cells is indicated by thick organic layers (Pl. 2/5). Each cell is enveloped by its own membrane, visible in corners of adjacent cells, and no calcification between the organic membranes is seen (Pl. 2/6). The aragonite is laid down in concentric layers of various thickness, sometimes having a lamellar appearance, but in most cases they show radial arrangement of acicular single crystals.

25.2.2.2 Calcite Precipitating Rhodophyta (Corallinaceae). In all specimens examined, three stages of calcification are recognizable: (a) the formation of a "Primary layer", (b) the formation of "Secondary layers", and (c) a final, ± inorganic precipitation of calcite in the remaining cell lumina.

a) The *Primary Layer* shows a very characteristic structure (Pl. 2/1–4). It is built up of numerous very small crystals, mainly in the longitudinal cell walls, their long axis arranged parallel to the cell wall (normally to the surface). The size of the single acicular crystals is very constant (up to 1 μ long, 0,15 μ thick). The horizontal cell walls are also calcified, but the shape of crystals is not recognizable.

The primary layer, by reason of its thinness very difficult to recognize in broken sections, appears as the first calcified structure in the weakly calcified upper cell layers.

◄ Pl. 2/1–6. Rhodophyta

Pl. 2/1. *Lithophyllum* sp. (Port Sudan/Red Sea); etched cross section; note primary layer (*arrow*) marking polygonal shape of cells, and concentric secondary layers; x 1350

Pl. 2/2. *Lithophyllum incrustans* Philip. (Banyuls/France, Mediterranean Sea); detail of etched longitudinal section; single crystals of primary layer clearly visible. secondary layers with crystal arrangement normally to middle primary layer; x 13 600

P. 2/3. *Lithophyllum* sp. (Port Sudan/Red Sea); horizontal etched section: view on tops of single crystals in primary layer; lamellar formation of secondary layers; x 6700

Pl. 2/4. *Lithophyllum* sp. (Tromsø/Norway); longitudinal etched section; primary layer in longitudinal cell walls clearly visible; x 3900

Pl. 2/5. *Peyssonelia polymorpha* Schm. (Banyuls/France); thick organic layers and lamellar aragonitic layers in etched section; x 1000

Pl. 2/6. Detail from Pl. 2/5; organic layers of adjacent cells; x 7900

It seems to be a calcification of the middle lamella region (Pobeguin, 1954). In contradiction to Baas-Becking and Galliher (1930), the single crystals show an orientation parallel to the cell walls.

The primary layer is characteristic for all calcitic *Rhodophyta* examined to date, from all different climatic regions. The well-ordered arrangement, the constant size of single crystals, and the independence of that structure from ecologic influences (especially temperature) suggest that the formation of the primary layer is due to biomineralization in a strict sense; the precipitation of crystals with well-defined morphology in a well-defined arrangement, probably controlled by organic matrix. It is the most complicated mode of precipitation of calcium carbonate in algae hitherto known.

The magnesia content of the primary layer seems to be constant in one colony and in the same climatic region (calculated from peak integrals of X-ray spectrometer; no absolute values can be given). The significant seasonal change in Mg content reported by Chave (1954) does not concern the primary layer, but is restricted to the secondary layers (described below).

The principal difference between aragonitic Squamariaceae and calcitic Corallinaceae is the formation of the primary layer by calcification of the middle lamella region in the latter group.

b) The *Secondary Layers* are comparable to the aragonitic deposits in Squamariacea; additional concentric layers of calcite are formed (Pl. 2/1–3). The single crystals, of about the same size as in the primary layer, exhibit, in the most cases, a perpendicular arrangement to the primary layer (Pl. 2/2). Only in few specimens, especially in *Goniolithon* but also in other genera, do the secondary layers show a concentric lamellar aspect, but by reason of the extremely small crystal size, a radial or tangential arrangement of single crystals is not recognizable (Pl. 2/3).

In fossil *Rhodophyta*, the diagenetic alteration, by reason of the very small crystal size, seems to occur very quickly. Even in species from the Upper Tertiary, the preservation of primary structures is the exception (Pl. 1/6). However, some well-preserved Solenopores from the Upper Triassic and Jurassic showed relict structures very similar to the primary structures of Corallinaceae. Besides a detailed description of structures etc., the phylogenetic aspects of the investigation of algal ultrastructures will also be discussed in a following paper.

25.3 Bibliography

Arnott, H.J., Pautard, F.G.E.: Calcification in plants. In: Biological Calcification. Schraer, H. (ed.). Amsterdam: North Holland Publ. Co., 1970, pp. 373–446.

Baas-Becking, L.G.M., Galliher, E.W.: Wall structure and mineralisation in coralline algae. J. Phys. Chem. **35**, 164–177 (1931).

Bayley, A., Bisalputra, T.: A preliminary account of the application of thin-sectioning, freeze-etching and scanning electron microscopy to the study of coralline algae. Phycologia **9**, 83–101 (1970).

Chave, K.E.: Aspects of the biochemistry of magnesium. I. Calcareous marine organisms. J. Geol. **62**, 266–283 (1954).

Lowenstam, H.A.: Aragonite needles secreted by algae and some sedimentary implications. J. Sed. Petr. **25**, 270–272 (1955).

Marszalek, D.S.: Skeletal ultrastructure of sediment producing green algae. Scanning Electron Microscopy. Johari, O., Corvin, I. (eds.). Chicago: 1971, pp. 273–280.
Pobeguin, T.: Contribution à l'étude des carbonates des calcium. Précipitation du calcaire par les végétaux. Comparaison avec le monde animal. Ann. Sci. Nat., Bot. **15**, 29–109 (1954).
Wilbur, K.M., Colinvaux, L.H., Watabe, N.: Electron microscope study of calcification in the alga *Halimeda* (order Siphonales). Phycologia **8**, 27–35 (1969).

26 Role of Calcareous Algae in Atlantic Continental Margin Sedimentation

J.D. MILLIMAN, Massachusetts, U.S.A.

26.1 Introduction

Much has been written about the habitat of modern benthic calcareous algae in the world oceans (e.g., Johnson, 1961; Ginsburg et al., 1972; Adey and MacIntyre, 1973). However, data concerning the role of these various algae in continental margin sedimentation are generally limited to regional studies, some of which have been summarized elsewhere (Milliman, 1974). Since many fossil algae are presumed to have grown in shelf-depth (20–200 m) environments (e.g., see papers by Flügel and by Conrad and others, this volume), calcareous algae and their sediments should be reconsidered with respect to modern continental margins. Data presented in this paper come primarily from three broad areas bordering the Atlantic Ocean (Fig. 1): eastern United States (Milliman, 1972; Milliman et al., 1972), Brazil (Milliman and Summerhayes, 1975; Vicalvi and Milliman, 1976), and western Africa (Summerhayes et al., 1976; Milliman, unpublished data); supplemental data have been used to extend these observations into an ocean-wide synthesis.

Contract support for these studies came from the U.S. Geological Survey (for the eastern United States). Projeto REMAC (a consortium of Brazilian agencies and companies) and the International Decade of Ocean Exploration (IDOE – National Science Foundation Grant GX–41960, Brazil) and IDOE (NSF Grand GX–28193, Africa). I thank K.O. Emery, O.H. Pilkey, and I.G. Macintyre for their comments on the manuscript. This is contribution number 3666 of The Woods Hole Oceanographic Institution.

26.2 Types of Calcareous Algae

Two types of benthic calcareous algae contribute predominantly to continental margin sediments: coralline algae (Rhodophyta) and the codiacean *Halimeda* (Chlorophyta). *Halimeda* generally occurs within unconsolidated sediments as discrete plates. It is found both within coral reef sediments (ranging in concentration, e.g., from 10 to 34% within the relict reef sediments of Florida; Milliman, 1972) and in shelf sands (see below).

Three general types of coralline algal sediments are recognized in the Atlantic, each composed of different growth forms:

1. "Maerl" is a sand or gravel in which more than half of the sediment is composed of the branches of freestanding corallines (Jacquotte, 1962); many maerls contain more than 75% branches (Fig. 2a). Although this sediment is widely recognized in the Medi-

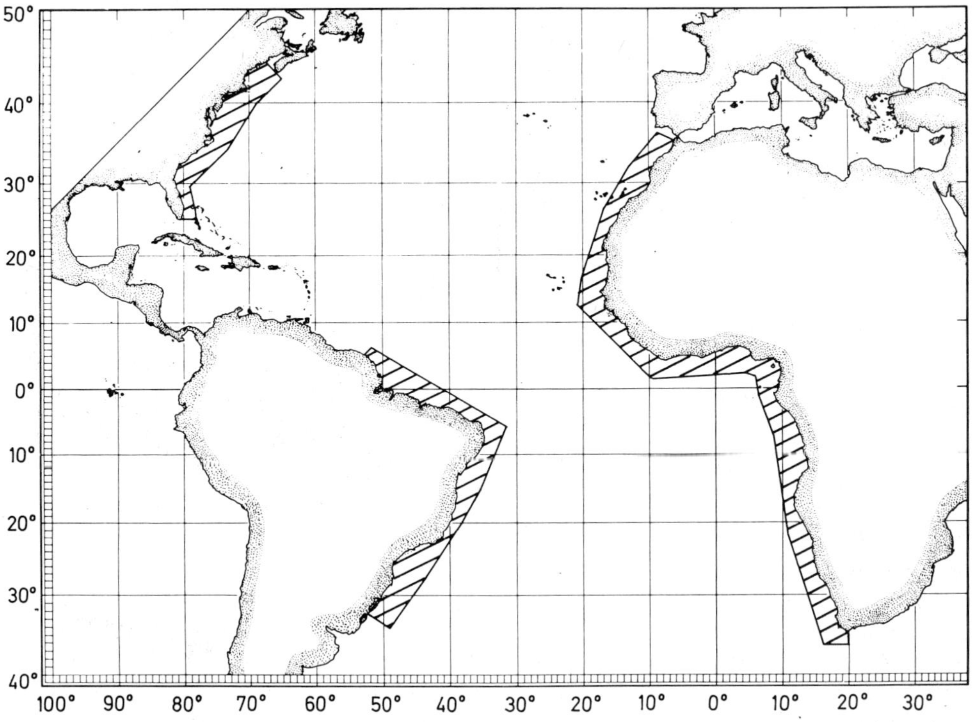

Fig. 1. Atlantic Ocean, showing three study areas upon which much of discussion based

terranean Sea (Blanc, 1972), similar deposits occur elsewhere in the eastern and western Atlantic (see below). Of those maerl deposits studied taxonomically, *Lithothamnium solutum, L. coralloides* and *Phymatolithon calcareum* are the major species (Jacquotte, 1962; Adey and Macintyre, 1973). The reefs from which these algal sands presumably are derived have not been found in our samples; most likely they occupy small areas relative to the wide-spread algal sands (Bosence, this vol.).

2. Encrusting corallines occur as reef limestones (such as coral reef or algal reef rock) and as prominent components within unconsolidated sediments (admixed usually with coral, barnacles and bryozoans). This sediment type is not dissimilar to the "biocenoses coralligenes" of Peres and Picard (1964). Encrusting algal sediments range in size from boulders (reef rock) to fine detritus (Fig. 2b), but in nearly all instances, the presence of nearby hard substrates, upon which the encrusting forms can attach, is implicit.

3. Rhodoliths include algal nodules and concretions in which the algal carbonate has grown around a nucleus (Bosellini and Ginsburg, 1971; Adey and MacIntyre, 1973) (Fig. 2c) as well as compact branching colonies, which can provide the branches for algal maerl (Bosence, this vol.). Sizes range from mm to cm in diameter. Form and shape of the rhodoliths depend upon the energy conditions and water depth as well as the species involved (Bosellini and Ginsburg, 1971; Adey and Macintyre, 1973). For instance, algal nodules are analogous to cyanophytic oncolites in that the concentric coatings presumably need turbulent environments by which new surfaces are continually presented for encrustation.

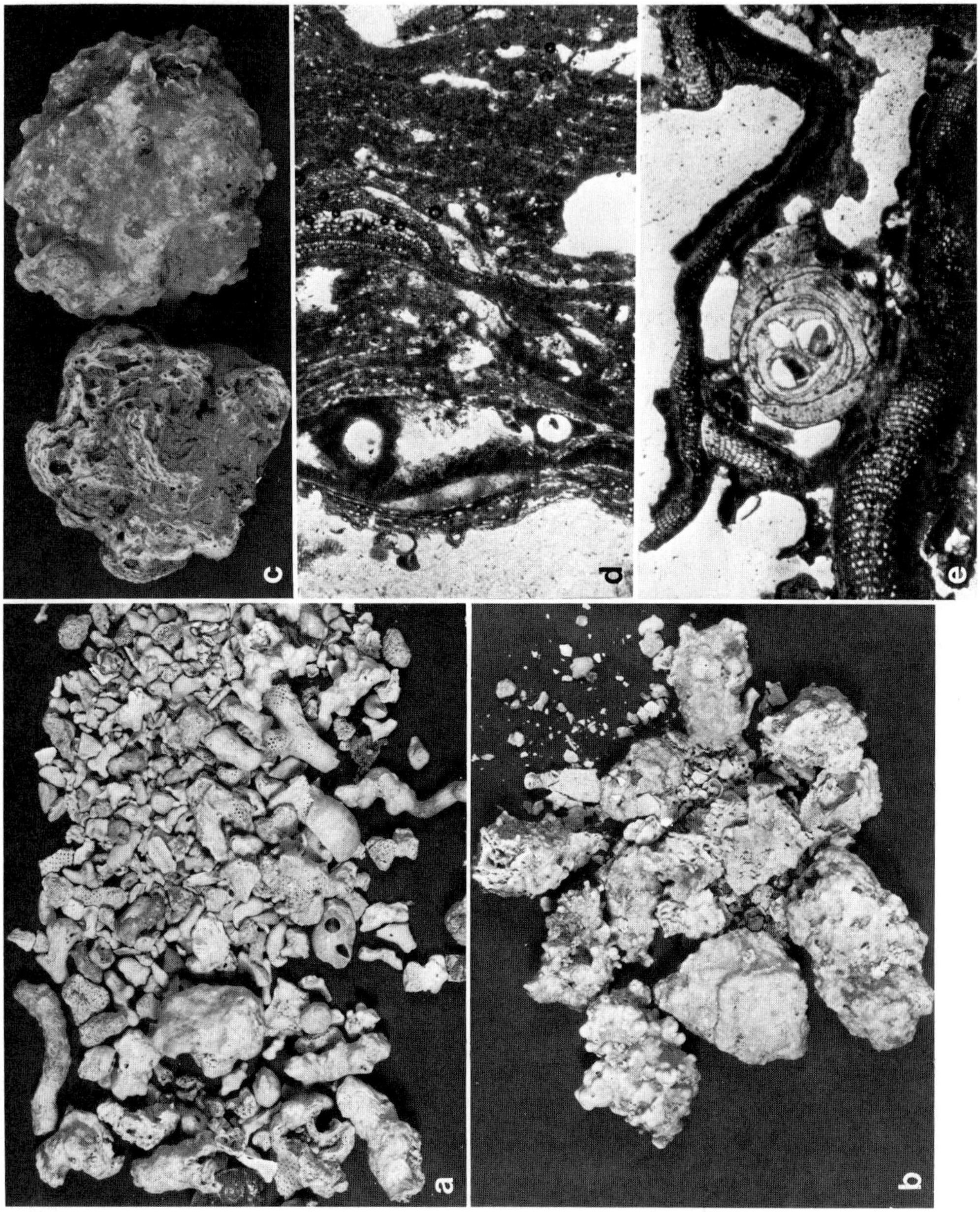

Fig. 2 a–e. Coralline algal types in the Atlantic Ocean. (a) Maerl, with significant amounts of bryozoan and molluscan fragments. (b) Coralline algal sediment, with both encrusting and nodular forms. (c) Algal nodules (rhodoliths), *left* one has been sliced to show concentric structure. (d) and (e) Photomicrographs of algal crusts from Moroccan shelf. (d) shows laminated structure of crust. (e) Illustrates incorporation other detritus between algal laminae, in this case hyaline foraminifera. Magnifications of (a–c) are 0.6, (d) and (e) are 12 x and 50 x, resp.

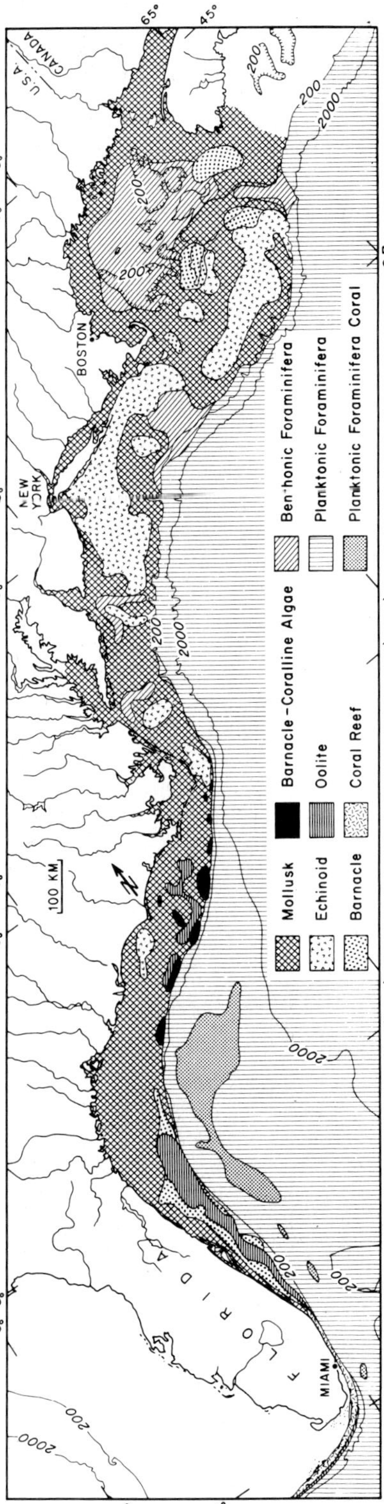

Fig. 3. Carbonate assemblages off eastern United States. Note assemblages containing coralline algae (Barnacle-Coralline Algae and Coral Reef) occur only south of Cape Hatteras (35° N). From Milliman et al. (1972)

26.3 Distribution

Western Atlantic. Coralline algae have been observed to coat entire rock surfaces in all climates, from arctic to tropical.(see Adey, 1966). However, as major components in shelf sediments, the distribution is far more restricted: calcareous algae are prominent constituents from Cape Hatteras, North Carolina (35° N) to southern Brazil (23° S). This is the area generally considered to encompass the tropical and subtropical oceanic environments in the western Atlantic (Vicalvi and Milliman, 1976). Corallines, being more cosmopolitan than *Halimeda*, occur throughout this entire zone, while *Halimeda* is strongly limited to the tropics (27° N to 20° S). In nearly all instances, however, calcareous algal sediments are restricted to those areas in which the influx of terrigenous sediment is low; generally this means the middle and (particularly) the outer shelf (see below).

Coralline algae off the southeastern United States generally constitute between 15 and 35 % of the modern and relict carbonates within the coral reef facies (*Halimeda* represents another 10–35%) and 10–45% within the barnacle-coralline algae facies (Fig. 3; Milliman, 1972). Similar sediments rich in encrusting corallines have been noted from the shelf in the Gulf of Mexico and Caribbean Sea (Gould and Stewart, 1955; Nota, 1958; Logan et al., 1969; Kornicker and Bryan, 1969). Although rhodoliths and maerl are locally present throughout this area, they do not appear to occur in the wide-spread assemblages seen for the encrusting forms.

A somewhat different situation exists off Brazil, where calcareous algae are *the* major sedimentary components on much of the northeast and southeast shelf. Coralline algal maerl occurs throughout much of the middle shelf, while the outer shelf is dominated by algal reef sediments (Fig. 4); Mabesoone and Coutinho, 1970; Kempf, 1972; Summerhayes et al., 1975; Melo et al., 1975; Vicalvi and Milliman, 1975). The maerl averages about 60% (by volume) coralline algae, with little or no algal limestone present. In contrast, encrusting coralline sediments contain about 75% coralline algae, of which about 30% is algal limestone (Milliman, 1975). In both sediment types, carbonates generally exceed 75% and often 95% of the sediment (Fig. 4). *Halimeda* sands, termed "maerl" because of their strong textural and environmental similarity to coralline algal maerl, are restricted to the middle shelf off northeastern Brazil; they contain an average of about 60% *Halimeda* (by volume) and an additional 20% coralline algae, even bryozoan-dominated sediments (Fig. 4) contain more than 25% calcareous algae (Milliman, 1975).

To this writer's knowledge, the vast distribution of algal-rich sediments off northeastern and southeastern Brazil is unmatched in the modern oceans: between Baia de São Marcos and Santos (Fig. 4), representing more than 3000 km of coastline and 25° of latitude, sediments on the middle and outer shelf average more than 40% calcareous algae and in many places appreciably more than 60%. This dominance results from the interaction of three prime factors. First, the shelf is shallow (generally less than 60 m deep) and the waters are warm. Second, the northeastern and southeastern Brazilian coastline and hinterland are predominatly semi-arid. Furthermore, because of the existing drainage patterns in Brazil, much of the hinterland is drained by the Amazon and Paraná rivers; along the intervening coast, only the São Francisco River can be considered important. As a result of these factors, sediment influx along northeastern and southeastern Brazil is slight, and hard substrates (upon which epifauna and epiphyta can attach) abound; carbonate content is correspondingly high. Third, most major species of hermatypic corals are notably lacking off Brazil (the result of being unable to migrate southward through the turbid conditions caused by the Orinoco and Amazon rivers, as well as the strong northward drift of the Guiana Current). Coral reefs, therefore, are absent. Freed from the competition of space with carbonate-producing epifauna (see Adey and MacIntyre, 1973, p. 894), and given the warm shallow waters, the absence of terrigenous influx and the many hard substrates, calcareous algae have assumed the dominant carbonate role. Only locally are sediments rich in epifaunal organisms (Fig. 4) and even bryozoan sediments contain appreciable algae (see above).

Another point of particular interest in the western Atlantic is the dominance of encrusting algal sediments and algal limestone on the outer shelf and upper slope along the southern United States, the Caribbean and northeastern South America. This feature has been particularly noted off the southern United States (e.g., Stetson, 1953; Parker and Curray, 1956; Ludwick and Walton, 1957; Menzies et al., 1966; MacIntyre and Milliman, 1970), although unquestionably this is partly a function of the more intense exploration of this area. This structure is not necessarily a "reef" in that it does not always rise above the adjacent bottom, nor need it be constructional: often the corallines encrust submerged coral reefs, oolitic dunes or even Tertiary outcrops (MacIntyre and Milliman, 1970). Also, this feature is not continuous geographically (being absent in areas with large terrigenous influx, such as river mouths) nor locally (reefs and sediments often occur as discrete clumps rather as a continuous ridge). Still, this algal assemblage represents one of the major sedimentologic facies on the outer shelf in the subtropical and tropical western Atlantic.

In subtropical waters, shelf-edge algal sediments are composed largely of corallines and barnacles, but in tropical waters, barnacles are replaced by coral and bryozoans (Milliman, 1972; MacIntyre and Milliman, 1970; Vicalvi and Milliman, 1976). Petrographic and radio-carbon data suggest that these "reefs" thrived during lower stands of sea level, although that does not necessarily infer that they formed intertidally (see below) nor that all organisms are relics of former lower sea levels. Many of the barnacles and bryozoans are probably modern, and are encrusting the relict coralline algae.

An interesting point of speculation is the fact that seismic profiles along much the same area presently covered by the algal facies show that the outer shelf and upper slope is underlain by deeply buried topographic highs, which Uchupi (Emery and Uchupi, 1972) has interpreted as organic reefs. Whether this buried reef, behind which much of the western Atlantic continental shelf seems to have accreted (Emery, 1968), is similar in composition and growth to the late Quaternary coralline algae facies must await future continental margin drilling.

Eastern Atlantic. Although data from the eastern Atlantic (particularly West Africa) are generally more localized (partly a function of the large number of nations which border this side of the ocean) and therefore less integrated, we can conclude that calcareous algae are far less important as sedimentary contributors than they are in the west.

Fig. 4. Carbonate content and assemblages in shelf sediments off Brazil. Note: broad ▶ extent of carbonate-rich sediments off much of northeastern and southeastern Brazil, and dominance of calcareous algae. Again, corallines decrease markedly south of Cabo Frio and are absent south of subtropical zone (approx. 25° S). From Milliman (1975)

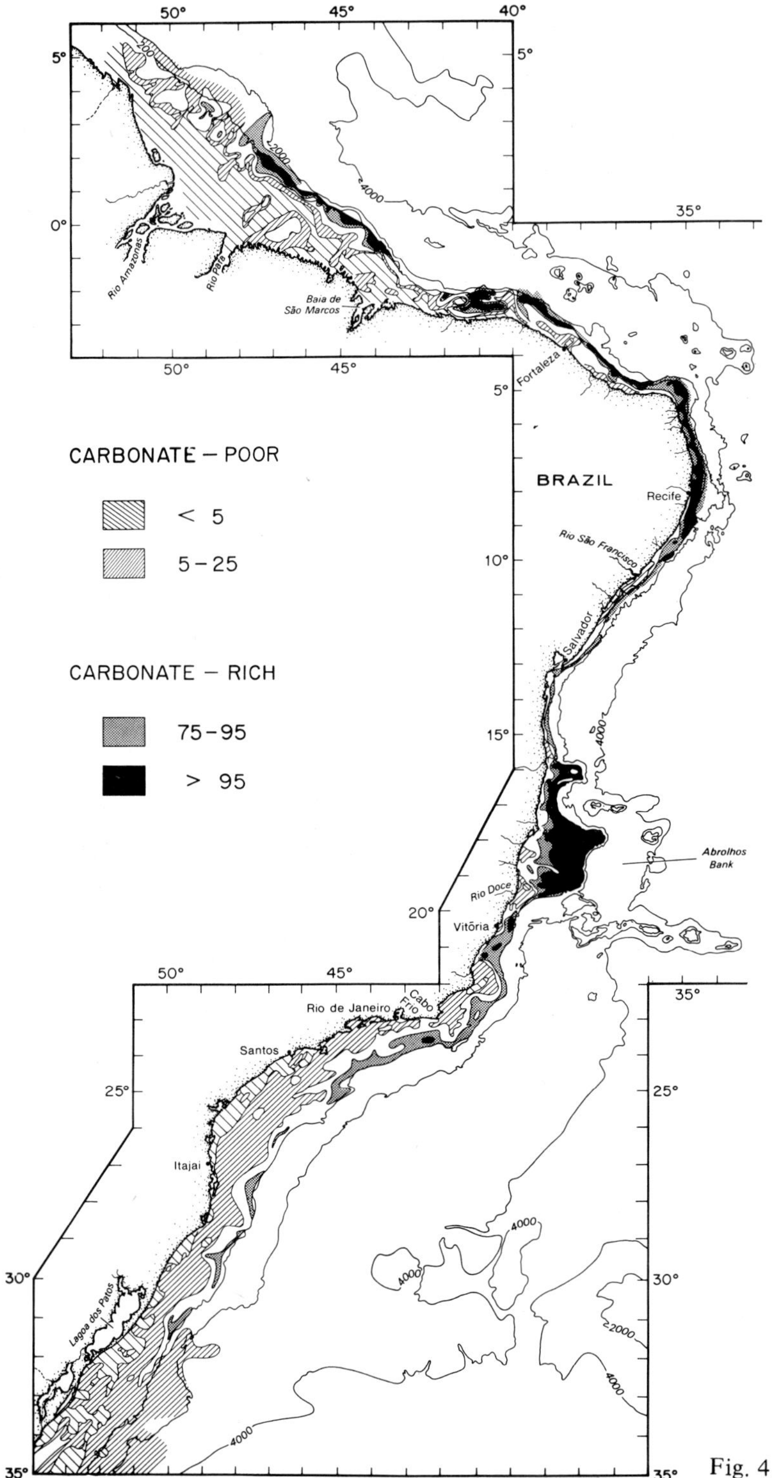

Fig. 4. Legend see p. 237

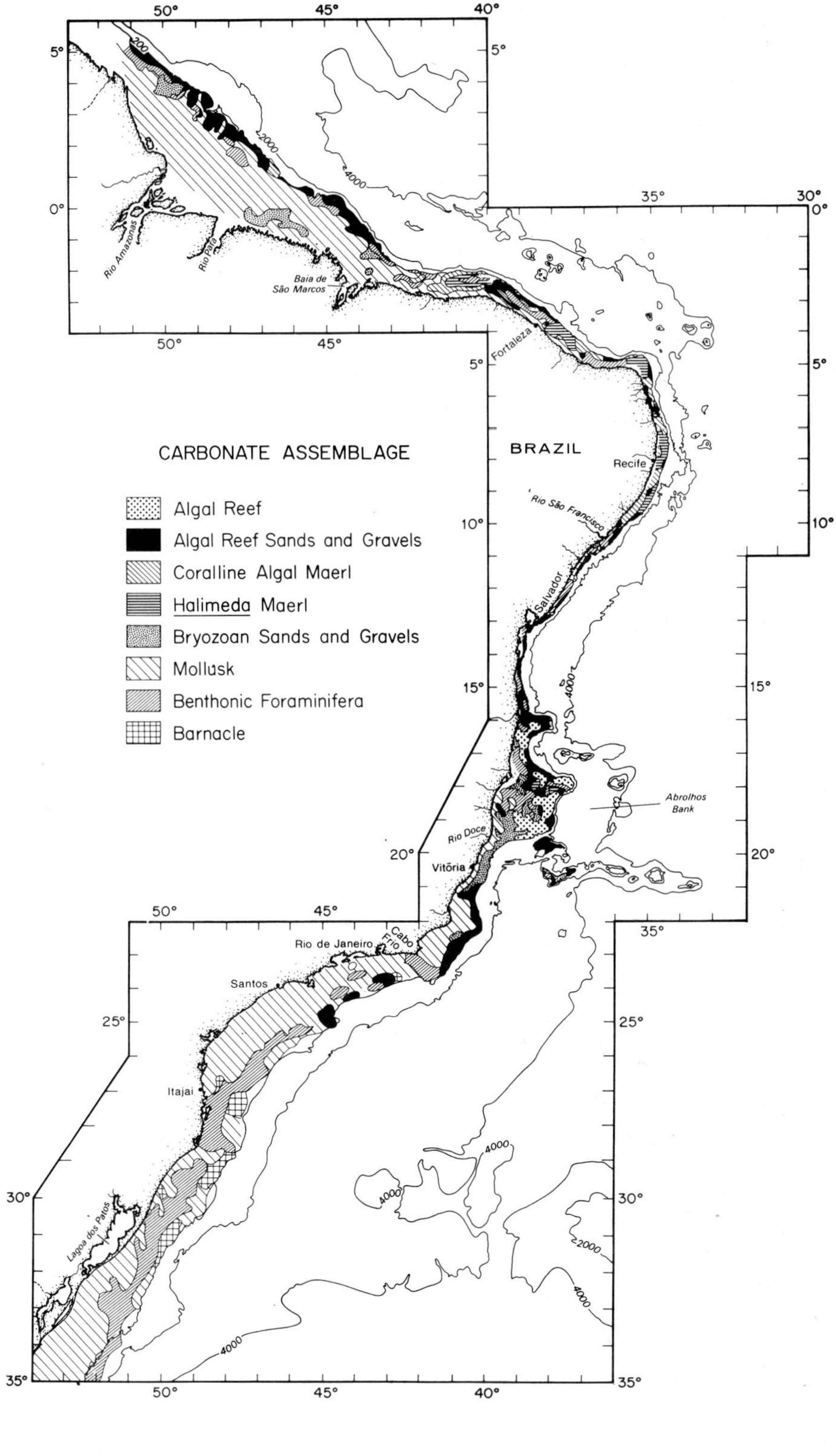
CARBONATE ASSEMBLAGE
Algal Reef
Algal Reef Sands and Gravels
Coralline Algal Maerl
Halimeda Maerl
Bryozoan Sands and Gravels
Mollusk
Benthonic Foraminifera
Barnacle
BRAZIL
Rio Amazonas
Rio Pará
Baia de São Marcos
Fortaleza
Recife
Rio São Francisco
Salvador
Abrolhos Bank
Rio Doce
Vitória
Cabo Frio
Rio de Janeiro
Santos
Itajaí
Lagoa dos Patos
200
2000
4000
50°
45°
40°
35°
30°
5°
0°
10°
15°
20°
25°

Corallines do form prominent facies off southern South Africa (Seisser, 1972), the Ivory Coast (Martin, 1973) and Senegal (McMaster et al., 1971), but these deposits are local and relatively small. This restriction results from several factors, such as the generally cooler temperatures (although as will be seen below, this factor is probably less important than others), the influx of modern fluvial sediments (e.g., off Nigeria and the Ivory Coast), the very deep shelf coincident with upwelling and resultant low oxygen content (southwestern Africa) and the lack in many areas of the necessary hard substrata upon which corallines can attach. As an example of this latter factor, the shelf sediments off Spanish Sahara are extremely rich in carbonate (generally greater than 90% $CaCO_3$), but the lack of hard substrates largely prevents the colonization of encrusting organisms (Summerhayes et al., 1976); as a result, sediments are predominantly a molluscan shell-hash.

Coralline algal development off western Europe can be prominent in many shallow-water locations (e.g., Brittany, Norway, and Ireland; Foslie, 1894; Boillot, 1964; Keary, 1967; Adey and MacIntyre, 1973; Bosence, this vol.), but in general, corallines do not appear to be significant sedimentary components in water depths greater than 20 m. Thus, the only prominent areas of algally-derived shelf carbonates appear to be off northwestern Africa (Morocco) and the Mediterranean. Off Morocco, corallines appear in two types of sediments. In nearshore waters (generally within 5 km of the coast and in water depths less than 30 m), prominent algal crusts, up to several cm thick, have grown over exposed rock outcrops (Fig. 2d, e). All indications are that these crusts are modern. The second type of algal-rich sediments is the *Miniacina miniacea* facies, in which both branching and encrusting forms occur (Fig. 5). Corallines within the *Miniacina* facies are particularly prominent in water depths shallower than 95 m; presumably most of the algal material is relict (Milliman, 1976).

A number of islands and sea mounts off northwest Africa also contain coralline algal sediments, primarily as encrusting and rhodolith assemblages (McMaster and Conover, 1966; Müller, 1969; von Rad, 1974). Corallines also are major sedimentary constituents on the shelf and banks throughout the Mediterranean Sea (Blanc, 1968; Caulet, 1972; Milliman et al., 1972). The three coralline facies in the Mediterranean (Blanc, 1972) correspond closely to those recognized in the western Atlantic; concretions (rhodoliths), maerl and coralligenous (encrustations). The concretions generally occur at greater depths (80–90 m) than either of the other two (maerl being restricted mainly to 10–40 m). Relict assemblages, however, occur to depths greater than 150 m (Caulet, 1972; Milliman et al., 1972).

26.4 Environmental Considerations

Water Temperature and Climate. Although both red and green algae contribute significant quantities of calcium carbonate to Atlantic continental margin sediments, greens are far more restricted, both taxonomically and climatically. The only green of any

Fig. 5. Distribution of calcium carbonate and carbonate assemblages in upper margin sediments off Morocco. Note: although carbonate contents high, abundance of corallines are markedly reduced compared to Brazil. To some extent probably reflects deeper and cooler waters as well as absence (particularly off southern Morocco and Spanish Sahara) of hard substrates. From Summerhayes et al. (1976)

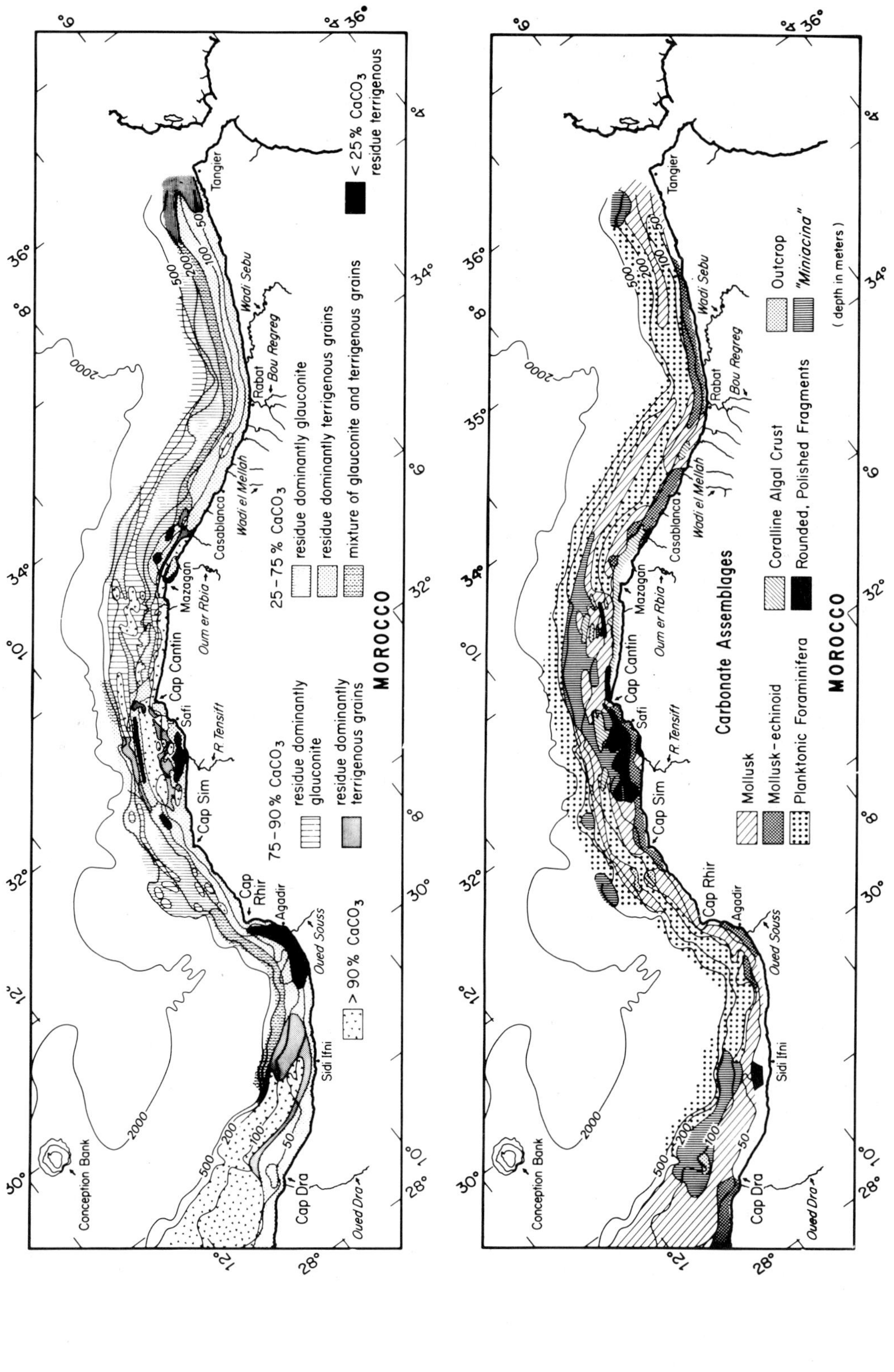
Tangier
Wadi Sebu
Bou Regreg
Rabat
Wadi el Mellah
Casablanca
Mazagan
Oum er Rbia
Cap Cantin
Safi
R. Tensift
Cap Sim
Cap Rhir
Agadir
Oued Souss
Sidi Ifni
Cap Dra
Oued Dra
Conception Bank
MOROCCO
< 25 % $CaCO_3$ residue terrigenous
25–75 % $CaCO_3$
residue dominantly glauconite
residue dominantly terrigenous grains
mixture of glauconite and terrigenous grains
75–90 % $CaCO_3$
residue dominantly glauconite
residue dominantly terrigenous grains
> 90 % $CaCO_3$
Carbonate Assemblages
Mollusk
Mollusk-echinoid
Planktonic Foraminifera
Coralline Algal Crust
Rounded, Polished Fragments
Outcrop
"Miniacina"
(depth in meters)

importance to the sedimentary budget appears to be the codiacean *Halimeda*. In the western Atlantic it occurs in many shelf sediments (although commonly in relatively shallow depths) in tropical areas (southern Florida to Brazil). In the cooler eastern Atlantic, *Halimeda* fragments do not occur as sedimentary components; although *H. tuna* is present in the Mediterranean, it apparently does not contribute to the sediment.

Coralline algae are capable of growing actively in nearly any marine climate. Corallines encrust rocks and pebbles in many high latitude areas and locally can even form coralligenous deposits (e.g., Brittany and Norway), but they do not appear capable of forming extensive carbonate deposits on high latitude shelves; this role apparently is reserved primarily for tropical and subtropical climates. In the western Atlantic, coralline facies extend between Cape Hatteras and southern Brazil, while in the eastern Atlantic similar facies are restricted primarily to northern Africa and Mediterranean. This restricted distribution in the eastern Atlantic, however, appears to be more the result of other environmental factors rather than temperature (see above).

Another major factor influencing calcareous accumulation is the importance of terrigenous sedimentation. Areas with high rates of terrigenous influx tend to have smaller concentrations of carbonate components (due to masking effects), but corallines seem particularly lacking. In part this reflects increased turbidity in the water column (thus lower levels of light penetration), but also it results in burial of potential substrata onto which corallines might attach. Thus, for instance, corallines do not occur off major rivers, such as the Mississippi, Amazon and Niger, although some topographic highs (such as off the Mississippi) may provide the necessary substrata upon which corallines can attach (Stetson, 1953; Parker and Curray, 1956). Martin (1973) has found modern fluvial muds overlying extensive algal carbonates, which apparently were deposited during more arid times at lower stands of sea level. It is in areas with minimal terrigenous influx, such as the Bahamas, Yucatan and northeastern and southeastern Brazil, where the waters are clear and sedimentation sufficiently low, that corallingenous deposits abound.

Water Depth. The depth to which calcareous algae can actively deposit calcium carbonate has been widely debated. Apparently certain species of *Halimeda* (such as *H. cryptica*) flourish at depths as great as 75 m in Jamaica (Goreau and Land, 1974). Off Brazil, modern *Halimeda* maerl seems to be restricted to much shallower depths (20–40 m), although in part this may reflect the generally shallow nature of the northeastern Brazilian shelf. If deeper habitats were studied, the vertical depth range might prove considerably greater.

Because coralline algae have been recovered from relatively great shelf depths, it is assumed that they can precipitate their carbonate in water depths as great as 200 m (Adey and MacIntyre, 1973), and living corallines have been noted from submersibles in depths exceeding 150 m (J.L. Wray, 1975, oral communication). However, accretion of coralline-rich deposits is more depth restricted: of more than 40 coralline samples that have been recovered from water depths greater than 75 m and subsequently age dated, only four show ages less than 1000 years. Nearly all the others are older than 10 000 years (Fig. 6), and presumably were deposited at lower stands of sea level. Moreover, most of the specimens capable of growing in depths greater than 50 m are restricted to the rhodoliths, although they also grow in shallower depths (Bosselini and Ginsburg, 1971); maerl and encrusting algae generally form in water depths shallower than 40 m (Blanc,

1972). Rhodolith growth, however, is slow in deeper waters, generally about 0.1 mm/yr (McMaster and Conover, 1967; Vogel, 1970).

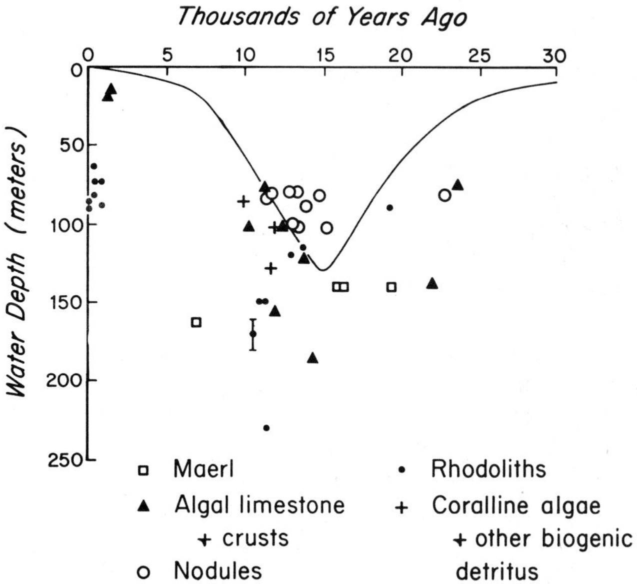

Fig. 6. Age of various coralline algae versus water depth from which recovered. Superimposed curve shows stand of sea level during past 30 000 years (from Milliman and Emery, 1968). Note: only modern corallines from shelf depths greater than 50 m are rhodoliths (concretions and nodules); other forms apparently restricted to significantly shallower depths. Data taken from Nota (1958), Koldewijn (1958), McMaster and Conover (1967), MacIntyre and Milliman (1970), Vogel (1970), Seisser (1972), Milliman et al (1972), Delibrias et al. (1972, 1974), Thommeret and Thommeret (1973) and Nair (1974)

26.5 Diagenetic Alteration

Both coralline algal and *Halimeda* fragments are susceptible to diagenetic alteration. In *Halimeda*, the utricles quickly fill with intragranular aragonite, and the grains alter into cryptocrystalline lumps (Fig. 7); often this can happen within individual plates within a living plant (Milliman, 1974).

Coralline algal alteration is somewhat more complex in that intragranular cements can occur as both aragonite and magnesian calcite, within both the cellular framework and the conceptacles (Fig. 7). Because of the widely divergent ages of corallines on various shelves in the western Atlantic, the role of temperature in controlling alteration and intragranular cementation is difficult to estimate. However, alteration seems to occur more slowly in cooler waters; practically all fragments found in Brazilian shelf sediments (some of which apparently are modern) are infilled with secondary carbonate, while many of the fragments in the cooler waters off Cape Hatteras and Morocco (many of

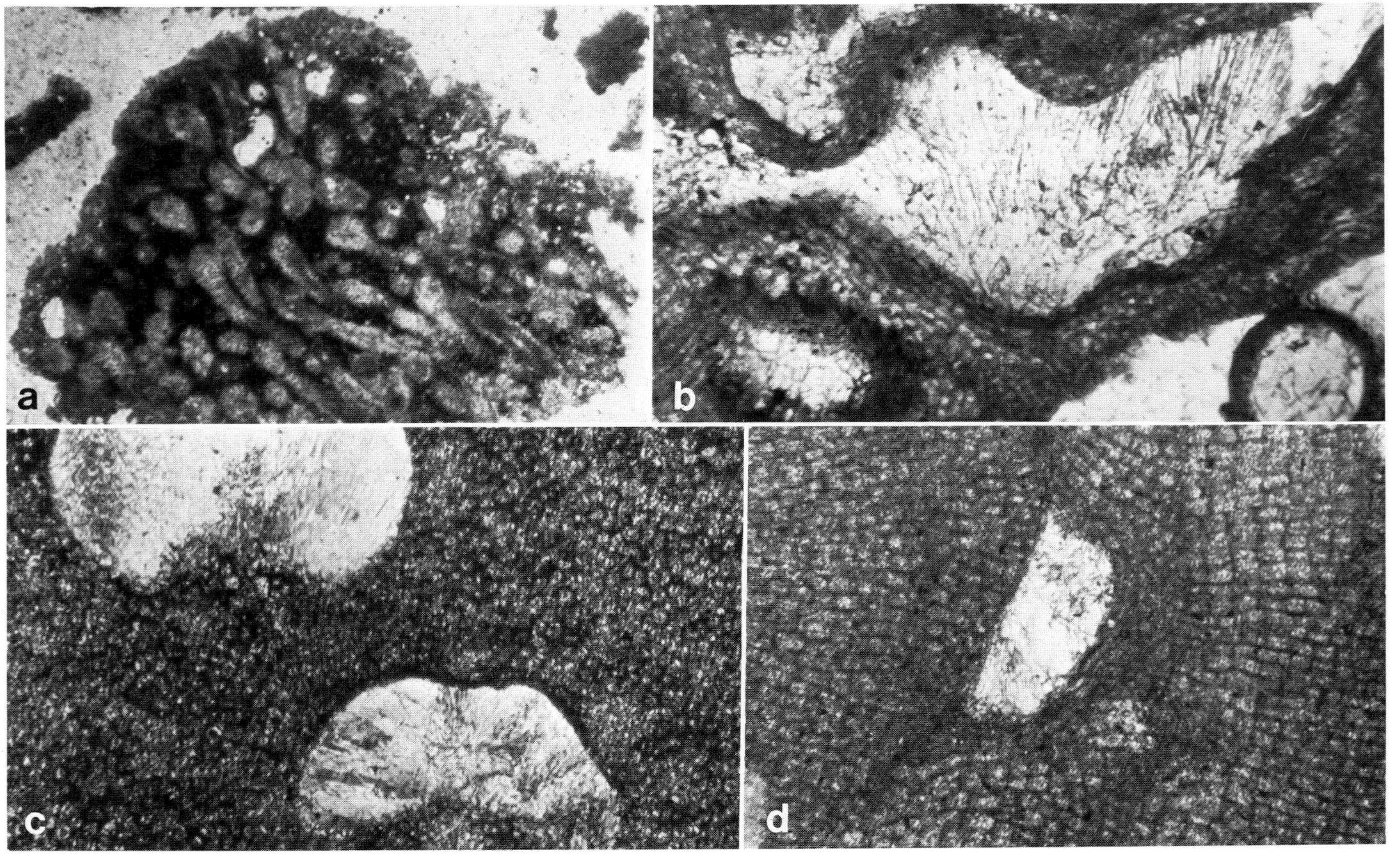

Fig. 7 a–d. Diagenetic alteration of calcareous algae. (a) Infilled *Halimeda* plate. (b) Intragranular cement between laminae in algal limestone from the inner shelf off Morocco. (c) and (d) Concepticle cements; note the cellular fillings in (d). Magnification in (a) is 50 x; in (b–d) 250 x these sediments most likely are relict) are relatively free from infilling. If the degree of primary infilling can be recognized in ancient sediments, it might provide a method by which the paleonotologist can differentiate climatic zones.

26.6 Bibliography

Adey, W.H.: Distribution of saxicolous crustose corallines in the northwestern North Atlantic. J. Phycol. **2**, 49–54 (1966).

Adey, W.H.: Effects of light and temperature on growth rates in boreal-subarctic crustose corallines. J. Phycol. **6**, 269–276 (1970).

Adey, W.H., MacIntyre, I.G.: Crustose coralline algae: a re-evaluation in the geological sciences. Geol. Soc. Am. Bull. **84**, 883–904 (1973).

Blanc, J.J.: Sedimentary geology of the Mediterranean Sea. Oceanogr. Marine Biol. Ann. Rev. **6**, 377–454 (1968).

Blanc, J.J.: Observations sur la sédimentation bioclastique en quelques points de la marge continentale de la Méditerranée. In: The Mediterranean Sea: A Natural Sedimentation Laboratory. Stanley, D.J. (ed.). Stoudsburg: Dowden, Hutchinson and Ross 1972, pp. 225–240.

Biollot, G.: Étude géologique de la Manache occidentale, fonds vocheux, dépôts Quaternaires, sédiments actuels. Ann. Inst. Océanogr. (Paris) **42**, 1–219 (1964).

Bosellini, A., Ginsburg, R.N.: Form and internal structure of recent algal nodules (Rhodolites) from Bermuda. J. Geol. **79**, 669–682 (1971).

Caulet, J.P.: Recent biogenic calcareous sedimentation on the Algerian continental shelf. In: The Mediterranean Sea: A Natural Sedimentation Laboratory. Stanley, D.J. (ed.). Stroudsburg: Dowden, Hutchinson and Ross 1972, pp. 261–277.

Coutinho, P.N., Morais, J.O.: Distribution de los sedimentos en la plataforma continental norte-nordeste de Brazil. Symp. Invest. Res. Caribb. Sea and Adjacent Regions, UNESCO, 273–284 (1971).

Delibrias, G., Guillier, M.R., Labeyrie, J.: Gif natural radiocarbon measurements VII. Radiocarbon **14**, 280–320 (1972).

Delibrias, G., Guillier, M.T., Labeyrie, J.: Gif natural radiocarbon measurements VIII. Radiocarbon **16**, 15–94 (1974).

Emery, K.O.: Shallow structure of continental shelves and slopes. Southeastern Geol. **9**, 173–194 (1968).

Emery, K.O., Uchupi, E.: Western North Atlantic Ocean: Topography, rocks, structure, water, life and sediments. Am. Assoc. Petroleum Geol. Memoir **17**, 532 pp. (1972).

Foslie, M.H.: The Norwegian forms of *Lithothamnion*. Norske Vidensk., Selsk. Skr. Kl. 29–208 (1894).

Ginsburg, R.N., Rezak, R., Wray, J.L.: Geology of calcareous algae. Comparative Sed. Lab., Univ. Miami (1972).

Goreau, T.F.: Calcium carbonate deposition by coralline algae and corals in relation to their roles as reef builders. Ann. N.Y. Acad. Sci. **109**, 127–167 (1963).

Goreau, T.F., Land, L.S.: Fore-reef morphology and depositional processes, North Jamaica. Soc. Econ. Paleontol. Mineral. Spec. Publ. **18**, 77–89 (1974).

Gould, H.R., Stewart, R.H.: Continental terrace sediments in the northeastern Gulf of Mexico. In: Finding Ancient Shorelines. Soc. Econ. Paleontol. Mineral., Spec. Publ. **5**, 2–19 (1955).

Hillis, L.W.: A revision of the genus *Halimeda* (order Siphonales). Inst. Marine Sci., Univ. Texas **6**, 321–403 (1959).

Huve, P.: Étude expérimentale de la réinstallation d'un "trottoir a Tenaera", en Méditerranée occidentale. C.R. Acad. Sci. Paris **239**, 323–325 (1954).

Huve, P.: Contribution a l'étude des fonds a *Lithothamnium (?) solutum* Foslie (*Lithophyllum solutum* (Foslie) Lemoine) de la region de Marseille. Red. Travaux Sta. Mas. d'Endoume Bull. **11**, 105–134 (1956).

Jacquotte, R.: Étude des fonds de maerl en Méditerranée. Rec. Travaux Sta. Mar. D'Endoume Bull. **26**, 141–230 (1962).

Johnson, J.H.: Limestone-building algae and algal limestones. Quart. Colo. School Mines 297 pp. (1961).

Johnson, H.W., Austin, L.F.: Growth rates in the articulated coralline, *Calliarthron* (Rhodophyta). Can. J. Botany **48**, 125–132 (1970).
Keary, R.: Biogenic carbonate in beach sediments of the west coast of Ireland. Sci. Proc. Roy. Dublin Soc., Ser. A **3**, 75–85 (1967).
Kempf, M.: A plataforma continental da costa leste brasileira, entre o Rio São Francisco e Ilha de São Sebastião (1o°30′–25° S): Notas sobre os principais tipos de fundo. Anais XXVI Congresso, Soc. Geo. Bras. **2**, 115–132 (1972).
Koldewijn, B.W.: Sediments of the Paria-Trinidad shelf. Rept. Orinoco Shelf Exped. **3**, 109 pp. (1958).
Kornicker, L.S., Bryant, W.R.: Sedimentation on continental shelf of Guatemala and Honduras. Am. Assoc. Petrol. Geol. Mem. **11**, 244–257 (1969).
Logan, B.W., Harding, J.L., Ahr, W.M., Williams, J.D., Snead, R.G.: Late Quaternary sediments of Yucatan shelf, Mexico. Am. Assoc. Petrol. Geol. Mem. **11**, 1–128 (1969).
Ludwick, J.C., Walton, W.R.: Shelf-edge calcareous prominences in northeastern Gulf of Mexico. Am. Assoc. Petrol. Geol. Bull. **41**, 2054–2101 (1957).
MacIntyre, I.G., Milliman, J.D.: Physiographic features on the outer shelf and upper slope, Atlantic continental margin, southeastern United States. Geol. Soc. Am. Bull. **81**, 2577–2598 (1970).
Martin, L.: Morphologie, sédimentologie et paléogéographie au quaternaire recent du plateau continental ivorien. Ph. D. thesis, Univ. Paris, 340 pp., 1973.
Mabesoone, J.M., Coutinho, R.N.: Littoral and shallow marine geology of northern and northeastern Brazil. Trab. Oceanogr. Univ. Fed. Pe., Recife **12**, 1–214 (1970).
McMaster, R.L., Conover, J.T.: Recent stromatolites from the Canary Islands. J. Geol. **74**, 647–652 (1967).
McMaster, R.L., Milliman, J.D., Ashraf, A.: Continental shelf and upper slope surface sediments off Portuguese Guinea, Guinea and Sierra Leone, West Africa. J. Sediment. Petrol. **41**, 150–158 (1971).
Melo, U. de, Summerhayes, C.P., Ellis, J.P.: Uppercontinental margin sedimentation off Brazil. Part 4, Salvador to Vitoria, Southeastern Brazil. Contr. Sedimentology **4**, 78–116 (1975).
Menzies, R.J., Pilkey, O.H., Blackwelder, B.W., Dexter, D., Huling, P., McCloskey, L.: A submerged reef off North Carolina. Intern. Rev. Ges. Hydrobiol. **51**, 393–431 (1966).
Milliman, J.D.: Atlantic continental shelf and slope of the United States. Petrology of the sand fraction northern New Jersey to southern Florida. U.S. Geol. Survey Prof. Paper **529**–J, 40 pp. (1972).
Milliman, J.D.: Marine Carbonates. Berlin–Heidelberg–New York: Springer 1974.
Milliman, J.D.: Upper continental margin sedimentation off Brazil. Part 6. A synthesis. Contrib. Sedimentol. **4**, 151–175 (1975).
Milliman, J.D.: *Miniacina miniacea*: modern foraminifera sands on the outer Moroccan shelf. Sedimentology **23**, 415–419 (1976).
Milliman, J.D., Pilkey, O.H., Ross, D.A.: Sediments of the continental margin off the eastern United States. Geol. Soc. Am. Bull. **83**, 1315–1334 (1972).
Milliman, J.D., Summerhayes, C.P. (eds.): Upper continental margin sedimentation off Brazil. Contrib. Sedimentology **4**, 175 pp. (1975).
Milliman, J.D., Weiler, Y., Stanley, D.J.: Morphology and carbonate sedimentation on shallow banks in the Alboran Sea. In: The Mediterranean Sea: A Natural Sedimentation Laboratory. Stanley, D.J. (ed.). Stroudsburg: Dowden, Hutchinson and Ross 1972, pp. 241–259.
Müller, J.: Mineralogisch-sedimentpetrographische Untersuchungen an Karbonatsedimenten aus dem Schelfbereich um Fuerteventura und Lanzarote (Kanarische Inseln). Unpubl. Ph. D. thesis, Univ. Heidelberg, 99 pp. (1969).
Nair, R.R.: Holocene sea-levels of the western continental shelf of India. Proc. Indian Acad. Sci. **79** (B), 197–203 (1974).
Nota, D.J.G.: Sediments of the western Guiana shelf. Mededel. Landbouwhogeschool Wagingen, Nederlands. **58**, 98 pp. (1958).
Parker, R.H., Curray, J.R.: Fauna and bathymetry of banks on continental shelf, northwest Gulf of Mexico. Am. Assoc. Petrol. Geol. Bull. **40**, 2428–2439 (1956).
Pérès, J.M., Picard, J.: Nouveau manuel de bionomie benthique de la Mer Méditerranée. Rec. Trav. Stn. Marine Endoume Bull. **14**, 5–137 (1964).

Rad, U. von: Great Meteor and Josephine Seamounts (eastern North Atlantic): Composition and origin of bioclastic sands, carbonate and pyroclastic rocks. Meteor. Forsch.-Ergebn. **C (19)**, 61 pp. (1974).

Siesser, W.G.: Relict algal nodules (Rhodolites) from the south African continental shelf. J. Geol. **80**, 611–616 (1972).

Stetson, H.C.: The sediments of the western Gulf of Mexico. Papers Phys. Oceanogr. Meteor. **12**. 1–45 (1953).

Summerhayes, C.P., Milliman, J.D., Briggs, S.R., Bee, A.G., Hogan, C.: Northwest African shelf sediments: Influence of climate and sedimentary processes. J. Geol. **84**, 277–300.

Thommeret, J., Thommeret, Y.: Monaco radiocarbon measurements IV. Radiocarbon **15**, 321–344 (1973).

Vicalvi, M.A., Milliman, J.D.: Carbonate sedimentation on the continental shelf off southern Brazil, with special reference to the benthonic foraminifera. Am. Assoc. Petrol. Geol. Memoir, in press (1976).

Vogel, J.C.: Gronigan radiocarbon dates IX, Radiocarbon **12**, 44–471 (1970).

27 Carbonate Construction and Decomposition by Epilithic and Endolithic Micro-organisms in Salt- and Freshwater

J. SCHNEIDER, Göttingen, W. Germany

27.1 Introduction

In the carbonate cycle micro-organisms such as Cyanophytes, Chlorophytes, Rhodophytes, fungi and lichens play an important and often decisive role. Some build up carbonate, some destroy carbonates and others can do both simultaneously.

27.2 The Build-up of Carbonates

The build-up of carbonates can occur in four different ways:

27.2.1 Trapping and Binding of Particles

Sediment particles of any material can be trapped mechanically and bound by algal filaments and mucus on the surface of algal mats and algal microreefs. This trapping and binding process is already well known from algal mats of subtidal to supratidal marine environments (e.g., Black, 1933; Logan, 1961; Monty, 1965, 1967; Gebelein, 1969; Golubić, 1973).

In the freshwater environment this process plays an important role in the build-up of travertine and lacustrine crusts in lakes and rivers.

27.2.2 Cementation by $CaCO_3$

By the assimilation of carbon dioxide and bicarbonate the level of $CaCO_3$-saturation in water can be raised and $CaCO_3$ is inorganically precipitated in microenvironments.

In microenvironments between algal filaments of marine and lacustrine milieus the pH can be raised to values higher than 10 by algal assimilation of CO_2 and HCO_3^- (Goldman et al., 1972). The CO_3^{2-}-concentration can thus be raised so far that $CaCO_3$ is precipitated. This is an inorganic precipitation of $CaCO_3$, which is enabled by the physiologic activities of micro-organisms. The $CaCO_3$ is precipitated as unoriented micrite within cavity systems (Pl. 2/4). In the marine environment the micrite consists of aragonite (Mg/Ca-ratio in seawater > 1) and in the freshwater environment of calcite (Mg/Ca-ratio in freshwater < 1). In seawater the precipitation is reduced by inhibitors and crystal poisons such as Mg^{2+} (Pytkowicz, 1969) or various organic substances (Chave and Suess, 1970; Simkiss, 1964).

Both possibilities, trapping and binding and cementation, occur in freshwater and in seawater.

27.2.3 Encrusting with $CaCO_3$

Blue-green algae, green algae and fungi can encrust their filaments and hyphae with $CaCO_3$.

This process is restricted to freshwater environments. Analogues from marine environments are not known, at least as far as recent algae and fungi are concerned (e.g., Pia, 1934; Monty, 1973; Golubić, 1973).

The encrustation consists of calcite, which grows outside the sheaths in a more or less oriented pattern (Pl. 1/2, Pl. 2/2 and 3).

The physiologic processes, which lead to the precipitation of micritic calcite on algal filaments and fungal hyphae, are not yet known in detail.

In the following, two examples for this type of encrustation will be given. In the dark parts of the cave of Škocjan in Slovenia (Yugoslavia) loose sinter crusts occur, which are built up by calcite-encrusted fungal hyphae (Pl. 1/1).

The percolating waters in this area contain not only sufficient dissolved $CaCO_3$, but also enough organic substances, which can be metabolized by heterotrophic fungi. The calcite is precipitated on the surface of the hyphae as oriented micrite.

Different species show encrustations of different thicknesses. At the cave entrance where the light just reaches the walls the fungi are accompanied by algae, which have filaments encrusted by coarser micritic calcite [Dr. S. Golubić, Boston, kindly showed me the locality and determined the cyanophyte as a *Geitleria calcarea* (Friedmann, 1955)]. The micritic calcite is arranged in an indistinct radial pattern around the filament (Pl. 1/2). The microflora of lacustrine crusts from the shallow sublitoral zone of carbonate-rich lakes consists to about 90% of *Schizothrix*-species (*S. lacustris, S. fasciculata*, see Kann, 1941, 1959). These contain several trichomes within one sheath. The outer surface of the sheath is encrusted by micritic calcite (Pl. 2/2 and 3). In this way the basic framework of the algal microreefs is built up (see also sect. 27.4).

27.2.4 Secretion of $CaCO_3$ within Mucus

The calcified colonies of *Rivularia haematites* (D. C.) AG. are important components of lacustrine crusts in lakes and rivers. The trichomes of the radially growing colonies are completely embedded in mucus. Calcite is precipitated along the sheaths of the trichomes within the mucus (Pl. 1/3 and 4). Wallner (1935) has described the pattern of the calcite in detail. References to older literature can be found in Pia (1934, pp. 12–15).

Oriented surficial encrustation of filaments and hyphae with $CaCO_3$, and secretion of $CaCO_3$ within mucus both occur only in freshwater. All these processes (trapping, binding, cementation, encrustation, calcification) produce characteristic biogenic carbonate structures, some of which have a high fossilization potential, for example: sinter crusts in caves, travertine, lacustrine crusts in lakes and rivers, onkoids and stromatolites.

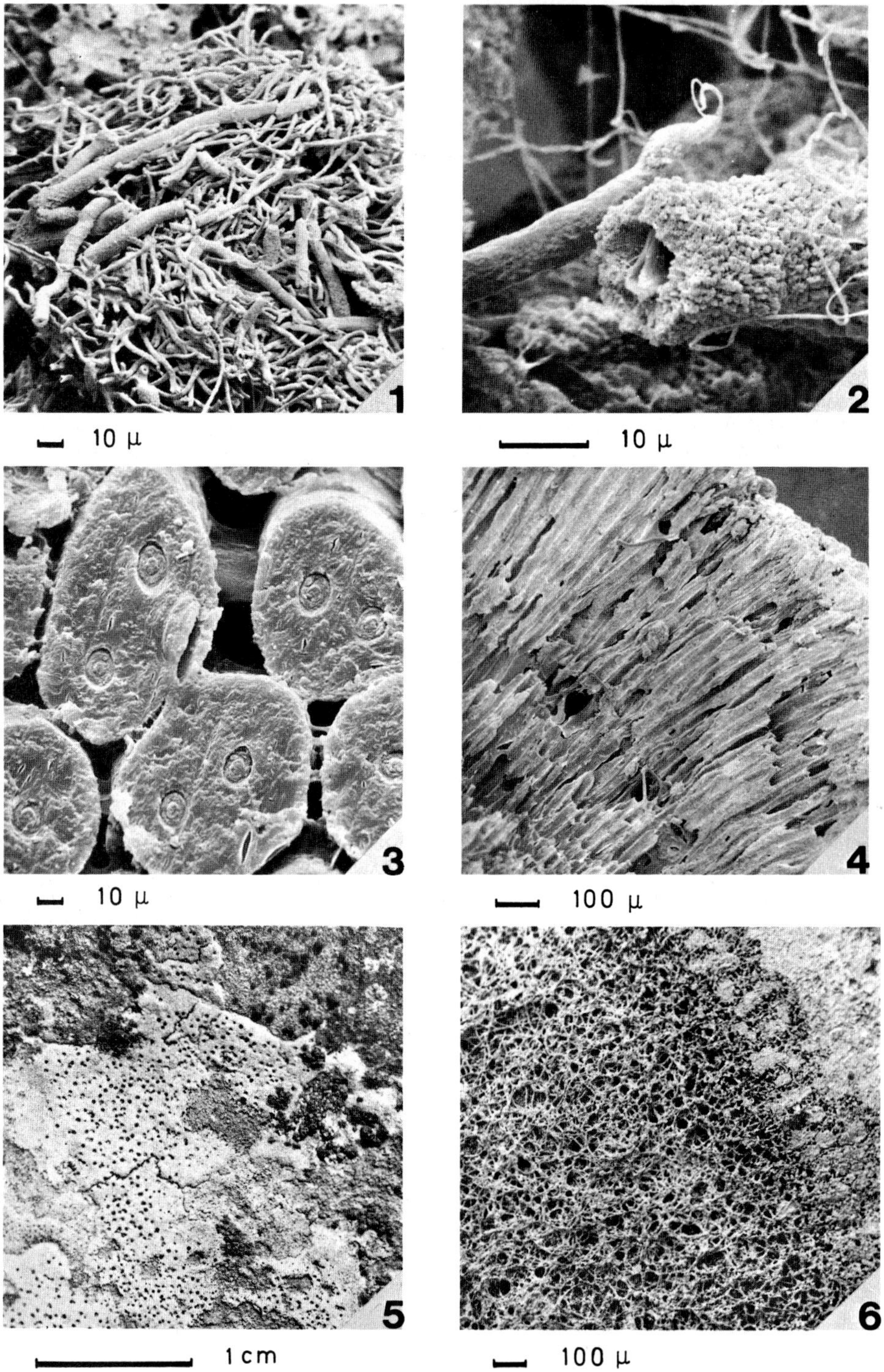

Plate 1. Legend see opposite page

27.3 The Decomposition of Carbonates

In the decomposition of carbonates in the terrestrial, lacustrine, and marine environments a decisive role is played by endolithic algae, fungi, and lichens. They can destroy carbonate substrates of all kinds whether rock (limestone and dolomite), shells or sediment grains. The endolithic micro-organisms affect the substrate in three different ways (27.3.1, 3.2, 3.3).

27.3.1 Biologic Corrosion by Endoliths

Endolithic bluegreen, green, and red algae, fungi and lichens penetrate actively into the carbonate substrate. They disintegrate the substrate in different ways, which are not yet completely understood. They thus carry out a biologic corrosion. Thereby the different micro-organisms produce partly specific boring patterns, which may be preserved in the fossil record (see Golubić, 1969, 1973; Golubić and Schneider, 1972; Schneider, 1973, 1976).

27.3.2 Preparation of Carbonate Surfaces for Any Inorganic Solution

Since the substrate surface is reduced by the endolithic micro-organisms to tiny fragments, it offers a larger active surface for any inorganic solution. In the terrestrial environment endolithic lichens participate in this process to an extraordinarily large extent. Endolithic lichens penetrate some millimeters deep into the carbonate substrate (see Pl. 1/5 and 6).

The small grains on the surface of the bored carbonate substrate can easily be dissolved by rain water run-off, which is undersaturated in respect to $CaCO_3$. Especially in the terrestrial environment this process plays an important role in the formation of such Karst-patterns as Karren.

◀ Pl. 1/1. Cave sinter crusts from dark part of Škocjan-caves (Yugoslavia). Crust formed by $CaCO_3$-encrusted hyphae of different fungi

Pl. 1/2. Close up view of sinter crust in cave entrance, at limit of light penetration. Next to encrusted fungal hypha (*background*) is micrite-encrusted filament of blue-green alga (*Geitleria calcarea*, Friedmann, 1955)

Pl. 1/3. Cross-section through some sheaths with trichomes of *Rivularia haematites* (D.C.) Ag. Sheaths and trichomes replaced by epoxy-resin. Interface between sheaths was filled with $CaCO_3$, dissolved after resin-embedding (locality: – 0.5 m depth, Attersee, Austria)

Pl. 1/4. Cross-section through colony of *Rivularia haematites,* showing radially arranged calcite-tubes. Sheaths and trichomes of alga were removed by chlorox

Pl. 1/5. Limestone surface with different thalli of endolithic lichens. Small black holes are perithecia of *Amphoridium calcisedum* (D.C.) Serv. (locality: limestone coast, uppermost supratidal, Northern Adriatic, south of Rovinj, Yugoslavia)

Pl. 1/6. Epoxy-resin-cast of endolithic lichen. Rock piece sliced perpendicular to surface (*right upper corner*). Immediately underneath surface, algal layer shown. Fungal hyphae of lichen bore 1–3 mm deep into limestone

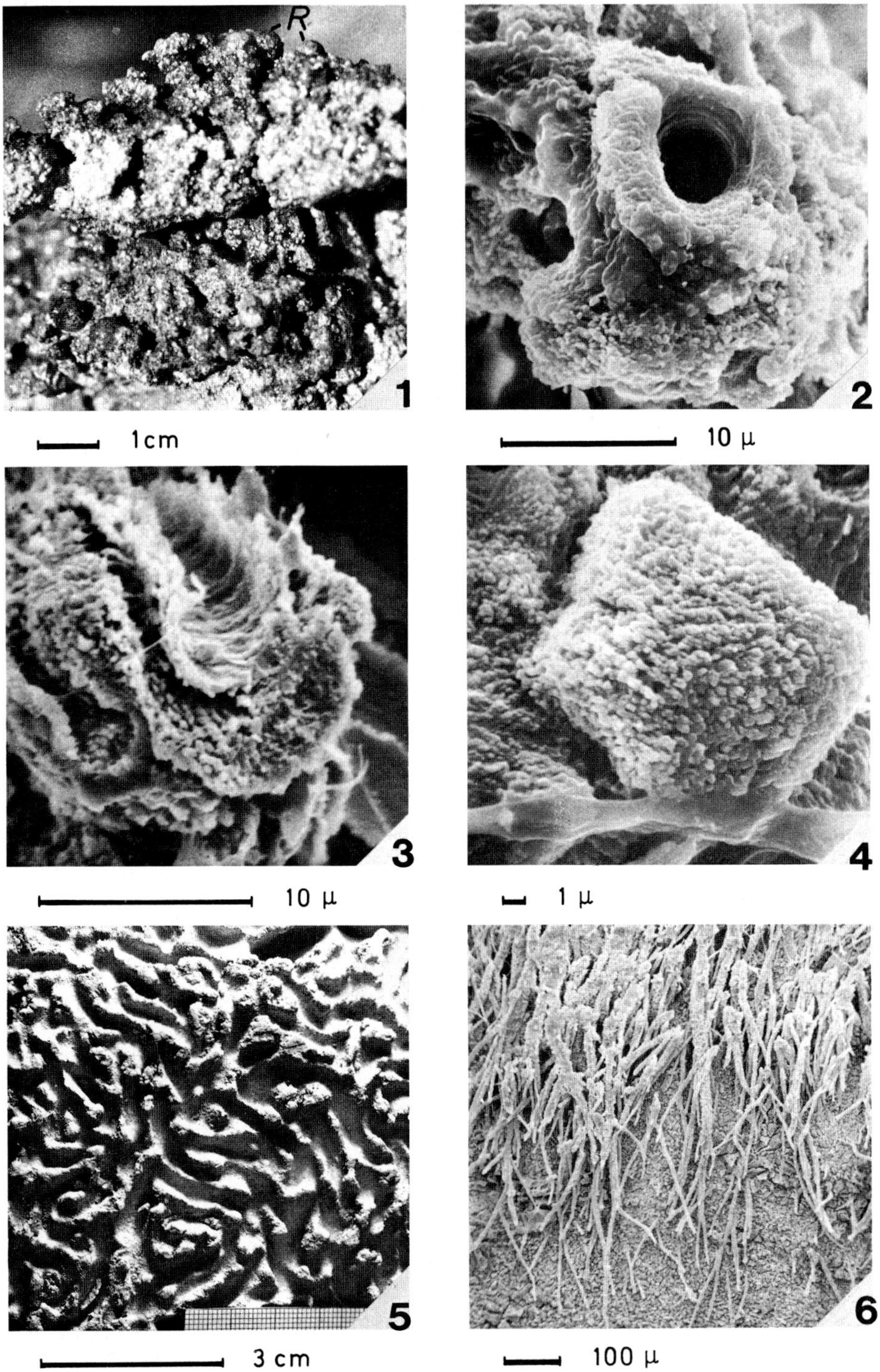

Plate 2. Legend see opposite page

27.3.3 Loosening of Carbonate Surfaces for Biologic Abrasion

The soft and loosened substrate is open to attack by grazing and rasping organisms. They eat the micro-organisms and the uppermost substrate layer and thus produce sediment. This sort of biologic erosion (or biologic abrasion) is carried out by various gastropods, echinoderms, fishes etc., especially in the marine environment, but also in lacustrine and terrestrial environments.

The resulting destruction forms, which I will call "biokarst", and the often characteristic borings of the endolithic micro-organisms have a high fossilization potential. They are therefore of significance to paleoecology and facies.

27.4 Simultaneous Build-up and Decomposition of Carbonates in Freshwater

Lacustrine crusts and furrows (Krusten- and Furchensteine) have been well known for 120 years. In 1857 Schimper described these forms from Lake Geneva. These observations were published by Forel (1878).

Lacustrine crusts and furrows occur in the shallow sublitoral zone of carbonate-rich rivers and lakes. Lacustrine crusts grow up on any substrate. The crusts comprise 30% algae, 60% trapped and inorganically or biogenically precipitated calcite and 10% trapped sediment. The algal microreefs show a characteristic cauliflower-like growth structure (Pl. 2/1).

Furrows occur only on carbonate rocks but always in connection with lacustrine crusts. The furrows have a brain-like pattern and show no regularities in direction (Pl. 2/5).

In particular Boysen Jensen (1909) worked on the formation of furrows (see also Kann, 1941), but up to the present day their genesis remains unsatisfactorily explained. The crusts originate from "patch-reef"-like algal microreefs, which grow in an irregular pattern on the surface of carbonate substrates (Fig. 1).

◀ Pl. 2/1. Lacustrine crust (locality: uppermost subtidal zone freshwater lake Attersee, Austria). Characteristic cauliflower-like growth structure of algal microreef. Round forms: colonies of *Rivularia haematites* (R)

Pl. 2/2. Micrite encrustation (calcite) of *Schizothrix*-filament. Trichomes and their common sheath removed by chlorox

Pl. 2/3. Remains of *Schizothrix*-sheath, encrusted by blocky micritic calcite. Most of algal material has been removed by chlorox

Pl. 2/4. Trapped sediment particle from lacustrine crust, encrusted by calcitic micrite. Micrite precipitated inorganically, caused by assimilation of CO_2 and HCO_3^-

Pl. 2/5. Surface of furrowed limestone from subtidal zone of a freshwater lake (Attersee, Austria). Lacustrine crust on furrows was removed by brush. Ridges between furrows are starting points for growth of lacustrine crusts. Ridges brown in color. Furrows have green color, due to endolithic bluegreen algae

Pl. 2/6. Epoxy-resin-cast of endolithic algae of furrows (Pl. 2/5) in perpendicular section. *Schizothrix* species bore initially with whole sheath. At certain depth single trichomes of *Schizothrix* leave sheath and bore separately into substrate

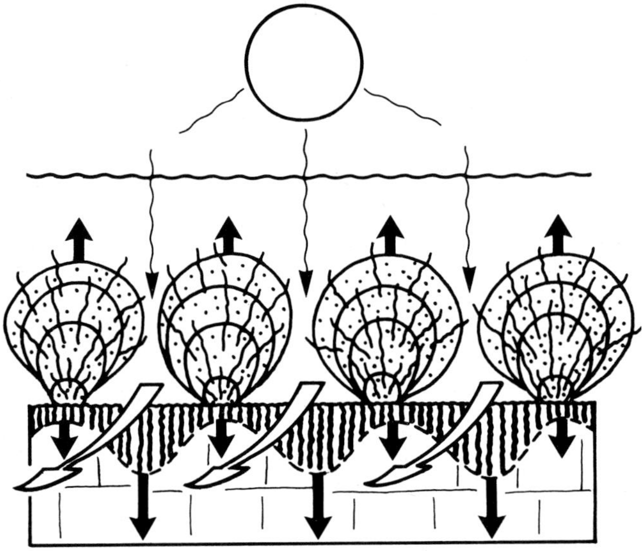

Fig. 1. Schematic genetic model of lacustrine crusts and furrows on carbonate substrates (shallow subtidal, freshwater). *Black arrows*: simultaneous biologic construction and decomposition of carbonates; *white arrows*: precolating water

The carbonate substrate (limestone or dolomite) is simultaneously attacked by endolithic algae and fungi. As soon as the crust has grown up to a height of a few millimeters, the intensive calcification prevents light penetration down to the substrate surface. Therefore endolithic algae can bore deeper only between the "patch-reefs" in spaces that are exposed to the light. Consequently biologic corrosion takes place preferentially there. Simultaneously tunneling and grazing organisms (*Tinodes*, nematods, copepods, and gastropods) invade the spaces between the calcified crusts and rasp away the loosened rocks surface. The biologic abrasion by the grazers leads to shallow furrows between the calcified algal microreefs (Fig. 1). Crusts can be smashed during autumn and winter storms. Pieces of them can thus form the nuclei of oncoids on the sediment bottoms or contribute to the carbonate sedimentation.

The following recolonization takes place preferentially on the ridges. The furrows are continuously deepened by simultaneous biologic corrosion and abrasion. As crusts become thicker and thicker (> 3 cm) an extreme situation can lead to caverns, in which microbial deterioration of the organic substances results in a raised CO_2-content and lowered pH-value. Inorganic solution can then take place as it is described by Golubić (1962, 1973, p. 443). However, for the formation of furrows (e.g., the biokarst), the simultaneous biologic corrosion and abrasion of the carbonate substrate between algal microreefs is of paramount importance. Detailed studies on lacustrine crusts and furrows are in progress in cooperation with Dr. Edith Kann (Vienna).

An example of a typical endolithic bluegreen alga from a furrowed limestone underneath crusts of *Schizothrix* and *Rivularia* from the Attersee (Austria) is shown in Plate 2/6.

Schizothrix bores initially with its whole sheath, contaning numerous trichomes. Below a certain depth the single trichomes of *Schizothrix* leave their sheath and bore separately into the substrate.

The fungi can be distinguished from the bluegreen algae by the different texture of the cast-surface (Pl. 3/1). The resin cast surfaces of bluegreen algae are rough, depending on the texture of the substrate; those of fungi are smooth. Bluegreen algae dissolve the matrix between the grains of the substrate (Pl. 3/5). The diameter of an algal bore hole is larger than that of the filament within. By contrast the fungi remove a minimum of substrate such that the diameter of the bore hole equals that of the hyphae. The boring mechanisms are not yet known. However, the different shape of the bore hole casts allows the distinction between fungal borings and those of Cyanophytes (Golubić and Schneider, 1972; Golubić et al., 1975; Golubić et al., 1970).

27.5 Destruction of Marine Carbonate Coasts

On marine limestone coasts there occurs a very characteristic color zonation and a specific morphologic zonation. The colors are caused by different species of epilithic and endolithic bluegreen algae (Ercegović, 1932; Le Campion-Alsumard, 1969; Schneider, 1973, 1976) and their zonation depends on the different ecologic conditions in each zone of the coast. The morphologic forms (rock pools, holes) are not a result of inorganic solution, but are due to simultaneous biologic corrosion caused by endolithic microorganisms and biologic abrasion carried out by grazing organisms (Schneider, 1973, 1976).

Examples of two characteristic endolithic cyanophytes from a marine limestone coast (Istria, Yugoslavia, intertidal zone) are shown on Plate 3/3 and 4: one is *Kyrtuthrix dalmatica* (Ercegović, 1929), which bores with its typical double-looped trichomes. The other is *Mastigocoleus testarum* (Lagerheim, 1886), which is characterized by its heterocysts.

The density of borings is in the order of 300.000 individuals/cm^2. The two species show different boring behaviors. *Kyrtuthrix* always bores perpendicular to the surface; *Mastigocoleus* bores irregularly into the substrate (see also Golubić and Le Campion-Alsumard, 1973) and can be identified by its characteristic heterocysts.

The boring tunnels of the endolithic Cyanophytes show a characteristic shape. On the walls of the borehole the matrix between the substrate grains is dissolved. Therefore the substrate grains are loosened by biologic corrosion. This leaves the substrate surface susceptible to attack by grazing organisms (Pl. 3/5).

The grazers (e.g., gastropods, see Pl. 3/6) eat the epilithic and endolithic algae and fungi together with the loosened carbonate grains. By this biologic abrasion the grazer produces carbonate sediment as fecal pellets (Schneider, 1976). Detailed investigations about the mass balance of these processes on carbonate coasts will be carried out in the future. The intensity of the destruction processes is dependent on the humidity of a

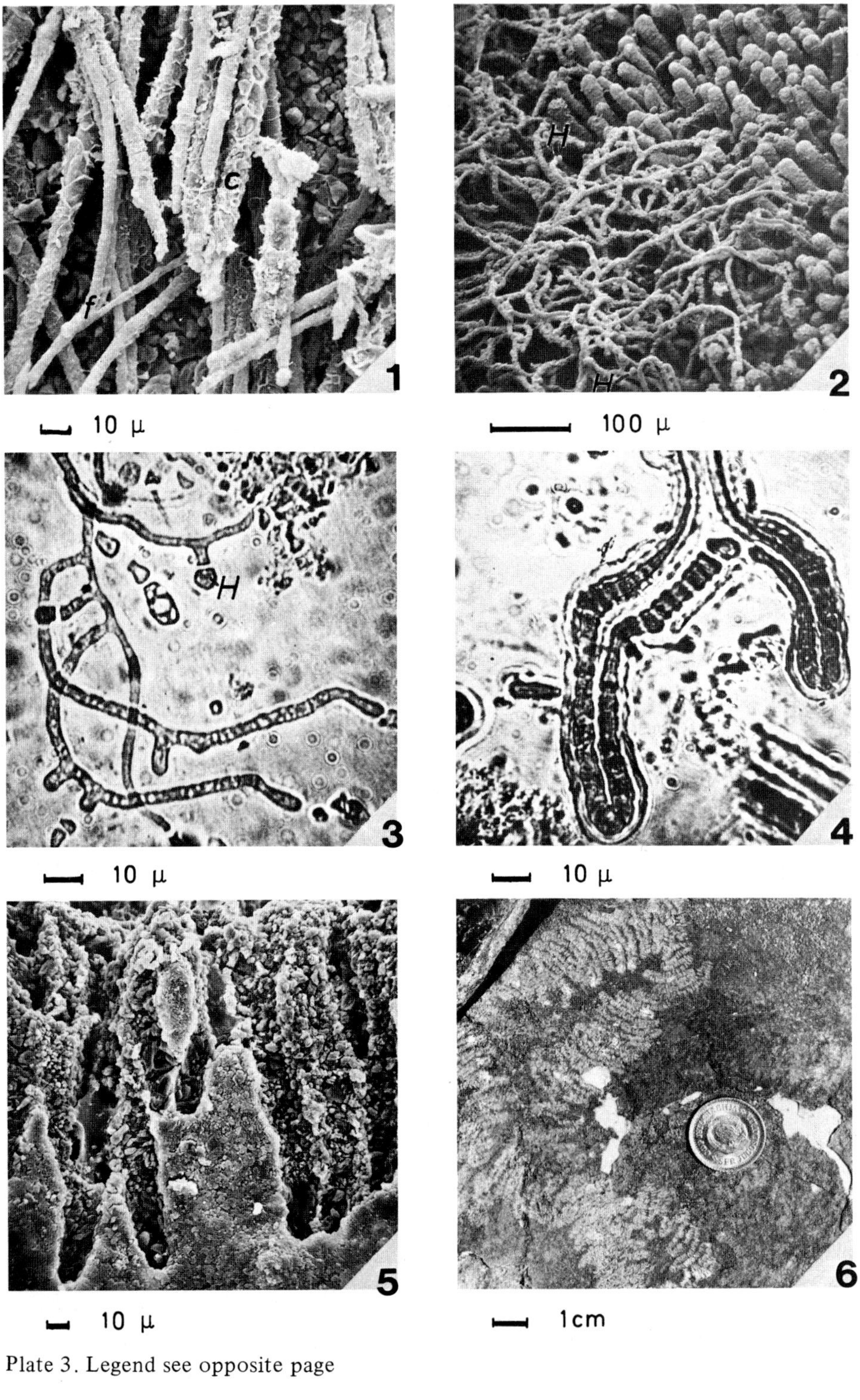

Plate 3. Legend see opposite page

given site. The higher the humidity at a given location (subtidal, intertidal or supratidal, rock pool or cavity) the higher is the intensity of biologic destruction. Therefore in the zone of highest contrast in humidity (e.g., wave zone, extreme high water line) and consequently in destruction rate, biokarst develops most rapidly and the characteristic morphology is very pronounced.

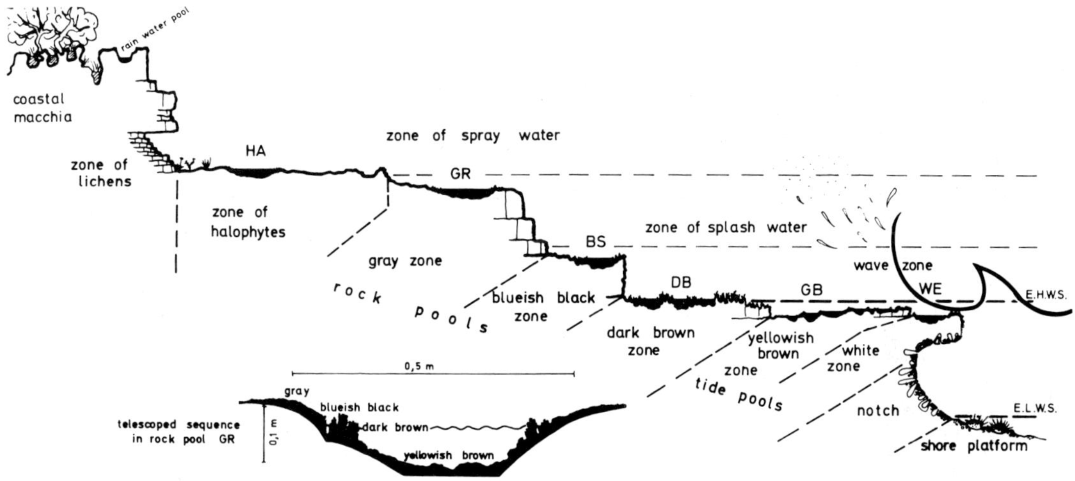

Fig. 2. Zonation at a limestone coast (Istria, Northern Adriatic, Yugoslavia). Telescoped sequence in rock pools gives example of relativity of bionomic and morphologic zonation

---

◀ Pl. 3/1. Epoxy-resin-cast of endolithic *Schizothrix* trichomes and fungal hyphae from a furrowed limestone. Cast surfaces of bluegreen algae (c) are rough, corresponding to substrate texture; those of fungi (f) are smooth (explanation see Sect. 4)

Pl. 3/2. Epoxy-resin-cast of two species *Kyrtuthrix* (*right*) and *Mastigocoleus* (*left*). Density of borings about 300.000 indiv./cm^2. Different boring behavior of two forms clearly visible. *Kyrtuthrix* bores perpendicular to surface, *Mastigocoleus* bores irregularly and can be identified by characteristic heterocysts (H)

Pl. 3/3. Typical endolithic bluegreen alga, *M. testarum* Lagerheim, from upper intertidal zone on limestone coast of Istria, Northern Adriatic (Yugoslavia). *Mastigocoleus* characterized by heterocysts (H).

Pl. 3/4. Typical bluegreen alga, *K. dalmatica* Ercegović, from upper intertidal zone on limestone coast of Istria, Northern Adriatic (Yugoslavia). *Kyrtuthrix* bores with looped trichomes.

Pl. 3/5. Broken limestone piece with boring tunnels of an endolithic bluegreen alga, *Hyella caespitosa* (Bornet and Flahault), wave zone of limestone coast of Istria, Northern Adriatic (Yugoslavia). Algae were removed by chlorox. On walls of boreholes matrix between substrate grains dissolved by boring activity. Substrate thus loosened by the biologic corrosion and substrate surface susceptible to attack of grazing organisms

Pl. 3/6. Meandering grazing traces of *Patella* on a limestone surface, colonized by epilithic and endolithic bluegreen algae. *Patella* grazes algae together with loosened carbonate grains. By this biologic abrasion grazer produces sediment in form of fecal pellets (wave zone of limestone coast, Istria, Northern Adriatic, Yugoslavia)

The biologic and the morphologic zonation of the coast is repeated in every rock pool in a telescoped sequence according to the degree of humidity (see Fig. 2). This accounts for the relativity of the bionomic and morphologic zonation (Schneider, 1973, 1976).

27.6 Conclusion

27.6.1 Carbonate Cycle in Lacustrine and Terrestrial Environment

Fig. 3 shows the carbonate cycle that arises in lacustrine and terrestrial environments, as for example in the Northern Alps.

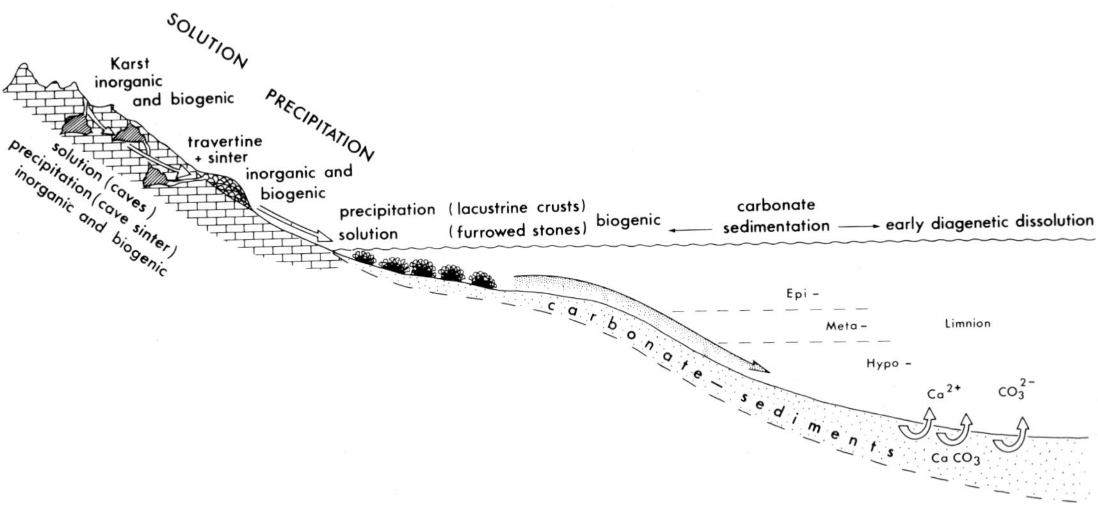

Fig. 3. Carbonate cycle in lacustrine and terrestrial environments (see Sect. 6.1)

Biogenic and inorganic decomposition of carbonates causes karst- and cave-formation. Biogenic carbonate precipitation occurs in the form of sinter crusts in caves, travertine, lacustrine crusts, onkoids, and carbonate-sediments.

Biologic corrosion and biologic abrasion leads to the formation of furrowed stones, i.e., biokarst. An early diagenetic solution can occur within the sediments during the summer-stagnation in the hypolimnion. This results in a recycling of $CaCO_3$ to the lake water. The dissolved carbonate can also be transported to the sea by rivers.

27.6.2 Carbonate Cycle in Marine Environments

Fig. 4 shows the carbonate cycle in the marine environment.

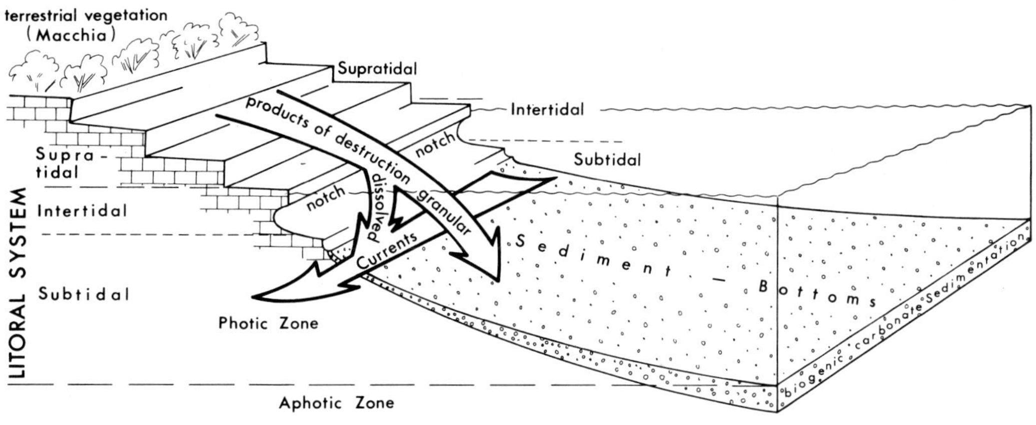

Fig. 4. Carbonate cycle in marine environments (see Sect. 27.6.2)

Biologic corrosion by micro-organisms produces dissolved carbonate. This is washed out to the sea by rain and waves. In the sea it is transported and distributed by currents and can be used by organisms with calcareous hard parts.

Biologic abrasion by grazing organisms produces fecal pellets containing carbonate particles. This particulate matter can be transported to the sediment bottoms in the subtidal areas, thus contributing to the sedimentation rate. Grain size reduction takes place on the sediment bottoms by biologic corrosion and by biologic abrasion. So, the grain size distribution, especially in areas with carbonate sedimentation, is influenced by the endolithic micro-organisms.

Acknowledgments. I wish to express my gratitude for helpful stimulation and discussions to the following colleagues: Dr. Thérèse Le Campion-Alsumard (Marseille), Dr. Edith Kann (Vienna), Dr. Stjepko Golubić (Boston). For much assistance in preparation techniques I thank Mrs. H. Peters. Dipl. Geol. H. Scholz (Göttingen) took the photographs on the Cambridge Scanning Electron Microscope. H. Grimme draughted the illustrations. Dr. Peter Sadler and Dipl. Geol. Thomas Junghans generously helped me with the english translation of the manuscript. The work at the Adriatic coast was supported by a grant and a fellowship from the Deutsche Forschungsgemeinschaft (Az.: Śchn 16/1–3).

27.7 Bibliography

Black, M.: The algal sediments of Andros Island, Bahamas. Phil. Trans. Roy, Soc. London, Ser. B **222**, 165–192 (1933).

Boysen Jensen, P.: Über Steinkorrosion an den Ufern von Furesö. Intern. Rev. Ges. Hydrobiol. **2**, 163–173 (1909).

Ercegović, A.: Ecological and sociological studies on lithophytic cyanophytes of the Adriatic coast (in serbocroatian). Jugosl. Akad. Znanosti Umijet., Zagreb **244**, 129–220 (1932).

Chave, K.E., Suess, E.: Calcium carbonate saturation in seawater: Limnology and Oceanography **15**, 633–637 (1970).

Forel, F.A.: Remarques sur la sculpture des galets des grèves des lacs Suisses. Act. Soc. Helv. Sci. Nat., Bern, 1878.

Friedmann, J.: *Geitleria calcarea* n. gen. n. sp., a new athmophytic lime-encrusting blue-green alga. Botan. Notiser **108**, 439–445 (1955).

Gebelein, C.D.: Distribution, morphology and accretion rate of recent subtidal algal stromatolites, Bermuda. J. Sediment. Petrol. **39**, 1, 49–69 (1969).

Goldman, J.C., Porcella, D.B., Middlebrooks, E.J., Toerien, D.F.: The effect of carbon on algal growth – its relationship to eutrophication. Water Res. **6**, 637–679 (1972).

Golubić, S.: Zur Kenntnis der Kalkinkrustation und Kalkkorrosion im Seelitoral. Schweiz. Z. Hydrobiol. **24**, 2, 229–243 (1962).

Golubić, S.: Cyclic and noncyclic mechanisms in the formation of travertine. Verh. Intern. Verein Limnol. **17**, 956–961 (1969).

Golubić, S.: The relationship between blue-green algae and carbonate deposits. In: The Biology of Blue-Green Algae. Carr, N.G., Whitton, B.A. (eds.). Oxford: Blackwell 1973, pp. 434–472.

Golubić, S., Brent, G., Le Campion-Alsumard, T.: Scanning electron microscopy of endolithic algae and fungi using a multipurpose casting-embedding technique. Lethaia **3**, 203–209 (1970).

Golubić, S., Le Campion-Alsumard, T.: Boring behaviour of marine blue-green algae *Mastigocoleus testarum* Lagerheim and *Kyrtuthrix damatica* Ercegovic as a taxonomic character. Schweiz. Z. Hydrol. **35**, 1, 157–161 (1973).

Golubić, S., Perkins, R.D., Lukas, R.: Boring micro-organisms and microborings in carbonate substrate. In: Study of Trace Fossils. Frey, R. (ed.). New York: Springer 1975, pp. 229–259.

Golubić, S., Schneider, J.: Relationship between carbonate substrate and boring patterns of marine microorganisms. Geol. Soc. Am., 1972 Ann. Meet. Abstr. Progr. **4**, 7, 518 (1972).

Kann, E.: Krustensteine in Seen. Arch. Hydrobiol. **37**, 504–532 (1941).

Kann, E.: Die eulitorale Algenzone im Traunsee (Oberösterreich). Arch. Hydrobiol. **55**, 2, 129–192 (1959).

Le Campion-Alsumard, T.: Contribution à l'étude des cyanophycées lithophytes des étages supralittorals et mediolittorals (Région de Marseille). Thetys **1**, 1, 119–172 (1969).

Logan, B.W.: *Cryptozoon* and associate stromatolites from the recent Shark Bay, Western Australia. J. Geol. **69**, 517–533 (1961).

Monty, C.: Recent algal stromatolites in the windward lagoon, Andros Island, Bahamas. Ann. Soc. Geol. Belg., Bull. **88**, 264–276 (1965).

Monty, C.: Distribution and structure of Recent stromatilitic algal mats, eastern Andros Island, Bah. Ann. Soc. Geol. Belg., Bull. **90**, 55–100 (1967).

Monty, C.: Precambrian background and phanerozoic history of stromatolitic communities, an overview. Ann. Soc. Geol. Belg. **96**, 3, 585–624 (1973).

Pia, J.: Die Kalkbildung durch Pflanzen. Beih. Bot. Zentralbl. **52**, Abt. A, 1, 1–72 (1934).

Pytkowicz, R.M.: Chemical solution of calcium carbonate in sea water. Am. Zoologist. **9**, 673–679 (1969).

Schneider, J.: Biologische und anorganische Faktoren beim Abbau einer Kalkküste. Habilitationsschrift, Göttingen, 329 S. (1973).

Schneider, J.: Biological and inorganic factors in the destruction of limestone coasts. Contr. Sedimentol. **6**, 112 (1976).

Simkiss, K.: Phosphates as crystal poisons in calcification. Biol. Rev. **39**, 487–505 (1964).

Wallner, J.: Zur Kenntnis der Kalkbildung in der Gattung *Rivularia*. Beih. Z. Bot. **54**, Abt. A, 151–155 (1935).

28 Carbonate Cementation in Recent Coralline Algal Constructions

T. ALEXANDERSSON, Uppsala, Sweden

28.1 Algal Constructions and Recent Submarine Cementation

Crustose coralline algae occur in close relationship with several forms of shallow-marine carbonate cementation in modern seas. Striking examples are: littoral and sublittoral algal constructions, such as the *Lithophyllum* corniche in the Mediterranean Sea (Pérès, 1967; Alexandersson, 1969), the algal ridges and reefs in Bermuda and the West Indies (Ginsburg and Schroeder, 1973; Adey, 1974; Adey and Vassar, 1975; Dean and Eggleston, 1975); and rhodoliths, unattached algal nodules that occur in virtually all seas (Bosellini and Ginsburg, 1971; Alexandersson, 1974). Internal carbonate cementation is often of great importance for the final composition and structure of such algal constructions, particularly where the frameworks are attacked by boring macroorganisms. In the case of *Lithophyllum* corniches and rhodoliths, as much as 50–70% of the biolith may consist of cement and internal sediment, with only scattered remains left of the initial algal skeletons (Fig. 1).

The crustose calcareous algae are essential for these constructions (1) as framework builders, this including the binding of foreign particles of widely different size, form, and origin, and (2) as producers of a multicellular organic blanket, sealing off the internal microenvironment from direct connection with the ambient seawater. Most of the skeletons serving as framework are dead, whereas the biochemically active organic blanket depends on living algal filaments.

A number of accessory organisms – bryozoans, serpulids, various noncalcareous algae – contribute to the living organic blanket, and thus to the isolation of the internal environment. On the other hand, a spectrum of boring macro-organisms – clionid sponges, polychaete worms, bivalves – may excavate tunnels and galleries into the calcareous framework and thus open up a variety of connections between the open seawater and the internal pore space. (Boring micro-organisms, such as endolithic algae and fungi, are of limited importance in the creation of this connective system. Rather, they are part of the isolating organic blanket.) The thickness of the organic blanket is in the order of millimeters, whereas the depth of the bored zone is in the order of centimeters.

It is worth pointing out that submarine cementation in present seas shows a conspicuous preference for living calcareous frames. Ten years ago, it was widely accepted that shallow marine cementation does not take place in the modern seas, and the change in opinion since then is to a great extent based on studies of living algal-coral frameworks. The cementation and lithification of shallow submarine sands is an exceptional case in comparison with the cementation of organic frameworks (cf. Taft et al., 1968; Shinn, 1969).

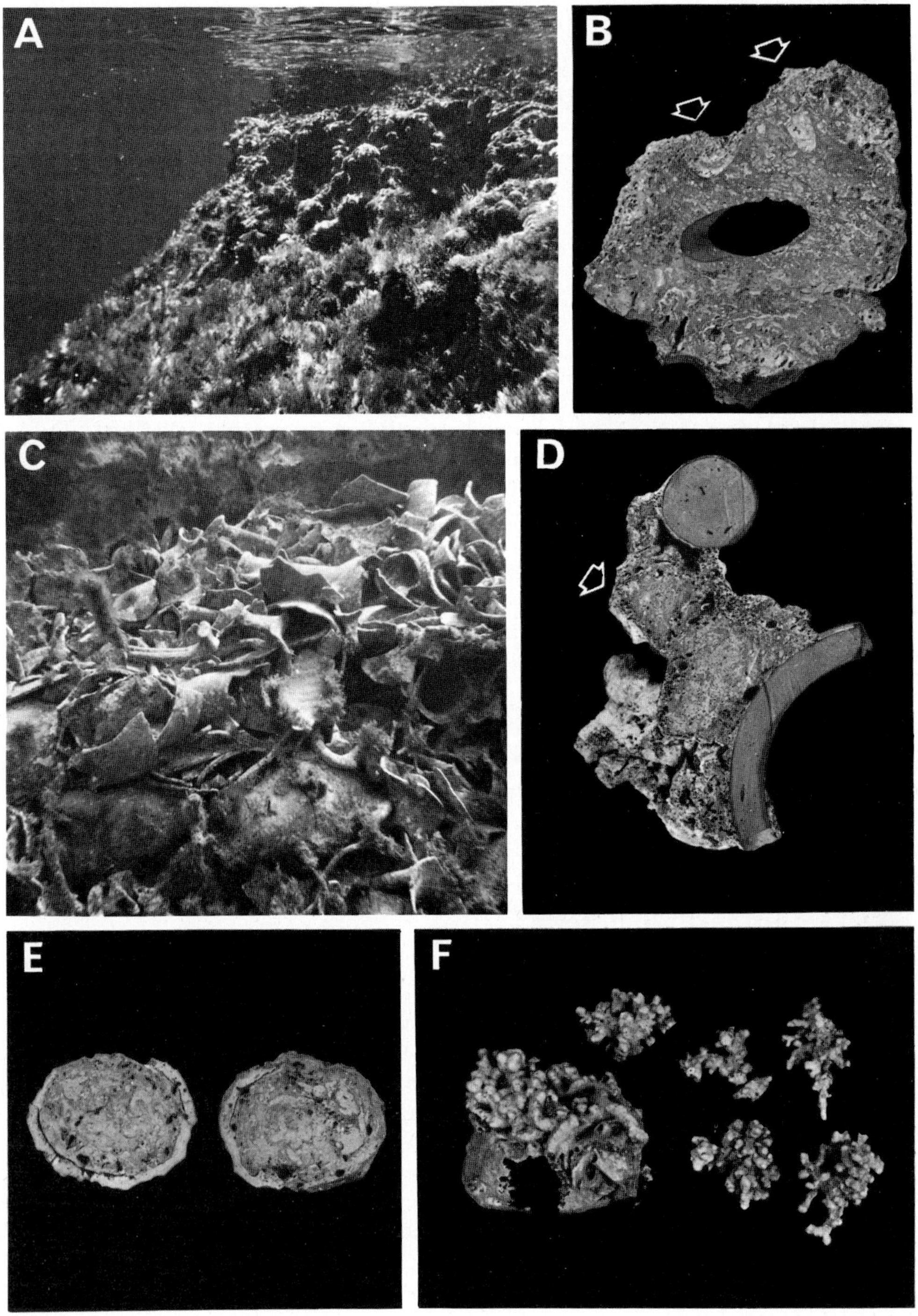

Fig. 1. Legend see opposite page

Why is submarine cementation so intimately linked with living calcareous frameworks? It is quite evident that the internal microenvironment of such frames is favorable to carbonate precipitation even in settings where the ambient seawater is barely saturated, or even undersaturated, with calcium carbonate (Alexandersson, 1974). This has been explained by two opposing hypotheses which, for emphasis and contrast, may be called the *dead-frame model* and the *living-frame model.*

The dead-frame model stresses processes related to the post-mortem decomposition of the various dead bodies in the frame, that is, mainly the bacterial breakdown of nitrogenous organic matter. The living-frame model, on the other hand, accentuates processes related to the life activites and growth of the frame-building organisms, and in the point of focus for this interest here are the crustose coralline algae.

(Obviously, the contradiction between the two models is partly contrived. Both are, in effect, concerned with the processes of living organisms, bacteria being no less alive than the rest of the frame-related community. In the scientific literature, however, the contradiction is real indeed. There is no doubt that the two models reflect fundamentally different views on the growth and early diagenesis of organic reefs, and the differences become particularly evident in attempts at paleoecologic reconstructions.)

This contribution is an argument in favor of the living-frame model, and the complexities of bacterial precipitation of marine calcium carbonate are left out of the discussion. For views on microbial precipitation, see e.g., Berner, 1969; Pautard, 1970; Di Salvo, 1973.

28.2 The Crustose Algal Skeleton and Algal Cement

28.2.1 The Skeleton

A transverse fracture through a crustose algal skeleton shows a great number of interconnected compartments that are joined into systems of closely spaced tubes with an inner diameter of about 5 μm (Fig. 2). The compartments are the empty cavities left by vegetative cells, and they are separated by double walls with a thickness of 2–8 μm. It appears that each cell secreted its own calcareous wall, one layer in thickness. Each layer consists of needles, 1–4 μm in length, which have an orientation normal to the wall surface.

◀ Fig. 1 A–F. Recent coralline algal constructions. (A) Underwater view of *Lithophyllum* corniche, Rhodes, Greece. Vertical notch is about 30 cm in height. (B) Slab of heavily bored and cemented algal crust, Rhodes. Central ellipsoidal cavity is after a boring mussel; *arrows* point to two cemented cavities with shells still in place. Slab is 8 cm across. (C) Sunken amphora cargo, perhaps 2000 years old, bound into rigid frame by crustose algae. Off Vurgundha, Karpathos, Greece. Depth 6 m. (D) Section through amphora neck and handle with heavily cemented algal framework. *Arrow* points to marginal bored zone. Diameter of circular handle, 30 mm. (E) Sectioned rhodolith from Mediterranean Sea. Patchy interior shows several generations of boring and filling, very little algal frame. Diameter 35 mm. (F) Rhodoliths (maerl) from the Skagerrak, North Sea. Nodules mostly algal skeleton with small amounts of intraskeletal cement. Depth 9 m, length of largest specimen 40 mm

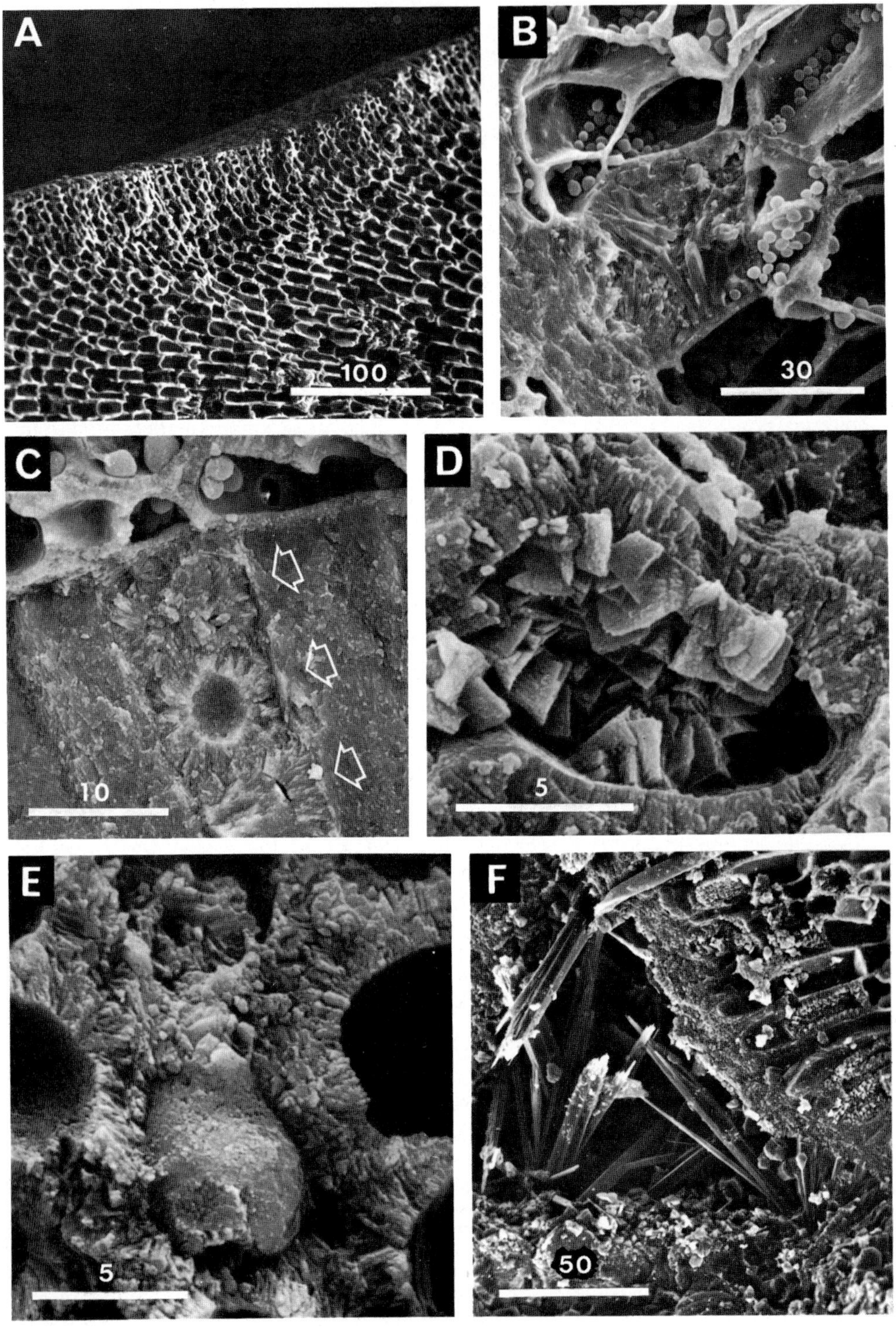

Fig. 2. Legend see opposite page

(Data presented by Dr. Gerd Flajs, Institut für Paläontologie, Bonn, during the algal symposium in Erlangen show that the crustose coralline algae have a more complex wall structure than the one apparent in simple fractures. See Flajs, 1975, and in preparation.)

The mineral form in the skeleton is Mg-calcite, usually with an Mg-content of 7–20 mole percent $MgCO_3$ (Chave, 1954; Moberly, 1970; Milliman et al., 1971). Schmalz (1965) found small amounts of an Mg-rich phase, probably brucite, as linings and fillings in vegetative cell compartments in the red alga *Goniolithon.*

Morphologically, the tissue of crustose coralline algae is differentiated into three portions: hypothallium, perithallium, and epithallium (e.g., Johnson, 1962). The hypothallium is the basal part, with elongated cells that initially grow horizontally and later curve upward (assuming a level crust). The perithallium is composed of vertical threads of cells and the calcite walls are as a rule thicker than in the hypothallium (Fig. 2). The epithallium cells are not calcified, and thus have no counterpart in the skeleton. Several layers of hypothallium – perithallium are usually seen in crusts, no more than a few mm in thickness.

The cavities of the reproductive cells, the conceptacles, are slightly flattened spheres with a diameter of 200–300 μm. As a rule, the spheres are divided into several smaller cavities by internal walls (Fig. 2).

28.2.2 The Cement

Cement – that is, nonskeletal crystals formed in voids in the structures – occurs (1) in primary intraskeletal openings, such as vegetative cell compartments and conceptacles, and (2) in virtually all sorts of secondary cavities, regardless of origin (Fig. 2). Common secondary openings are: the sheet cavities between a skeleton and its substrate, and between successive skeletal laminae; micro- and macroscopic borings; and cavities within incorporated foreign skeletons and shells, such as bryozoans, foraminifers, and boring mussels.

Mg-calcite is by far the predominant cement polymorph, the first to form within the algal skeleton proper, and the one forming the bulk of the rock in *Lithophyllum* corniches and in many rhodoliths. In the Mediterranean Sea, the calcite cement in algal constructions has about 15 mole percent $MgCO_3$ in solid solution (Alexandersson, 1969), and, interestingly enough, petrographically identical Mg-calcite cement is also predominant in beachrock in the same region (Alexandersson, 1972a).

Aragonite cement occurs as a minor, but very persistent, constituent in coralline algal constructions. Evidently, there are always some microscopic sites in the framework where the conditions for crystal growth favor the aragonite lattice. At a given site, these

---

◀ Fig. 2 A–F. SEM micrographs of algal cement. Length of scale bars in μm. (A) Vertical fracture through algal skeleton with hypothallium and perithallium. (B) Part of conceptacle with cement in one cavity. Round bodies are nestling unicellular green algae, ubiquitous in algal skeletons. (C) Hypothallium on top of older, heavily cemented perithallium. *Arrows* point to three cemented perithallium compartments. (D) Mg-calcite cement in vegetative cell compartment. (E) Intraskeletal microdruse, forming precise cast of internal space in cell compartment. (F) Aragonite needle cement in interskeletal cavity. At a distance of a few μm, Mg-calcite cement grows in algal cell compartments conditions may change through time – a common sequence of cement is from Mg-calcite to aragonite – but in the frame as a whole both polymorphs seem to grow simultaneously.

Field oberservations on marine carbonate precipitation suggest that the mineralogy of the precipitate is independent of the major physiochemical parameters of the seawater, e.g., salinity, temperature, pH, and ionic ratios. Virtually each supersaturated sublittoral environment contains both aragonite and Mg-calcite as intragranular cement (Glover and Pray, 1971; Alexandersson, 1972b; Boyer, 1972), and the two polymorphs duplicate each other as cement under various conditions, and also as marine ooliths (Füchtbauer, 1969; Bricker, 1971; Milliman and Barretto, 1975; Marshall and Davies, 1975).

The condition that Mg-calcite is the predominent cement in coralline algal constructions is hardly a coincidence. A general correspondence in mineralogy between biogenic substrates and cement in marine cementation is the rule, rather than the exception. It seems probable that the control depends to a great extent on stereochemically active organic substances, originally formed during the processes of biomineralization. Such substances are active in the organic matrices governing the mineralogy and crystal organization of hard tissues during biologic calcification (e.g., Wilbur and Watabe, 1963), and their derivatives retain some ability to control polymorphic form and fabric, even during inorganic precipitation (Suess and Fütterer, 1972).

28.3 Crustose Algae and Microenvironmental Control

28.3.1 Architecture

The general organization of an algal framework suggests that the opportunities for an algal control of the internal microenvironment are very great. There are few connections between the ambient seawater and the pore space in the frame, except through the blanket of living algal filaments. Direct connections almost exclusively consist of macroscopic borings *empty of their borers.* In a living crust, however, vacated borings are rapidly overgrown by laterally spreading algal laminae so the majority of empty borings are in fact part of the sealed-off internal environment. A living algal surface has a considerable ability to withstand and repair attacks by borers, and also by grazing fish, limpets, and echinoids (Adey and Vassar, 1975).

In conclusion: as long as a living film of algae is present on the surface of an organic framework, the interior of this framework remains an isolated, biologically controlled microregime. Free circulation is limited, and most of the communication with the ambient seawater must take place through the blanket of living cells.

28.3.2 Extracellular Production

Biochemic data lend further support to the hypothesis that the metabolism of the algae may be a major factor in an organic control of the microenvironment. It is known that extracellular products – that is, soluble organic substances liberated from healthy as distinct from injured or decomposing cells – are produced by algae in great profusion and variety (Fogg, 1966; Sieburth and Jensen, 1968, 1969; and others). In red algae, the total amount of dissolved organic matter released per year has been estimated to 38% of the gross production (Khailov and Burlakova, 1969). The compounds in external metabolites show a broad diversity in molecular weights, at least from 200 to 5000. It is evident that the extracellular products of algae are of great ecologic importance, since more that 1/3 of the total algal production may contribute directly to this mechanism for microenvironmental control (cf. Lucas, 1961).

The release of algal extracellular products provides a mechanism that seems exceptionally well suited for influencing the chemistry of the internal environment in the framework. Where fluids are pumped through the framework, e.g., in a turbulent nearshore situation, the external metabolites may be flushed through most of the openings available to the seawater.

Clearly, the data on extracellular production in algae are at best circumstantial evidence in the case of cementation. However, they do support the petrographic evidence in a general way by suggesting a mechanism for organic environmental control in exactly those organisms where a great deal of influence was expected.

28.4 Frame-Builders and Cementation: Conclusions

The theory that the frame-building organisms in reefs and algal crusts contribute directly to the genesis of cavity-filling cement within the frames accomodates several characteristics of modern cemented reefs. For example: (1) large-scale cementation of biogenic frameworks may take place in areas where little or no carbonate precipitation occurs in the surrounding sediments; (2) within the frames, cementation may be varied and ubiquitous, proceeding even in the absence of obvious triggering agents; (3) since the processes are extracellular, biologic isotopic fractionation should be limited; (4) the local factors most clearly related to submarine cementation should be a living frame and a pumping mechanism.

However, it must be stressed that even biochemic cementation is dependent on the ambient state of carbonate saturation. In supersaturated seas, like the Mediterranean and the Caribbean, the cement is accumulated in frameworks that last indefinitely in their environment of formation. On the other hand, in undersaturated regions, like the Skagerrak in the North Sea, the algae must work against an unfavorable chemical gradient. They manage to produce small amounts of cement in restricted frames only, and after the death of the organisms both skeleton and cement are attacked by dissolution.

In short, submarine carbonate cementation of organic frameworks leads to geologically significant constructions only in situations where the marine environment as a whole is favorable to carbonate deposition.

28.5 Bibliography

Adey, W.H.: A survey of red algal biology and ecology with reference to carbonate geology, and the role of reds in algal ridge and reef construction. In: Recent Advances in Carbonate Studies. Abstract Volume 6, St. Croix: West Indies Laboratory 1974, pp. 3–6.

Adey, W.H., Vassar, J.M.: Colonization, succession and growth rates of tropical crustose coralline algae (Rhodophyta, Cryptonemiales). Phycologia **14**, 55–69 (1975).

Alexandersson, E.T.: Recent littoral and sublittoral high-Mg calcite lithification in the Mediterranean. Sedimentology **12**, 47–61 (1969).

Alexandersson, E.T.: Mediterranean beachrock cementation: marine precipitation of Mg-calcite. In: The Mediterranean Sea: A Natural Sedimentation Laboratory. Stanley, D.J. (ed.). Stroudsburg, Penns.: Dowden, Hutchinson and Ross, 1972a, pp. 203–223.

Alexandersson, E.T.: Intragranular growth of marine aragonite and Mg-calcite: evidence of precipitation from supersaturated seawater. J. Sed. Petrol. **42**, 441–460 (1972b).

Alexandersson, E.T.: Carbonate cementation in coralline algal nodules in the Skagerrak, North Sea: biochemical precipitation in undersaturated waters. J. Sed. Petrol. **44**, 7–26 (1974).

Berner, R.A.: Chemical changes affecting dissolved calcium during the bacterial decomposition of fish and clams in seawater. Marine Geol. **7**, 253–274 (1969).

Bosellini, A., Ginsburg, R.N.: Form and internal structure of Recent algal nodules (rhodolites) from Bermuda. J. Geol. **79**, 669–682 (1971).

Boyer, B.W.: Grain accretion and related phenomena in unconsolidated surface sediments of the Florida reef tract. J. Sed. Petrol. **42**, 205–210 (1972).

Bricker, O.P. (ed.): Carbonate Cements. Johns Hopkins University Studies in Geology, Vol. 19. Baltimore–London: The Johns Hopkins Press 1971.

Chave, K.E.: Aspects of the biogeochemistry of magnesium. J. Geol. **62**, 266–283, 587–599 (1954).

Dean, W.E., Eggleston, J.R.: Comparative anatomy of marine and freshwater algal reefs, Bermuda and central New York. Geol. Soc. Am. Bull. **86**, 665–676 (1975).

Di Salvo, L.H.: Microbial ecology. In: Biology and Geology of Coral Reefs. Jones, O.A., Endean, R. (eds.). Vol. II. New York–London: Academic Press 1973, pp. 1–15.

Flajs, G.: Submikroskopische Strukturen von Kalkalgen. In: Program and Abstracts. Intern. Symp. Fossil Algae. Flügel, E. (ed.). Erlangen: Institut für Paläontologie 1975, pp. 8–9.

Fogg, G.E.: The extracellular products of algae. Oceanogr. Marine Biol. Ann. Rev. **4**, 195–212 (1966).

Füchtbauer, H. (ed.): Lithification of Carbonate Sediments I. Sedimentology (Special Issue) **12**, 7–159 (1969).

Ginsburg, R.N., Schroeder, J.H.: Growth and submarine fossilization of algal cup reefs, Bermuda. Sedimentology **20**, 575–614 (1973).

Glover, E.D., Pray, L.C.: High-magnesian calcite and aragonite cementation within modern subtidal carbonate sediment grains. In: Carbonate Cements. Bricker, O.P. (ed.). Johns Hopkins Univ. Studies Geol., Vol. 19. Baltimore–London: The Johns Hopkins Press 1971, pp. 80–87.

Johnson, J.H.: The algal genus *Lithothamnium* and its fossil representatives. Colorado School of Mines Quart. **57**, 1–111 (1962).

Khailov, K.M., Burlakova, Z.P.: Release of dissolved organic matter by marine seaweeds and distribution of their total organic production to inshore communities. Limnol. Oceanogr. **14**, 521–527 (1969).

Lucas, C E.: On the significance of external metabolites in ecology. Symp. Soc. Exp. Biol. **15**, 190–206 (1961).

Marshall, J.F., Davies, P.J.: High-magnesian calcite ooids from the Great Barrier Reef. J. Sed. Petrol. **45**, 285–291 (1975).

Milliman, J.D., Barretto, H.T.: Relict magnesian calcite oolite and subsidence of the Amazon shelf. Sedimentology **22**, 137–145 (1975).

Milliman, J.D., Gastner, M., Müller, J.: Utilization of magnesium in coralline algae. Geol. Soc. Am. Bull. **82**, 573–580 (1971).

Moberly, R., Jr.: Microprobe study of diagenesis in calcareous algae. Sedimentology **14**, 113–123 (1970).

Pautard, F.G.E.: Calcification in unicellular organisms. In: Biological Calcification. Schraer, H. (ed.). Amsterdam: North-Holland Publ. Co. 1970, pp. 105–201.
Pérès, J.M.: The Mediterranean benthos. Oceanogr. Marine Biol. Ann. Rev. **5**, 449–533 (1967).
Schmalz, R.F.: Brucite in carbonate secreted by the red alga *Goniolithon* sp. Science 149, 993–996 (1965).
Shinn, E.A.; Submarine lithification of Holocene carbonate sediments in the Persian Gulf. Sedimentology **12**, 109–144 (1969).
Sieburth, J. McN., Jensen, A.: Studies on algal substances in the sea I. Gelbstoff (humic material) in terrestrial and marine waters. J. Exp. Marine Biol. Ecol. **2**, 174–189 (1968).
Sieburth, J. McN., Jensen, A.: Studies on algal substances in the sea II. The formation of Gelbstoff (humic material) by exudates of phaeophyta. J. Exp. Marine Biol. Ecol. **3**, 275–289 (1969).
Suess, E., Fütterer, D.: Aragonite ooids: experimental precipitation from seawater in the presence of humic acid. Sedimentology **19**, 129–139 (1972).
Taft, W.H., Arrington, F., Haimovitz, F., MacDonald, C., Woolheater, D.: Lithification of modern marine carbonate sediments at Yellow Bank, Bahamas. Bull. Marine Sci. **18**. 762–828 (1968).
Wilbur, K.M., Watabe, N.: Experimental studies in calcification in molluscs and in the alga *Coccolithus huxleyi*. New York Acad. Sci. Ann. **109**, 82–112 (1963).

29 Ecological Studies on two Carbonate Sediment-Producing Algae

D. BOSENCE, London, Great Britain

29.1 Environment Setting and Lithofacies

The bathymetry and location of Mannin Bay are shown in Fig. 1. A submarine valley leads into the bay and up to a slope that levels off at about 8 m to form a platform with occasional rocky outcrops. The shore is mainly rocky (Dalradian metasediments) but some carbonate beaches and dunes are present. In sheltered lagoons a muddy shore is found. The sea water is normal marine with seasonal water temperature varying from 5 to 15° C. Salinity varies seasonally from 33.5 to 35‰ (Lees et al., 1969). The normal marine environment is the result of mixing by tidal currents which, however, are not strong enough to transport the algal clasts (Buller, 1969). Study of the algal bed forms shows that the main hydraulic forces are the oscillatory currents produced by waves.

The sediments of Mannin Bay are mainly carbonate as is the case with much of the littoral and sublittoral of the coast of western Eire (Keary, 1970; Lees et al., 1969). The Mannin Bay carbonates can be divided into four facies (Fig. 1) based on particle size, composition and mineralogy of the carbonate grains (Buller, 1969; Gunatilaka, 1972; Bosence, 1976).

The distribution of the facies can, in the main, be explained in terms of the input of hydraulic energy into the bay. The tidal wind and fetch directions indicate that the maximum energy is derived from about 230°N (Fig. 1). The area most exposed to this direction is represented by the coarse sand facies. Where slight protection is afforded fine sand accumulates, and in the very sheltered bay heads mud is deposited (Buller, 1969). The *Lithothamnium* (or maerl) facies, which is the principle concern of this paper, does not fit this pattern as it is basically an autochthonous deposit and the grain size is related to the size of the thallus and its comparative rate of destruction.

29.2 Algae

29.2.1 The Free-Living Corallines

The *Lithothamnium* facies of Buller (1969) is composed of the free-living coralline algae (subfamily Melobesioideae) *L. corallioides* Crouan and *Phymatolithon calcareum* (Pallas) Adey and McKibbin (1970).

The main form of reproduction observed in Mannin Bay is vegetative. The thallus is broken by strong bottom currents and by organic reworking of the maerl. The branches of the thallus are of multiaxial construction with filaments of unlimited growth forming

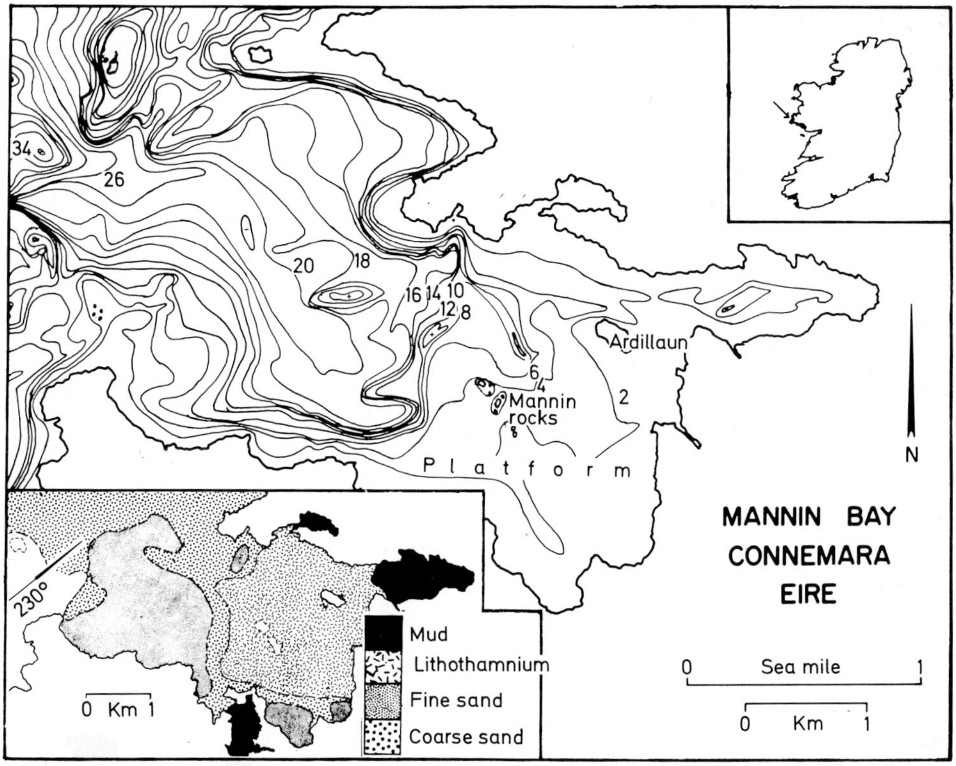

Fig. 1. Location, bathymetry and lithofacies of Mannin Bay (after Buller, 1969)

a central medulla. These axial filaments branch and grow laterally as filaments of limited growth forming a cortex and a branch width typical for the species.

Growth is periodic as illustrated by layers of different cell size. These growth layers cover the apex and some way down each branch (Pl. 1/4). The growth layers illustrate that growth may be patchy and occasionally abrasion and rehealing can be found.

Branching can occur in two different ways. In dichotomous branching the axial filaments diverge to form two equal branches with a plane of symmetry between them (Pl. 1/4). Intercalary branching occurs through rejuvenation of the lateral filaments. This is recognized by the initiation of new medulla from pre-existing lateral filaments (Pl. 1/5) resulting from rejuvenation of the intercalary meristem.

The living maerl species cover up to 100% of the bottom and live in both quiet and exposed environments within the depth range of low water spring tides to 16 m (Fig. 2). Where growth is prolific an in situ three-dimensional growth structure is formed by the interlocking algal thalli, which may be stacked to a height of 30 cm and covering areas measured in thousands of square meters, so forming a bank (Pl. 1/1 and 2). These banks can consist of one or both maerl species and they support a diverse biota including borers, encrusters, nestlers, grazers, infaunal deposit and suspension feeders and epiphytic algae.

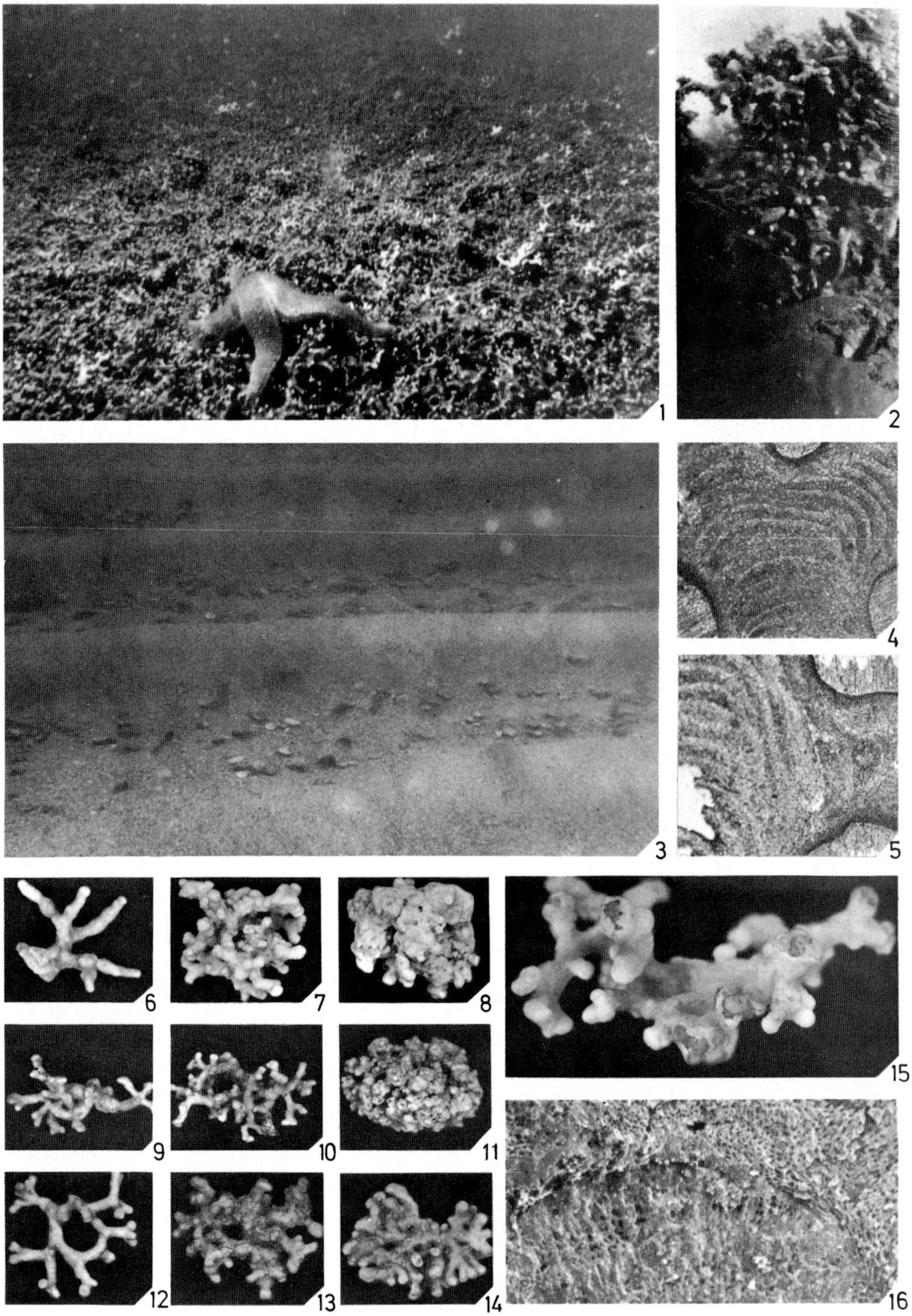

Plate 1. Legend see opposite page

The banks are broken down locally by physical and biologic erosion to form a gravel-sized debris (Pl. 1/3). This is either transported into symmetric wave ripples by wave induced currents with some living corallines in the ripple troughs, or it is a bioturbated bottom.

29.2.2 Distribution of Corallines

Fig. 2 shows the percentage cover of living maerl in Mannin Bay. The distribution is discontinuous and three main concentrations can be noted: a northern central bank, a central eastern bank and a small southern bank. (Dense growth is not found lower than 8 m or above 1 m below LWST.) This distribution is discussed below with respect to the physical and chemical parameters operating in the bay.

The salinity of the bay water cannot be limiting as it is normal marine and does not vary at one point of time over the bay (Gunatilaka, 1972). Nutrient distribution has not been studied but the analyses for salinity, Ca, Mg, and pH indicate a well-mixed open marine environment and in the absence of any rivers bringing in concentrated nutrients there are not likely to be any areas of nutrient enrichment. The availability of O_2 and CO_2 reflected in the Eh is not limiting as this is constantly positive (Gunatilaka, 1972). Grazers on the algae are unimportant on this scale. Although abundant, gastropods only brush the surface mucous and debris. There is no grazing by echinoids. Although within the *Laminaria* zone, there is no competition between the maerl species and these brown algae as they are restricted to the hard substrate areas of the bay. However, epiphytic filamentous algae cover the living corallines especially in quiet waters. In addition to this the filamentous algae trap mud and eventually smother the unattached corallines to form oncolites. Thereby restricting their occurrrence in the sheltered areas of the bay where the mud facies is found.

Light and temperature have been shown to be the most important controlling factors in the depth and geographic distribution of corallines in the northern Atlantic by Adey

---

◀ Pl. 1/1. Surface of algal bank of *P. calcareum* with *Asterias rubens*, x 1/4

Pl. 1/2. Core through bank of *L. corallioides*. Note interlocking of thalli and grain supported sediment with muddy matrix, x 1

Pl. 1/3. Wave-rippled algal debris with dead infaunal bivalve shells in troughs, x 1/40

Pl. 1/4. Longitudinal section of *L. corallioides* illustrating growth layers and dichotomous branching (acetate peel), x 20

Pl. 1/5. Longitudinal section of *L. corallioides* illustrating intercalary branching (acetate peel), x 20

Pl. 1/6–8. *L. corallioides* spheroidal II, III, and IV, x 2

Pl. 1/9–11. *L. corallioides* ellipsoidal II, III, and IV, x 1 1/2

Pl. 1/12–14. *L. corallioides* discoidal II, III, and IV, x 1 1/2

Pl. 1/15. *L. corallioides* showing healing of dead apices, x 4

Pl. 1/16. Section through preserved apex of *L. corallioides* showing dead apex grown over by new tissue (2^{ry} hypothallus and perithallus from both sides (SEM micrograph), x 100

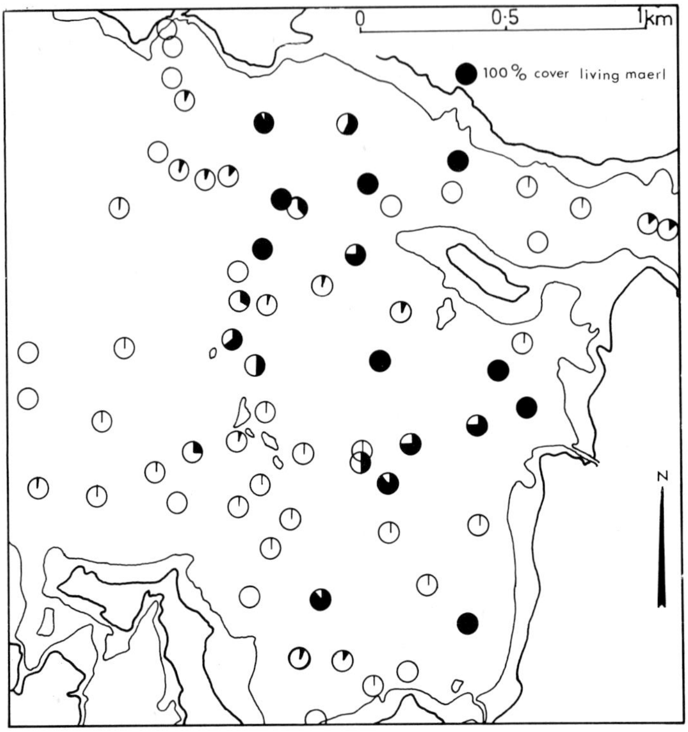

Fig. 2. Percentage cover of living maerl, Mannin Bay

(1970), Adey and McKibbin (1970) and Adey and Adey (1973). *P. calcareum* and *L. corallioides* both range from Norway to the Mediterranean. A comparison of experimental work by Adey and McKibbin (1970) and measurements in Mannin Bay show that optimum temperature for growth exists for the maerl species from 1 to 20 m depth. However living maerl is not found over this range and the algae are thought to be restricted by light which is only 180 lux at 20 m, in the summer and may reach as low as 160 lux in 3 m. Adey and McKibbin (1970) have shown that there is little growth at these luminosities under optimum temperatures. At the depth limit of the algae (16 m) 30% of the surface light is reaching the bottom (Bosence, 1976).

The parameter used for estimating the relative bottom current velocities was the presence or absence of wave ripple marks (Fig. 3). Wave ripples are found in the algal debris subfacies where waves meet the edge of the platform and where waves are refracted and converge on islands and submarine ridges. Where waves pass over the platform or diverge into bays they are not capable of forming ripples (Fig. 3).

Figs. 2 and 3 show the presence and orientation of ripples and the distribution of the living maerl species. There is a marked negative correlation between these two parameters: 82% of those stations with more than 50% living maerl do not have ripples. This suggests an inverse relationship between the oscillating bottom currents and the living algal thallus, which is supported by observations made whilst diving, that the growing apical tips are abraded when the algae are rolled on the sea bed.

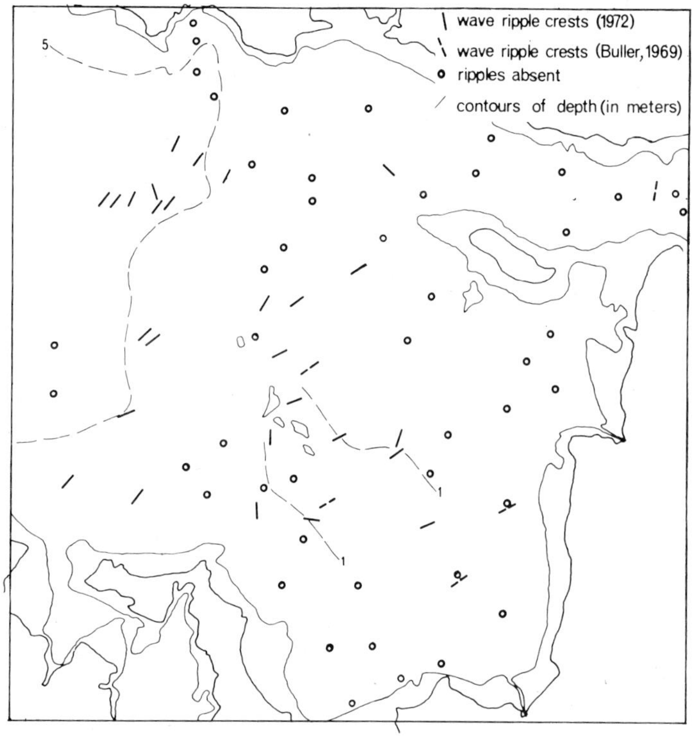

Fig. 3. Presence and orientation of wave ripples, Mannin Bay

29.2.3 Variation of Growth Form

Both maerl species show a parallel series of growth forms (Pl. 1/6–14) (Foslie, 1894; Cabioch, 1966; Bosence, 1976). They are a continually varying morpholical sequence. A scheme is suggested (Bosence, 1976) which describes the variation found in the free-living corallines incorporating the systems of Cabioch (1966) and Bosellini and Ginsburg (1971) by relating forms to shape and branching density classes. Three shape classes can be recognized: spheroidal, ellipsoidal, and discoidal. Within these shape classes four groups of branching density can be seen; (I) one single main branch, (II) a few branches, (III) frequent branching, and (IV) dense and solid intergrown thallus or rhodolith.

Figs. 4 and 5 show the distribution of thallus shape and branching density classes respectively for *L. corallioides*. The commonest shapes are spheroidal and ellipsoidal but these two growth forms show no obvious pattern of distribution. However, discoidal forms show a concentration in the southern part of the bay. The substrate in this area is a moderately sorted sand (Fig. 1) in contrast to the poorly sorted coarse algal gravel substrate of the rest of the platform. The distribution of the branching density classes illustrates a close correlation with exposure to wave-induced currents as shown by the presence of ripples: 72% of stations with more than 5% branching density IV are rippled. This is also noted from observations that the densely branched forms are found growing in the ripple throughs.

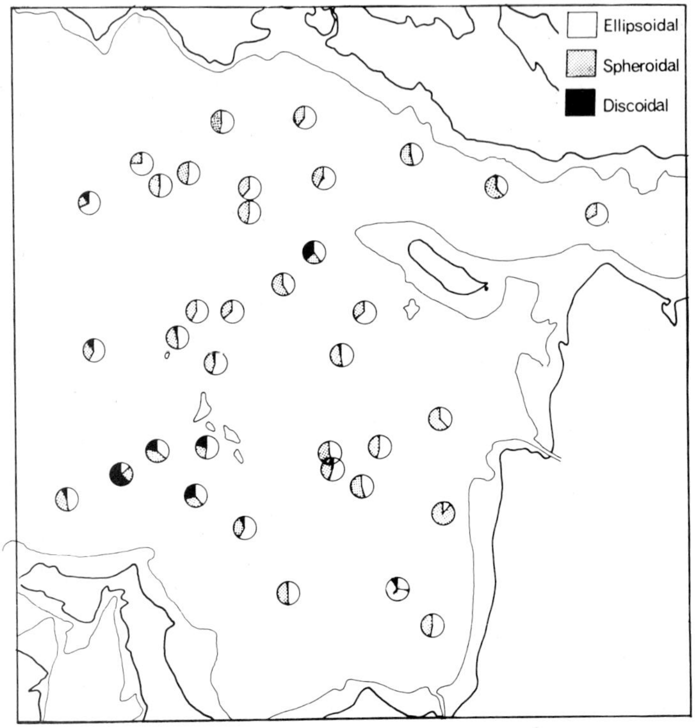

Fig. 4. Distribution of shape classes, Mannin Bay

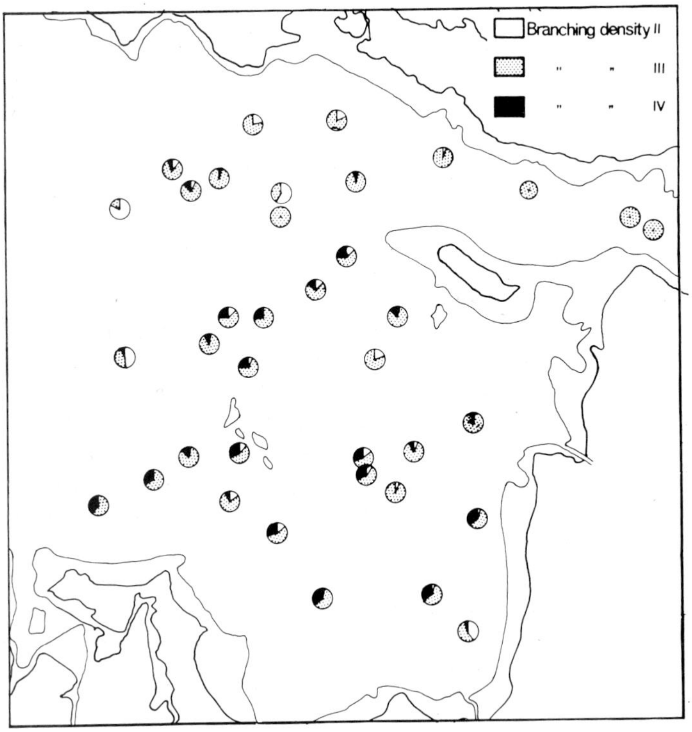

Fig. 5. Distribution of branching density classes, Mannin Bay

29.2.4 Hydrodynamics of the Thallus of *L. corallioides*

Because of the field relationship between branching density and exposure and bank development and exposure, experiments were designed to establish the velocity required for transport of the algal thallus. To establish this both actual specimens and plasticene replicates were studied in a wave tank (Bosence, 1976). Ellipsoidal forms are more easily transported than spheroidal forms, and discoidal forms are the most stable. Open-branched forms are found to be more stable than densely branched forms of the same shape and size.

29.2.5 Discussion and Results of Growth Form Study

29.2.5.1 Distribution of Shape Classes. In Mannin Bay the origin of the discoidal forms can be interpreted as a response to growth on a moderately sorted sand substrate in a relatively strong current area. A thallus living on sand will be prevented from growing downwards into the sand by the lack of light and will therefore grow upward and laterally. If the thallus is frequently turned only lateral growths will be possible. In the maerl this will not apply as the thallus is well supported by its neighbours (Pl. 1/2). The spheroidal and ellipsoidal forms do not appear to be related to any regional parameters and the development of these forms must involve the detail of its position with respect to the substrate and the nearest neighbors throughout the life of the plant. Shape is also effected by breakage during transport.

29.2.5.2 Distribution of Branching Density Classes. Comparing the results from the wave tank studies with the distribution of the forms in the bay shows that the open-branched forms are found in the quiet waters and the dense forms in the exposed waters. The most stable forms are therefore living in quiet waters and the most unstable forms in the exposed waters. This may be explained as follows: when the thallus is rolled the apical tips of the branches are observed to be abraded and perhaps broken; further growth is restricted from these apical meristems. If the abrasion is not prolonged the apex may heal from lateral undamaged cells growing over the tip (Pl. 1/15 and 16). However, if abrasion continues, growth may occur by resumption of previously dormant cells to produce an intercalary meristem within the protection of the main branches. This process will lead to an increase in branching density and also to a stronger form, because the branches may join, which will resist breakage but will also be more easily entrained. This ties in well with the relative proportions of dichotomous and intercalary branching found in the different growth forms; the densely branched forms having intercalary branching and the open-branched forms dichotomous branching. The frequency of intercalary branching in the dense forms would indicate that the dense forms (rhodoliths) are a response by the alga to wave motion.

Acknowledgments. I thank Dr. R. Goldring for supervision of this work and colleagues at Reading University and the British Museum (N.H.) for their help and criticism. Professor Francis, Imperial College, London, kindly allowed access to a wave tank. I gratefully acknowledge financial help from Goldsmiths' College, Reading University, and the Royal Society.

29.3 Bibliography

Adey, W.H.: The effects of light and temperature on growth rates in boreal-subarctic crustose corallines. J. Phycol. **6**, 269–276 (1970).

Adey, W.H., Adey, P.J.: Studies on the biosystematics and ecology of the epilithic crustose Corallinaceae of the British Isles. Brit. Phycol. J. **8**, 343–407 (1973).

Adey, W.H., McKibbin, D.L.: Studies on the maerl species *Phymatolithon calcareum* (Pallas) nov. comb. and *Lithothamnium corallioides* Crouan in Ria de Vigo. Botan. mar. **13**, 100–106 (1970).

Bosellini, A., Ginsburg, R.N.: Form and internal structure of recent algal nodules (Rhodolites), Bermuda. J. Geol. **79**, 6, 669–682 (1971).

Bosence, D.W.J.: Ecological studies on two unattached coralline algae, Connemara, Eire. Palaeontology **19** (1976) (in press).

Buller, A.T.: Source and distribution of calcareous skeletal componants in recent marine carbonate sediments, Mannin and Clifden Bays area, Connemara, Ireland. Unpubl. Ph. D. thesis, Reading Univ., 1969.

Cabioch, J.: Contribution à l'étude morphologie anatomique et systematique de deux Melobesiées: *Lithothamnium calcareum* (Pallas) Areschoug et *Lithothamnium corallioides* Crouan. Botan. mar. **9**, 33–53 (1966).

Foslie, M.: The Norwegian forms of *Lithothamnium*. D. Kgl. Norske Vidensk. Selsk. Skrift. **2**, 1–203 (1894).

Gunatilaka, H.A.: A survey of the geochemistry and diagenesis of the recent carbonate sediments from Connemara, western Eire. Unpubl. Ph. D. thesis, Reading Univ. 1972.

Keary, R.: Coastal climate and the shelf bottom sediments, a comment. Marine Geol. **8**, 363–365 (1970).

Lees, A., Buller, A.T., Scott, J.: Marine carbonate sedimentation processes Connemara Ireland. Univ. Reading Geol. Dept. Rep. **2** (1969).

30 The Coralline Algae from the Miocene Ziqlag Formation in Israel and Their Environmental Significance

B. BUCHBINDER, Jerusalem, Israel

30.1 Introduction

The Tortonian Ziqlag Formation consists of coral reefs and biocalcarenites. Remnants of this formation occur as isolated outcrops along the western foothills of the Judea-Samaria mountains (Fig. 1), and as scattered patches in the subsurface of the Mediterranean coastal plain and in the offshore area (Gvirtzman and Buchbinder, 1969; Gvirtzman, 1970; Buchbinder, 1975). The Ziqlag Formation forms the transgressive part of a sedimentary cycle which was terminated by the desiccation of the entire Mediterranean during Messinian times. The Ziqlag Formation unconformably overlies Miocene marls (Ziqim and Bet Guvrin formations) or Eocene to Cretaceous chalks, limestones and dolomites. In the subsurface the formation is covered by Late Miocene evaporites (Mavqiim Formation), or by Pliocene marls (Yafo Formation) and sands (Pleshet Formation). Two major facies groups were differentiated in the Ziqlag Formation (Fig. 1) (Buchbinder, 1975): (1) back reef platform and beach biocalcarenites; (2) coral reefs which are distributed mainly in the trough and slope area.

Coralline algae are highly abundant in the Ziqlag Formation as skeletal grains and as in situ crusts. The crustose corallines are the main skeletal contributors to the Ziqlag reef complex amongst the coralline algae, and are the most useful for environmental interpretation. The articulated corallines are represented by a single genus – *Corallina* – and they play only a limited role as environmental indicators. They are relatively scarce, equally distributed among the different environments, and are too poorly preserved to facilitate detailed taxonomic determinations. Green algae are very rare in the Ziqlag Formation. A few segments of *Halimeda* were found mainly in the Ziqlag hills area. Dasycladaceae are practically absent from the Ziqlag Formation.

The following twenty-one taxa of coralline algae from the Ziqlag Formation were determined and described (Buchbinder, 1975): 1. *Archaeolithothamnium* cf. *A. figiensis* Johnson and Ferris; 2. *Archaeolithothamnium* sp. A; 3. *Archaeolithothamnium* sp. B; 4. *Lithothamnium* cf. *L. betieri* Lemoine; 5. *Lithothamnium* cf. *L. saxorum* Capender; 6. *Lithothamnium* sp. A; 7. *Lithothamnium* sp. B; 8. *Mesophyllum laffittei* Lemoine; 9. *Mesophyllum lemoineae* Souaya; 10. *Mesophyllum* cf. *M. sancti dionysii* Lemoine; 11. *Mesophyllum* cf. *M. savornini* Lemoine; 12. *Mesophyllum* sp. A; 13. *Mesophyllum* sp. B; 14. *Lithophyllum albanense* Lemoine; 15. *Lithophyllum bonyense* Johnson; 16. *Lithophyllum ghorabi* Souaya; 17. *Lithophyllum* cf. *L. glangeaudi* Lemoine; 18. *Lithophyllum prelichenoides* Lemoine; 19. *Lithophyllum* cf. *L. pseudoamphiroa* Johnson; 20. *Lithoporella melobesoides* Foslie; 21. *Corallina* sp.

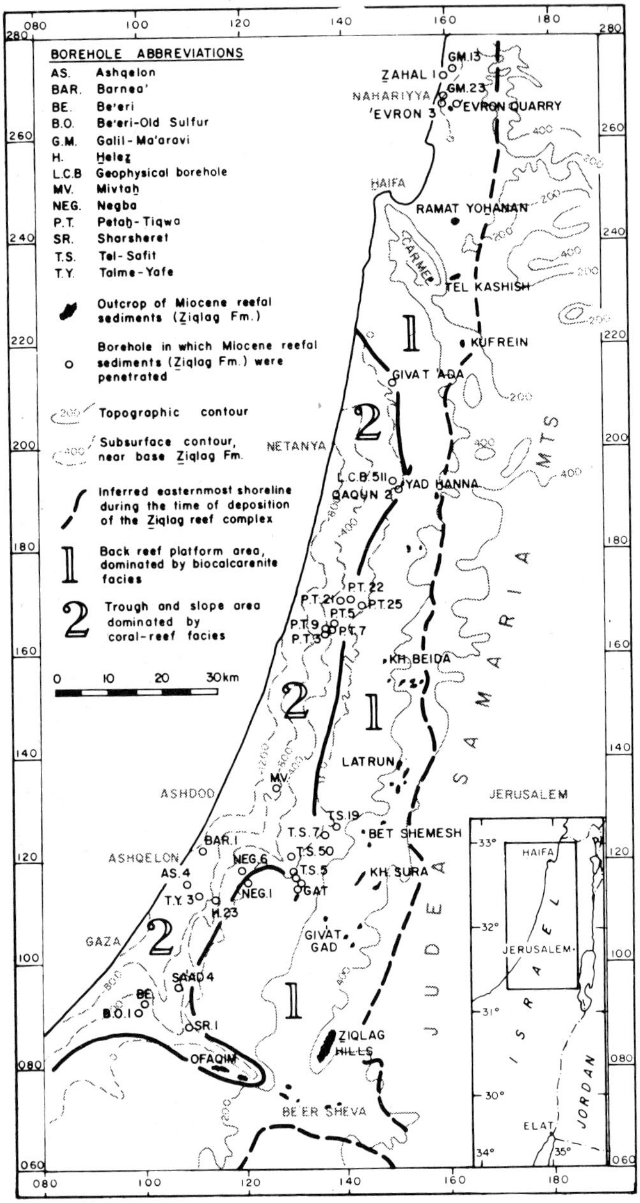

Fig. 1. Areal distribution and major facies groups of the Miocene Ziqlag Formation in Israel

30.2 Genera and Species Distribution

An attempt has been made to describe the frequency distribution of the various taxa in the Ziqlag Formation based mainly on their number in the samples taken. The results must be regarded as approximate since the number of samples and thin sections in which a certain taxon occurs also depends on factors other than its original distribution, e.g., degree of exposure, or the weathering state of the outcrop.

Blake (1935) and Daniel (1959) described abundant *Lithothamnium* from sediments which were later designated as the Ziqlag Formation. Blanckenhorn (1914) named the Miocene bioclastic sediments as „Lithothamnienkalke". However, a detailed taxonomic study (Buchbinder, 1975) has shown that *Lithothamnium* is a relatively rare genus in the Ziqlag Formation; only two specimens were found in the Ofaqim Reef area and other rare specimens were found in Barnea' 1 boreholes.

Lithophyllum is the dominant algal genus in the Ziqlag Formation. Its various species inhabited the reefs and the back reef platform environments. *L. prelichenoides* (Fig. 2) is the most abundant species, outnumbering any other *Lithophyllum* species. *L. ghorabi* (Fig. 3) is the second most abundant species. It is mainly associated with coral colonies in the Ofaqim and Bet Shemesh areas. *L.* cf. *glangeaudi* is also found in the Ofaqim Reef, but is much less abundant than *L. prelichenoides* and *L. ghorabi*. Other species of *Lithophyllum* are much less frequent and are represented by a few specimens only.

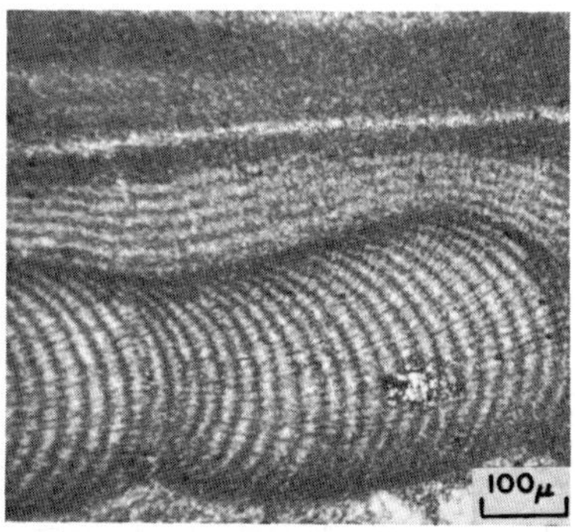

Fig. 2.

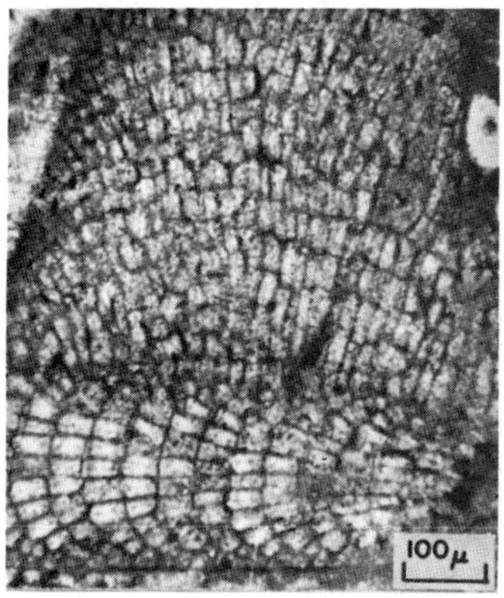

Fig. 3.

Fig. 2. *Lithophyllum prelichenoides.* Vertical section through a crust showing basal coaxial hypothallium and perithallium

Fig. 3. *Lithophyllum ghorabi.* Vertical section through a crust showing basal coaxial hypothallium and perithallium

The monostromatic crusts of *Lithoporella melobesoides* are highly abundant in the coral reefs of the Ziqlag Formation. They encrust corals, other crustose corallines, and even encrusting foraminifera.

The genus *Archaeolithothamnium* is relatively rare in the Ziqlag Formation. Only a few fragmented specimens of this genus were found; one fragment of a branch was detected in the Ofaqin Reef, and other specimens were encountered in the Helez 23 and Tal Safit 5 boreholes.

The genus *Mesophyllum* occurs chiefly in the Ofaquim Reef area. The single species *M. laffittei* (Fig. 4) is exclusively restricted to the forereef area where it built inclined algal beds (Buchbinder, 1975). It also occurs in the Helez 23 and Mivtah 1 wells, probably under similar forereef conditions. *M. lemoineae* also occurs in the Ofaquim Reef, but only in the reefcore belt. Single specimens of other species of *Mesophyllum* were occasionally found in the Ofaqim Reef and the Ziqlag hills areas.

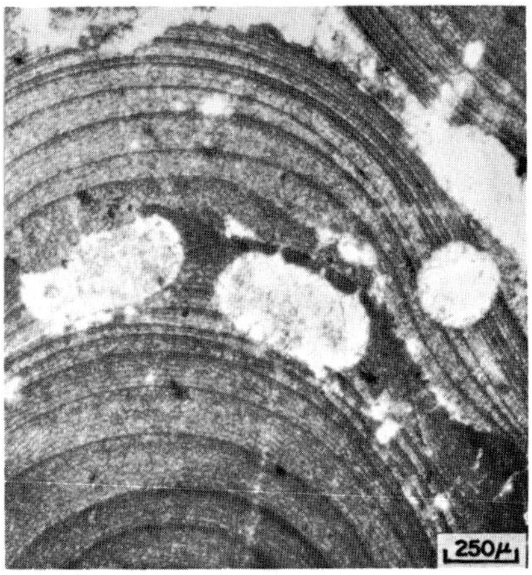

Fig. 4. *Mesophyllum laffittei.* Section through knobby protuberance showing zoned perithallium and multipored conceptacles

Most of the algal taxa found in the Ziqlag Formation were described also from other Miocene sediments in the Tethys belt and the Pacific Ocean (Lemoine, 1939; Souaya, 1963a, b; Johnson, 1957, 1964a, b; and many others). However, the species *Lithophyllum ghorabi* and *Mesophyllum laffittei* were described from the Mediterranean region only (Egypt and Algeria). Since most publications on algae do not include discussions of the relative abundance of the different taxa, it is impossible to compare the relative abundance of the algal assemblage of the Ziqlag Formation to other Miocene occurrences. However, comparison with Recent Mediterranean flora (Pérès, 1967; Adey, 1970) shows that the Ziqlag Formation demonstrates: (1) a notable absence of the heterocysted genus *Neogoniolithon*; (2) paucity of *Lithothamnium*; (3) a relatively high abundance of the genus *Lithoporella*.

Judging from the available literature, the genus *Neogoniolithon* was absent from the entire Mediterranean area during Miocene times. However, Johnson (1964b, 1965) recorded Miocene *Goniolithon* (= *Neogoniolithon*) from Guam Island in the Pacific and from Guatemala. It is probable, however, that although this genus appeared in the Pacific and Caribbean provinces in Miocene times. It only spread to the Mediterranean during later periods, and therefore its absence from the Ziqlag Formation has no environmental significance.

Based on the global distribution of modern genera (Johnson, 1961; Adey, 1970), the relatively high abundance of the genus *Lithoporella* and the scarcity of *Lithothamnium* in the Ziqlag Formation presumably indicate tropical marine conditions during Ziqlag times. This assumption is supported by the dominant distribution of *Lithophyllum* in the Ziqlag Formation and agrees with the conclusion already deduced on the basis of the presence of hermatypic coral fauna (Buchbinder, 1975).

30.3 Morphology and Mode of Occurrence of the Crustose Coralline Algae

Three morphological types of algal occurrences are distinguished: (1) attached crusts characterizing reefcore area; (2) free of semifree crusts and nodules characterizing forereef area; (3) attached or free crusts and branches characterizing back reef platform area.

30.3.1 Attached Crusts (Reefcore Area)

The algal crusts consist mainly of *Lithophyllum prelichenoides* and *Lithophyllum ghorabi* which form superimposed undulating crusts 200–600 μ thick, occasionally alternating with encrusting foraminifera; these crusts tend to form knobby protuberances and short branches. The algal crusts bind the coral colonies into a dense and rigid framework. The reefcore system represents a continuous struggle for space between corals and coralline algae. In many cases coralline algae completely surround coral colonies, presumably following the death of the colonies.

30.3.2 Free or Semi-Free Crusts and Nodules (Forereef Area)

This type was found mainly in the forereef sediments of the Ofaqim Reef. The algal deposits are composed almost exclusively of the genus *Mesophyllum laffittei*; however, rare *Lithothamnium* specimens (cf. *L. betieri*) were also detected. Detailed examination has revealed that most of the algal forms vary from coarse rhodoliths (4–8 cm in size, Fig. 5) to free crusts and branches a few millimeters in size, which originally formed coarse algal sand. Rhodoliths are known today from a 2–150 m depth (Bosellini and Ginsburg, 1971; Adey and Macintyre, 1973). They are a characteristic element of the sea floor at depths of about 50–150 m on many oceanic offshore banks and continental margins (Adey and Macintyre, 1973).

Fig. 5. Algal nodules (rhodoliths) composed of *M. laffittei*. Ofaqim forereef

In the Ofaqim forereef the algal deposits dip away from the reefcore displaying an inclination of 22° adjacent to the reef core and 20° at 200 m off the reefcore. Assuming that these dips represent the original forereef slope, a depth of about 73 m is calculated for a distance of 200 m off the reefcore, if we consider the algal layers adjacent to the reefcore to be at about 0 m depth. However, since the algal layers next to the reefcore were deposited at the lower parts of the reef wall at depths exceeding 0 m (as described from Recent reefal occurrences, Logan et al., 1969; Hottinger, 1973), the calculated depth of 73 m is a minimal figure for the algal deposits located 200 m off the reefcore.

The restriction of *M. laffittei* to the forereef beds in the Ofaqim Reef gives this alga a significant role as an environmental indicator.

30.3.3 Attached or Free Crusts and Branches (Back Reef Platform Area)

In the back reef platform area, algae commonly occur as abraded grains in calcarenitic sediments and are either micritized or too small to allow taxonomic identification. Solid crusts are relatively scarce. They are generally found associated with isolated coral colonies (e.g., in Bet Shemesh and Tel Moreshet Gat, Gvirtzman and Buchbinder, 1969) or forming small algal banks which pave the hard substrates (e.g., near Latrun). The crusts consist chiefly of *Lithophyllum prelichenoides* and occasionally *L. ghorabi.*

The origin of the abraded algal grains is not quite clear; they might derive from the attached crusts. However, these crusts are either too small or too scanty to account for the high amount and uniform distribution of the algal grains. The reefs at the western edge of the platform supplied algal and other reef derived debris to their immediate vicinity only, and not to the entire platform. Therefore, there must have been some other source for the algal grains within the platform area. Abundant growth of algal colonies as easily disintegrating aggregates of loosely packed branches or semiattached crusts are known from Recent shallow platforms or back reef environments, e.g., Shark Bay, Australia (Davies, 1970), Florida Keys (Ginsburg, in Ginsburg et al., 1971) and from back reef environments in Ras Muhammad, Red Sea (personal observation). Similar occurrences in the Ziqlag Formation could have accounted for the high frequency and uniform distribution of the algal grains in the back reef platform area.

Acknowledgments. The present study constitutes part of a Ph. D. thesis submitted to the Hebrew University, Jerusalem. Special thanks are due to Dr. R. Rezak, of Texas A & M University, for his valuable advice in the study of the algae, to Dr. W.H. Adey, of the Smithsonian Institution, for his comments on some taxonomical problems, and to Prof. Z. Reiss, of the Hebrew University, for his helpful comments and suggestions.

30.4 Bibliography

Adey, W.H.: A revision of the Foslie crustose coralline herbarium. Norske Vidensk-Selsk Akr. Kl. **1**, 1–46 (1970).

Adey, W H., Macintyre, I.G.: Crustose coralline algae: a re-evaluation in the geological sciences. Geol. Soc. Am. Bull. **84**, 883–904 (1973).

Blake, G.S.: The Stratigraphy of Palestine and its Building Stones. Jerusalem: Govt. Printing and Stationary Office 1935, pp. 1–133.

Blanckenhorn, M.: Syrien, Arabien und Mesopotamien. Handbuch der regionalen Geologie, Bd. 5. Heidelberg 1914, pp. 1–159.

Bosellini, A., Ginsburg, R.N.: Form and internal structure of Recent algal nodules (rhodolites) from Bermuda. J. Geol. **79**, 669–682 (1971).

Buchbinder, B.: Lithogenesis of Miocene reef limestones in Israel with particular reference to the significance of the red algae. Unpublished Ph. D. thesis, Hebrew Univ., Jerusalem 1975, pp. 1–170.

Daniel, E.J.: Lexique Stratigraphique International – Palestine (unpublished draft), 1959.

Davies, G.R.: Carbonate bank sedimentation, eastern Shark Bay, Western Australia. Am. Assoc. Petrol. Geol. Mem. **13**, 85–168 (1970).

Ginsburg, R., Rezak, R., Wray, J.L.: Geology of calcareous algae (notes for short course). Rosenstiel School Marine Atmosph. Sci. Miami, Fla., 1–15 (1971).

Gvirtzman, G.: The Saqiye Group (Late Eocene to Early Pleistocene) in the Coastal Plain and Hashephela regions, Israel. Geol. Surv. Isr. Rep. OD/5/67 (in Hebrew, English abstract), 1–170 (1970).

Gvirtzman, G., Buchbinder, B.: Outcrops of Neocene formations in the central nd southern coastal plain, Hashephela and Be'er Sheva regions, Israel. Geol. Surv. Isr. Bull. **50**, 1–73 (1969).

Hottinger, L.: Larger foraminifera and selected other sediment producing organisms in the Gulf of Elat, Red Sea. Rep. Activity Marine Biol. Sta. Elat (May–June, 1973).

Johnson, J.H.: Calcareous algae of Saipan. U.S. Geol. Surv. Profess. Papers **280–E**, 209–246 (1957).

Johnson, J.H.: Limestone-building algae and algal limestones. Colorado School Mines Spec. Publ. 1–297 (1961).

Johnson, J.H.: Miocene coralline algae from northern Iraq. Micropaleontology **10**, 477–485 (1964a).

Johnson, J.H.: Fossil and Recent calcareous algae from Guam. U.S. Geol. Surv. Profess. Papers **403–G**, 1–40 (1964b).

Johnson, J.H.: Fossil algae from Guatemala. Colorado School Mines Profess. Contrib. **1**, 1–152 (1965).

Lemoine, M.: Les algues calcaires fossils de L'Algérie. Matériaux pour la Carte Géologie de L'Algérie, Sér. 1, Paleont. **9**, 1–128 (1939).

Logan, B.W., Harding, J.L., Wayne, M.H., Williams, J.D., Snead, R.G.: Carbonate sediments and reefs, Yucatan Shelf, Mexico. Am. Assoc. Petrol. Geol. Mem. **11**, 1–196 (1969).

Pérès, J.M.: The Mediterranean benthos. Oceanogr. Marine Biol. Ann. Rev. **5**, 449–553 (1967).

Souaya, F.J.: Micropaleontology of four sections south of Quseir, Egypt. Micropaleontology **9**, 233–266 (1963a).

Souaya, F.J.: On the calcareous algae (Melobesioideae) of Gebel Gharra (Cairo–Suez road) with a local zonation and some possible correlations. J. Paleontol. **37**, 1204–1216 (1963b).

31 Paleogeographic Significance of Rhodolites: Some Examples from the Miocene of France and Turkey

F. ORSZAG-SPERBER, A.-F. POIGNANT, and A. POISSON, Paris, France

31.1 Introduction

Bosselini and Ginsburg (1971) have defined the term rhodolite: it consists of algal nodules, principally composed of red algae (Corallinaceae). These structures are frequent in the marine Neogene formations. We have tried to determine the environmental conditions of the deposits for a few examples mainly located in France (Aquitaine and Corsica) and in Turkey.

31.2 Rhodolites from the Miocene of the Southern Aquitaine

These deposits, described by Buge et al. (1973), show rhodolites imbedded in grey shales from the Helvetian stage (pro-parte Seravallian). Most of the rhodolites are spherical in shape, with diameters of about 10 cm and weight of about 500 g. We have observed burrows and borings and several species of Corallinaceae can be recognized:

Archeolithotamnium intermedium Raineri is not the most frequent form, but where it appears it clearly dominates. The tissue is constituted of cells arranged in rows; hypothallus is relatively reduced in size (12–14 x 25 μ) and the perithallus (12–15 x 12–15 μ) has rectangular cells. Sporangia are imbedded in the tissue (60–70 x 90–120 μ av.). The preservation of the tissue is generally poor, but a determination as *A. intermedium* (formerly known from the Cirenaica) is possible.

Palaeothamnium archaeotypum Conti. Encrusting algae, with many tubes can constitute the main part of some rhodolites. Various stages of the development of the reproduction cells can be seen (size 100–120 x 500–700 μ). The thickness of the hypothallus is 50–100 μ; the perithallus cells (15 x 10 μ) are set in rows, occasionally in lines. The algae closely resemble *Palaeothamnium oligocenicum* Mastrorilli and appears to be a synonym of the Miocene form met in the Vienna basin.

Lithothamnium pseudoramossissimum Poignant. This alga which always constitutes the main part of a nodule, grows in a bush-like form, its size being slightly different from that of the type described from the Turkish Miocene. The hypothallus is 60–80 μ thick, can exceptionally reach 100–120 μ. Cells size is about 20 μ, but often can be as small as 8–10 μ. The perithallus has a clear zonation, each zone being an average of six rows of cells (8–10 x 5 μ). The ill-ordered and confused aspect already noticed for this type is confirmed here. Conceptacles are frequent, being ca. 200–250 μ in height and 700 μ in width. The roof show six to eight pores.

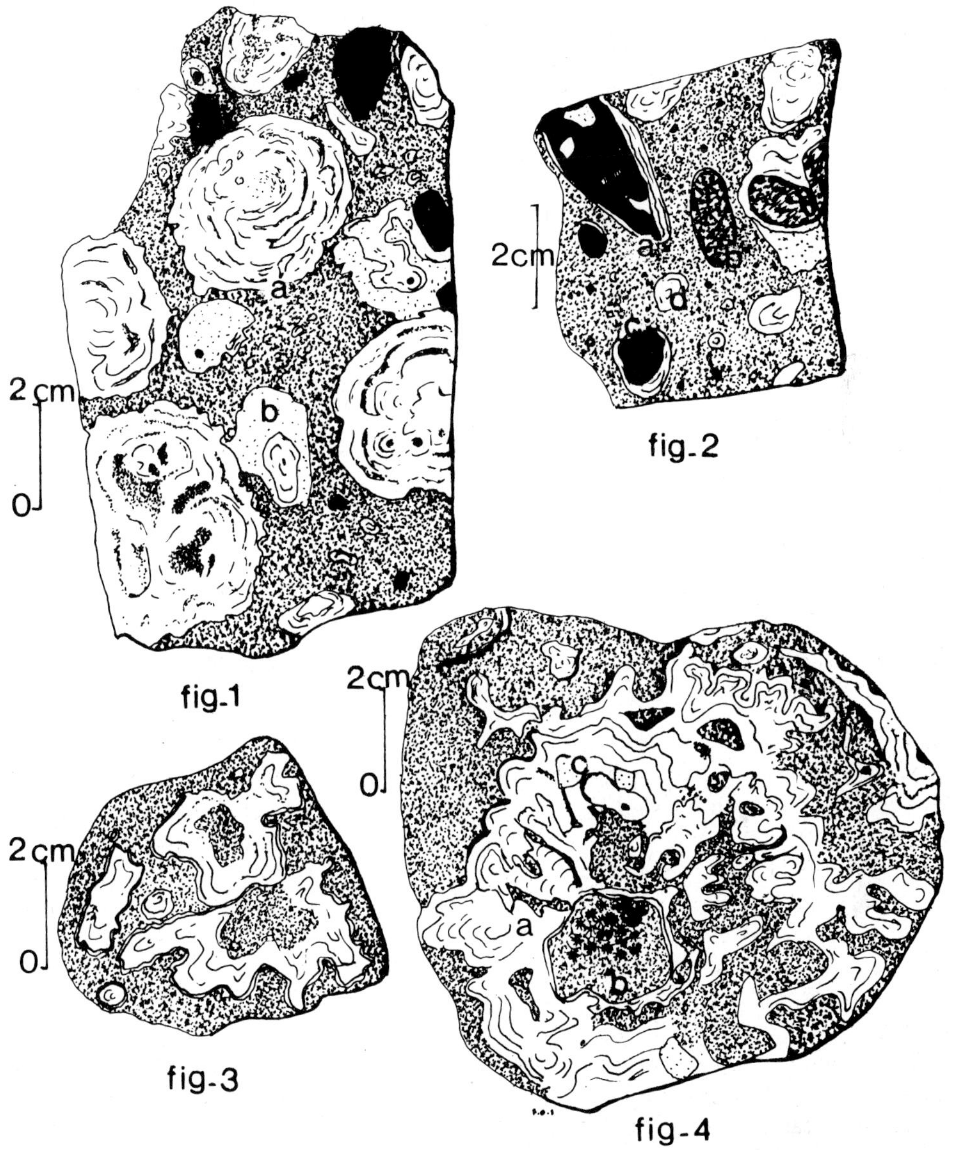

Pl. 1/1. Polished section of a Corallinaceae calcarenite (Saint-Florent, Corsica). Calcareous algae constitute a large nodule (*a*), often these nodules are found encrusted by Bryozoa (*b*). Here the substrate pebbles are not encrusted but imbedded in a calcareous sandstone

Pl. 1/2. Polished section of a conglomerate (Saint-Florent, Corsica). Pebbles from the substrate are encrusted by Corallinaceae. Encrustment is irregular (*a*) showing the pebble stayed in the same site for some time, before being moved and imbedded, together with other pebbles (*b*) and (*c*) in a cement made of quartz, feldspath and various bioclasts. Nodules entirely made of Corallinaceae and quite spherical in shape can also be found (*d*)

Pl. 1/3. Polished section of a nodule (Eastern Corsican plain). This type of rhodolite is extremely frequent in reef formations

Pl. 1/4. Polished section of a rhodolite. Corallinaceae (*a*) are encrusting a fragment of Coral (*b*). Also a few debris of Bryozoa can be seen (*c*)

Lithophyllum microsporum Maslov is relatively rare. The basal hypothallus is unknown, the perithallus is made of aligned cells (10–14 μ). This aspect results from the disposition of the septa, which are not of the same height. Small flattened conceptacles (80 x 30 μ) are seen frequently.

Lithophyllum albanense Lemoine is also rare, and always developed as crusts. The basal hypothallus is strongly reduced, the perithallus extremely irregular and the conceptacles are oval with sharp black boundaries (350 x 150 μ); they are found as fragments only. The hypothallus is made of concentric rows and the perithallus is zoned. The cells are small and conceptacles have many pores and generally measure 150–300 μ. The morphology of these nodules from Bearn are similar to those described from Sainte-Marie-de-Gosse (Landes) from the Lutetian, but the same algal species are not found: in particular we observed that there is a complete lack of *Pseudolithothamnium, Solenomeris,* and *Lithoporella* (Boulanger and Poignant, 1969) in the Miocene nodules.

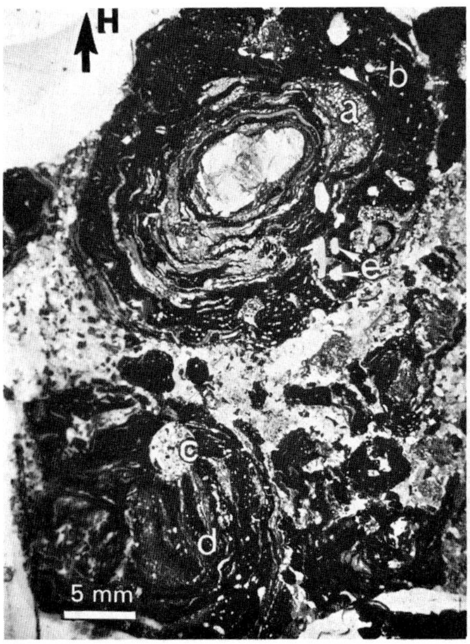

1

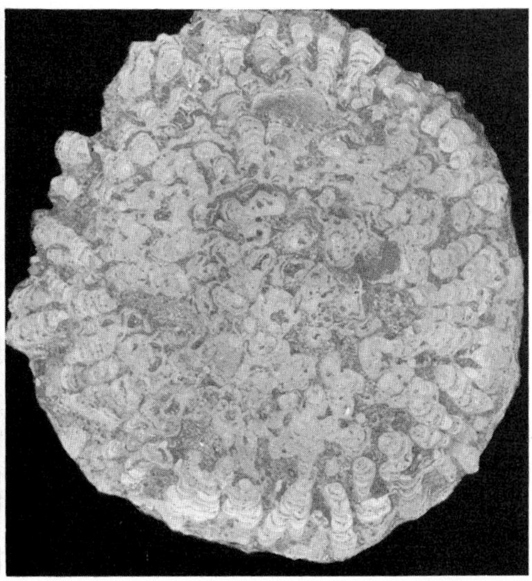

2

Pl. 2/1. Thin section from a calcarenite with rhodolites (Saint-Florent, Corsica). A pebble has been encrusted by Bryozoa (*a*) but mainly by Corallinaceae (*b*). On that same nodule one can see the oriented filling (micrite-sparite) of cavities (*e*) bored by lithophages. Another cavity (*c*) has been filled, it intersects a fragment of *Lithothamnium* (*d*) also incrusted by algae

Pl. 2/2. Rhodolite (15 cm) from Karabayir (Turkey). This nodule almost perfectly spherical in shape, reveals only Corallinaceae of the columnar type. The growth appears regular. No nucleus is to be seen, nor any trapped debris. Also to be noticed is the lack of bioturbation (indicating a fast imbedded process)

31.3 Rhodolites from the Turkish Miocene

The Karabayir-Korkuteli formation is characterized by the association of two very different facies:

1. coarse limestones, with massive beds rich in algae (algal limestones), and littoral fauna.
2. micritic limestones and shales containing a rich planctonic microfauna.

Various organisms are associated with these algae, but they seldom can be identified. Most often only some rounded fragments with rather constant grain size are found. Among those essentially benthic organisms are *Balanus*, Corals and Foraminifera and from Aquitanian (*Lepidocyclina, Operculina,* etc.). This association is characteristic for a shallow marine environment, not very far from shore.

The Karabayir formation is a typical level in the limestone series of the western part of the Bay Daglari; the facies being very uniform. Algal nodules can reach a size of 15 cm. Rhodolites are presented either as radiating, dichotomous branches, or as encrusting columnar structures with concentric layers around a pelecypod or gastropod bioclast. All the large nodules consist of four crustose alga species: *Lithothamnium pseudoramossissimum* Poignant is by far the most abundant, closely interwined with *Pseudolithothamnium album* Pfender; they are often associated with *Solenomeris douvillei* Pfender and less frequently with *Lithoporella melobesioides* (cells: 50 x 16–18 μ). Very often bryozoa are of importance. The very large, zoned, columnar or dendroid (bushes of 3 cm) nodules are very similar to the present *Lithothamnium var. subspherica* Foslie.

Lithothamnium pseudoramossissimum Poignant presents a central hypothallus of only 60 μ thick, built of lines of long cells (12–14 x 20–25 μ). The cell boundaries are sharply defined in the lower part of zones, which consist of a maximum of five to eight rows. Conceptacles are frequent, sometimes very abundant (180–250 x 400–700 μ). The roof often shows eight to ten pores. This species is nearly always associated with crustose forms of *Solenomeris douvillei* Pfender. The structure of these columnar nodules indicates a rather quiet environment (see Bosellini and Ginsburg, 1971).

Other small rhodolites are characterized by a striking laminar aspect. The following species have been determined:

Lithophyllum capederi Lemoine: thalli are abundant, undulated, mainly consisting of perithallus with regular rows of cells. Conceptacles present a very marked conical structure; cells from the colwalls are sharply raised.

Lithophyllum lacroixi Lemoine appears in thick crusts, with a zoned perithallus. Conceptacles which are numerous, are not very high but large (150–450 μ).

Lithophyllum microsporum Maslov appears as an homogenous and relatively thick tissue. Cells are aligned. Conceptacles are conical in shape (80–100 x 50–60 μ), and present blurred outlines.

Lithothamnium cf. macrosporangium Mastrorilli differs from the Calabrian form only by its smaller size. The smaller nodules, about 1 cm in size, are made of *Pseudolithothamnium album*, which is extremely abundant, *Dermatholithon* and to a lesser extent *Lithoporella*. Fragments of *Archaeolithothamnium intermedium* with clearly structured perithallus are also to be found.

In conclusion, it can be said that only the large columnar rhodolite are made of *Lithothamnium pseudoramossissimum.*

31.4 Corsican Miocene Rhodolites (France)

The rhodolites we have examined come from two different basins. From a stratigraphic point of view these basins are bioclastic formations above upper Burdigalian level in the Saint-Florent basin (North Corsica), and between Langhien and Tortonian formations in the Eastern plain basin. We did not find any fundamental difference, so far as the concerned species were, or between rhodolites from either origin, in spite of the fact that, as will be shown, deposition modes were different.

Lithothamnium ramossissimum Gümbel (Conti), zoned, has always been found sterile. This species is well known in Austria, Iraq, and Italia Miocene.

Mesophyllum commune Lemoine is also abundant. This polymorphic species is often poorly preserved. Being generally sterile, this alga has mainly been identified by the general morphology of the tissue.

Mesophyllum koritzae Lemoine appears most often as superposed crustal fragments. Conceptacles, typical, are rather abundant. This species is not unfrequent in Albania, Spain, Algeria, and in the Miocene of the Ukrania.

Lithophyllum duplex Maslov is relatively abundant with encrusting thallus and perithallus with cells set in bending lines or fan-like structure.

The characteristic aspect of conceptacles described by Orszag-Sperber and Poignant (1972) appears again very clearly. In fact, most conceptacles present a deeply indented bottom, almost like a tuber. In place of the pore, we find a hemispheric lodge opening inside the conceptacle cavity.

Lithophyllum microsporum Maslov is identified by its rather thick thallus, and also by its numerous small-sized conceptacles.

Lithophyllum johnsoni Ishijma infrequent; seldom appears and does not seem to play any part in the rhodolites formation.

Lithothamnium pseudoramossissimum Poignant is identified. It will be interesting to ascertain the stratigraphic value of this species.

Archaeolithothamnium intermedium is very rare. The cells, rectangular, are set in rows. A few sporanges make it possible to confirm diagnosis.

Palaeothamnium archaeotypum Conti is also rare, and found only as fragments. The few pieces examined contained conceptacles at various development stages.

Leptolithophyllum platicarpum (Maslov) Poignant is by far the most original among Corsican nodules. This variety has been described by Maslov as being *Lithothamnium,* and this from Ukrainia and Crimea Tortonian. The hypothallus is very thin, and made of lines of long cells (12 x 20 μ). Perithallus has an irregular grid (10 x 10–15 μ). Conceptacles are very characteristic: extremely flat, wiry, with only one pore (200 but only 40 μ in height). All observations seem to confirm that the variety described by Maslov is in fact *Leptolithophyllum*. Ours would be, thus, the first Miocene *Leptolithophyllum* identified outside Ukrainia.

1

2

3

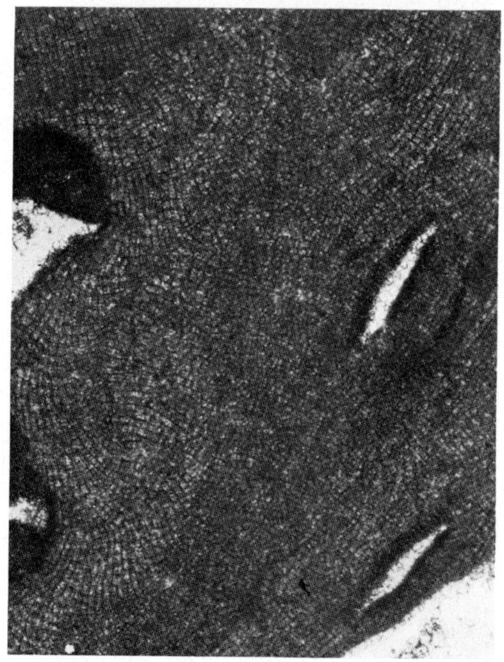

4

Pl. 3/1. *Archaeolithothamnium intermedium*, Salies-de-Béarn, France (x 90)

Pl. 3/2. *Lithophyllum microsporum*, Salies-de-Béarn, France (x 150)

Pl. 3/3. *Palaeothamnium archaeotypum*, Corsica (x 160)

Pl. 3/4. *Leptolithophyllum platicarpum*, Corsica (x 90)

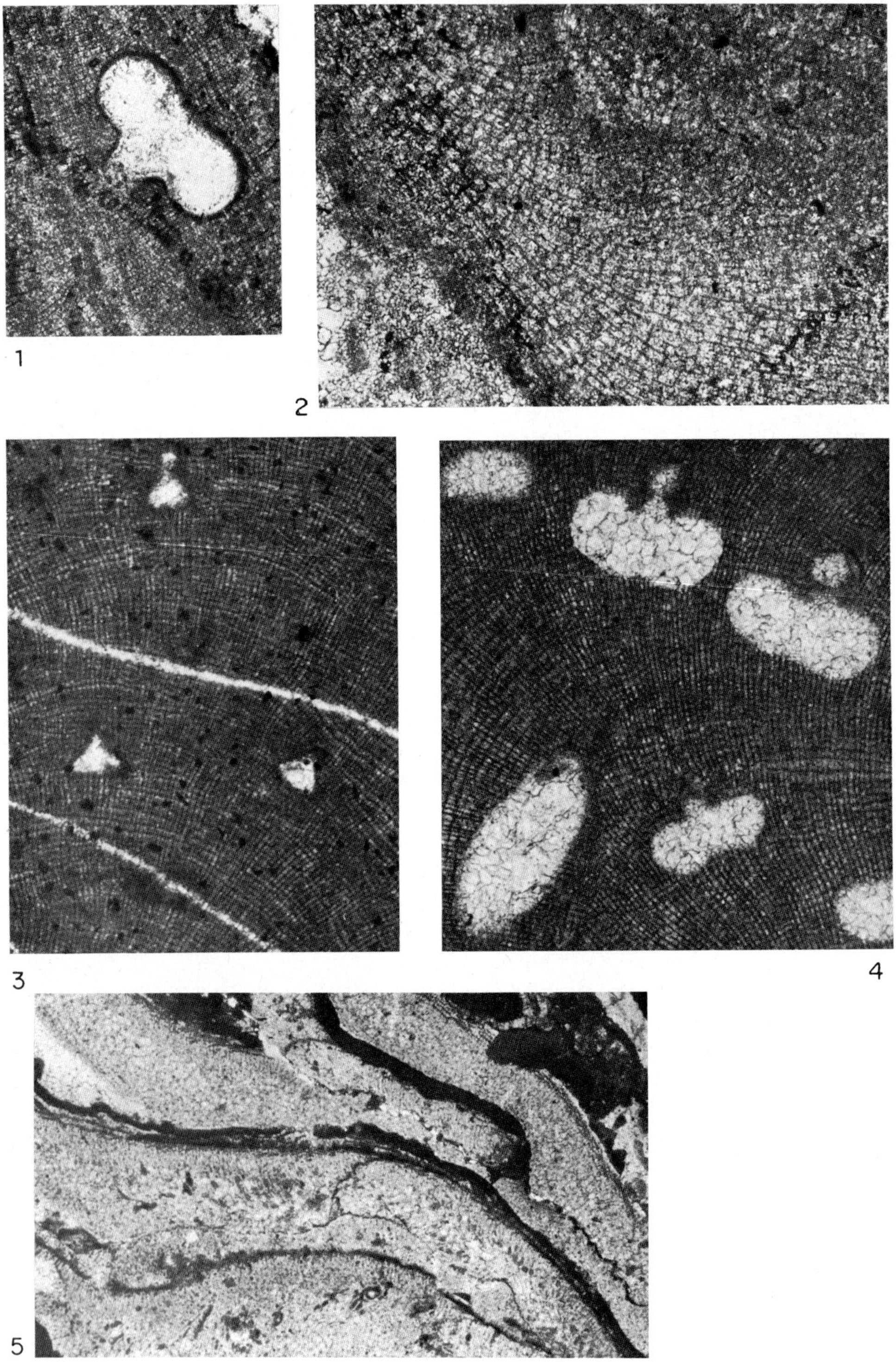

Plate 4. Legend see opposite page

In the eastern plain, rhodolites are small, with diameters ranging from 2–5 cm, and on exception 10 cm. Generally the Corallinaceae have grown around a fragment of coral, occasionally around a granite or schists pebble of the substratum. These rhodolites are found on sandy bottoms, and support the growth of coral bioherms.

Most rhodolites appear non-symmetric, which indicates little turning. We also find fecal pellets in the filling of some burrows, which again points toward quiet deposition.

In the Saint-Florent basin we find no reef associated with rhodolites. The sections reveal an alternance of levels either rich in rhodolites or almost empty of algal nodules. These levels are about 1.50–2 m thick, and two to three alternances are usually to be found. The rhodolites levels are not lenticular: their regular thickness, together with the abundance of cement imbedding the rhodolites, make it possible exclude storm deposits.

The nodules are either entirely made of Corallinaceae (Pl. 1/1), or have grown around a nucleus, generally some pebble of the substrate. The examination of this section shows that these nodules have been displqced (Pl. 2/1).

We found very few Bryozoa, although more than in the Eastern plain. This can result from an unfavorable environment, but also the considerable growth of the Corallinaceae may have exluded them.

31.5 Discussion

The study of rhodolites provides us with very important information on the various deposition modes of Miocene formations.

A first conclusion can be drawn from the rhodolites we have examined: algae are set according to two main types; a columnar type (branched or not) and a laminar type. This is in accordance with Bosselini and Ginsburg (1971) results.

In our case, the laminar rhodolites have also been overturned on sandy bottom, as can be seen from their disposition in the eastern Corsica plain and in Turkey, but on the contrary, the columnar type seems to be charcteristic of immobile and irregular rhodolites, like the substrate of patch-reefs in Corsica. The use of rhodolites alone as bathymetric indicators seems to lead to uncertain results unless appropriate precautions are taken. In fact, rhodolites have been found in shallow water (3–5 m), as in Bermuda. In order to explain their presence at greater depths, Siesser (1972) thinks that these algal nodules, found between 100 and 200 m of depth, probably developed in a shallow water environment. Then the sea level has risen (glaciation). The original site was thus probably in much shallower water than where nodules are found now. As far as the rhodolites we examined are concerned, it is fairly certain they grew in shallow water: the sediment contains various fragments of Echinoderms, *Ostrea,* and Foraminifera such

---

◄ Pl. 4/1. *Lithophyllum albanense*, Salies-de-Béarn, France (x 90)

Pl. 4/2. *Lithothamnium pseudoramossissimum*, Salies-de-Béarn, France (x 150)

Pl. 4/3. *Lithophyllum microsporum*, Corsica (x 90)

Pl. 4/4. *Lithophyllum duplex*, Corsica (x 90)

Pl. 4/5. *Pseudolithothamnium album*, Corsica (x 37)

as *Heterostegina*, which means the deposition zone was near the shore. Even though, it is known that rhodolites are frequently to be found in a cold climate (Jacquotte, 1962; Cabioch, 1966), those we studied were certainly formed in warm regions. In the first place they are found associated with warm region fauna, in the second place, the occurrence, even when infrequent, of *Archaeolithothamnium* in all deposits proves the temperature to have been rather high. The establishment of reefal formations on the rhodolites (Corsica) also confirm this hypothesis.

A more detailed study of present rhodolites should reveal whether one of the existing difference between actual and fossil rhodolites lies in the monospecificity of the former. It would also be of interest to compare living maerl (Cabioch, 1970; Pinot, 1974) with actual rhodolites, in order to ascertain the influence of temperature on their development.

31.6 Bibliography

Bosselini, A., Ginsburg, R.M.: Form and internal structure of recent algal nodules (Rhodolites) from Bermuda. J. Geol. **79**, 6, 669–682 (1971).

Boulanger, D., Poignant, A.F.: Sur les nodules algaires du Lutétien supérieur de Sainte-Marie-de-Gosse (Landes). C.R. Soc. Géol. Fr. 109–110 (1969).

Buge, E., Debourle, A., Deloffre, R.: Gisement miocène à nodules algaires (Rhodolites) à l'Ouest de Salies-de-Béarn (Aquitaine Sud-Ouest). Bull. Centre Rech. Pau, S.N.P.A. **7**, 1, 1–51 (1973).

Cabioch, J.: Le maërl des côtes de Bretagne et le problème de sa survie. Penn ar Bed, Brest, Vol. VII, no. 63, 421–429 (1970).

Jacquotte, R.: Etude des fonds de maërl de Mediterranée. Rec. Traux. Station Mar. Endoume **26**, 141–235 (1962).

Maslov, V.P.: Algues rouges fossiles d'U.R.S.S. et leur rapport avec les faciès. Trad. B.R.G.M., Paris (1962).

Orszag-Sperber, F., Poignant, A.F.: Corallinacées du Miocène de la plaine orientale de Corse. Rev. Micropaléontologie **15**, 2, 115–124 (1972).

Pinot, J.P.: Le précontinent breton entre Penmarc'h, Belle-Ile et l'escarpement continental. Lannion, Impram. 1974, 256 p.

Poisson, A., Poignant, A.F.: La formation de Karabayir, base de la transgression miocène dans la région de Korkuteli (Départment d'Antalaya-Turquie). *Lithothamnium pseudoramossissimum*, nouvelle espèce d'algue rouge de la formation de Karabayir. Bull. Min. Expl. Inst. Turkey **82**, 67–72 (1974).

Siesser, W.G.: Relier algal nodules (Rhodolites) from the South African continental shelf. J. Geol. **80**, 611–615 (1972).

32 The Lower Cretaceous Calcareous Algae in the Area Surrounding Geneva (Switzerland): Biostratigraphy and Depositional Environments

M.A. CONRAD, Geneva, Switzerland

32.1 Introduction

This work is based on samples collected by the following authors: Mouty (1966), Steinhauser (1970), Steinhauser and Charollais (1971), Jaquet (1973a, b) and Conrad (1969). In addition, samples belonging to the Swiss National Foundation were studied. All together, more than 3000 slides were investigated.

32.2 External Infralittoral Environment

This is characterized by the simultaneous occurrence of photophil algae and large bryozoans (Conrad, 1969). The following *subenvironments* are distinguished and summarized in Fig. 2:

32.2.1 Shallow Water Outer Platform

Well-sorted and rounded grainstones containing commonly mixed-up and transported algal-bryozoans assemblages (Alleves, Fontanil, Guiers and Lower Urgonian formations). Variant: corals-sponges-encrusting algae-bryozoans biostromes and/or small bioherms.

32.2.2 Former Sea-Grasses on Soft Substrates

Wackestones and boundstones with encrusting algae and encrusting bryozoans assemblages. *Lithocodium* and *Bacinella* are prominent. This association is completed by large dasyclads species bearing tight whorls of numerous branches: *Griphoporella aff. piae* and *Pseudocymopolia pluricellata* within the Berriasian-Valanginian interval, *Acroporella? parva*, n. sp., in the Lower Barremian. Let us not forget that similar forms were described together with *Lithocodium aggregatum: Pseudocymopolia aff. anadyomenea* (see Praturlon, 1964, p. 172) and *Pseudocymopolia orientalis* (see Elliott, 1970, p. 325). The rareness of other types of dasyclads, in our material, suggests the possible existence of an environment encouraging the development of forms belonging to different genera and species, but having common morphological features. This subenvironment is common in the upper and lower parts of infralittoral formations surrounded by deeper water, circalittoral facies (Upper Chambotte, Allèves and Lower Urgonian formations).

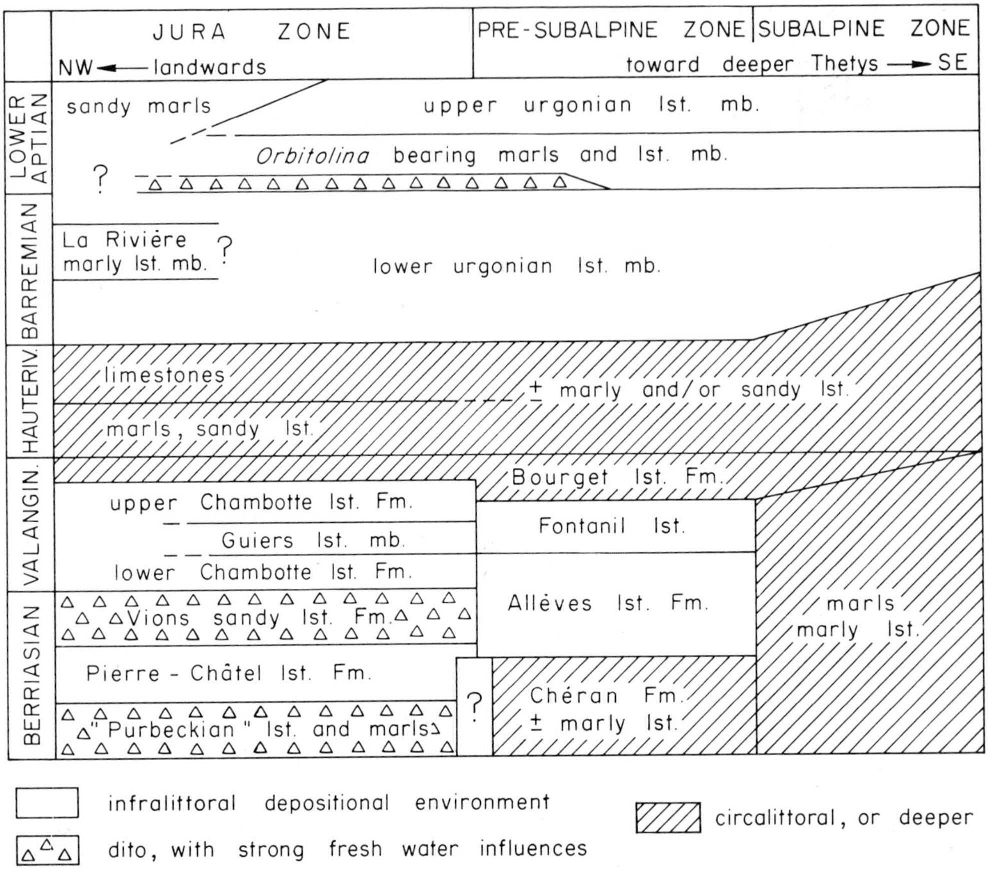

Fig. 1. The Lower Cretaceous in the area surrounding Geneva. Main lithostratigraphical features. After Steinhauser and Lombard (1969), Charollais and Lombard (1966) and Conrad (1969). No scale

32.2.3 Sea-Grasses on Banks

Strong skeletal production (?), abundant, diversified and heavily calcified dasyclads + large encrusting and/or ramose bryozoans. Pro parte coated grainstones. Lower Urgonian 1st. mb., with *Salpingoporella genevensis* (indicator organism?).

32.3 Middle Infralittoral Environments

Rather protected marine environment, lacking some suspension-feeders such as large bryozoans and sponges. Several subenvironments, probably depending on substrates. *Thaumatoporella* is a general indicator alga. Strong development of *Bacinella* within rudistid biostromes. Dasyclads are seldom abundant. Middle infralittoral environments are common in the "Purbeckian", Pierre-Châtel, Allèves, Chambotte, and Urgonian formations.

32.4 Internal Infralittoral Environment

Depending on brackish or hypersaline influences, it develops typical algal and/or animal assemblages. It is found in the "Purbeckian", Pierre-Châtel, Vions, and Urgonian formations. Lacking space, only dasyclads occurring within the Vions formation will be considered here. The Vions formation was sampled at regular 30 cm intervals and studied by Jaquet (1973a and b), using mathematical analysis. The material consists of 770 slides corresponding to seven field sections. Dasyclads were grouped as one bulk category. In reviewing the slides, the dasyclad *Macroporella embergeri* were separated, being easy to recognize in thin sections thanks to the special structure of its calcareous envelope.

The following relative frequencies have been observed; the theoretical relative frequencies computed assuming statistical independence, are indicated within brackets. (1) Charophycophyta (vegetative and/or fertile parts) alone: 6.4 (9.8) %; (2) Dasyclads alone (without *M. embergeri*): 67.3 (48.5) %; (3) *M. embergeri* alone: 14.5 (16.8) %; (4) Charoph. + dasyclads: 1.4 (5.8) %; (5) Charoph. + *M. embergeri*: 6.8 (1.0) %; (6) Dasyclads + *M. embergeri*: 3.6 (15.4) %; (7) Charoph. + dasyclads + *M. embergeri*: 0 (2.6) %. A check was made (with the kind help and assistance of E. Davaud, Department of Geology, University of Geneva) by chi square tests as to whether theoretical frequencies differed significantly from those observed. The value of chi square indicates that the null hypothesis should be rejected. Since the mixing of flora should be retained only for a small number of cases, this means that Charophycophyta, *Macroporella embergeri* and other dasyclads developed in separated living environments. In addition, the vertical distribution within field sections of algae shows the existence of separated facies with dasyclads and *M. embergeri* + Charophycophyta; the latter corresponding to the peak of the Berriasian regression. *M. embergeri's* ecological niche was therefore akin to Charophycophyta's paleo-environment, that is subject to strong fresh water influences. On the contrary, the "facies with dasyclads" seems to correspond to a hypersaline environment, as shown by the absence of brachiopods, some foraminifers and *Salpingoporella annulata*; the foregoing dasyclad being commonly found in the middle infralittoral environment.

32.5 Circalittoral Environment

Only very rare, possibly in situ, crusts of tiny tubes probably belonging to *Girvanella* sp. were found on ramose bryozoans in the upper Hauterivian. Otherwise, occasional algal bioclasts of infralittoral origin were transported in deeper circalittoral depositional environments. Sciaphil calcareous algae seem hardly to have developed in the Neocomian.

32.6 Palaeoecology Versus Chronostratigraphy

Species mentioned in Fig. 3 are found in the following infralittoral environments. Internal (brackish): *Macroporella embergeri* Bouroullec and Deloffre, 1968. Middle and/or internal: *Clypeina jurassica* Favre, 1927, *Clypeina? solkani* Conrad and Radoičić, 1972,

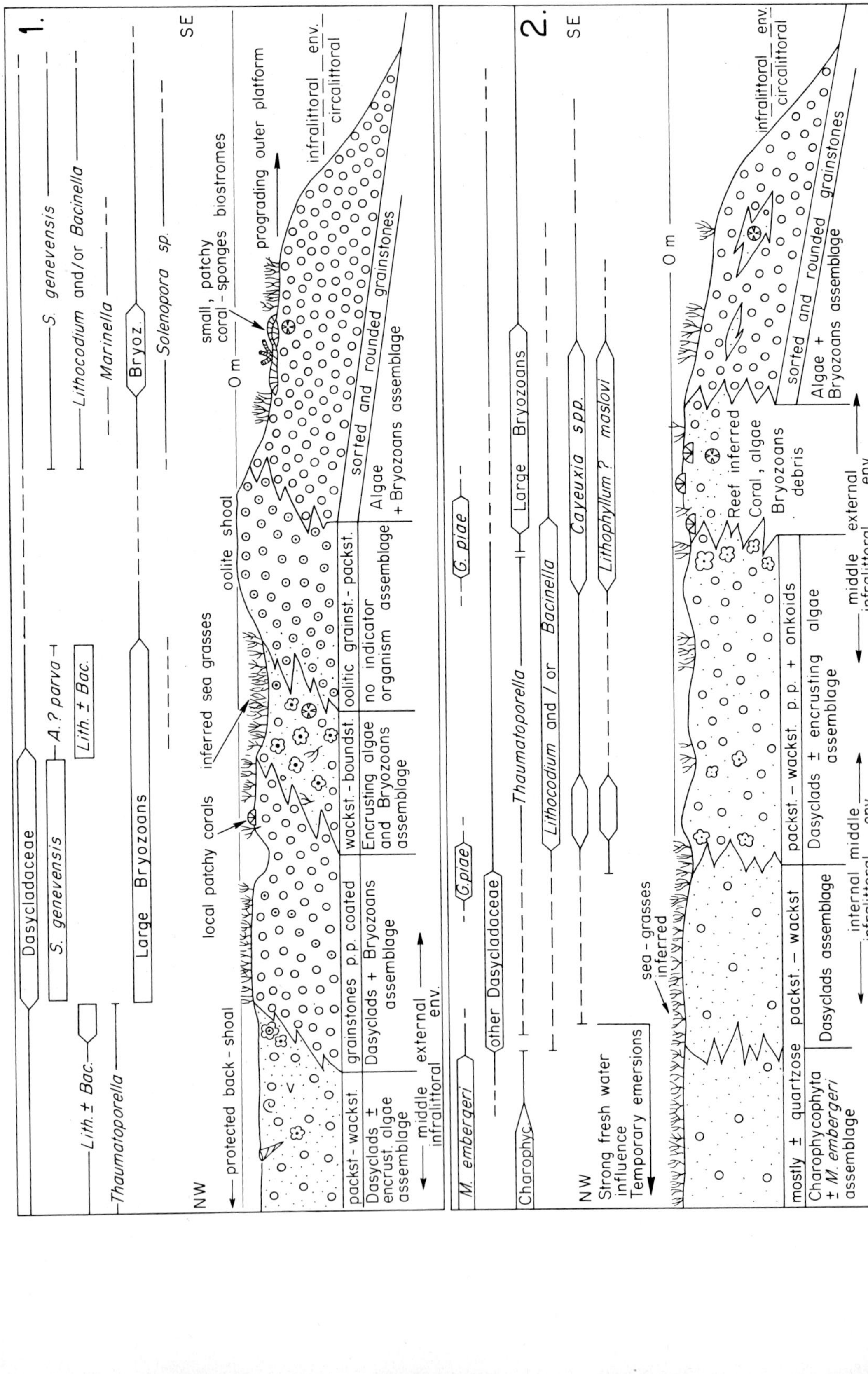
1.
NW
SE
Dasycladaceae
S. genevensis
A.? parva
S. genevensis
Lith. ± Bac.
Lith. ± Bac.
Lithocodium and/or Bacinella
Thaumatoporella
Marinella
Large Bryozoans
Bryoz.
Solenopora sp.
protected back-shoal
local patchy corals
inferred sea grasses
oolite shoal
small, patchy coral-sponges biostromes
0 m
prograding outer platform
infralittoral env.
circalittoral
packst.-wackst.
grainstones p.p. coated
wackst.-boundst.
oolitic grainst.-packst.
sorted and rounded grainstones
Dasyclads ± encrust. algae assemblage
Dasyclads + Bryozoans assemblage
Encrusting algae and Bryozoans assemblage
no indicator organism assemblage
Algae + Bryozoans assemblage
middle infralittoral
external env.
2.
NW
SE
M. embergeri
G. piae
G. piae
other Dasycladaceae
Charophyc.
Thaumatoporella
Large Bryozoans
Lithocodium and / or Bacinella
Cayeuxia spp.
Lithophyllum ? maslovi
Strong fresh water influence
Temporary emersions
sea-grasses inferred
0 m
Reef inferred
Coral, algae Bryozoans debris
infralittoral env.
circalittoral
mostly ± quartzose
packst. – wackst
packst. – wackst. p.p. + onkoids
sorted and rounded grainstones
Charophycophyta ± M. embergeri assemblage
Dasyclads assemblage
Dasyclads ± encrusting algae assemblage
Algae + Bryozoans assemblage
internal infralittoral
middle env.
middle infralittoral
external env.

Heteroporella lemmensis (Bernier), 1971, n. comb., *Kopetdagaria? spp. aff. K. iailensis* Maslov, 1965, *Salpingoporella steinhauseri* Conrad et al., 1973, *Salpingoporella aff. muehlbergii* (Lorenz), 1902, *Triploporella? neocomiensis* Radoičić, 1963 (present within a narrow stratigraphical interval; see also Radoičić, 1963, p. 144), *Likanella campanensis* Azema and Jaffrezo, 1972 (rare), *Linoporella aff. L. kapelensis* Sokac and Nikler, 1973 (rare), *Cayeuxia? anae* Dragastan, 1971 (an incertae sedis). Middle and external: *Salpingoporella annulata* Carozzi, 1954, *Likanella? danilovae* Radoičić, 1968, *Heteroporella? paucicalcarea* Conrad, 1970, *Lithophyllum? maslovi* Dragastan, 1971 (an incertae sedis). External: *Griphoporella sp. aff. G. piae* Dragastan, 1971, *Pseudocymopolia pluricellata* Bakalova, 1973, *Acroporella? parva*, n. sp. which was previously described and illustrated (Conrad, 1970, p. 74, Pl. XI), *Salpingoporella genevensis* (Conrad), 1969. Ubiquitous infralittoral: *Actinoporella podolica* (Alth) 1878, *Pseudoactinoporella fragilis* Conrad, 1970, *Salpingoporella muehlbergii* (Lorenz), 1902, *Salpingoporella melitae* Radoičić, 1965, *Vermiporella? tenuipora* Conrad, 1970 (an incertae sedis).

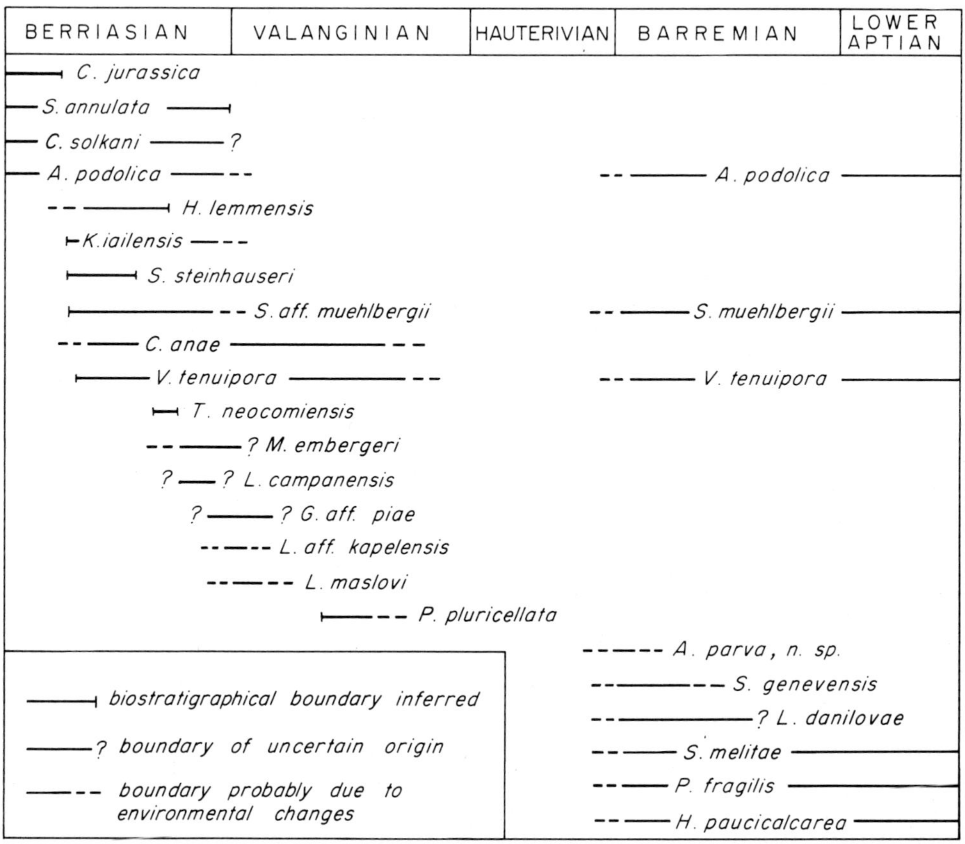

Fig. 3. Time stratigraphic chart of dasyclads and some incertae sedis of algal origin. See text for generic names

◀ Fig. 2. Main infralittoral depositional environments. 1. Barremian interval (Lower Urgonian 1st. mb.), 2. Upper Berriasian – Lower Valanginian interval (Vions, Lower Chambotte and Allèves 1st. Fm.

These data are still insufficient. For example, information on the external infralittoral environment of Lower and Middle Berriasian age is as yet lacking. However, it appears that dasyclads, for example, were photphil algae sensitive moreover to other factors influencing different infralittoral subenvironments. This represents a major obstacle for their large scale use in stratigraphy.

An important renewal of the flora takes place from the Middle Berriasian onwards. Several little known species occur in the middle infralittoral environment of Valanginian age which, chronostratigraphically, is still ill-characterized. Finally, let us note that the Hauterivian was entirely deposited in circalittoral environment and deprived of algae.

32.7 Conclusions

Calcareous algae, including dasyclads, are a useful and sensitive tool for detailed palaeoenvironmental investigations. However, their specific ecology should always be taken into account before using them for stratigraphical purposes.

32.8. Bibliography

Bernier, P.: Deux nouvelles algues dasycladacées du Jurassique supérier du Jura méridional. Geobios **4**/3, 173–184 (1971).

Charollais, J., Lombard, A.: Stratigraphie comparée du Jura et des chaînes subalpines aux environs de Genève. Arch. Sci. Genève **19**/1, 49–81 (1966).

Conrad, M.A.: Les calcaires urgoniens dans la region entourant Genève. Eclogae Geol. Helv. **62**/1, 1–79 (1969).

Conrad, M.A.: Barremian and Lower Aptian Dasycladaceae in the area surrounding Geneva (Switzerland). Geol. Rom. **9**, 63–100 (1970).

Elliott, G.F.: Pseudocymopolia, a mesozoic tethyan alga (Family Dasycladaceae). Palaeontology **13**/2, 323–326 (1970).

Jaquet, J.-M.: Définition de microfacies par classification automatique et comparaison des descripteurs. Arch. Sci. Genève **26**/2, 137–171 (1973a).

Jaquet, J.-M.: Analyse statistique de la variation verticale des descripteurs dans la Formation de Vions (Berriasien supérieur, Jura méridional). Arch. Sci. Genève **26**/3, 247–283 (1973b).

Mouty, M.: Le Néocomien dans le Jura méridional. Thesis 1369. Damas Univ.. Syria (1966).

Praturlon, A.: Calcareous algae from Jurassic-Cretaceous limestone of Central Apennines (Southern Latium-Abruzzi). Geol. Rom. **3**, 171–203 (1964).

Radoičić, R.: Triploporella neocomiensis spec. nov. des couches valanginiennes – hauteriviennes du Montenegro. Vesnik **31**, ser. A (1963).

Steinhauser, N.: Recherches stratigraphiques dans le Crétacé inférieur de la Savoie occidentale (France). Thesis 1506, Geneva: Médecine et Hygiène, 1970.

Steinhauser, N., Charollais, J.: Observations nouvelles et réflexions sur la stratigraphie du "Valanginien" de la région neuchâteloise et ses rapports avec le Jura méridional. Geobios **4**/1, 7–59 (1971).

Steinhauser, N., Lombard, A.: Définition de nouvelles unités lithostratigraphiques dans le Crétacé inférieur du Jura médridional (France). C.R. SPHN Genève, NS **4**/1, 100–113 (1969).

33 Relationships of Algae with Depositional Environments and Faunal Assemblages of the Panormide Carbonate Platform, Upper Triassic, Northwestern Sicily

B. ABATE, Palermo, R. CATALANO, Palermo, B. D'ARGENIO, Naples,
P. DI STEFANO, Palermo, and R. RICCOBONO, Palermo, Italy

33.1 Introduction

In northwestern Sicily large bodies of Triassic carbonate rocks crop out. They display several lithofacies corresponding from shelf to basin environments. Facies analysis during the last three years has allowed the differentiation and the mapping of different lithofacies (Abate and Catalano, 1974; Catalano et al., 1974a, b).

Algae are present in almost every facies unit. They form in some places large bodies, either in growth position, or as cemented detritus. The present paper is a schematic description of the relationships between the algae and the sedimentary environments. An attempt is made to correlate the distribution of the main algal groups with the most important sedimentological features and faunal assemblages.

33.2 Regional Setting

The Upper Triassic carbonates of northwestern Sicily form large thrust sheets, which are parts of the Alpidic chain in Sicily. Paleotectonic events are Upper Triassic up to Lower Tertiary in age. Tectogenetic deformations took place in the external zones during the Upper Tertiary, detaching the carbonate bodies from their basement and displacing them southward in a pile of thrust sheets. In the last events the paleogeographic units lost their internal coherence as well as their relationships with the adjacent ones (e.g., Panormide Carbonate Platform and Imerese Basin).

33.3 Facies Analysis

The areal distribution of the sedimentary facies was used as a tool in tracing the lateral continuity of the overthrusted rocks for a better understanding of the regional significance of the structures.

In the Panormide Carbonate Platform some belts, characterized by identity of facies (*isopic zones*), can be traced from North to South (Fig. 1):

1. Tidal flat zone
2. Back reef lagoon zone
3. Reef complex zone
4. Slope zone (not considered here).

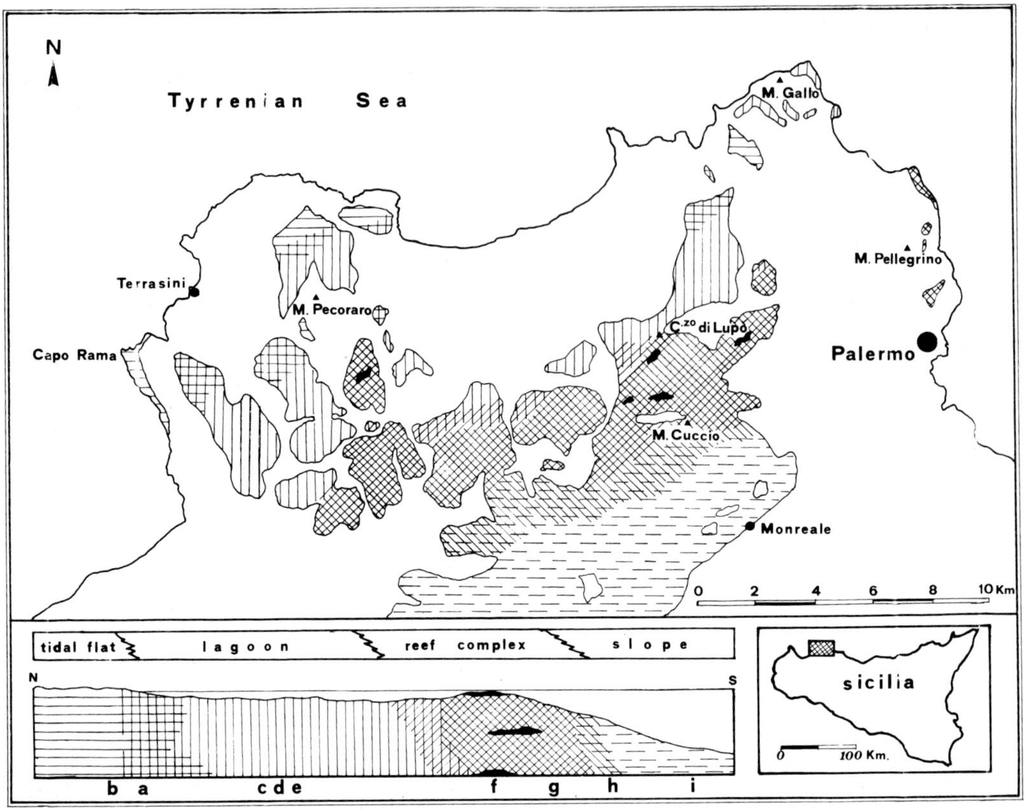

Fig. 1. Facies map of the upper Triassic carbonate deposits (Panormide Carbonate platform) in the Palermo Mountains

Environment	*Lithofacies*
Tidal flat	A) Algal mat dolomitic limestones
	B) Loferitic breccias
Back reef lagoon	C) Dolomitic limestones and calcareous dolomites with Megalodonts
	D) Coral biolithites
	E) Algal biolithites
Reef complex	F) Dolomitized spongid biolitithes
	G) Dolomitic biocalcirudites and biocalcarenites
	H) Intrabioclastic dolorudites
Slope	I) Slope deposits

These isopic zones have been identified in terms of lithofacies (or groups of lithofacies). They trend from ENE to WSW and crop out along about 20 km.

33.3.1 Back Reef Deposits

Back reef lagoon and tidal facies also alternate vertically in regular cyclothems. They are characterized by three major lithofacies:

Lithofacies A: Algal Mat Dolomitic Limestones

Deposits are formed by laminar algal mats (loferites and stromatolites) containing various desiccation structures; they are considered to be supratidal (tidal flats) (Pl. 1/4 and 6).

Faunal content: Foraminifera, gastropods, pelecypods, small ammonites.

Floral content: laminar algal mats, oncolites and thalli of *Cayeuxia* type.

Lithofacies B: Loferitic breccias

Breccias with large elements sometimes arched to form tepee-like structures with a chaotic textures and open spaces, filled with mechanical and/or chemical deposits. These rocks are derived from the alteration (including desiccation) processes of the stromatolitic and loferitic deposits in a phreatic and vadose hydrologic regime. These tepee-like structures and collapse breccias (Pl. 1/3) are strongly modified by caliche processes.

The interfaces between the nonaltered and the overlying altered (i.e. brecciated) sediments are undulated and resemble erosional surfaces.

Faunal content: Foraminifera, sponges (rare), hydrozoa [*Heterastridium conglobatum* (Reuss)], bryozoa, gastropods, pelecypods (small megalodonts), ammonites.

Floral content: thalli and fragments of *Cayeuxia* type, algal mats and Solenoporaceae fragments. These fossils may be transported by storms on the tidal flat (Fig. 2).

Fig. 2. *Loferitic breccias.* Thin section of bioclastic rudites and arenites filling the cavities of the tepee and the collapse breccias structures. The well-preserved test of the molluscs and the presence of oncolites point out their origin of lagoonal deposits, later transported by storm into the exposed desiccated tidal flats. Inner lagoon. Capo Rama, Palermo Mountains. Negative print

Lithofacies C: Dolomitized Limestones and Calcareous Dolomites with Megalodonts

White to grey doloarenites and dolorudites, biocalcarenites, and biocalcilutites characterized by the presence of a rich Megalodont fauna.

Plate 1. Legend see opposite page

Faunal content: encrusting foraminifera, hermatypic corals, bryozoans, brachiopods, gastropods, pelecypods (*Megalodon* sp., *Dicerocardium* sp.).

Floral content: Dasycladaceae [*Diplopora tubispora* Ott, *Heteroporella zankli* (Ott), *Diplopora* sp.], Cyanophyceae (*Cayeuxia* sp.), Solenoporaceae (*Solenopora* sp., *Parachaetetes* sp.), very large oncolites and rare stromatolites, rare spores (Fig. 3 and 4).

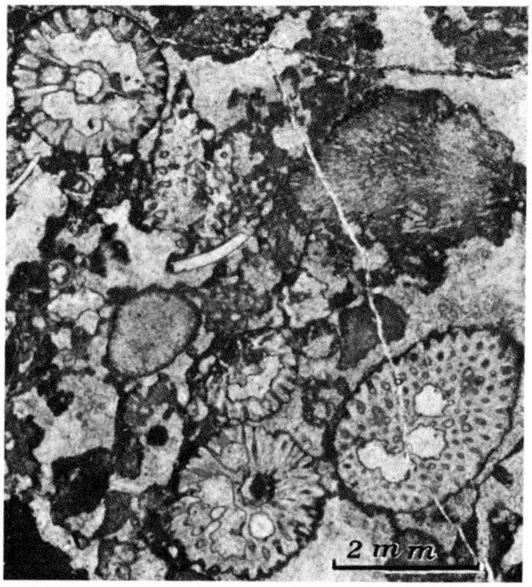

Fig. 3.

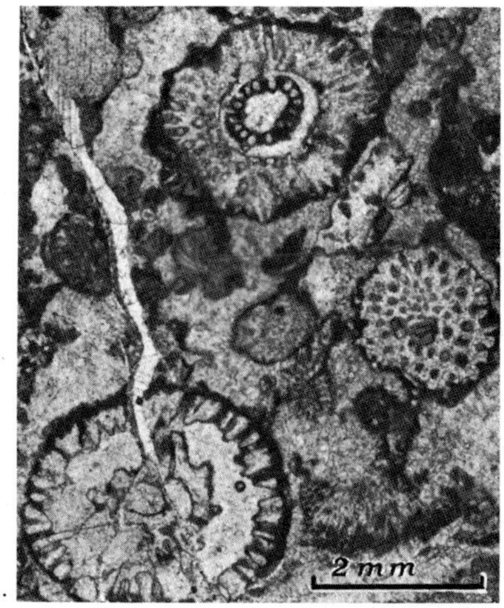

Fig. 4.

Fig. 3. Algal biolithites. Algal assemblage. In association *Heteroporella zankli* (variously sectioned) and *Cayeuxia* sp. Outer lagoon. Cozzo di Lupo, Palermo Mountains

Fig. 4. Algal biolithites. Algal assemblage. In association *Diplopora cf. Diplopora tubispora*. Outer lagoon. Cozzo di Lupo, Palermo Mountains

---

◀ Pl. 1/1. Rhythmic sequence of the three lithofacies in the inner lagoon; from the bottom: *a*: algal mat dolomitic limestones, *1*: alteration surface, *b*: loferitic breccias, *2*: erosion surface, *c*: dolomitized limestones and calcareous dolomites with Megalodonts. Top left, a tepee-like structure. Capo Rama, Palermo Mountains

Pl. 1/2. Dolomitized limestones and calcareous dolomites with Megalodonts. Megalodonts appear pervaded by vadose pisolites. Inner lagoon. Capo Rama, Palermo Mountains

Pl. 1/3. Loferitic breccias. The picture shows two types of structures to the top and collapse breccias to the bottom. The effects of the vadose diagenesis are very diffused mainly in the internal sediments forming the matrix of the breccias. Inner lagoon. Capo Rama, Palermo Mountains

Pl. 1/4. Algal mat dolomitic limestones. Loferites with aligned shrinkage pores and sheet cracks. Thickness about 40 cm. Inner lagoon. Capo Rama, Palermo Mountains

Pl. 1/5. Dolomitized limestones and calcareous dolomites with Megalodonts. Vadose pisolites. Siltitic concentric laminae alternate with microsparry envelopes. Sometimes the laminae vanish in the surrounding sediment. Some nuclei are coated grains. Inner lagoon. Capo Rama, Palermo Mountains. x 4, negative print

Pl. 1/6. Algal mat dolomitic limestones. Early cemented, bioclastic arenites later covered by algal mats. Note inversely graded peloidal laminae and loferites with desiccation structures. Inner lagoon, Capo Rama, Palermo Mountains. x 2

The lithofacies types attributed to a tidal flat environment prevail in this area in contrast to those of back-reef lagoon. The rocks formed by these cyclothems represent however the inner part of the back-reef area in Palermo Mountains (Pl. 1/1).

The outer part is formed by rocks in which the lithofacies C is prevailing (together with the following lithofacies D and E), A is very rare, while the lithofacies B is lacking.

This environment (outer lagoon) was characterized by low energy and soft bottoms in which the large Pelecypods (Megalodontidae and Dicerocardiidae) lived and tabular masses of corals and algal biolithites flourished; the lagoon was bisected by tidal channels with areas colonized by stromatolotic mats.

Lithofacies D: Coral Biolithites

This lithofacies consists of limestones and dolomitic limestones, tabular masses of hermatypic corals (mostly of the *Thecosmilia* group) up to 20 m in length and approximately 1 m in height (Pl. 2/1–3).

Faunal content: encrusting foraminifera, corals (cf. *Thamnasteria* sp. and *Palaeastrea* sp.), small gastropods.

Floral content: Dasycladaceae (*Diplopora* sp.), Solenoporaceae (*Solenopora* sp.) and rare fragments of the *Cayeuxia* type.

Lithofacies E.: Algal Biolithites

This lithofacies is characterized by colonies of *Solenopora* and *Cayeuxia* (20–70 cm thick) (Pl. 2/5 and 6).

The cavities within the biolithites are filled with dasycladacean and other bioclastic elements cemented by calcite (Fig. 5).

Faunal content: Foraminifera, gastropods, ostracods.

Floral content: Dasycladaceae (*Heteroporella zankli* (Ott), *Diplopora* cf. *tubispora* Ott, *Diplopora* sp.), Cyanophyceae (*Cayeuxia* type sp.), Solenoporaceae (*Solenopora* sp., *Parachaetetes* sp.) (Pl. 3/3 and 6).

Pl. 2/1. Coral biolithites and Megalodont limestones in detail the tabular masses of corals (*Thecosmilia* group) passing downward to Megalodont levels. Outer lagoon. Cozzo di Lupo, Palermo Mountains ▶

Pl. 2/2. Coral biolithites – Tabular masses of hermatypic corals (*Thecosmilia* group). Outer lagoon. Cozzo di Lupo, Palermo Mountains

Pl. 2/3. A sequence of three different lithofacies typical of the outer lagoon. Dolomitized limestones with Megalodont, at the bottom of the picture, are followed to the top by coral biolitithes and finally by algal biolithites. Outer lagoon. Cozzo di Lupo, Palermo Mountains

Pl. 2/4. Algal biolitithes. Solenoporaceae (detail). Outer lagoon. Cozzo di Lupo, Palermo Mountains

Pl. 2/5. Algal biolithites. Tabulate masses of Solenoporaceae. Outer lagoon, Cozzo di Lupo, Palermo Mountains. x 2,2

Pl. 2/6. Algal biolithites (to the top) and coral biolithites. The first lithofacies is intersected by paleokarstic solution fissures and cavities filled with fibrous calcitic cements. Outer lagoon. Cozzo di Lupò, Palermo Mountains

Pl. 2/7. Algal biolithites. Dasycladaceae and other bioclastic elements cemented by calcite. Outer lagoon. Cozzo di Lupo, Palermo Mountains

Plate 2. Legend see opposite page

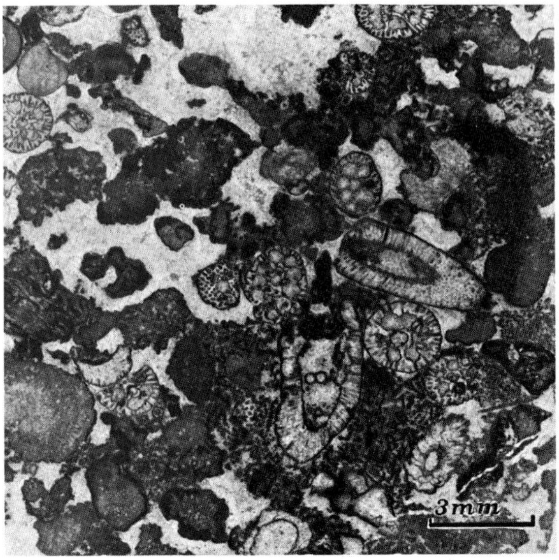

Fig. 5. Algal biolithites. Algal assemblage. In association *Heteroporella zankli, Diplopora tubispora, Cayeuxia* sp. Outer lagoon. Cozzo di Lupo, Palermo Mountains

33.3.2 Marginal Deposits

The rims of the lagoon (reef complex) are formed by alternations of either reef (patch-reef and inter-reef deposits) or fore reef deposits (chiefly consisting of bioclastic materials derived from the erosion of the biolithites).

In the *reef complex zone* the following lithofacies have been recognized:

Lithofacies F: Dolomitized Sponge Biolithites (patch reef)

These consist of two main lithotypes: grey biolothites with calcisponges and biomicritic arenites and rudites with small intraclasts and bioclasts.

The framework of the biolithites (more than 50%) is made by typical framebuilding, organisms such as calcisponges (*Colospongia* sp., *Peronidella* sp., *Molengraafia* sp.,

---

Pl. 3/1. Dolomitized spongid biolithites (patch reef). Calcareous sponges (*Molengraafia* group). Reef-complex. Cozzo di Lupo, Palermo Mountains ▶

Pl. 3/2. Dolomitized spongid biolithites. Sponges and Hydrozoa assemblages. Reef complex. Cozzo di Lupo, Palermo Mountains

Pl. 3/3. Algal biolithites. Fragments of *Parachaetetes*. Outer lagoon. Cozzo di Lupo, Palermo Mountains. Negative print

Pl. 3/4. Dolomitized spongid biolithites (patch reef). Close view of a large sponge. Reef-complex. Cozzo di Lupo, Palermo Mountains

Pl. 3/5. Dolomitized biocalcirudites and biocalcarenites. Bioclastic and lithoclastic compontents in the reef complex. Cozzo di Lupo, Palermo Mountains

Pl. 3/6. Algal biolithites. In association *Cayeuxia*-type thalli and Solenoporaceae. Outer lagoon. Cozzo di Lupo, Palermo Mountains. Negative print

Pl. 3/7. Dolomitized biocalcirudites and biocalcarenites. Algae (Dasycladaceae mostly), sponge fragments and other bioclastic grains in the inter-reef deposits. Reef complex. Cozzo di Lupo, Palermo Mountains. Negative print

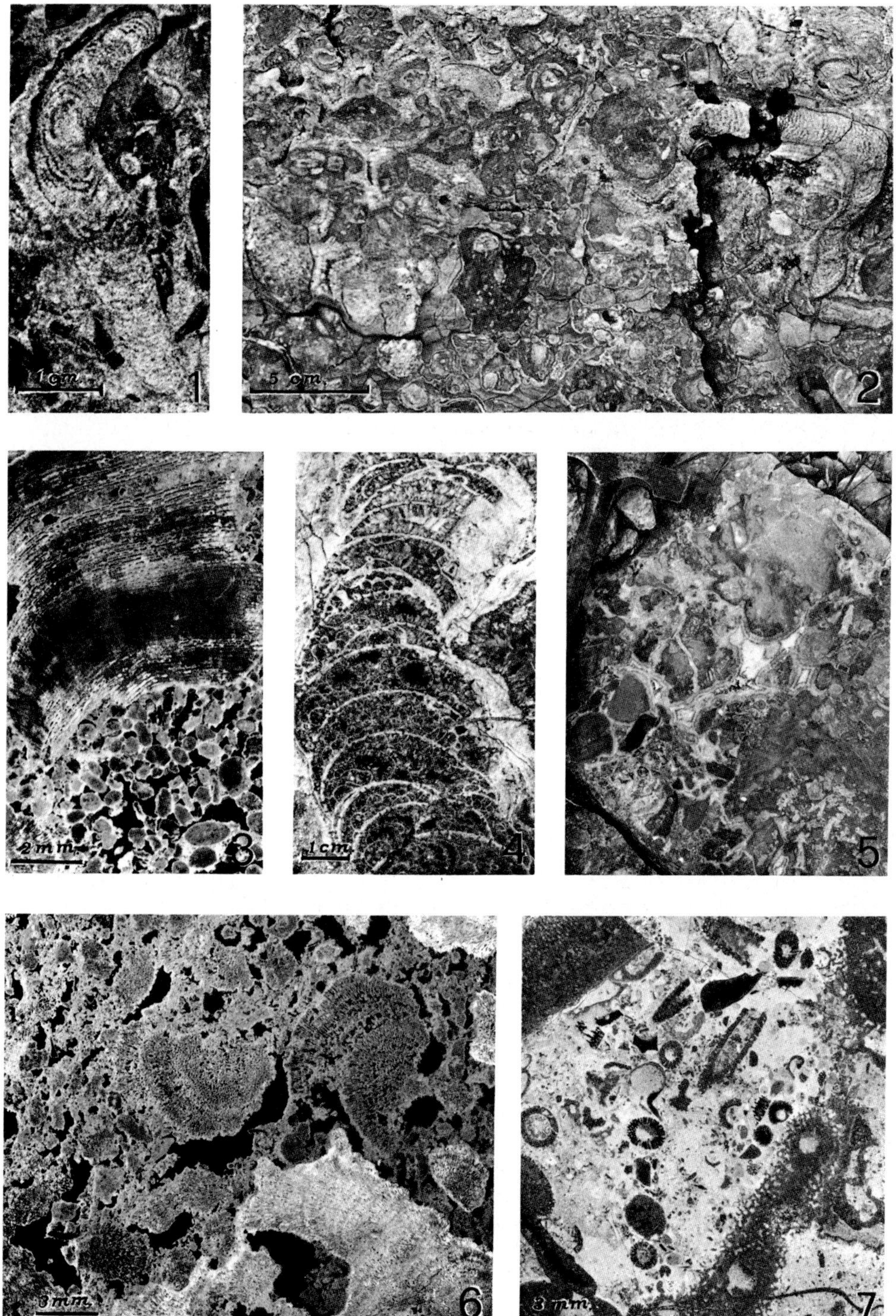

Plate 3. Legend see opposite page

Polytholosia sp.) (Pl. 3/1, 2 and 4). – Other *faunal content*: encrusting foraminifera, hydrozoans, stromatoporoids, corals (*Thecosmilia* type), brachiopods, very rare gastropods, echinoderm fragments, abundant problematica (*Cheilosporites tirolensis, Microtubus communis*).

Floral content: Solenoporaceae (*Solenopora* sp.), Cyanophyceae (*Cayeuxia* sp.), rare Dasycladaceae and very abundant oncolites.

The open spaces are filled with mudstone and fine grained fossil debris (mainly organisms living in the reef cavities; brachiopods, gastropods, echinoderms) and with fibrous calcitic cement, typical of a phreatic marine hydrological regime.

Lithofacies G: Dolomitized Biocalcirudites and Biocalcarenites

It is possible to distinguish the following lithotypes: calcareous dolomitized rudites with irregular intraclasts (biomicritic arenites) and small bioclasts (calcareous sponges), intramicritic rudites and arenites with subangular grains and large bioclasts. This lithofacies is considered as a typical inter-reef deposit (Pl. 3/5).

Faunal content: Foraminifera, fragments of calcareous sponges, stromatoporoids, corals, brachiopods (*Rhynchonellina bilobata, Rhynchonella* sp., *Terebratula* sp.) small ammonites, echinoderms and their fragments (*Cidaris* sp., *Pentacrinus* sp.).

Floral content: Dasycladaceae (*Diplopora* sp.), Solenoporaceae (*Solenopora* sp.), Cyanophyceae (*Cayeuxia* type fragments) (Pl. 3/7).

Early diagenetic cements, typical of a marine phreatic and meteoric phreatic regime, indicate a quick lithification.

The reefs are laterally graded into the inter-reef deposits. The latter are roughly bedded.

Lithofacies H: Intrabioclastic Dolorudites

This lithofacies is characterized by lithotypes with a more or less coarse ruditic texture: (1) Biomicritic rudites and arenites with prominent bioclasts (more or less angular grains composed of large fragments of sponges, encrusting hydrozoans, corals, and echinoderms). (2) Dark grey rudites, mostly constituted of smaller fragments of the reef and back reef deposits, bound by a dolomicritic matrix; primary, rare small cavities are partially filled with fibrous cement of marine phreatic environment. (3) Coarse calcirudites, with clasts reaching 80 cm in diameter, and smaller and angular fragments filling the intergranular cavities are bound by dark grey or yellowish-grey dolomicritic and doloarenitic cements, with the same textural features of the previously described deposits (grains and cements show dolomitization which often erases the original structure of the single elements). A period of synsedimentary faulting activity, clearly evident in lower parts of the scarp (not described in this paper) is also proposed as an important mechanism for the origin of a part of the larger blocks.

Faunal content: Foraminifera, abundant fragments of sponges, colonial and encrusting hydrozoans, fragments of stromatoporoids, corals, small brachiopods, ammonites (*Orthoceras billiemensis, Rhacophyllites billiemensis, Orthoceras* sp.), small pelecypods, fragments of crinoids.

Floral content: Oncolites and *Cayeuxia* type thalli fragments.

Textural features (particularly the large fragments from the inner zones), the presence of micrite, the absence of a distinct bedding and the lateral relationships with the reef facies indicate a fore reef environment.

33.4 Conclusions

Lithofacies herein described occupy well defined and mappable areas of the Panormide Carbonate Platform (Northwestern Sicily). They can be assigned to different sedimentary environments, representing tidal flat to slope environments (Pl. 4).

33.4.1 Tidal Flat Zone

This zone is characterized by large tidal flat deposits, with algal mats displaying desiccation and alteration features in their upper parts, alternating with sublittoral deposits. The only evidence of "autochthonous" algal activity is shown by the algal mats which form the bulk of the tidal flat deposits.

Other algae (*Solenopora* and *Cayeuxia*) are scattered by storms and are found interbedded in the stromatolites, together with other non-algal fossil fragments.

In the desiccated and altered tidal flats, with their typical tepee-like structures, algal fragments (*Cayeuxia, Solenopora,* algal mats) are carried in from the adjacent lagoon.

33.4.2 Back Reef Lagoon Zone

The sediments of this zone are here most widely exposed. Their algal content shows the highest diversity and abundance. Lagoon physiography, as shown by sediment distribution and grain size variation, was irregular and there is evidence of channels filled by rudites and coarse arenites displaying various sedimentary structures (point bars, cross laminations, etc.).

Solenoporaceae biolithites, about 1 m thick and extending laterally for several tens of meters, are very frequent. These biolithites also contain *Cayeuxia* type thalli and populations of *Diplopora* and *Heteroporella.*

The rare stromatolites observed do not show desiccation structures. The oncolites of algal-foraminiferal nature increase in size toward the tidal flat. It is of some interest that *Baccanella floriformis* is also present.

33.4.3 Reef Complex Zone

The algal content of the sediments in this zone is very poor. Crenulate oncolites often occur together with few *Diplopora* and rare Solenoporaceae. The observed diversity of algae in the sediments of the upper Triassic of the Panormide Carbonate Platform is the highest in the back reef lagoon. The lowest diversity obviously occurs in the tidal flat,

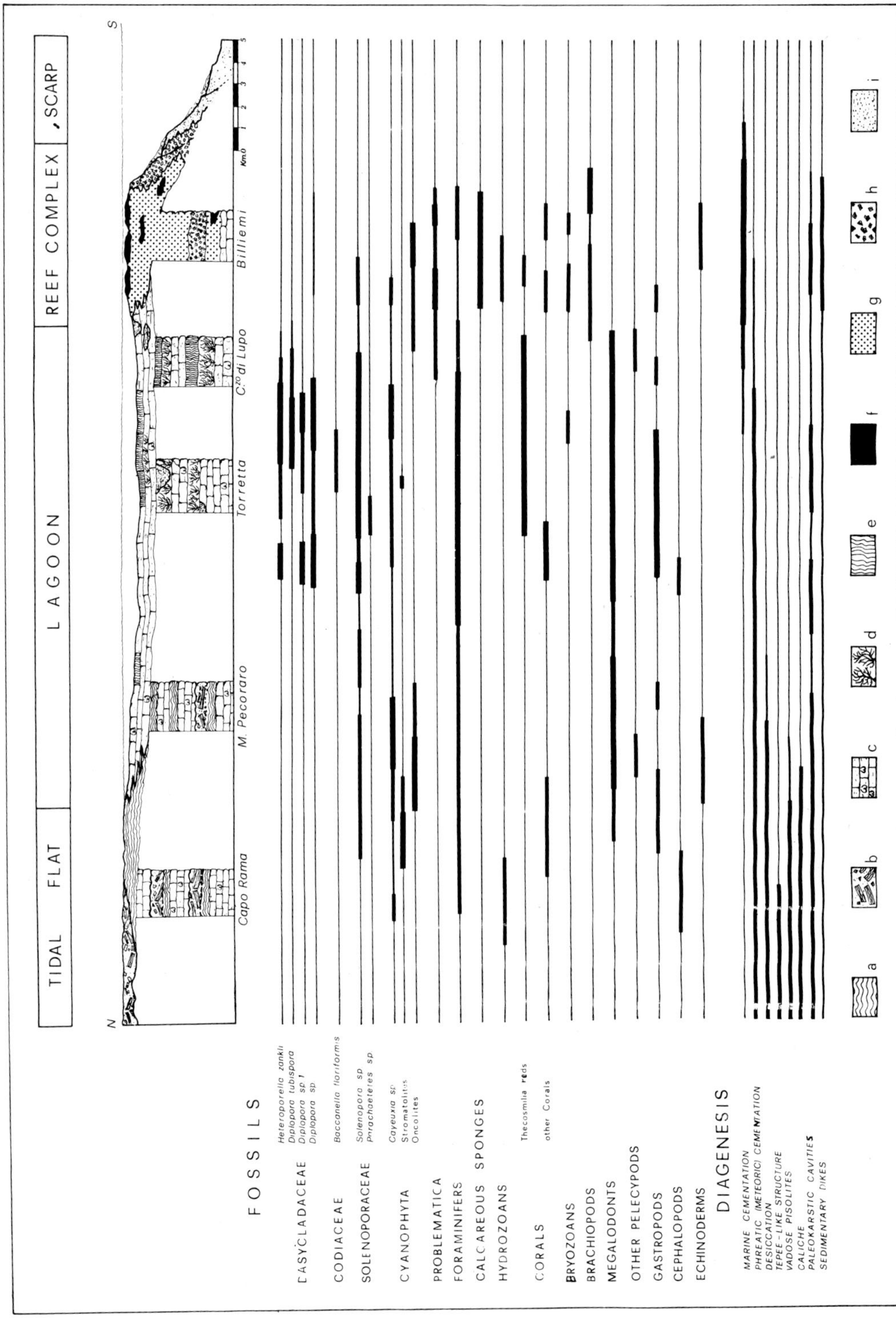

Plate 4. Legend see opposite page

and intermediate conditions have been observed in the reef complex zone. In contrast, the volumes of sediments produced are highest in the tidal flat zone (taking the stromatolites into account) and are gradually decreasing toward the reef complex zone.

33.5 Bibliography

Abate, B., Catalano, R.: Il margine della Piattaforma Carbonatica Panormide triassica nei monti di Palermo. Boll. Soc. Nat. Napoli **83**, 1–16 (1974).

Bystricky, J.: Die obertriadischen Dasycladaceen der West-Karpaten. Geol. Sbornik **18**, 285–309 (1968).

Catalano, R., D'Argenio, B., Lo Cicero, G.: Ciclotemi triassici di Capo Rama (Monti di Palermo). Geol. Romana **13**, 125–149 (1974a).

Catalano, R., D'Argenio, B., Lo Cicero, G.: Ritmi deposizionali e processi diagenetici nella successione triassica di piattaforma carbonatic dei Monti di Palermo. Soc. Geol. Ital. **93** (1974b).

Fischer, A.G.: The lofer cyclothems of the Alpine Triassic. Kansas Geol. Surv. Bull. **169**, 107–149 (1964).

Flügel, E.: Kalkalgen aus Riffkomplexen der alpin-mediterranen Obertrias. Verh. Geol. 297–346 (1975).

Ott, E.: Dasycladaceen (Kalkalgen) aus der nordalpinen Obertrias. Mitt. Bayer. Staatssamml. Paläontol. Hist. Geol. 7, 205–226 (1967).

Zankl, H.: Der hohe Göll, Aufbau und Lebensbild eines Dachsteinkalk-Riffes in der Obertrias der nördlichen Kalkalpen. Abh. Senckenberg. Naturforsch. Ges. **519**, 1–123 (1969).

◀ Pl. 4. Relationships of algae with depositional environments and faunal assemblages of the Panormide Carbonate Platform, upper Triassic, Northwestern Sicily. (a) algal mat dolomitic limestones; (b) loferitic breccias; (c) dolomitic limestones and calcareous dolomites with Megalodonts; (d) coral biolithites; (e) algal biolithites; (f) dolomitized spongid biolithites; (g) dolomitic biocalcirudites and biocalcarenites; (h) intrabioclastic dolorudites; (i) slope deposits

34 Environmental Models for Upper Paleozoic Benthic Calcareous Algal Communities

E. FLÜGEL, Erlangen, West Germany

Recent calcareous algae appear to be restricted to specific depositional facies (see Klement, 1966; Milliman, this vol.), they are most important therefore for the interpretation of Ancient sedimentary environments.

In connection with the study of the marine Permian of the Southern Alps, specific recurrent algal associations have been recognized, which are characterized by the predominance of one or a few abundant genera and species. These communities occur in limestones of different stratigraphic age; they can be found in different parts of the Paleotethys not only in the Alpine area but also in Yugoslavia, Anatolia, Japan, and in North America.

The purpose of this paper is

1. the typification of some common and widely distributed Permian Algal Communities
2. the description of distribution models, which can be used for the facies interpretation of Upper Paleozoic shelf and reef environments.

34.1 The Permian Algal Communities of the Southern Alps

34.1.1 Stratigraphy and Paleogeographical Situation

The algal material used in this study has been collected in Lower Permian and Upper Permian deposits of the Southern Alps (see Table 1). The algal-bearing localities are situated in the Sexten Dolomites (Fig. 1/1), in the Carnic Alps at the Austrian/Italian boundary (Fig. 1/2–4), in the Karawanken mountains (Fig. 1/5), in Slovenia (Fig. 1/6), and in the Velebit mountains, Croatia (Fig 1/7).

Paleontological, sedimentological and geochemical studies of the Permian sediments from the Carnic Alps and its neighboring areas south of the Periadriatic lineament reveal the following paleogeographic development during the Permian (see Buggisch et al., 1976; Flügel, 1975):

The Lower Permian deposits are characterized by a shelf- and shelf-edge sedimentation:

The Lower *Pseudoschwagerinas* Limestone (UPK of Table 1) of the Rattendorf formation were deposited cyclically in a near-coast inner-shelf area during alternating regressive and transgressive phases. Four lithological cycles with basal clastic beds and overlying predominately carbonates are known. The carbonate members consist of irregularly bedded biomicritic limestones and massive limestones, which sometimes correspond with organic buildups, formed by different phylloid algae (*Ancicodium, Eugonophyllum*).

Table 1. Permian stratigraphy of the Southern Alps. Algal-bearing formations are marked by an asterisk. UPK: Lower *Pseudoschwagerina* limestone; GB: Grenzland group; OPK: Upper *Pseudoschwagerina* limestone

		Carnic Alps	Yugoslavia	Sexten Dolomites	*Carnic Alps* Forni Avoltri	*Carnic Alps* Naßfeld	*Carnic Alps* Tarvis	Karawanken Mountains	*Yugoslavia* Sava Folds, Slovenia	*Yugoslavia* Velebit Mountains, Croatia
Upper Permian	Tatarian	Bellerophon stage		*Bellerophon formation* *	*Bellerophon formation* *	*Bellerophon formation* *	*Bellerophon formation* *	*Bellerophon formation* *	Zazar formation *	Velebit formation *
Middle Permian	Murghabian	Gröden stage	Velebit formation	*Gröden formation*	*Gröden formation*	*Gröden formation*	*Gröden formation*	*Gröden formation*		Velebit formation *
	Kungurian	Gröden stage	Velebit formation	*Gröden formation*	*Gröden formation*	*Gröden formation*	*Gröden formation*	*Gröden formation*		Velebit formation *
Lower Permian	Artinskian	Trogkofel stage	Kosna formation	Gröden Conglomerate *	Tarvis breccia * / Trogkofel ls. *	Tarvis breccia * / Tresdorf ls. *	Tarvis breccia * / Goggau ls. *	Tarvis breccia * / Trogkofel ls. * / Kosma formation *	Kosna formation	
	Sakmarian	Trogkofel stage	Kosna formation	Quartzporphyr	Rattendorf formation	Trogkofel ls. *	Rattendorf formation	(OPK) *	Kosna formation	
	Asselian	Rattendorfer stage	Rattendorfer stage	Quartzporphyr	Rattendorf formation	Rattendorf formation: OPK * / GB * / UPK *	Rattendorf formation	GB * / (UPK) *	Kosna formation	

Fig. 1. Lower Permian algal localities in the Southern Alps and in Yugoslavia. *1*: Sexten Dolomites (Kreuzberg area, Trogkofel limestones); *2*: Carnia (Forni Avoltri, Trogkofel formation); *3*: Carnic Alps (Nassfeld area and Schulterkofel area, Rattendorf and Trogkofel formation; *4*: Tarvis area (Trogkofel formation); *5*: Karawanken Mountains (Austrian and Slovenian parts, Rattendorf formation and different facies of the Trogkofel formation); *6*: Save folds (Rattendorf formation and clastic Trogkofel facies); *7*: Velebit Mountains (and Lika area; Rattendorf formation, and Middle Permian algal limestones)

Low diverse invertebrate communities (smaller foraminifera, fusulinids, corals, brachiopods) and mostly rather monomict algal associations indicate the existence of restricted, shallow-marine and subtidal environments. The total thickness of this group is about 160 m.

The Grenzland Group (GB of Table 1) consists of up to 130 m sandstones and silty shales with thin intercalations of biomicrites with fusulinids, echinoderms, and different algal groups (phylloid algae; crustose forms like *Ramovsia*; oncoids of the *Osagia* type). Biota, sediment structures, and grain-size analysis show the Grenzland beds to be deposited in a near-coast high-energy environment with intertidal to subtidal conditions.

In contrast to the Lower *Pseudoschwagerina* Limestones and to the Grenzland beds, the predominately carbonatic beds of the Upper *Pseudoschwagerina* Limestones (OPK of Table 1) include middle to high diverse biota, characterized by dasycladaceans, codiaceans, oncoid-forming algae, many paleotextulariid foraminifera, and monomict pseudoschwagerinid fusulinids as well as colonial corals. A regular recapitulation of microfacies sequences in stratigraphic sections may indicate a repeated shifting of ecological zones from very shallow water areas to more off-shore environments in an open-marine shelf lagoon with normal water circulation.

The Trogkofel Formation consists of three facies belts:

1. Organic buildups with encrusting organisms (*Tubiphytes*) and stromatolite-lined cavities as well as with fenestellid bryozoans, brachiopods and crinoids are developed in the Sexten Dolomites and in the Carnic Alps. This boundstone facies is very similar to the reef core facies described from the Lower and Middle Permian of Texas and New Mexico (Malek-Aslani, 1970; Dunham, 1972). A shelf-edge position of some of these *Tubiphytes* biomicrites and biosparites is probable.

2. Well-bedded Trogkofel limestones of the western Karawanken Mountains and of the Italian part of the Carnic Alps (region of Forni Avoltri, see Fig. 1) show typical biotas of outer shelf platforms. The algal communities are very similar to those of the Upper *Pseudoschwagerina* limestones.

3. A clastic Trogkofel facies, consisting of up to 2000 m quartz sandstones, conglomerates, and some limestones, is known from Slovenia, Croatia and from the Austrian Carnic Alps. Lithoclastic limestones in the Western Karawanken Mountains and in the Ortnek area, Slovenia (see Kochansky-Devidé et al., 1973; Ramovs and Kochansky-Devidé, 1965) give hints to a long-lasting period of synsedimentary re-sedimentation. Relatively small limestone lenses with monotonous coral-, brachiopod-, and echinoderm-faunas may have been formed in an inner-shelf basin.

A widely distributed erosion during the end of the Lower Permian is manifested in breccias and conglomerates (Tarvis breccia, Gröden conglomerate), underlying the Gröden formation. These sediments have been formed in areas of intensive, intra-Permian block-faults. A tectonic uplift of some sedimentary regions seems to be responsible for a uniform submarine and subaerial destruction of the Trogkofel platform and of the Trogkofel reefs in high-energy environments.

The red bed facies of the Gröden formation was deposited as continental clastics in South Tyrol during the first phase of sedimentation and as predominately marine sediments later on (Carnic Alps, Karawanken). Only a few possible algal structures are known from this facies.

The increasing transgression during the Upper Permian resulted in the deposition of bituminous sediments and basal evaporites of the Bellerophon formation. In the southwestern near-coast area evaporites are dominating (facies fiammaza, Accordi, 1959), whereas gypsum and open-marine carbonates (facies badiotica) rich in monomict algal floras with gymnocodiaceans and dasycladaceans occur in northeastern Tyrol.

34.1.2 Algal Floras

The calcareous algae from the Lower Permian Rattendorf and Trogkofel formations of the Carnic Alps have been investigated by E. Flügel (1966) and Homann (1972), the

Table 2. Distribution of calcareous algae in Lower Permian shelf environments of the Southern Alps

	Inner shelf			Outer shelf			
	Near-coast area		Buildups	Platform		Buildups	
	UPK	GB	UPK	OPK	TK	TK	Clastic TK (inner shelf?)
Blue-Greens							
Porostromata							
Garwoodia gregaria (Nicholson)				x			
Girvanella ducii Wethered	x	x	x	x	x		
Girvanella cf. *kordeae* Güvenc					x		
Girvanella permica Pia		x		x	x		
Girvanella texana Johnson	x						
Ortonella morikawai Endo					x		
"Osagia" sp.		x		x	x		
Problematic Blue-Greens							
Ramovsia limes Kochansky-Devidé		x					
Tubiphytes obscurus Maslov	x	x		x	x	x	
Greens							
Dasycladaceae							
Anatoliopora carbonica Konishi	x			x			
Anthraporella spectabilis Pia	x	x	x	x	x		x
Anthracoporella vicina Kochansky and Herak					x		
Atractyliopsis carnica Flügel	x	x		x	x		
Gyroporella igoi Endo				x	x	x	
Gyroporella intusannulata Kochansky-Devidé					x		
Gyroporella microporosa Endo					x		
Gyroporella nipponica Endo and Horiguchi					x	x	
Gyroporella symmetrica Johnson	x			x			
Gyroporella? tenuimarginata Endo					x		

Table 2 (continuation)

	Inner shelf			Outer shelf			
	Near-coast area		Buildups	Platform		Buildups	
	UPK	GB	UPK	OPK	TK	TK	Clastic TK (inner shelf?)
Likanella sp.					x		x
Macroporella maxima Endo				x			
Mizzia cornuta Kochansky and Herak	x			x	x		
Mizzia velebitana Schubert	x			x			x
Mizzia yabei (Karpinsky)	x	x		x	x		x
Pseudogyroporella mizziae-formis Endo		x					
Salopekiella sp.				x			
Problematic dasycladaceans							
Epimastopora alpina Kochansky and Herak	x	x	x	x	x		x
Epimastopora hunzaensis Zanin-Buri	x	x		x			
Epimastopora kansaensis Johnson			x	x			
Epimastopora kanumai Endo and Kanuma				x			
Epimastopora ketini Bilgütay	x	x		x			
Epimastopora piae Bilgütay	x	x		x	x		x
Pseudoepimastopora iwaizakiensis (Endo)	x			x			
Pseudoepimastopora japonica (Endo)	x	x		x			
Pseudoepimastopora kroatica Homann	x			x			
Pseudoepimastopora likana (Kochansky and Herak)	x	x		x	x		
Vermiporella nipponica Endo	x			x	x		
Codiaceae							
Succodium duisbergi Homann				x			
Problematic codiaceans							
Neoanchicodium cate-noides Endo	x	x		x	x	x	x

Table 2 (continuation)

	Inner shelf			Outer shelf			
	Near-coast area		Buildups	Platform		Buildups	
	UPK	GB	UPK	OPK	TK	TK	Clastic TK (inner shelf?)
Reds							
Solenoporaceae							
Solenopora centuriones Pia				x			
Solenopora texana Johnson		x					
Gymnocodiaceae							
Gymnocodium bellerophontis Rothpletz	x						
Permocalculus sp.	x			x	x		
Ancestral Squamariaceae?							
Anchicodium fukujiense Endo and Horiguchi			x	x	x		
Anchicodium magnum Endo	x	x	x				
Eugonophyllum johnsoni Konishi and Wray	x	x	x	x	x	x	x
Ancestral corallines?							
Archaeolithophyllum lamellosum Wray						x	
Archaeolithophyllum missouriense Johnson	x		x	x	x		
Cuneiphycus johnsoni Flügel	x	x		x	x		x
Problematic red algae							
Archaeolithoporella sp.				x	x	x	
Lithostroma oklahomaense Mamay	x						
Ungdarella uralica Maslov	x		x	x			
Brown Algae?							
Permopadina fallax Ramovs							x

Permian algal floras of the Karawanken mountains by Kochansky-Devidé (1970) and Kochansky-Devidé et al. (1973). Information on the algae from the clastic Trogkofel facies of Slovenia can be found in papers by Ramovs and Kochansky-Devidé (1965) and Ramovs (1968). Upper Permian algae from the Bellerophon formation were studied by Accordi (1956) and Praturlon (1963).

Table 2 and 3 show the algal inventory recognized in the Lower and Upper Permian of the Southern Alps. The classification follows the proposals made by Wray and by Riding (this vol.).

Table 3. List of calcareous algae, found in the Bellerophon formation, Upper Permian

Greens

Dasycladaceae

Atractyliopsis lastensis Accordi
Macroporella preromangica Praturlon
Mizzia velebitana Schubert
Physoporella laevis Praturlon

Reds

Gymnocodiaceae

Gymnocodium bellerophontis Rothpletz
Gymnocodium nodosum Ogilvie-Gordon
Permocalculus fragilis (Pia)
Permocalculus monoliformis (Pia)
Permocalculus solidum (Pia)
Permocalculus tenellus (Pia)

34.1.3 Algal Communities

Calcareous algae are the most important part of the biofraction of the Lower Permian shelf carbonates in the Southern Alps. Table 4 characterizes the frequency of algae in samples, which have been taken from several geological sections in the Carnic Alps and in the Karawanken mountains.

Considering the most abundant genera in algal-bearing samples 19 "Algal Communities" can be separated. These communities are characterized by

1. one or two quantitatively predominating algal genus or species,
2. a repeatedly recognized association with special invertebrates like Smaller Foraminifera, fusulinids, bryozoans, gastropods, etc.,
3. a correlation between some algal communities and certain texture types of the limestones.

The algal communities are named after the most frequent genus. They can be attributed to several adaption groups according to their dominating growth forms (see Pia, 1928):

Table 4. Frequency of algae. A.: P. 1997, cycle IV, Carnic Alps; B: P. 1997, cycle III; C: Tresdorfer Höhe, Nassfeld area, Carnic Alps; D: Schulterkofel, Carnic Alps; E: Zottachkopf, Carnic Alps; F: Zweikofel, Carnic Alps; G: Col di Mezzodi, Forni Avoltri, Carnia; H: Teufelsschlucht, Slovenian Karawanken Mountains

	Inner shelf environment				Outer shelf environment			
	Lower *Pseudoschwagerina* ls.				Upper *Pseudo-schwagerina* ls.		Bedded Trogkofel ls.	
	A	B	C	D	E	F	G	H
Total sum of the samples	48	71	39	42	204	44	105	42
Percentage of samples with algae	98	40	62	77	84	88	91	81

A. Crustose forms (e.g. *Pseudovermiporella* community, *Tubiphytes* community, *Ramovsia* community, *Archaeolithophyllum lamellosum* community)

B. Phylloid forms (e.g., phylloid algae community with *Eugonophyllum* and *Anchicodium*, perhaps *Neoanchicodium* community)

C. Nodular forms (e.g., *Girvanella* community, oncoid communities with blue-green algae or red algae)

D. Cylindrical forms (e.g., different dasyclad communities, *Anthracoporella* community, *Gynmocodium/Permocalculus* community)

With regard to the general diversity of the biotas and to the diversity of the algal associations, it is possible to distinguish:

I. Low biotic diversity (many algal thalli together with only a few invertebrate groups)

A. Low algal diversity with generally one or two algal species and rich to abundant populations

a) Green algae:

1. *Anthraporella* community (Pl. 1/1)

Biomicrites with abundant populations of the dasyclad *Anthracoporella spectabilis* Pia, together with some fragments of phylloid algae or some rare other dasyclads. Fauna: smaller foraminifera, some single corals.
Distribution: Lower *Pseudoschwagerina* ls.

Plate 1. Low diverse algal communities ▶

Pl. 1/1. *Anthracoporella* community with *Anthracoporella spectabilis* Pia, and some smaller foraminifera. Lower *Pseudoschwagerina* limestone, Tresdorfer Höhe, Carnic Alps. x 11

Pl. 1/2. *Gymnocodium* community, with *Gymnocodium bellerophontis*, Rothpletz. Bellerophon limestone, Upper Permian; St. Ulrich, Gröden valley, South Tyrol. x 5.4

Pl. 1/3. Phylloid algal community, with recrystallized plates of *Eugonophyllum* and *Archaeolithophyllum*. Lower *Pseudoschwagerina* limestone, P. 1997, Carnic Alps. x 5.6

Pl. 1/4. Phylloid algal community, with plates of *Eugonophyllum johnsoni* Konishi and Wray. Lower Rattendorf formation, Dolzanova soteska near Trzic, Karawanken Mountains. x 7.5

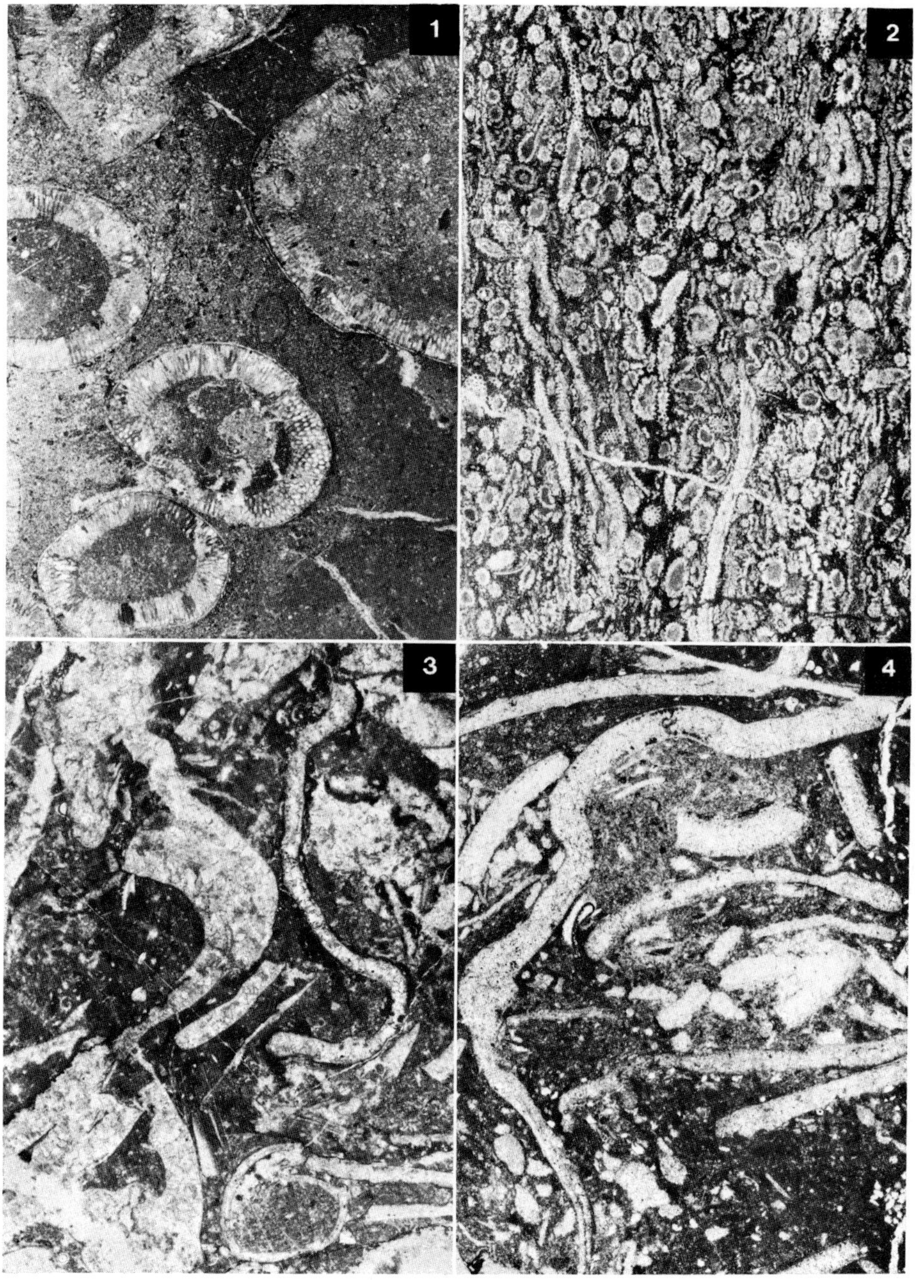

Plate 1. Legend see opposite page

2. *Tauridium* community

   Biomicrites with abundant populations of the codiacean algae *Tauridium cuvillieri* Güvenc and *Tauridium fragilis* Güvenc, together with only a few other algae. Fauna: smaller foraminifera, some fusulinids.
   Distribution: known from the Middle and Upper Permian of Anatolia and from the Bellerophon formation of the Dolomites.

3. *Pseudovermiporella* community

   Biomicrites with rich to abundant populations of the problematic dasyclad *Pseudovermiporella sodalica* Elliott, mostly as encrusted forms. Fauna: many to abundant specimens of miliolid foraminifera, and some fusulinids.
   Distribution: in the Bellerophon formation, often together with many gymnocodiacean algae.

4. *Gymnocodiacean* community (Pl. 1/2)

   Biomicrites with abundant populations of one or several species of *Gymnocodium* Pia and *Permocalculus* Elliott, together with abundant to many specimens of *Pseudovermiporella*, and some dasyclads. Fauna: miliolid and nodosariid foraminifera, sometimes a few gastropods.
   Distribution: Bellerophon formation.

5. Phylloid algal community (Pl. 1/3 and 4)

   Biomicrites and biosparites with many to abundant fragments of "ancestral Squamariaceae" (see Wray et al., 1975), which may belong to several species and genera (*Eugonophyllum, Anchicodium, Archaeolithophyllum*). Other algae: encrusted forms of *Vermiporella* and *Girvanella*. Fauna: sessile foraminifera upon algal thalli (*Tuberitina, Apterinella*), paleotextulariid and miliolid foraminifera, rare corals, some gastropods.
   Distribution: widespread in the Lower *Pseudoschwagerina* limestones (see Fig. 3).

6. *Komia* community

   Biomicrites with many specimens of the problematic red alga *Komia* Korde. Fauna: some smaller foraminifera with ammodiscid and miliolid species.
   Distribution: very rare in Lower Permian deposits, better known from Pennsylvanian shelf margin facies (see Freeman, 1964; Wilson, 1975).

b) Blue-green algae:

7. *Girvanella* community (Pl. 2/1)

   Biosparites with abundant nodules, consisting of parallel arranged tubes of *Girvanella ducii* Wethered. Fauna: some smaller foraminifera, fusulinids.
   Distribution: only rare in the Upper *Pseudoschagerina* limestone of the Karawanken Mountains.

8. *Tubiphytes* community (Pl. 2/2 and 3)

   Biosparites and biomicrites as well as intrabiosparites with many to abundant, encrusted and free thalli of the problematic blue-green alga *Tubiphytes*

obscurus Maslov. Different stages of preservation may be responsible for different sizes of the thalli and for dense or open microstructures; isolated tubes also seem to belong to *Tubiphytes.* Many thalli are linked with low stromatolites of the types LLH-C and LLH-S. These stromatolites consist of very densely arranged, crenulated micrite layers and sparry-filled interspaces. No significant vertical growth has been observed. Fauna: as epizoans on *Tubiphytes* some foraminifera; bryozoans, echinoderms and brachipods are common.
Distribution: widespread in the organic buildups of the Trogkofel formation.

9. *Ramovsia* community

Biomicrites and biosiltites which abundant encrusting forms of the microproblematicum *Ramovsia limes* Kochansky-Devidé. Fauna: encrusting foraminifera.
Distribution: known from the Grenzland group of the Rattendorf formation only.

10. Micrite-oncoid community (Pl. 3/1)

Biosparites with many small, oncoid-like foraminiferal-algal consortia. These probably encrusted organisms are characterized by micritic, very densely zoned "walls", several sparry-filled caverties, and sometimes a finely perforated periphery. Similar fossils have been described as problematic algae, problematic foraminifera, or as an intergrowth of encrusting foraminifera with blue-green algae. Some of these forms include sessile foraminifera and algae like *Cuneiphycus johnsoni* Flügel. Fauna: rare fusulinids, paleotextulariid foraminifera, and echinoderms.
Distribution: known from the well-bedded Trogkofel limestone of the Teufelsschlucht/Karawanken Mountains, and from the Rattendorf formation of the Lika area/Velebit Mountains, Croatia.

B. High algal diversity with several species or different genera, generally rich populations

a) Green algae:

11. *Epimastopora* community (Pl. 3/2)

Biosparites and biomicrites with many fragments of the problematic dasycladaceans *Epimastopora* Pia and *Pseudoepimastopora* Endo; usually several species can be recognized. Other dasycladaceans (*Gyroporella, Mizzia*) as well as oncoids are common. Fauna: gastropods common, several groups of Smaller Foraminifera, echinoderms.
Distribution: widespread in sediments of the outer shelf (Upper *Pseudoschwagerina* limestone, well-bedded Trogkofel limestone of Forni Avoltri and of the Karawanken Mountains), but also known from the inner shelf environment (Lower *Pseudoschwagerina* limestone).

b) Red algae:

12. *Solenopora* community

Biomicrites with many thalli of *Solenopora* sp., together with rare Smaller Foraminifera, echinoderms, and bryozoans.
Distribution: very scarce in the Rattendorf formation.

13. *Archaeolithophyllum* community (Pl. 2/4)

Biomicrites and biosparites with many crustose thalli of *Archaeolithophyllum lamellosum* Wray, together with stromatolites, and *Tubiphytes* sp. Fauna: bryozoans and echinoderms common.
Distribution: framebuilders in organic buildups of the Trogkofel formation (Sexten Dolomites) and in the clastic Trogkofel facies (Tresdorfer limestone, Carnic Alps).

II. High biotic diversity (algae associated with several groups of invertebrates, "mixed associations")

A. Low algal diversity

a) Green algae:

14. Low diverse dasyclad community

Biosparites with many algae which usually belong to one or two dasycladacean species. Oncoids may be present. Fauna: Smaller Foraminifera, fusulinids, bryozoans, corals, brachipods, gastropods, and echinoderms. Distribution: widespread in the Inner-shelf and Outer-shelf environments with open-marine water circulation.

15. *Neoanchicodium* community

Biosparites with many to abundant specimens of the problematic codiacean algae *Neoanchicodium catenoides* Endo, together with plates of *Archaeolithophyllum, Epimastopora, Mizzia, Atractyliopsis.* Fauna: paleotextulariid foraminifera, fusulinids, mollusks, corals, etc.
Distribution: common in outer shelf environments (Upper *Pseudoschwagerina* limestones and well-bedded Trogkofel limestones).

Plate 2. Low diverse algal communities ▶

Pl. 2/1. *Girvanella* community, with many nodules composed of *Girvanella ducii*, Wethered. Upper *Pseudoschwagerina* limestone, Pristovnik Alm near Trögern/Eisenkappel, Karawanken Mountains. x 7.5

Pl. 2/2. *Tubiphytes* community, with the problematic blue-green alga *Tubiphytes obscurus* Maslov. Trogkofel formation, organic buildup; Matzenboden, Sexten Dolomites, South Tyrol. x 7.5

Pl. 2/3. *Tubiphytes* community, with stromatolite-lined cavities, and *Tubiphytes* sp.; Matzenboden, Sexten Dolomites. x 5.7

Pl. 2/4. *Archaeolithophyllum/Tubiphytes* community, with plates of *Archaeolithophyllum lamellosum* Wray, and some sections of *Tubiphytes obscurus*, Maslov; Trogkofel limestone, Matzenboden, Sexten Dolomites. x 7.5

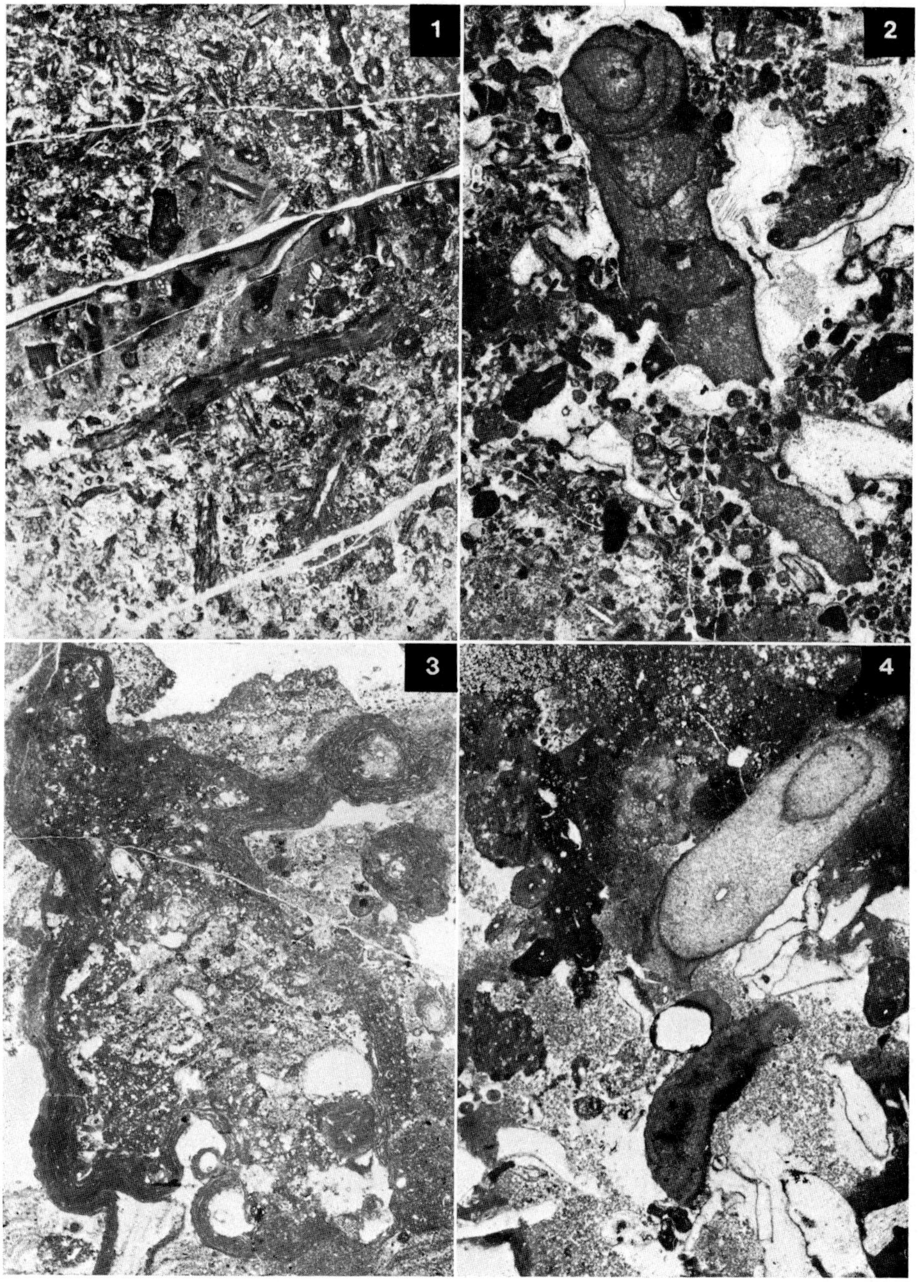

Plate 2. Legend see opposite page

b) Oncoid assemblage

16. *Osagia*-oncoid community (Pl. 4/1 and 2)

Small bean-shaped oncoids without distinct laminae. Nuclei are any available skeletal grain, which is often corroded and bored by girvanellid algae. Laminae can be recognized only by concentrations of *Girvanella* and encrusting foraminifers (*Hedraites, Apterrinella*). Fauna: many groups of invertebrates, and several dasycladadean and codiacean algae.
Distribution: oncoids are widely distributed in shallow, near-coast but also in off-shore environments (Grenzland group; Upper *Pseudoschwagerina* ls., bedded Trogkofel ls.).

17. *Archaeolithophyllum*-oncoid community (Pl. 4/3)

Relatively large (up to 10 cm), subrounded oncoids with concentrically arranged thalli of the red alga *Archaeolithophyllum lamellosum* Wray, together with many encrusting foraminifers like *Hedraites, Tuberitina*, and *Apterrinella*. Fauna: similar to community 16.
Distribution: off-shore, outer-shelf environments, and near-coast areas (see Toomey, 1974).

B. High algal diversity

a) Green algae:

18. High diverse dasyclad community

Similar to community 14, but usually with several dasycladacean species and genera.
Distribution: widespread in outer-shelf environments with normal marine, open-water circulation.

19. Algal sporen assemblage (Pl. 4/4)

Biosparites with numerous sparry spheres, which have been interpreted as sporen of dasycladaceans (*Atractyliopsis, Mizzia* etc.), together with several types of dasycladaceans, *Neoanchicodium, Epimastopora,* oncoids. Fauna: similar to community 14 and 18, Gastropods and *Epimastopora* relatively common.
Distribution: widely distributed in very shallow areas of the outer-shelf platform (Upper *Pseudoschwagerina* limestone, bedded Trogkofel limestone) in the Carnic Alps, and in the Karawanken Mountains.

Pl. 3/1. Micritic oncoid community, with problematic encrusting organisms, together with some fusulinids. Trogkofel formation, bedded limestone; Dolzanova soteska near Trzic, Karawanken Mountains. x 7.5 ▶

Pl. 3/2. *Epimastopora* community, with many fragments of *Epimastophora alpina*, Kochansky and Herak, and *Pseudoepimastopora* sp. Upper *Pseudoschwagerina* limestone, Eustachiushütte near Trögern, Karawanken Mountains. x 5.2

Pl. 3/3. Low diverse dasyclad community, with many thalli of *Gyroporella igoi*, Endo, and *G. symmetrica*, Johnson. Tresdorf limestone (clastic Trogkofel facies), Tresdorf Alm, Carnic Alps. x 5.6

Pl. 3/4. High diverse dasyclad community, with *Mizzia velebitana*, Schubert, *Mizzia cornuta*, Kochansky and Herak, *Clavaphysoporella* sp., and other dasycladaceans. Middle Permian Velebit, Croatia. x 5.6

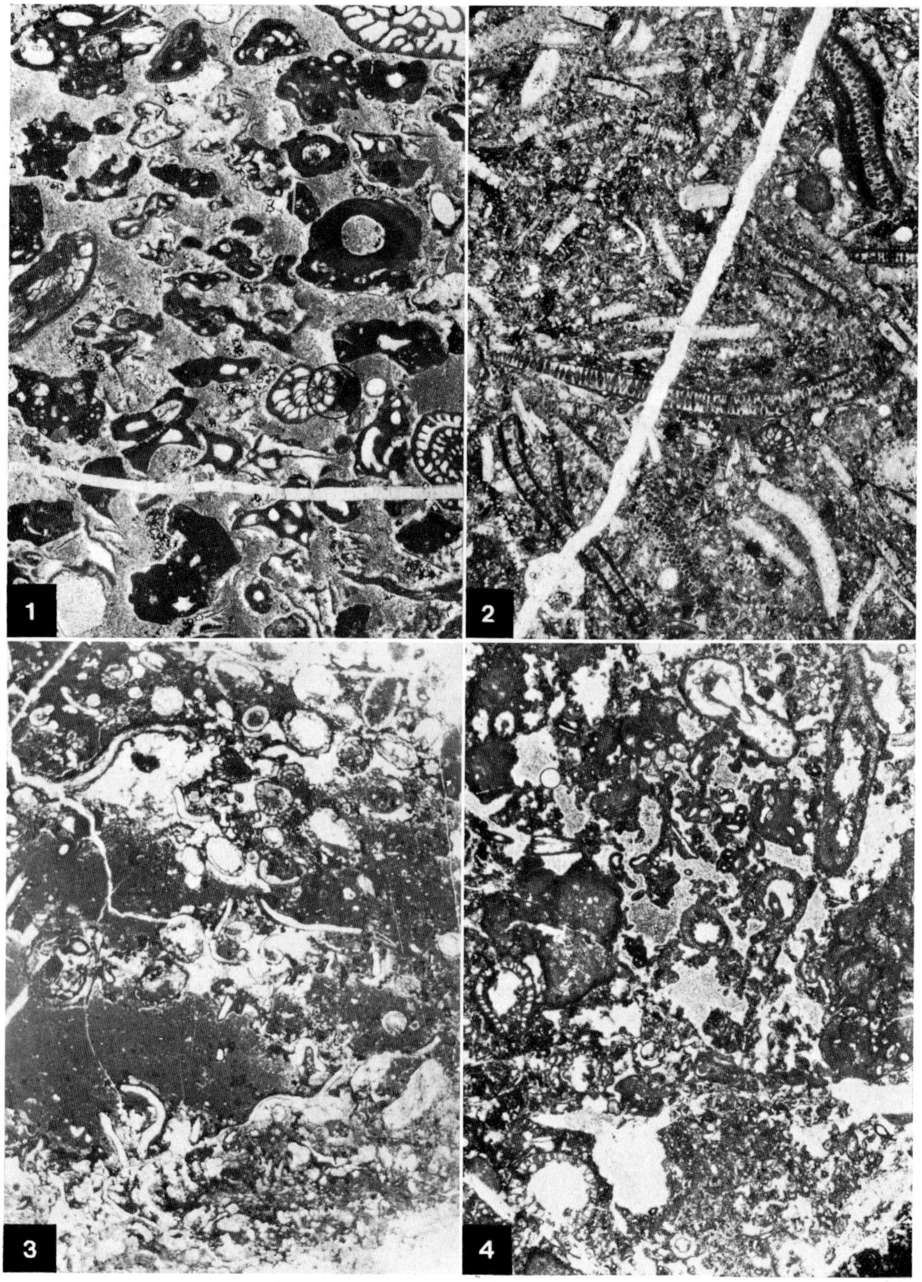

Plate 3. Legend see opposite page

31.1.4 Distribution of the Algal Communities

Using the Jaccard Coefficient of Community (see Chetham and Hazel, 1969) the algal floras of the different Lower Permian facies belts can be compared in order to determine similarities between distinct shelf environments.

Table 5. Relationships between the algal floras in different parts of the Lower Permian shelf platform in the Southern Alps. High values of the Jaccard Coefficient indicate a large number of joint species

	UPK	GB	UPK buildup	OPK	TK	TK buildup	cl.TK
UPK	–						
GB	0.360	–					
UPK belt	0.233	0.218	–				
OPK	0.490	0.410	0.194	–			
TK	0.326	0.370	0.200	0.465	–		
TK belt	0.090	0.125	0.250	0.135	0.206	–	
cl. TK	0.218	0.304	0.200	0.216	0.027	0.133	–

An interpretation of Table 5 must consider the fact that most of the algal species known from the different facies units of the Lower Permian have a wide stratigraphic range. Therefore it seems possible to compare not only time equivalents like the Lower *Pseudoschwagerina* limestones and the organic buildups of this facies of the different facies of the Trogkofel formation, but also all stratigraphic units.

Nearly all stratigraphic units except the organic buildups and the clastic Trogkofel facies show high Jaccard Coefficients with regard to the algal floras. The different parts of the shelf area therefore were inhabited by predominately eurytopic species. Inner and outer parts of the shelf platform are characterized not by the distribution of species but by the distinct predominance of some species within algal communities and by various degrees of biotic diversity. The low Jaccard Coefficients of the phylloid algal buildups and the *Tubiphytes* reefs indicate the existence of special ecological factors (water depth and water circulation) in these areas.

Pl. 4/1. *Osagia*-oncoid community, with bean-shaped oncoid. The laminae are marked by *Girvanella* sp., and by different encrusting foraminifera; a fragment of *Epimastopora* is circumcrusted. Trogkofel formation, bedded limestones; Creta di Rossa near Forni Avoltri, Carnia, Italy. x 7.5 ▶

Pl. 4/2. *Osagia*-oncoid community, with *Epimastopora* and *Neoanchicodium* sp. Upper *Pseudoschwagerina* limestone, Zottachkopf section, Carnic Alps. x 3.7

Pl. 4/3. *Archaeolithophyllum*-oncoid community. Large oncoid, composed of distinct layers of *Archaeolithophyllum lamellosum*, Wray. Upper *Pseudoschwagerina* limestone, Zottachkopf section, Carnic Alps. x 15

Pl. 4/4. Algal sporen association. Many dasycladacean sporen, together with *Atractyliopsis carnica*, Flügel. Upper *Pseudoschwagerina* limestone, Carnic Alps. x 5.6

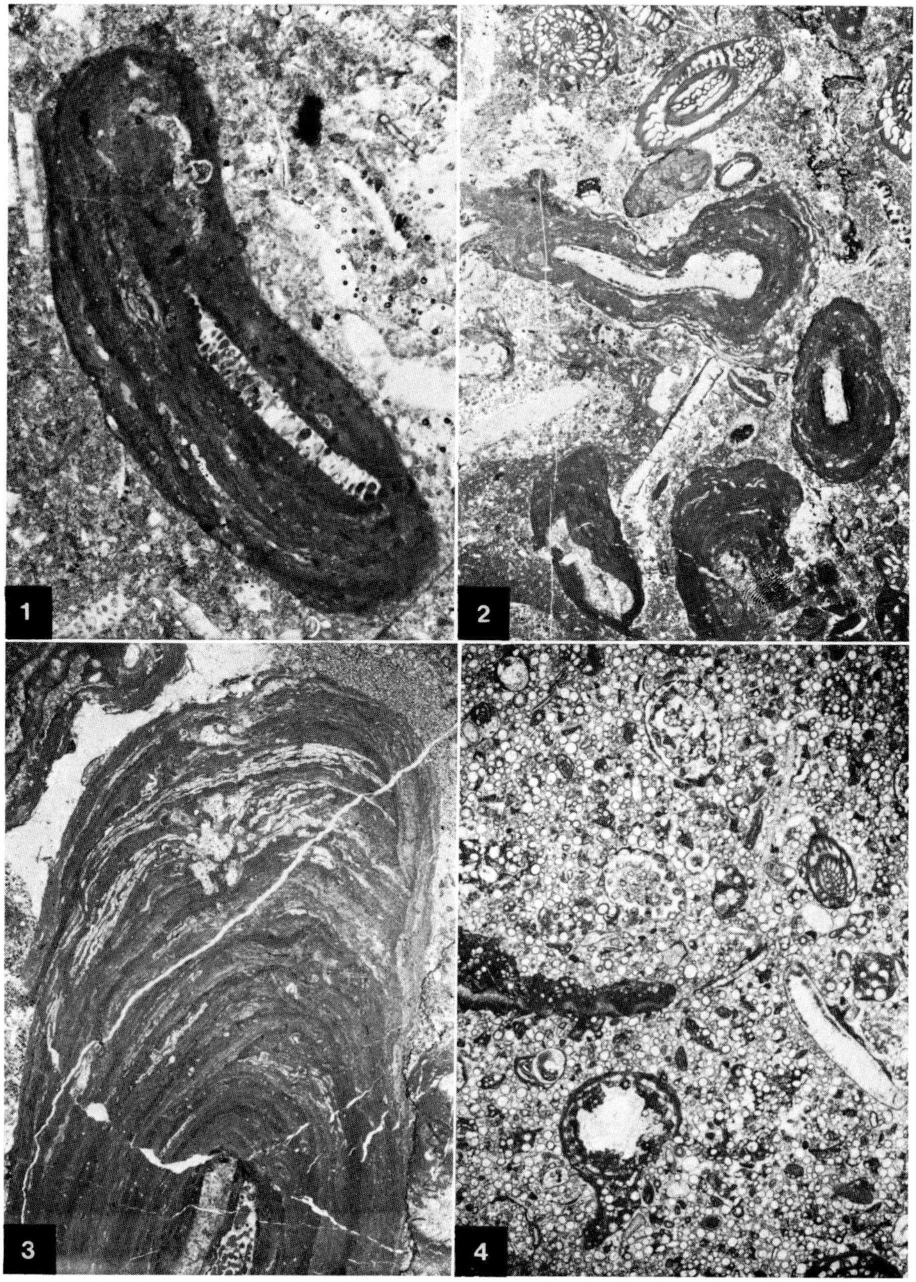

Plate 4. Legend see opposite page

The possible distribution of the algal communities on the Lower Permian shelf of the Southern Alps is shown in Table 6. The classification of the shelf environments follows Wilson (1975).

Table 6. Distribution of algal communities in the Lower Permian of the Southern Alps

	Inner shelf			Outer shelf		Shelf-edge (or shelf position)
	Restricted platforms			Open Platforms	Sands on platform	Organic reefs
	Restricted marine shelf-lagoons	Restricted marine shoals	Relatively open circulation	shelf facies open circulation	edge. Shoal environment in agitated waters	
Osagia-oncoid C.	-----				-----	
Girvanella community	-----					
Archaeolithophyllum C.	-----					
Anthracoporella C.		-----				
Epimastopora community		-----	-----	-----		
Low diverse dasyclads		-----				
Ramovsia C.			-----			
Micritic oncoid C.			-----			
High diverse dasyclads				-----	-----	
Neoanchicodium C.				-----	-----	
Algal sporen A.					-----	
Phylloid algal C.						-----
Tubiphytes community						-----
Archaeolithophyllum C.						-----

A semi-quantitative survey of the more frequent skeletal grains of the Lower Permian limestones from the Carnic Alps and adjacent areas has resulted in the recognition of 16 bioassociation types (see Flügel, 1975):

Group A, high diversity, at least three, mostly five and more biota groups

1. Biomicrite with diverse organisms (Smaller Foraminifera, fusulinids, bryozoans, brachiopods, gastropods, crinoids, echinoids, dasyclads); the bioclasts may be micritized. Homogenization of the texture by bioturbate burrowing is common. This type corresponds to the Standard Microfacies Types 9 and 10 (shelf facies with open circulation).
   Distribution: Common in the Lower *Pseudoschwagerina* limestones, and in the Upper *Pseudoschwagerina* limestones (here together with oncoids).
2. Bioarenosparite resp. grainstone with diverse organisms. Echinoderms, gastropods and bryozoans more frequent than in type 1; coated grains common. Corresponds to SMF 11 (shoal environment in agitated water).
   Distribution: common in the Lower *Pseudoschwagerina* limestones.

Group B, low diversity, dominance of one or two biota groups

3. Biomicrite resp. wackestone and bafflestones with platy (phylloid) algal fragments (= algal community no. 5). Corresponds to SMF 7. Widely distributed in micritic mud mounds and in horizontally bedded strata of the Lower *Pseudoschwagerina* limestones; rare in the Grenzland group, very scarce in the Upper *Pseudoschwagerina* ls. and in the bedded Trogkofel limestone.
4. Biomicrite resp. wackestone and packstone with abundant thalli of *Anthraporella* (= algal community no. 1) and encrusting foraminifers. Corresponds to SMF 18 (restricted marine shoals). Lower *Pseudoschwagerina* ls.
5. Biomicrites resp. wackestone and packstones with many gastropods and abundant fragments of *Epimastopora* (= algal community no. 11). Corresponds to SMF 18.
   Distribution: widespread in the Lower and Upper *Pseudoschwagerina* limestones, and in the bedded Trogkofel ls.
6. Biomicrite resp. packstone with numerous fusulinids. Corresponds to SMF 10 (shelf facies, open circulation). Known from all members of the Lower-Permian sections.
7. Biosparite and biomicrite resp. pack- or floatstone with abundant fragments of echinoderms. Corresponds to SMF 12 (shoal environment in agitated waters, possible slope sediment). Lower *Pseudoschwagerina* ls.
8. Microbioclastic biosiltite and biopelmicrite resp. encrusting bindstone with abundant encrusting foraminifers (*Ammovertella* etc.), together with paleotextulariid foraminifers, and other biota. Corresponds to SMF 7. Not rare in the Lower *Pseudoschwagerina* and Upper *Pseudoschwagerina* limestones.
9. Microbioclastic biosiltite resp. bindstones with numerous encrusting organisms (mostly *Ramovsia* = algal community no. 9). Corresponds to SMF 7. Known from the Grenzland group only.
10. Oncobiomicrites resp. oncoidal wacke-packstone, with large oncoids of the *Osagia*- and *Archaeolithophyllum*-type (= algal communities nos. 16 and 17), together with diverse other biotas. Corresponds to SMF 22 (restricted marine

shelf lagoons, protected environment). Common in the Grenzland group, in the Upper *Pseudoschwagerina* limestone, and in the bedded Trogkofel limestone.

11. Biomicrite resp. packstone with numerous, densely packed echinoderms, together with gastropods. Corresponds to SMF 12. Known from the Grenzland group only.
12. Biosparite resp. grainstone with abundant dasyclad sporen, and dasycladaceans (*Atractyliopsis* etc. = algal community no. 19). Corresponds to SMF 12.
13. Biomicrite resp. wackstone with a diverse fauna and flora, and a high percentage of *Neoanchicodium* (= algal community no. 15). Corresponds to SMF 11 and 13 (shoal environment in agitated waters).
    Distribution: common in the Upper *Pseudoschwagerina* limestones and in the bedded Trogkofel limestones.
14. Biosparite resp. grainstone with numerous dasycladaceans, gastropods, and foraminifers. Corresponds to SMF 12 and to the algal community no. 18).
    Distribution: bedded Trogkofel limestones.
15. Biomicrites and intramicrites resp. encrusting boundstone with abundant encrusting *Tubiphytes*, bryozoans and brachiopods; resedimentation is common. Corresponds to SMF 7. Typical microfacies of Trogkofel buildups.
16. Biosparites resp. encrusting bindstone, with abundant *Tubiphytes* and algal stromatolites (= algal community no. 8). Corresponds to SMF 7. Typical microfacies of the Trogkofel buildups.

Some of these bioassociations seem to be significant for outer shelf areas with open circulation upon open platforms and with shoal environments in agitated waters near the edge of the platform. A typical outer shelf sequence is shown in Table 7. The sequence is characterized by an alternation of 10 bioassociations (and of 7 algal communities).

Whereas the open platform environments yield generally large quantities of calcareous algae, the more restricted parts of the inner-platform shoals show much more distinct differences in the amount of algae (see Fig. 2). Variable portions of algae, smaller foraminifera, and fusulinids in many sections of the Lower *Pseudoschwagerina* limestones may reflect changing bathymetrical situations: Point-counter analysis reveals five groups within the biofractions according to the relative frequency of the calcareous algae (mainly phylloids like *Eugonophyllum, Archaeolithophyllum*), fusulinids, and smaller foraminifera. Group A is characterized by low algal content and high content in smaller foraminifera (15–20% of the biofraction), fusulinids are generally rare (mostly less than 10%). Group B has 20–50% algae, about 15% smaller foraminifera, and only a few or no fusulinids. Group C includes samples with up to 50% algae, about 10–15% smaller foraminifera, and up to 20% fusulinids. In group D samples with abundant algae are gathered, together with about 10–20% smaller foraminifera, and fusulinids. In the last group E calcareous algae are abundant too whereas fusulinids are lacking; smaller foraminifera occur with 8–10%. The biofraction groups A and B seem to indicate relatively deep deposition areas (according to the predominance of paleotextulariid foraminifers; see Stevens, 1966). Shallowing is shown by the increase in algal content within the groups C, D, and E.

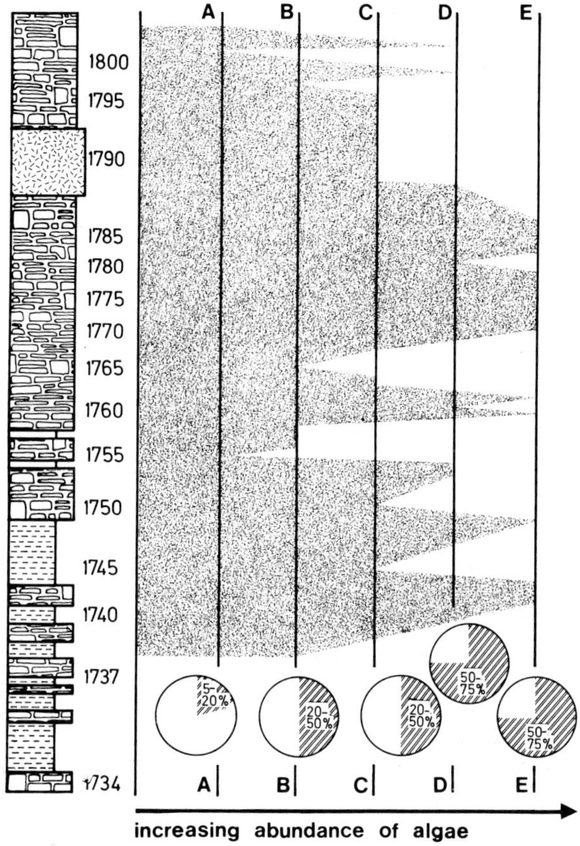

Fig. 2. Distribution and frequency of calcareous algae reflect changing bathymetrical situations within shoals of the inner platform. Lower *Pseudoschwagerina* limestones, sedimentation cycle IV, P. 1997, near the Rattendorf Alm, Carnic Alps. 5 biofraction groups (A–E) are characterized by a different amount of algae, smaller foraminifera, and fusulinids. see p. 334 and Fig. 3. Height of section about 15 m

A combined microfacial and geochemical investigation of a section from the Lower *Pseudoschwagerina* limestone (see Fig. 3) indicates smaller cyclically alternating environmental conditions (shown by the terrigenous residuals, and, perhaps, by some rare elements) within the sedimentary cyle. The causal connections between these alternations and the different amount of algae is not clear.

34.2 Distribution Models

The distribution of recent calcareous algae is controlled by water depth, water temperature, substrat, and sedimentary regime. Milliman (1974, and this vol.) has given excellent descriptions of the importance of calcareous algae in continental margin sedimentation. Today the codiacean alga *Halimeda*, some dasycladaceans, and coralline red alga are essential for the shelf sedimentation. According to the original growth form three general types of algal sediments have been recognized: (a) Maerl (sand or gravel with more than 50% made by algal fragments, branches of free-standing corallines or green-algae),

Table 7. Bioassociations and algal communities of an Outer Shelf sequence. Upper *Pseudoschwagerina* limestone, Zottachkopf section, Carnic Alps. Rapid changes of more open and more restricted water circulation. Correlation of the bioassociations with the Standard Microfacies Types of Wilson (1975); 3: SMF 7; 6: SMF 10; 7, 11, 12, 14: SMF 12; 8, 9, 10: SMF 9; 13: SMF 11, 13; 15: SMF 7. The thickness of the section is 73 m

Bioassociation	Algal communities	Interpretation			
		Open platform		Shoal environments in agitated waters	Organic buildups
		restricted circulation	open circulation		
Top of the section					
15	8				x
10	16	x			
8			x		
6			x		
10	16	x			
13	15			x	
10	16	x			
13	15			x	
10	16	x			
13	15			x	
10	16	x			
13	15			x	
10	16	x			
6			x		
10	16	x			
12	19			x	
10	16	x			
13	15			x	
12	19			x	
5	11	x			
13	15			x	
10	16	x			
12	19			x	
13	15			x	
10	16	x			
12	19			x	
13	15			x	
12	19			x	
8			x		
6			x		
12	19			x	

Table 7 (continuation)

Bioasso-ciation	Algal communities	Interpretation			
		Open platform		Shoal environ-ments in agi-tated waters	Organic buildups
		restricted circulation	open circulation		
10	16	x			
8			x		
6			x		
8			x		
6			x		
11			x		
13	15			x	
	12	x			
6			x		
12	19			x	
6	12	x			
12	19			x	
6	12	x			
12	19			x	
10	16	x			
11				x	
10	16	x			
11				x	

(b) encrusting algae such as encrusting corallines, and (c) algal nodules (including rhodoliths and cyanophytic oncolites).

Some of the Permian algal sediments may be compared with these recent types: Maerl is represented by the *Epimastopora* community, and by other dasyclad communities as well by the *Gymnocodium/Permocalculus* community of the Upper Permian limestones. Encrusting algae are known with the *Tubiphytes* and the *Archaeolithophyllum lamellosum* community. The *Osagia*- and *Archaeolithophyllum*-oncoid communities belong to the algal nodules group.

According to Milliman (this vol.) recent algal sediments of the tropics are restricted to those areas, in which the influx of terrigenous sediment is low (middle and particularly outer shelf). This pattern may be in agreement with the general distribution of the Lower Permian algae of the Upper *Pseudoschwagerina* limestone and of the Trogkofel limestone, but the phylloid algal community of the Lower *Pseudoschwagerina* limestones seems to have flourished in the inner, sublitoral marine environment, where strong non-carbonate sedimentation is evident.

A comparison of recent calcareous algae and of the Permian algal floras seems rather ambiguous because of the problematic systematic position of many Upper Paleozoic algae (see Table 2). Only the dasycladaceans can be used as paleoecological indicators in

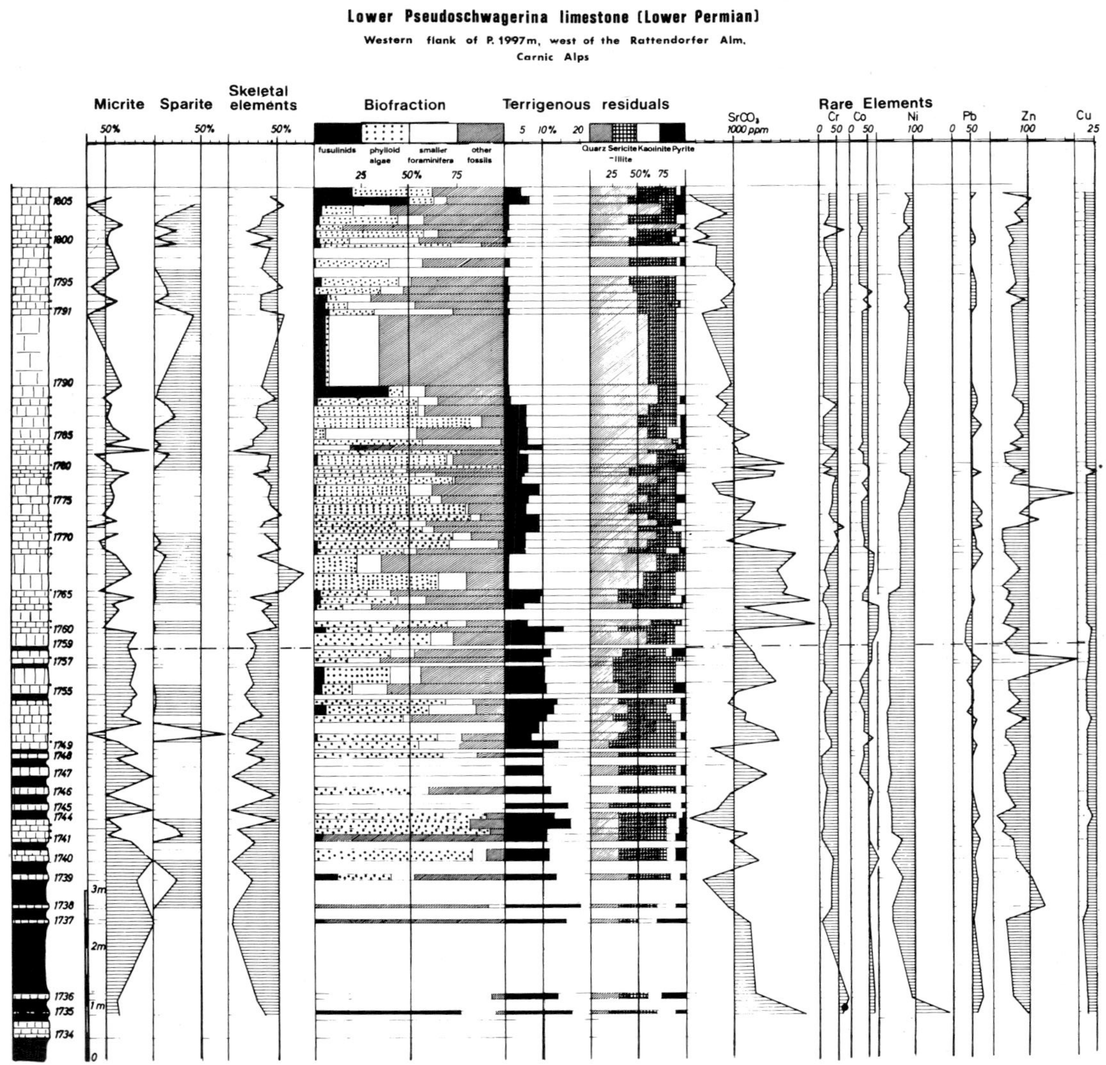

Fig. 3. Inner shelf facies with strongly fluctuating organic and inorganic parameters. Lower *Pseudoschwagerina* limestones, section near P. 1997, Carnic Alps. The distribution of the algae does not seem to be controlled by the cyclically alternating terrigenous influx

a rather general way. Rich autochthonous dasycladacean floras are widely distributed today in coastal, tropical, and subtropical marine environments. Most recent dasycladaceans seem to prefer very shallow, sheltered waters (very often less than 5 m, see Senes, 1967; Klement, 1966). In the Florida Keys area the most profilic growth is found on mud banks (together with sediment-baffling sea-grass), and in open marine shoreline lagoons (together with many codiaceans, and with *Thalassia*). Restricted areas with lacking connections to the open ocean are characterized by rather sparse dasycladacean populations.

Similar distributions may be inferred for the Permian dasyclad communities of different diversity. The "high diverse dasyclad community" seems to have flourished in the open-marine shelf lagoons whereas dasyclad floras with low diversity but high pupulation density may indicate mud bank areas in quiet waters. For these communities a possible transport of dissociated debris of dasyclads (see Elliott, 1968, 1975) can be excluded.

Many authors have given information on the paleogreographical patterns of Upper Paleozoic calcareous algae. Using the valuable contributions by Achauer (1969), Crowley (1969), Freeman (1964), Frost (1975), Heckel and Cocke (1969), Kendall (1969), Kotila (1973), Laporte (1962), Malek-Aslani (1970), McCrone (1963), Ross (1968, 1969, 1971), Schenk (1967), Stevens (1971), Tillman (1971), Toomey (1969, 1974, 1976), Toomey and Winland (1973), Troell (1969), J.L. Wilson (1967), and Wray (1964, 1971, this vol.) some generalizations can be made concerning the distribution of the algal communities which have been recognized in the Lower Permian shelf environments of the Southern Alps:

The phylloid algal community is found within algal mounds in different parts of the shelf (near-coast position, see Stevens, 1971; Kotila, 1973; as barriers between an inner-shelf, more restricted facies and an open-marine outer-shelf facies, see Tomey and Winland, 1973), shelf-edge (see J.L. Wilson, 1974; Malek-Aslani, 1970), and in shallow basins (see Tillman, 1971; Wilson, 1975). Generally this community seems to be widely distributed in the clear water, photic zone with sheltered as well with more open conditions. In the Southern Alps the community is known from the restricted platform environments of the inner shelf only.

The *Tubiphytes* community seems to be predominately distributed within true organic buildups [reef-wall facies (Malek-Aslani, 1970); sponge-algal facies (Achauer, 1969; Kendall, 1969)]. In the Trogkofel formation of the Southern Alps this boundstone facies is developed on exposed crests and (possibly) on seawards flanks of large carbonate buildups. The shelf-edge position of this community recognized by Wilson (1975) for many Upper Paleozoic *Tubiphytes* reefs in the United States cannot be seen in the distribution of all Trogkofel buildups; some of these reefs may have been formed in the inner or middle part of the shelf platform (Sexten Dolomites, see Fig. 1).

The oncoid assemblages have been used widely for paleoecological indications such as remarks on water depth and the relative nearness to the coast (see Toomey, 1974; Troell, 1969; Tillman, 1971). Most authors agree in a deposition area close to the shorelines, quiet waters with generally slow sedimentation, and sometimes possibly brackish environments. Differences may be inferred for the *Osagia*-oncoids, and for the *Archaeolithophyllum*-oncoids (see Toomey, 1974) with regard to water movements and main areas of oncoid development. According to the situation found in the Lower

Table 8. Distribution patterns of Upper Paleozoic shallow-water biota

	Inner shelf		Outer shelf		
	Restricted platforms				
	Restricted shelf-lagoons	Restricted shoals	Open-marine platforms	Shoals in agitated waters, platform-edge	Organic buildups
Diversity of algal communities	generally low	low	low and high	generally high	generally low
Characteristical algal communities	*Girvanella* C.	low-diverse dasyclad C. *Anthracoporella* C.	high-diverse dasyclad C.	algal sporen C.	*Tubiphytes* C. phylloid algal C.
Widely distributed algal communities	*Osagia*-oncoid C. *Archaeolithophyllum*-oncoid C.	*Epimastopora* C.	*Osagia*-oncoid C. *Archaeolithophyllum*-oncoid C. *Epimastopora* C. *Neoanchicodium* C.	*Osagia*-oncoid C. high-diverse dasyclad C. *Neoanchicodium* C.	*Tubiphytes* C.
Relatively rare algal communities		micritic oncoid C. *Ramovsia* C. *Solenopora* C.		*Komia* C.	*Archaeolithophyllum*/*Tubiphytes* C.
Dominant limestone textures	mudstone wackestone	wackestone packstone bindstone	wackestone floatstone grainstone	grainstone rudstone wackestone	bindstone or bafflestone resp.
Other biota:					
Encrusting forams	abundant	common	common	present	common
Paleotextulariids	present	present	common	abundant	rare
Nodosariids	present	present	present	common	rare
Fusulinids	common	rare	abundant	rare	rare
Corals	rare	rare	common	rare	rare
Bryozoans	rare	rare	rare	rare	common
Brachiopods	rare	rare	common	rare	common
Gastropods	rare	common	common	common	rare
Echinoderms	rare	present	common	present	common

Permian of the Southern Alps *Osagia*-oncoids as well as *Archaeolithophyllum*-oncoids seem to have had a wide distribution in near-coast environments (like the Grenzland group), but also in shoal environments in agitated waters of the outer shelf area (bedded Trogkofel limestones of Forni Avoltri and the Karawanken Mountains).

The dasyclad communities of the Upper Paleozoic cyclic shelf deposits of the United States have been interpretated as biotas of very shallow, well-lit shoal waters in a back-reef or back-bank lagoon (see Wilson, 1975). Similar conditions may be supposed for the dasyclad communities, for the Algal sporen assemblage, and perhaps for the *Epimastopora* community of the Upper *Pseudoschwagerina* limestones and of the Trogkofel formation of the Southern Alps.

Table 8 shows some of the most important biotic features which can be recognized in Upper Carboniferous and in Lower Permian shelf- and shelf-edge environments. More studies of quantitative distributional data of the lower-level taxa of the Upper Paleozoic algae together with a careful re-evaluation are needed for the improvement of these models.

Acknowledgments. Sincere thanks are due to the Deutsche Forschungsgemeinschaft, which has assisted the paleontological studies of the author in the Southern Alps by grants (Projects Fl 42/19 and Fl 42/22).

34.3 Bibliography

Accordi, B.: Il Permiano superiore delle Dolomiti e le sue relazioni con l'orogenesi ercinica. Estud. Geol. **15**, 1–17 (1959).

Achauer, C.W.: Origin of Capitan Formation, Guadalupe Mountains, New Mexico and Texas. Am. Ass. Petrol. Geol. Bull. **53**, 2314–2323 (1969).

Buggisch, W., Flügel, E., Leitz, F., Tietz, G.-F.: Die fazielle und paläogeographische Entwicklung im Perm der Karnischen Alpen und in den Randgebieten. Geol. Rundschau **65**, 649–690 (1976).

Cheetham, A.H., Hazel, J.E.: Binary (presence-absence) similarities coefficients. J. Paleontol. **43**, II/1, 1130–1136 (1969).

Crowley, D.J.: Algal-Bank Complex in Wyandotte Limestone (Late Pennsylvanian) in Eastern Kansas. State Geol. Surv. Kansas, Bull. **198**, 1–52 (1969).

Dunham, R.J.: Capitan Reef, New Mexico and Texas: Facts and Questions to Aid Interpretation and Group Discussion. Permian Basin Sec., Soc. Econ. Paleontol. Mineral. Publication 72–14 (1972).

Elliott, G.F.: Permian to Palaeocene Calcareous Algae (Dasycladaceae) of the Middle East. Bull. Brit. Mus. (Nat. Hist.) Geol., Suppl. **4**, 1–111 (1968).

Elliott, G.F.: Transported algae as indicators of different marine habitats in the English middle Jurassic. Palaeontology **18**/2, 351–366 (1975).

Flügel, E.: Algen aus dem Perm der Karnischen Alpen. Carinthia II, Sonderh. **25**, 76 p. (1966).

Flügel, E.: Fazies-Interpretation der unterpermischen Sedimente in den Karnischen Alpen. Carinthia II, **164/84**, 43–62 (1975).

Freeman, T.: Algal limestone of the Marble Falls Formation (Lower Pennsylvanian), Central Texas. Geol. Soc. Am. Bull. **75**, 669–676 (1964).

Frost, J.G.: Winterset algal-bank complex, Pennsylvanian, Eastern Kansas. Am. Ass. Petrol. Geol. Bull. **59**/2, 265–291 (1975).

Heckel, P.H., Cocke, J.M.: Phylloid algal-mound complexes in outcropping Upper Pennsylvanian rocks of Mid-Continent. Am. Ass. Petrol. Geol. Bull. **53**, 1058–1074 (1969).

Homann, W.: Unter- und tiefmittelpermische Kalkalgen aus den Rattendorfer Schichten, dem Trogkofelkalk und dem Treßdorfer Kalk der Karnischen Alpen (Österreich). Senckenbergiana Lethaea **53**, 3/4, 135–313 (1972).

Kendall, Ch.G.S.C.: An environmental re-interpretation of the Permian evaporite/carbonate shelf sediments of the Guadalupe-Mountains. Geol. Soc. Am. Bull. **80**, 2503–2526 (1969).

Klement, K.W.: Studies on the ecological distribution of lime-secreting and sediment-trapping algae in reefs and associated environments. N. Jb. Geol Paläontol. Abh. **125** (1966).

Kochansky-Devidé, V., Buser, St., Cajhen, J., Ramovs, A.: Podroben Profil Skozi Trogkofelske Plasti v Potoku Kosnutnik v Karavankah. (Detailliertes Profil im Perm der Trogkofelschichten am Bache Kosutnik in den Karawanken). Razprave, Diss. XVI/4, Ljubljana (1973).

Kotila, D.A.: Algae and Paleoecology of Algal and Related Facies. Morrow Formation, Northeastern Oklahoma. Diss. Univ. Oklahoma (1973).

Laporte, L.F.: Paleoecology of the Cottonwood limestone (Permian), northern Midcontinent. Geol. Soc. Am. Bull. **73** (1962).

Malek-Aslani, M.: Lower Wolfcampian Reef in Kemnitz Field. Lea County, New Mexico. Am. Ass. Petrol. Geol. Bull. **54**, 2317–2335 (1970).

McCrone, A.W.: Paleoecology and biostratigraphy of the Red Eagle Cyclothem (Lower Permian) in Kansas. State Geol. Surv. Kansas Bull. **164** (1963).

Milliman, J.D.: Marine Carbonates. Berlin–Heidelberg–New York: Springer 1974.

Pia, J.: Die Anpassungsformen der Kalkalgen. Palaeobiol. **1**, 211–224 (1928).

Praturlon, A.: Dasycladaceae from Upper Permian of the Dolomites (Italy). Geol. Romana **2**, 119–150 (1963).

Ramovs, A.: Biostratigraphic der klastischen Entwicklung der Trogkofelstufe in den Karawanken und Nachbargebieten. N. Jb. Geol. Paläontol. Abh. **131**, 72–77 (1968).

Ramovs, A., Kochansky-Devidé, V.: Die Entwicklung des Jungpaläozoikums in der Umgebung von Ortnek in Unterkrain. Rapr. Diss. Slov. Akad. **8**, 323–389 (1965).

Riding, R.: Problems of Affinity in Paleozoic Calcareous Algae. This vol. (1976).

Ross, C.A.: Paleoecology of Fusulinaceans. Proc. Intern. Paleontol. Union Intern. Geol. Congr. 23, Sess. 2, 5, 301–318 (1968).

Ross, C.A.: Paleoecology of *Triticites* and *Dunbarinella* in Upper Pennsylvanian strata of Texas. J. Paleontol. **43**, 298–311 (1969).

Ross, C.A.: Paleoecology of Late Pennsylvanian Fusulinids (Foraminiferida), Texas. C.R. 6th Congr. Intern. Strat. Geol. Carbonifere, Sheffield, 1429–1440 (1971).

Schenk, P.E.: Facies and phases of the Altamont limestone and Megacyclothem (Pennsylvanian), Iowa to Oklahoma. Bull. Geol. Soc. Am. **78**, 1369–1384 (1967).

Senes, J.: Répartition bathymétrique des algues fossilisables ou mediterranées. Geol. Sbornik **18**, 141–150 (1967).

Stevens, C.H.: Paleoecologic implications of early Permian fossil communities in Eastern Nevada and Western Utah. Bull. Geol. Soc. Am. **77**, 1121–1130 (1966).

Stevens, C.H.: Distribution and Diversity of Pennsylvanian marine faunas relative to water depth and distance from shore. Lethaia **4**, 403–412 (1971).

Tillman, R.W.: Petrology and Palaeoenvironments, Robinson Member, Minturn Formation (Desmoinesian), Eagle Basin, Colorado. Am. Ass. Petrol. Geol. Bull. **55**/4, 593–620 (1971).

Toomey, D.F.: The biota of the Pennsylvanian (Virgilian) Leavenworth Limestone, midcontinent region. Part 2. Distribution of Algae. J. Paleontol. **43**, 1313–1330 (1969).

Toomey, D.F.: Algally coated-grains from the Leavenworth Limestone (U. Pennsylv., Midcontinent Region, USA). N. Jb. Geol. Paläontol. Mh. **1974**, 175–191.

Toomey, D.F.: A Permian Marine Plant Controlled Community, Northern Franklin Mountains, West Texas. El Paso Geol. Soc. Symp. Franklin Mountains, 67–76 (1976).

Toomey, D.F., Winland, H.D.: Rock and biotic facies associated with Middle Pennsylvanian (Desmoinesian) Algal Buildup, Nena Lucia Field, Nolan County, Texas. Am. Ass. Petrol. Geol. Bull. **57**/6, 1053–1074 (1973).

Troell, A.R.: Depositional Facies of Toronto Limestone Member (Oread Limestone, Pennsylvanian), Subsurface Marker Unit in Kansas. Bull. State Geol. Surv. Kansas **197**, 1–29 (1969).

Wilson, J.L.: Cyclic and reciprocal sedimentation in Virgilian strata of Southern New Mexico. Geol. Soc. Am. Bull. **78**, 805–818 (1967).
Wilson, J.L.: Characteristics of carbonate platform margins. Ass. Petrol. Geol. Bull. **58**/5, 810–824 (1974).
Wilson, J.L.: Carbonate Facies in Geologic History. Berlin–Heidelberg–New York: Springer 1975.
Wray, J.L.: Archaeolithophyllum, an abundant calcareous alga in limestones of the Lansing group (Pennsylvanian), southeastern Kansas. Kansas Geol. Surv. Bull. **170**, (1964).
Wray, J.L.: Algae in Reefs through Time. Proc. N. Am. Paleontol. Conv., 1969, Part J., 1358–1373 (1971).
Wray, J.L., James, N.P., Ginsburg, R.N.: The puzzling Paleozoic phylloid algae: a Holocene answer in Squamariacean calcareous red algae. Am. Ass. Petrol. Geol. Ann. Meetings Abstr. Dallas **2**, 82–83 (1975).

35 Devonian Calcareous Algae from the Dinant and Namur Basins, Belgium

H.H. TSIEN and E. DRICOT, Louvain-la-Neuve, Belgium

35.1 Introduction

The area of Devonian outcrops in Belgium is one of those important, relatively small basins which are very complex and from which the ecological variations of the organism and the reefs can be easily deduced. Reef development begins in the Couvinian and continues until the Upper Frasnian. Five types of reefs can be distinguished in the Devonian reef complexes of Belgium (Figs. 1 and 2, R1 to R5). Most of the reefs are developed during relatively stable phases. Mud mounds are developed during transgressive phases (Tsien, 1974, 1975).

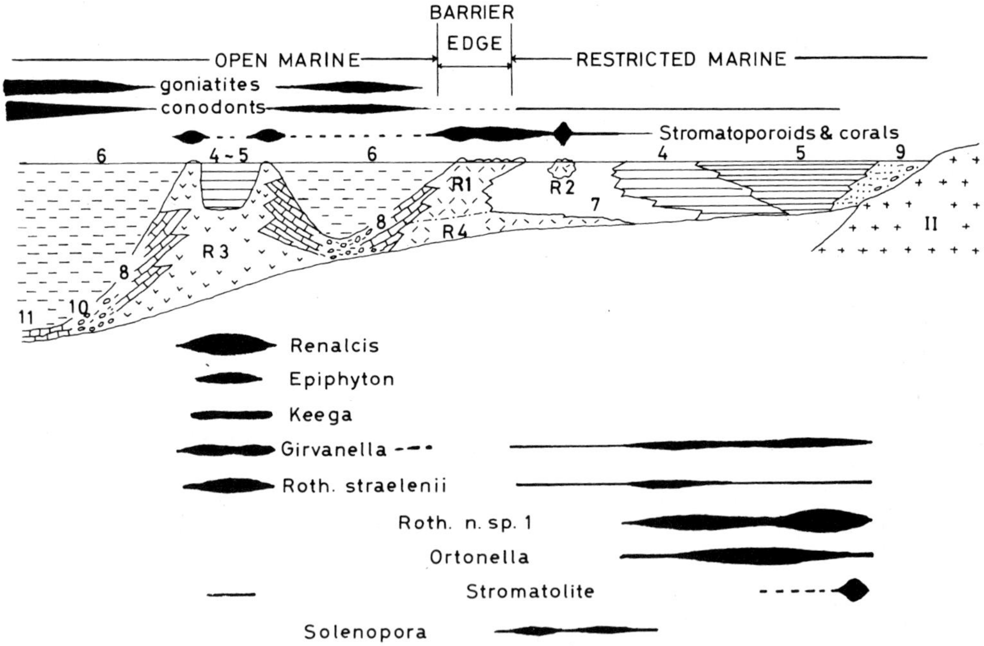

Fig. 1. Environmental distribution of calcareous algae and their relative abundance in the reef complex developed during a relatively stable phase. *R1:* barrier reef; *R2:* dome or patch reef; *R3:* biohermal complex; *R4:* biostrome; *7:* back-reef facies; *4:* sub-lagoonal facies; *5:* lagoonal facies. (Simplified model of the reef complex after Tsien, 1974)

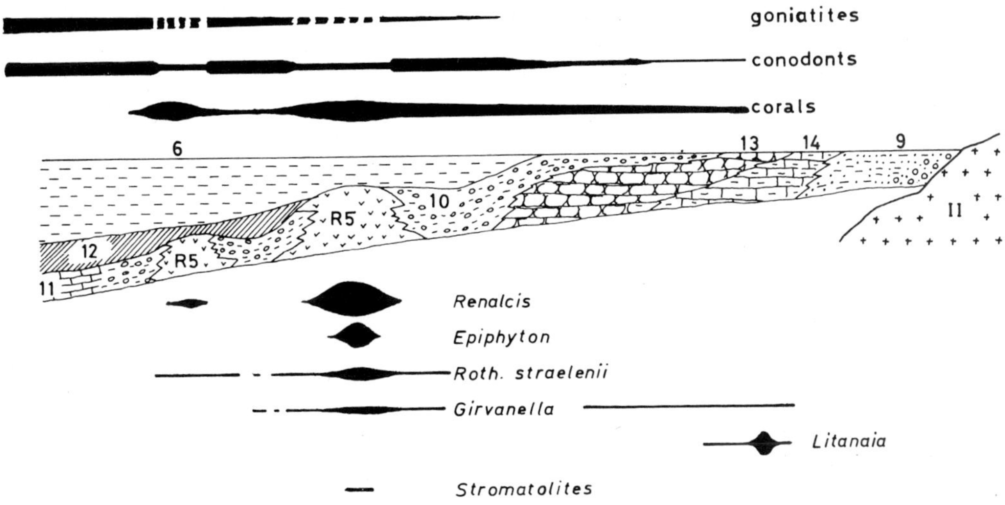

Fig. 2. Environmental distribution of calcareous algae and their relative abundance in the reef complex developed during the transgressive phase. *R5:* Mud mound. (Simplified model of the reef complex after Tsien, 1974)

Bioherm complexes (R3) are always encountered where subsidence is relatively rapid. Biostromes (R4) occur on a stable shelf and presumed barrier reefs (R1) are located at the border of the shelf, with dome reefs or patch reefs (R2) occurring behind the barriers. During development of these reefs open marine conditions occurred in the basin to the south, and at the same time restricted conditions occurred on the shelf to the north (Fig. 1). During transgressive phases, as long as subsidence is not too rapid, mud mounds tended to develop.

Calcareous algae are rare in Couvinian time; but they are common in Givetian and Frasnian reef complexes, both in the reef facies of the mud mounds (R5) and bioherm complexes (R3), and also in the sublagoonal or lagoonal facies.

35.2 Roles of Calcareous Algae

Algae, in some cases, are the principal frame builders and in other cases, are subordinate to stromatoporoids and corals in the reef complexes (Tsien, 1974). During the relatively stable phase, in bioherm complexes (R3), algae occur in protected areas or in the central lagoon. The principal role of the algae is to trap sediments to bind organic debris, and probably to secrete carbonate between the larger organisms. They are subordinate to stromatoporoids and corals.

During the transgressive phase, in the mud mound (R5), the main organisms are algae, stromatactis, and corals. Stromatactics forms a considerable proportion of the mounds and hence represents some sort of constructive organism, here interpreted to be of algal origin. All these organisms can trap and bind sediments. In these mounds, algae (*Renalcis, Epiphyton, Rothpletzella*) are the principal frame builders.

Algae are rare in R1, R2, R4 reefs; however, some encrusting *Rothpletzella* which binds organic debris, is found in R2 and R4 reefs.

Ortonella, Rothpletzella, Litanaia are the main rock-forming organisms in the sub-lagoonal facies and lagoonal facies.

35.3 Distribution

Calcareous algae do not occur in all facies of the reef complexes. They are ecologically restricted. Most can be regarded as indicators of particular environments. In Belgium, they are abundant in the bioherm complexes (R3), mud mounds (R5), and the sub-lagoonal and lagoonal facies. They are occasionally found in the back-reef facies, and are rare in the R1, R2, and R4 reefs. The growth forms (outlined below) vary from simple, small, floating bodies to large, columnar colonies in both the reef facies and sub-lagoonal or lagoonal facies. As environmental indicators, the algal assemblages are probably more significant than the algal growth forms. The *Renalcis, Epiphyton, Rothpletzella, "Keega"*, and *Girvanella* assemblage is indicative of the reef facies. The *Rothpletzella, Ortonella, Girvanella* assemblage is indicative of the shallow sublagoonal facies. *Ortonella,* together with stromatolites, is indicative of the shallow lagoon facies with higher salinity. The distribution and relative abundance of the algae in the reef complexes are summarized in Figs. 1 and 2.

35.3.1 Cyanophyceae or Chlorophyceae

1. *Renalcis* Vologdin, 1932

   Renalcis is one of the most important and distinctive reef-building algae, especially in the Upper Frasnian mud mounds (R5). Simple small floating colonies are abundant around the coral debris encrusted by *Rothpletzella* in the bioherm complexes (R3) and mud mounds (R5). Columnar growth forms are found in lagoons of the bioherm complexes (R3). Significantly large masses of *Renalcis* are found in the protected part of the mud mound; occasionally, *Renalcis* is associated with *Epiphyton* in rocks with fenestral texturies of the mud mounds.

   Renalcis devonicus Johnson, 1964
   R. turbitus Wray, 1967
   R. sp.

Pl. 1/1. *Girvanella ducii*, x 11.8, Gi2β, Mont d'Haurs, 799 (Louvain) ▶

Pl. 1/2. *Rothpletzella* sp. 1, x 4.2, Gi2β, Alvaux 14, 12506 I (Louvain)

Pl. 1/3. *Ortonella* sp., x 6.3, Gi2β, Alvaux, 10151 (Louvain)

Pl. 1/4. *Litanaia* sp., x 6.3, Fr2, Lavoir, 10152 (Louvain)

Pl. 1/5. *Rothpletzella straeleni*, x 11.8, Fr2α, Solre, St.-Géry, 2878, 681A' (Louvain)

Pl. 1/6. *Renalcis devonicus*, x 8.5, Fr2β, Car. Beau Château, Senzeille, 6842 (Louvain)

Pl. 1/7. *Epiphyton* sp., x 8.5, Fr2α, Car. des Croisettes, Vodecée, 13710 (Louvain)

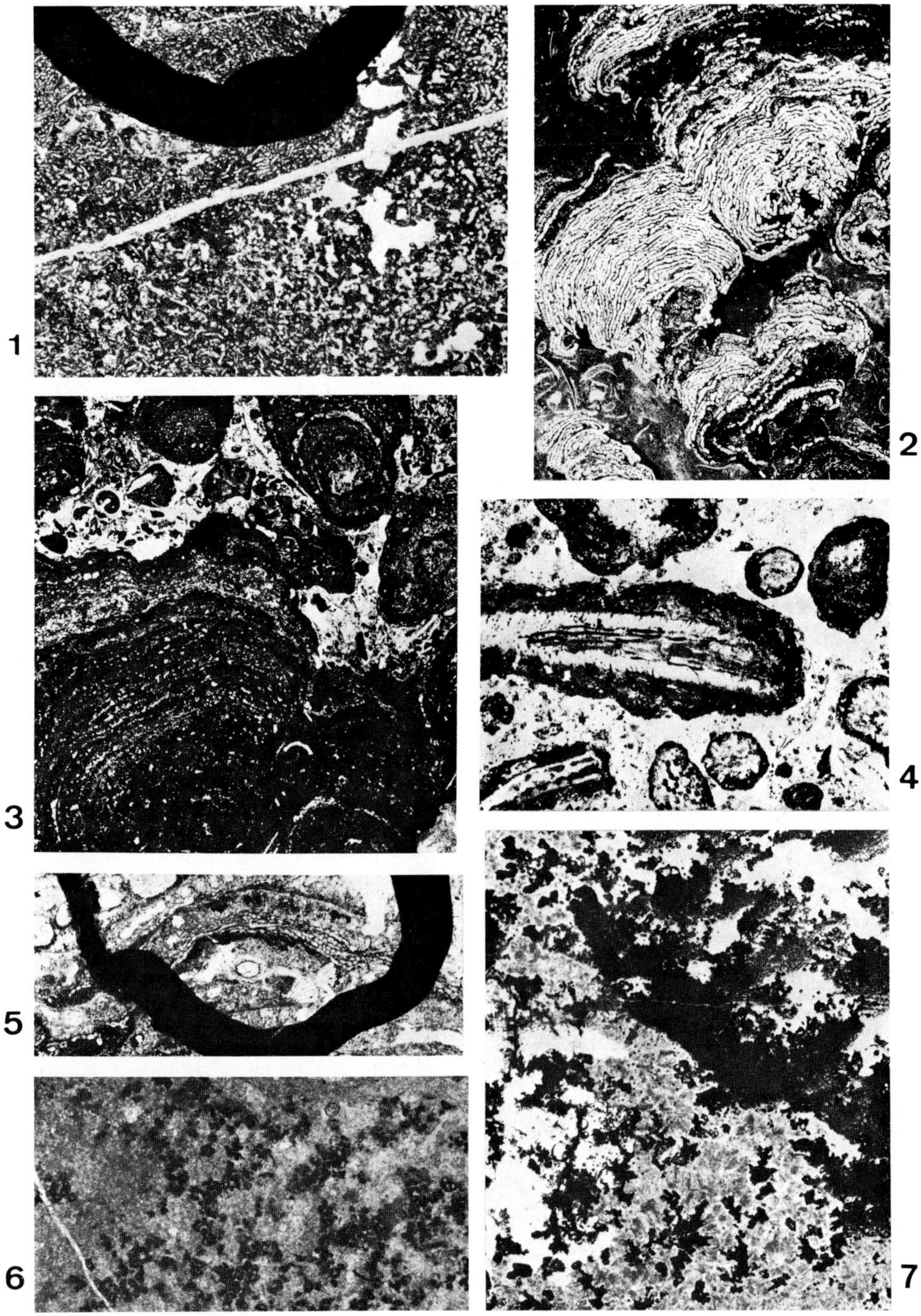

Plate 1. Legend see opposite page

2. *Epiphyton* Bornemann, 1886

   Epiphyton is generally associated with *Renalcis* and common in the mud mounds (R5).

3. *Rothpletzella* Wood, 1948 (= *Sphaerocodium* auct.)

   Rothpletzella is an important reef-building alga in more protected environments. Growth forms vary considerably in shape and size, ranging from encrustations of only a few filament layers to masses of several centimeters thick. Nodular and columnar forms are common in the lagoonal facies. Successive *Rothpletzella* encrustations of corals or stromatoporoids are common in the reef facies.

   Rothpletzella straelenii (Lecompte, 1936)
   R. sp. 1
   R. sp. 2

4. *Girvanella* Nicholson and Etheridge, 1880

   Girvanella is very common in the reef facies and it also occurs in the restricted carbonate facies and open marine shale facies. In the reef facies it coats organic fragments and also occurs in spherical growth forms. In the sublagoonal facies, it forms thin lamellar or spheroidal structures (oncolite). Only small – up to 3 mm – spherical forms occur sporadically in the open sea shale facies.

   Girvanella ducii Wethered, 1890
   G. wetheredi Chapman, 1908
   G. nicholsoni (Wethered, 1886)

35.3.2 Rhodophyceae

5. *Solenopora* Dybowski, 1877

   Solenopora is common in the back-reef facies, rare in the sublagoonal facies, and generally forms very thick masses.

6. "*Keega*" Wray, 1967

   Keega is common in the reef facies and forms thick masses (± 2 cm). It encrusts corals, stromatactis, and occasionally sediments in the reefal environment. *Keega* is sometimes difficult to distinguish from certain stromatoporoids, as noted and discussed by Wray and Playford (1970). In the sublagoonal facies, it is rare and occurs as an isolated body. *Keega* has been recently reinterpreted to be a stromatoporoid by Riding (1974) and accepted by Wray (this volume). The reinterpretation by Riding and Wray suggests that *Keega australe*, a type of *Keega*, is not an alga but a stromatoporoid. But the internal structures and external structures of our *Keega* specimens, such as assymmetric pattern of growth line, finer structures, encrusting nature, and growth forms etc., show that it is not a *Stachyodes*. Therefore, we still keep "*Keega*" in the algal group and use "*Keega*" as an informal name.

35.3.3 Chlorophyceae

7. *Ortonella* Garwood, 1914

   Nodular (oncolite), columnar, and laminar forms are common in the sublagoonal facies. In the lagoonal facies, they often alternate with stromatolites.

8. *Litanaia* Maslov, 1956

   Litanaia is common in the sublagoonal facies and sometimes forming rock beds.

35.4 Stratigraphic Value

Calcareous algae have long geologic ranges (Wray, 1971, Fig. 20); consequently, they have limited value for age determinations (Wray, 1971). But as Chuvashov (1967) stated "the range of each species or genus in each region is dependent not on the duration of their natural existence but on the duration in the region of conditions favourable for their life activity" and "Dating of the deposits is only possible with reference to the whole assemblage and to the historical development of each particular part of the basin". This point of view is also held by the present authors.

Some algae have stratigraphic value only within well-studied sedimentary basins. In the Dinant and Namur basins, the fauna in the lagoonal and sublagoonal facies is poor but in contrast algae are abundant. In the lagoonal and sublagoonal facies, only algae can be used for biostratigraphic correlations, as has been demonstrated by Tsien (1973). The following species are useful for biostratigraphic correlations in the Dinant and Namur basins:

Rothpletzella sp. 1; Upper Middle Givetian
Rothpletzella sp. 2; Upper Frasnian
Renalcis sp.; Upper Frasnian
Litanaia (when abundant): Upper Frasnian

Acknowledgment. We are grateful to Dr. Eric W. Mountjoy for reading the manuscript and for many valuable comments.

35.5 Bibliography

Bornemann, J.G.: Die Versteinerungen des cambrischen Schichtensystems der Insel Sardinien. Nova Acta Kgl. Leop. Carol. Dtsch. Akad. Naturforsch. **41**, 1–83 (1886).

Chapman, F.: On the relationship of the genus *Girvanella* and its occurrence in the Silurian limestones of Victoria. Rept. Austral. Ass. Adv. Sci. **1907**, 377–386 (1908).

Chuvashov, B.I.: Algae from the Devonian, Carboniferous and Lower Permian of the Middle and Southern Urals, their ecology and stratigraphic value. In: Fossil Algae of the USSR.Vozzhennikova, T.F., Glezer, Z.I., Jouze, A.P., Saks, V.N., Chechoukova-Poretskaia, V.S. (eds.). Akad. Nauk, SSSR, Sibirsk. Otd.; Inst. Geol. Geophys. Izd. Nauka, Moscou, 1967. (In russian). Translated by NLL, pp. 206–214.

Dybowski, V.B.: Die Chaetetiden der Ost-Baltischen Silurformation. Verh. K. Russ. min. Gesellsch., T. 14, 1877, 134 pp.
Garwood, E.: Some new rock-building organisms from the lower Carboniferous beds of Westmorland. Geol. Mag., Decade 6, vol. 1, no. 6, pp. 265–271 (1914).
Johnson, J.H.: Lower devonian algae and encrusting foraminifera from New South Wales. J. Paleontol. **38**, 98–108 (1964).
Lecompte, M.: Contribution à l'étude des "récifs" du Frasnien de la Belgique. Mém. Inst. Géol. Louvain **10**, 31–112 (1936).
Maslov, V.P.: Les algues calcaires fossiles de l'URSS (in Russian). Trudy inst. geol. Nauk, SSSR **160**, 301 pp. (1956). Trad. francaise BRGM no. 3517.
Nicholson, H.A., Etheridge, R.: A monograph of the Silurian fossils of the Girvan district, Ayrshire, Part 1. Scotland geol. Surv. Mem. **23**, 23 (1880).
Riding, R.: The Devonian genus *Keega* (Algae) reinterpreted as stromatoporoid basal layer. Palaeontology **17**, 565–577 (1974).
Tsien, H.H.: Polypiers et algaes. In: Le Sondage de Boussu. Dejonghe, L. (ed.). Serv. Géol. Belg., Prof. Pap. no. 3, 1973, pp. 86–94.
Tsien, H.H.: Paleoecology of Middle Devonian and Frasnian in Belgium. Intern. Symp. Belg. Micropaleont. Limits, Namur, Pub. no. 12, 1974, 53 pp.
Tsien, H.H.: Introduction to the Devonian reef develpment in Belgium. In: Guide Book of Excursion C, Second Intern. Symp. Cnidarian Fossils, Paris, Serv. Geol. Belg., 1975.
Vologdin, A.G.: Archaeocyathids of Siberia , Vol. 2. Moscou–Leningrad: State Scient. Techn. Explorat. Publ., 1932 (in Russian).
Wethered, E.: On the structure and organisms of the lower limestone shales, carboniferous limestone and upper limestone of the forest of Dean. Geol. Mag., N.S., Decade 3, vol. 3, no. 12, 1886, pp. 529–540.
Wethered, E.: On the occurrence of the genus *Girvanella* in oolitic rocks, and remarks on oolitic structure. Quart. J. Geol. Soc. **46**, 270–293 (1890).
Wood, A.: "Sphaerocodium" a misinterpreted fossil from the Wenlock limestone. Proc. Geol Ass. **59**, 9–22 (1948).
Wray, J.L.: Upper Devonian calcareous algae from the Canning basin, Western Australia. Prof. Contr. Colorado School Mines **3**, 1–76 (1967).
Wray, J.L.: Algae in reefs through time. Proc. N. Am. Pal. Convention, 1358–1373 (1971).
Wray, J.L., Playford, P.E.: Some occurrences of Devonian reef-building algae in Alberta. Bull. Canad. Petrol. Geol. **18**, 544–555 (1970).

36 Some Ordovician and Silurian Algae from Selected Areas of the Southwestern United States

D.F. TOOMEY and D. LeMONE, Texas, U.S.A.

36.1 Introduction

Algae are relatively common biotic constituents in some marine carbonate units of Ordovician and Silurian age in the southwestern United States.

Two types of floral assemblages are conspicuous. The more widespread, and volumetrically dominant, is found associated with algal/sponge organic buildups of Lower Ordovician age. This is a diverse floral assemblage consisting of: (1) varied algal stromatolitic structures, (2) blue-green algae of the genera *Renalcis, Epiphyton*, and the perforating and encrusting *Girvanella*, and (3) the problematical algal form *Nuia*. Most of these algal components aided in binding together the skeletal elements of the organic buildup, and conversely, through the boring and perforating action of *Girvanella*, of converting skeletal and nonskeletal grains to micrite – a function of profound importance. This algal assemblage is characteristic of the Arbuckle and El Paso Groups of southern Oklahoma, and western Texas (Fig. 1).

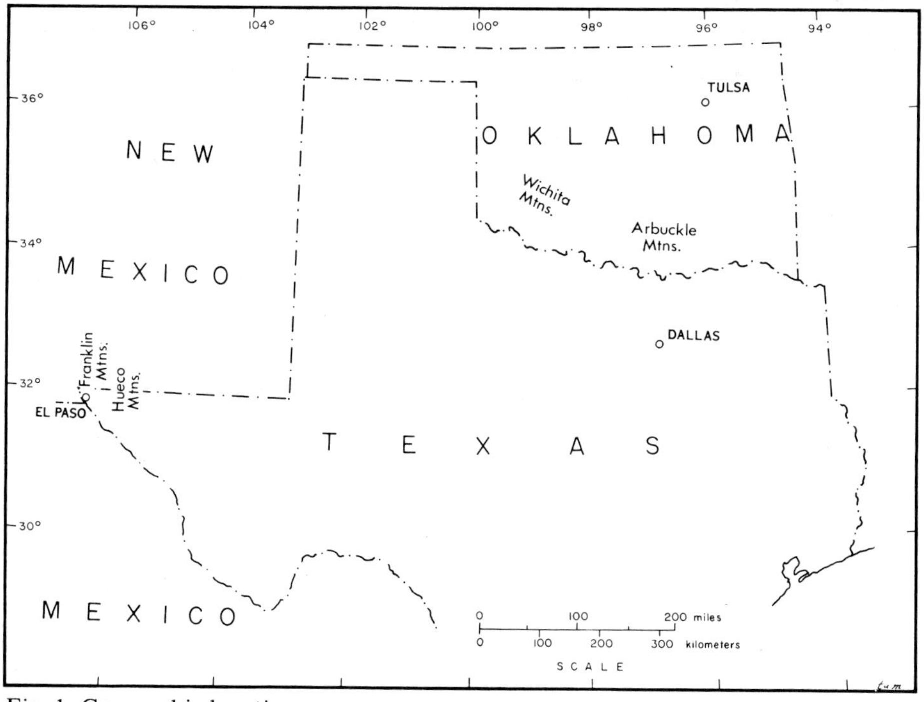

Fig. 1. Geographic location map

The second distinctive Lower Paleozoic algal assemblage found in the southwestern United States, occurs in rocks interpreted to have been laid down on slowly subsiding shallow-marine carbonate platforms. This assemblage consists of algally coated-grains, composed principally of *Girvanella* and *Hedstroemia*, algal/borings, recrystallized nodules ot the red alga *Solenopora*, and tuffs of an *Epiphyton*-like form. Within this assemblage algal stromatolites are not common. This floral assemblage is found in the Middle Ordovician portion of the Simpson Group of southern Oklahoma, and in the upper Ordovician Montoya Group and Silurian Fusselman Dolomite of western Texas.

36.2 Floral Assemblages Present in Lower Ordovician Organic Buildups

Carbonate mounds are a conspicuous sedimentary feature in a number of Lower Ordovician stratigraphic horizons in southern Oklahoma, west Texas, and southern New Mexico. The best studied are those present in the McKelligon Canyon Formation of the southern Franklin Mountains of western Texas (Toomey, 1970). The mounds of the El Paso Group (McKelligon Canyon Formation) vary markedly in size. Mostly, they are rather small features, 3–5 feet in length and 2–5 feet in height. The largest organic buildup found to date is 45 feet long and 19 feet high. Some of the larger mounds may contain conspicuous channels filled with coarse calcarenitic debris; sometimes the channel fill is cross-bedded. The mounds are massive nonbedded muddy carbonates containing a distinctive biota, and enclosed and surrounded by thinner, well-bedded carbonate units containing much clastic skeletal debris.

The organic buildups are composed principally of mud with relatively common skeletal components. The skeletal framework is dominated by three biotic framework builders. These are: (1) *Pulchrilamina spinosa* (Toomey and Ham, 1967), a laminated spinose organism, with probable coelenterate affinities, (2) the siliceous lithistid sponge *Archaeoscyphia annulata* Cullison, and (3) *Calathium* Billings, a problematical organism, that may possibly be a dasyclad alga. It is estimated that the mud-filling that supports

Fig. 2 A–G. Thin section photomicrographs (x 80) of algal or algally altered grains from the organic buildups in the Lower Ordovician, McKelligon Canyon Formation, southern Franklin Mountains, west Texas ▶

(A, B). Fragments of the questionable alga *Nuia siberica* Maslov; note algal corrosion and algal boring effects along the grain periphery of grain in (A) along with the distinctive multilayering of *Nuia*

(C, E). Pelmatozoan ossicles showing a micrite envelope and the effects of algal borings on skeletal grains; (E) (under crossed-nichols) shows ossicle with calcite overgrowth and the upper surface of the ossicle with algal borings filled with dark micrite. Both photomicrographs are from rocks filling in erosional channels within the organic buildup

(D). Marginally altered brachiopod fragment with dark micrite corrosion rim, probably the result of algal borings

(F). Slightly recrystallized algally bored corrosion rim circumscribing an intraclast; tubules appear to be those of the perforating blue-green alga *Girvanella*

(G). An almost totally bored intraclast showing a mass of well-developed *Girvanella* tubules

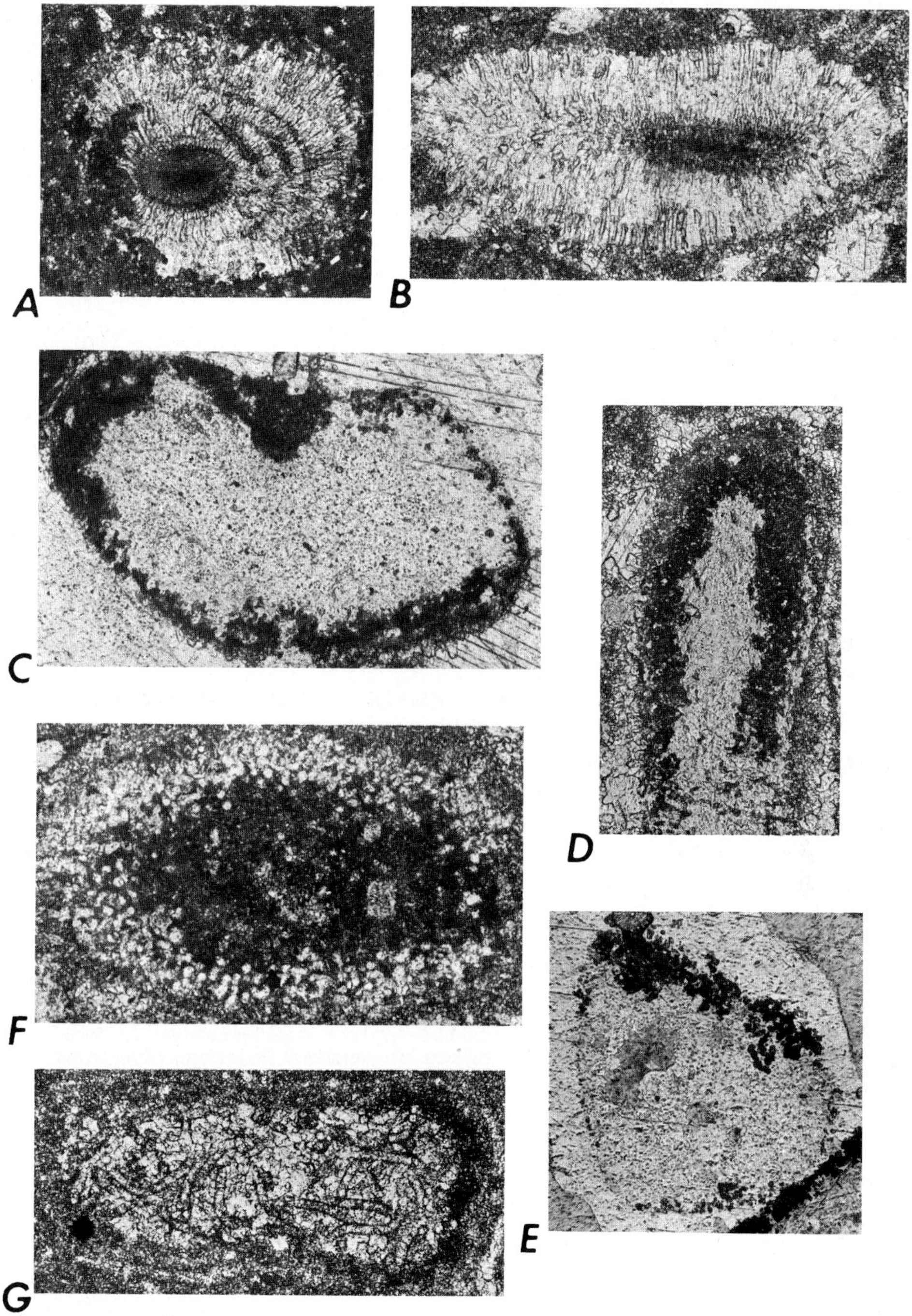

Fig. 2. Legend see opposite page

the three principal framework builders constitutes 60–75% of what may be thought of as a "normal mound".

Petrologically, the mound rock is primarily a skeletal wackestone, that can grade into skeletal packstone and boundstone. Intraclasts are common, as are conspicuous organism burrows. True indication of boundstone is usually present only near the top surface of the mound where much *Pulchrilamina* is found.

Algal stromatolites are most common only near the base of the larger mounds. These are digitate forms with individual "fingers" averaging about 1/2 inch in diameter, and colonies approximately 3 inches in height. *Girvanella* tubules, ranging in diameter from 10 to 16 microns commonly entwine the stromatolitic laminae. The algal laminae are delineated by fine quartz silt grains (Fig. 4D). In addition, marginal grain corrosion. so well developed on many of the skeletal and nonskeletal grains of the mound carbonates, has been shown to be the result of the boring and perforating action of the filamentous blue-green alga *Girvanella* (Klement and Toomey, 1967) (Fig. 2E, G). Particle alteration within the algal corrosion rim leads to micritization of the grain (Fig. 2C–D). Partial recrystallization of the micrite envelope, or corrosion rim, frequently occurs (Fig. 2F). This may indicate prior leaching and thus remobilization of the carbonate matter during the penetrative process of the algal filaments, or during their decay. The work of grain-boring blue-green algae appears to be a major agent in carbonate alteration, destruction, and ultimate micrite formation. It is suggested that local blooms of blue-green algae may be an important factor in the early diagenesis of carbonate sediments.

The organic buildups of the El Paso Group contain abundant remains of a problematical micro-organism referred to *Nuia sibirica* Maslov. Morphologically, this problematical fossil is a straight to slightly curved tubular form with a conspicuous, dark central canal, and a distinctive radial-hyaline wall microstructure (Figs. 2 A–B). The wall is usually single-layered, but multi-layered walls are known. Soviet workers consider *Nuia* to be an alga, although they do not agree on the major algal group to which *Nuia* should be assigned. Toomey and Klement (1966), after study of the El Paso material, concluded that *Nuia* should be considered as a problematical micro-organism, since the peculiarities

---

Fig. 3 A–E. Lower Ordovician algal structures ▶

(A). "Mini-herm" exposed in the Kindblade Limestone, 450 feet above the base, at Mill Creek, Murray County, Arbuckle Mountains of southern Oklahoma; length of chisel 7 1/2 inches. Organic structure built by stromatolitic-forming algae

(B). Gray dolomite containing conspicuous vertically stacked hemispheroid stromatolites (measuring tape in inches). Interareas between stromatolite colonies contain sandy intraclastic debris. Photograph from the upper part of the José Formation, El Paso Group, southern Franklin Mountains, west Texas

(C). A small stromatolitic algal "head" in the lower Kindblade Limestone, Longhorn Mountain, Kiowa County, Wichita Mountains, southwestern Oklahoma; length of hammer is 14 inches

(D). Digitate stromatolite colonies with a height of 16 inches, exposed in the Kindblade Limestone 450 feet above the base on Unap Mountain, Kiowa County, Wichita Mountains, southwestern Oklàhoma. Intraclasts and flat pebble conglomerate surround stromatolite colonies

(E). A dome-shaped pod of digitate stromatolites exposed laterally from the stromatolites shown in (D): length of chisel 7 1/2 inches

Fig. 3. Legend see opposite page

of its internal structure do not permit its assignment within one of the established groups of fossil algae.

Many nonmound horizons within the El Paso Group contain conspicuous intervals with vertically stacked hemispheroid stromatolites (Fig. 3B). These are especially interesting because the interareas between the stromatolite colonies contain much sandy intraclastic debris, and the stromatolites may be capped with thin layers of flat pebble conglomerate.

Approximately 500 miles to the northeast (see Fig. 1) in coeval sediments of the Arbuckle Group, exposed in both the Arbuckle and Wichita Mountains of southern Oklahoma, similar organic buildups occur. The skeletal framework of these mounds contains the same three biotic framework builders: *Pulchrilamina, Archaeoscyphia*, and *Calathium*, as present in the west Texas organic buildups. Morphologically and petrologically the Arbuckle mounds are also identical, although maximum mound size is greater. The largest organic buildup found to date in the Wichita Mountains, has a measured height of 65 feet and a width of 200 feet at the base. The only major difference noted in the Arbuckle organic buildups is the absence of the problematical micro-organism *Nuia* and replacement with an abundance of the blue-green algae *Renalcis* and *Epiphyton* (Figs. 4 B, C).

The Arbuckle mounds contain rather dense growths of *Epiphyton* and *Renalcis*, and although there is no evidence that they were mutually attached, despite attachment of *Renalcis* to the sponge *Archaeoscyphia*, the presence of numerous small voids between the two algal forms suggests that they formed a grain-supported sediment locally. Both *Epiphyton* and *Renalcis* tend to have elongated growth forms adjacent to skeletons and grains. This may reflect morphological adaptation to an attached habit (Riding and Toomey, 1972).

In the Arbuckle organic buildups, especially in the Wichita Mountains, *Epiphyton* and *Renalcis* were not independent frame-builders. They were, however, capable of colonizing the fine-grained mound substrate and produced relatively thin layers that do not

---

Fig. 4 A–E. Algal forms from the southwestern United States ▶

(A). Algally coated-grains (x 3) from the Silurian Fusselman Dolomite, Bean Ranch locality, Hueco Mountains, west Texas. Coated-grains contain abundant *Girvanella* tubules. Matrix contains *Girvanella* along with *Solenopora* and *Epiphyton*-like tuffs (not all seen at this magnification)

(B). *Renalcis* sp. (x 52) showing irregular morphology and branching habit; organic buildup horizon 450 feet above the base of the Kindblade Limestone, Unap Mountain section, Kiowa County, Wichita Mountains, southwestern Oklahoma

(C). Compact forms of *Epiphyton*? sp. (x 47) from the same interval and locality as (B)

(D). Digitate stromatolite colony (x 4) with distinctive laminations. Laminae delineated by quartz silt grains (*tiny white flecks*). Dense border along right side of colony consists of *Girvanella* tubules. Calcarenitic channels on either side of the stromatolite contain shell debris and lithoclasts. Algal structure from Kindblade Limestone, 462 feet above the base at the Highway 77 section, Murray County, Arbuckle Mountains, southern Oklahoma

(E). Scattered algal stromatolite "heads" exposed on outcrop surface in the Kindblade Limestone 450 feet above the base, Joins Ranch section, Murray County, Arbuckle Mountains, southern Oklahoma; length of hammer head 7 1/2 inches

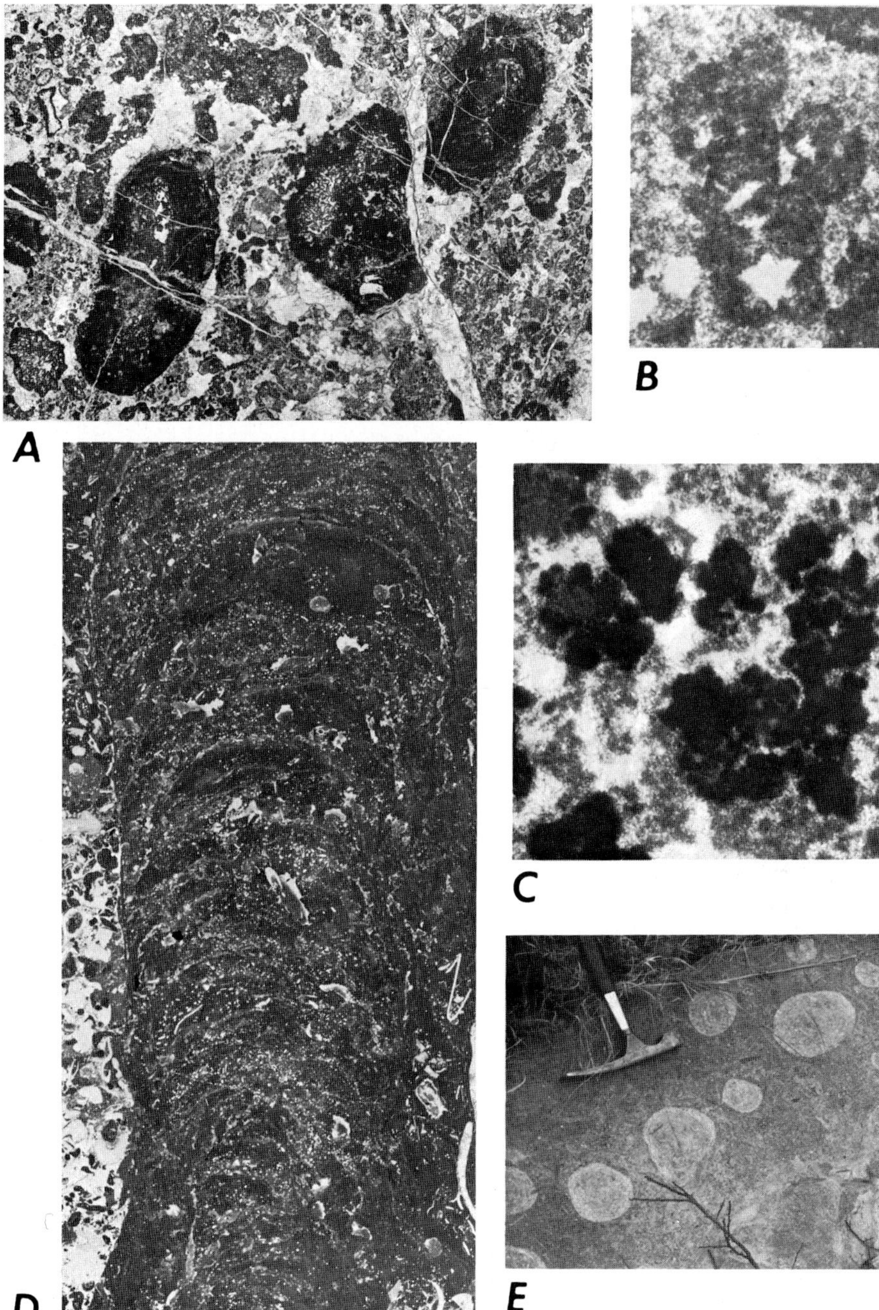

Fig. 4. Legend see opposite page

appear to have raised themselves above the surrounding sediment. Seemingly they preferred the firm substrate offered by other skeletons or grains. Their basic contribution to the mass of the organic buildup forming organisms was relatively minor, for their principal sedimentological role was a binding and sediment stabilizing effect produced by their local veneering of sponges and mound sediments alike.

Small stromatolitic "mini-herms" (Fig. 3A, C) are rather abundant on bedding planes associated with the mound horizon and, scattered stromatolitic algal "heads" are also common within this horizon (Fig. 4E). In addition, there are some Arbuckle mound horizons that contain pods of digitate stromatolite colonies off the main mound masses (Fig. 3E); there are also scattered, very well-developed, individual digitate stromatolite colonies within the same horizon (Fig. 3D).

36.3 Floral Assemblages Present in Middle Ordovician to Silurian Platform Carbonates

Platform carbonates, interpreted to have been deposited on slowly subsiding shallow-marine carbonate shelves, contain limited algal assemblages. Most of these carbonate units have been dolomitized with resultant algal destruction. Some carbonate units, however, have escaped dolomitization and these carry representative algal suites that are dominated by algally coated-grains.

In the Middle Ordovician portion of the Simpson Group (McLish Formation), exposed in the Arbuckle Mountains of southern Oklahoma, algally coated-grains are common. These oncolites contain tubules of the blue-green alga *Girvanella*, associated with tubes of the green alga *Hedstroemia*. These algally coated-grains range in diameter from 8 to 30 mm, with an average diameter of 18 mm (n = 7). The algae have indiscriminately coated, readily available skeletal and nonskeletal grains. Many of the skeletal grains seen in the McLish Formation have pronounced micrite envelopes. As Bathurst (1966) noted, most micrite envelopes are probably the result of the actions of boring algae.

In the Upper Ordovician Upham Dolomite of the Montoya Group, exposed in the Franklin Mountains of western Texas, algae are relatively rare. The Upham Dolomite is a massive, highly burrowed dark carbonate unit containing abundant *Receptaculites* (possible dasyclad alga; see Nitecki, 1972), large gastropods and cephalopods, and abundant corals. The algal assemblage is composed totally of scattered recrystallized nodules thought to be the alga *Solenopora*, many of which appear to be coated with encrusting stromatoporoids.

At the Bean Ranch locality in the Hueco Mountains of western Texas, the Silurian Fusselman sequence is light-colored, thick-bedded dolomite and dolomitic limestone. There are many fossiliferous horizons within the Fusselman of the Hueco Mountains, but dolomitization and recrystallization have obliterated many fossils leaving only ghost impressions. Pelmatozoan fragments are abundant in most of the units, and casts and molds of pentamerid brachiopods are locally abundant. Tabulate corals, especially *Favosites* and *Halysites*, are common, and sponges along with nautiloid cephalopods are also present. The algal assemblage present in the Fusselman consists primarily of algally coated-grains. These coated-grains (Fig. 4A) range in diameter from 3 to 22 mm, and average approximately 8.8 mm in diameter (n = 18). The Fusselman oncolites contain abundant *Girvanella* tubules, along with fragments of the red alga *Solenopora*, and tuffs of an *Epiphyton*-like algal form. Many of the other skeletal and nonskeletal grains seen in thin section have micrite envelopes and show clear evidence of algal (*Girvanella*?) borings. The age of the Fusselman sequence at this location is probably Middle Silurian (McGlasson, 1967).

All of the above algal assemblages, briefly noted from a few selected areas in the southeastern United States, appear to be indigenous to carbonates that were laid down in shallow warm, clear, normal marine environments. Paleogeographically, all of these occurrences can be related to deposition on relatively shallow, slowly subsiding marine shelves.

36.4 Bibliography

Bathurst, R.G.C.: Boring algae, micrite envelopes and lithification of molluscan biosparites. Geol. J. **5**, 15–32 (1966).

Klement, K.W., Toomey, D.F.: Role of the blue-green alga *Girvanella* in skeletal grain destruction and lime-mud formation in the Lower Ordovician of west Texas. J. Sed. Petrol. **37**, 1045–1051 (1967).

McGlasson, E.H.: The Siluro-Devonian of west Texas and southeast New Mexico. In: Symp. Silurian-Devonian Rocks of Oklahoma and Environs. Toomey, D.F. (ed.). Tulsa Geol. Soc. **35**, 148–164 (1967).

Nitecki, M.H.: North American Silurian receptaculitid algae. Fieldiana, Geol. Field Mus. Nat. Hist., Chicago **28**, 120 pp. (1972).

Riding, R., Toomey, D.F.: The sedimentological role of *Epiphyton* and *Renalcis* in Lower Ordovician mounds, southern Oklahoma. J. Paleontol. **46**, 509–519 (1972).

Toomey, D.F.: An unhurried look at a Lower Ordovician mound horizon, southern Franklin Mountains, west Texas. J. Sed. Petrol. **40**, 1318–1334 (1970).

Toomey, D.F., Ham, W.E.: *Pulchrilamina*, a new mound-building organism from Lower Ordovician rocks of west Texas and southern Oklahoma. J. Paleontol. **41**, 981–987 (1967).

Toomey, D.F., Klement, K.W.: A problematical micro-organism from the El Paso Group (Lower Ordovician) of west Texas. J. Paleontol. **40**, 1304–1311 (1966).

37 Early Cambrian Algal Carbonates in Southern Spain

I. ZAMARREÑO, Oviedo, Spain

37.1 Introduction

Algae played an important role in the formation of Cambrian carbonate sequences in southern Spain. The purpose of this paper is to describe these early Cambrian algal carbonates, their distribution, controlling organisms, associated lithofacies, and depositional environments.

For this purpose, an area near Cordoba (Sierra Morena) was selected. The sections studied occur along the roads from Cordoba to Villaviciosa (between km 7.5 and 9), and from Cordoaba to Zafra (between km 2 and 3) (Fig. 1).

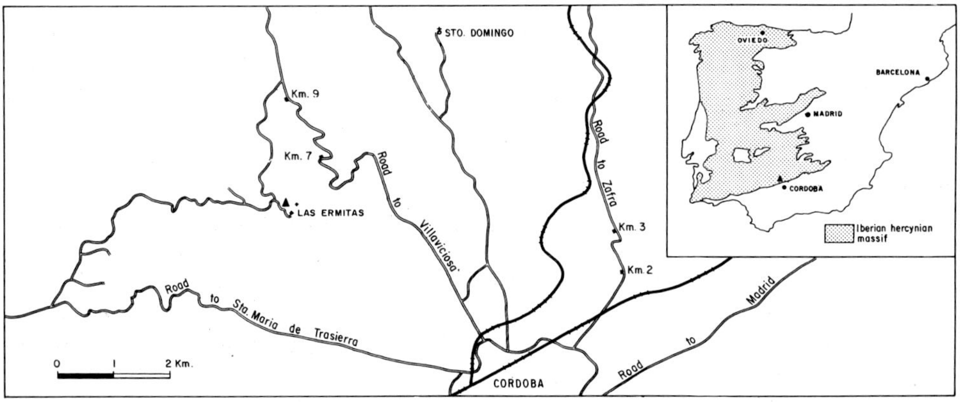

Fig. 1. Index map with localities cited in the text

The carbonate sequence consists mainly of limestones and interbedded siltstones, shales, and sandstones, the two former exhibiting nodular textures. This sequence, several hundred meters thick, rests on an andesitic foundation and is overlaid by a thick succession of terrigenous clastic deposits, of Cambrian age.

Two chief algal carbonate rock types are found: mounds and tabular units. The mounds are discrete carbonate bodies, moundlike or lenslike in shape, embedded in rocks of different lithology (siltstones), referred to as algal mounds in this paper. The tabular units are thick massive carbonate units displaying a clotted fabric and they constitute repetitive units within the section (cyclic algal limestones).

The algal mounds occur at several horizons within a 15 m interval, while the cyclic algal limestones are located higher in the succession, where they spread laterally over long distances.

Due to the tectonic complexity of the area, only incomplete successions are to be found in each locality. Therefore the algal mounds and cyclic algal limestones provide good criteria for correlation.

37.2 Lithofacies and Depositional Environment

Several lithofacies can be recognized in the carbonate sequence studied: (1) algal wackestones, (2) oolitic grainstones, (3) pelletal grainstones, and (4) intraclastic grainstones. This paper deals only with the algal wackestones and oolitic grainstones, which are in fact the most abundant lithofacies.

37.2.1 Algal Wackestones

Algal wackestones are the most widespread lithofacies in this area. Two types of algal assemblages can be distinguished corresponding to the two main limestone morphologies in the field: (1) algal mounds, and (2) cyclic algal limestones.

37.2.1.1 Algal Mounds. The algal mounds crop out as distinct grey, nonlaminated carbonate bodies surrounded by thin-bedded siltstones, green or pink in color. They range in height from 20 cm to 50 cm, and in breadth from 50 cm to 1 m. Most of the algal mounds are completely enveloped by siltstones. In some instances, the base of the mounds rest on an oolite bed, up to 25 cm thick, and they are uniform in size and regularly spaced (Fig. 2B). At one locality (along the road from Cordoba to Zafra) the algal mound is established upon a foundation of oncolitic and intraclastic packstone (Fig. 2A).

Dominant biotic constituents of this lithofacies are the alga *Epiphyton* and the problematical *Renalcis* (Fig. 2 D, E). Occasionally *Girvanella* is also present.

The framework of the mounds consists of clusters of *Epiphyton* and *Renalcis*, which are responsible for the clotted macroscopic fabric (Fig. 3A). The spaces between the clusters are filled with calcareous mudstone along with minor amounts of detrital silt-sized quartz. In some mounds archaeocyathid debris and *Chancelloria* are also present.

Epiphyton is always found in growth position. Its fragile dendritic structure could not have withstood a high degree of current and wave action; therefore it probably thrived in a quiet water environment.

The algal mounds surrounded by siltstones indicate that carbonate deposition only took place where organic mounds developed. Thus, the establishment of algal thickets on the sea floor probably creates a microenvironment favorable to the accumulation of algal-derived carbonates, for the trapping of particles (archaeocyathids, *Chancelloria*, and terrigenous detritus), and even for the precipitation of calcium carbonate. The calcium carbonate could have contributed to the cementation of tha algal thickets now present as mounds in the stratigraphic record.

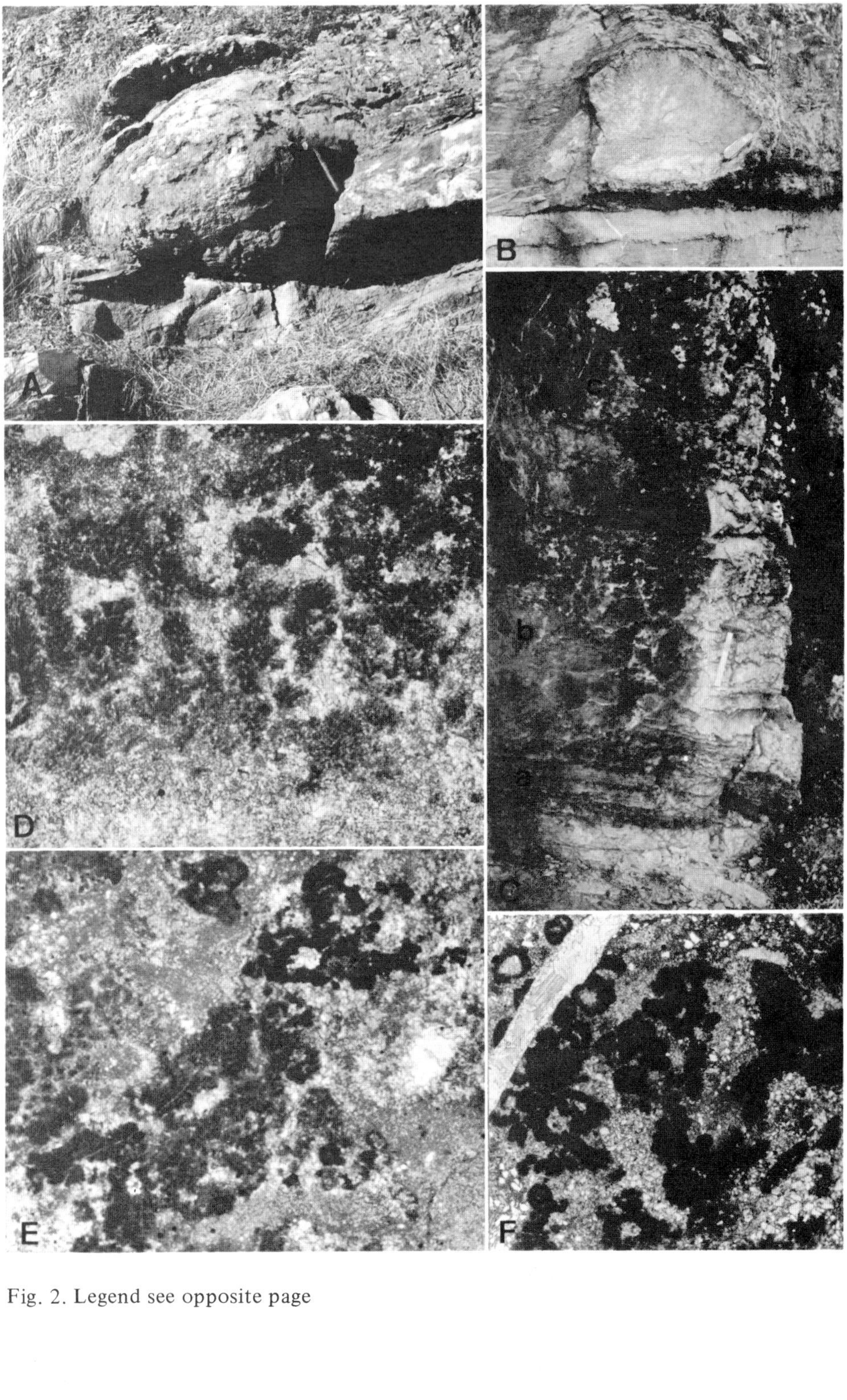

Fig. 2. Legend see opposite page

In some cases, recrystallization or dolomitization has partly obliterated the original texture and the algae are not preserved.

In the locality of Las Ermitas, an algal mound attains such size and displays such a complexity of associated lithofacies as to be called a bioherm (Debrenne and Zamarreño, in press).

37.2.1.2 Cyclic Algal Limestones. These constitute thick tabular cyclic carbonate units, each unit comprising oolitic grainstone at the base, followed by massive algal limestone grading upward into nodular siltstones (Fig. 2C). Cyclic units range from 2.5 m to 8 m in thickness.

The dominant algal limestone-forming organisms are the problematical *Renalcis* and the alga *Epiphyton*, the former being predominant (Fig. 2F, 3B). The interstices between algae are filled with calcareous mudstone, silt-sized quartz, and abundant debris of trilobites and archaeocyathids; some *Chancelloria* is also present. In comparison with the algal mounds, the cyclic algal limestones show a more abundant and diverse nonalgal biota.

The cyclic algal limestones always rest on an oolite bed, suggesting a change from agitated to more quiescent conditions in the environment of deposition allowing the profilic growth of algae. Quiet-water conditions are also indicated by the abundance of lime mud, unbroken *Renalcis* and *Epiphyton*, and lack of bedding within the algal limestones which indicates little transportation of sediments. Nevertheless, pockets of bedded oolitic and intraclastic grainstones also occur within the algal limestones, suggesting that at times they were disturbed by strong waves and currents.

Presumably algal limestone development ceased because of increase of terrigenous clastic sediment.

Dolomitization or recrystallization has affected some of the cyclic algal limestones where the original texture has been completely obliterated.

Much debate has arisen concerning the origin of cyclic sequences (Duff et al., 1967), which were probably in this area controlled by sedimentary distributive mechanisms.

37.2.1.3 Oolitic Grainstones. This lithofacies consists of oosparites in which the well-preserved ooids with micrite nucleii exhibit concentric and/or radial structures. Detrital quartz, peloids, and trilobite fragments also occur in association with the ooids.

The oolitic grainstones, although not volumetrically important, are common throughout the studied section. They are in general located at the base of the cyclic algal limestones and algal mounds, but also exist in an interval 8 m thick. In the latter case, thin to medium-bedded oolites up to 0.70 m thick are interbedded with sandstones; both

---

◀ Fig. 2 A–F. Algal carbonate morphologies and algal assemblages.

(A) Algal mound on the road from Cordoba to Zafra.

(B) Algal mound resting on an oolitic grainstone bed, road from Cordoba to Villaviciosa.

(C) Typical algal limestone cycle showing (*a*) oolite bed at the base, followed by (*b*) massive algal limestone and capped by (*c*) nodular siltstones, road from Cordoba to Villaviciosa.

(D) Dense growth of *Epiphyton* in algal mound texture, x 25

(E) *Epiphyton* and *Renalcis* in algal mound texture, x 17

(F) *Renalcis* in cyclic algal limestone, x 13

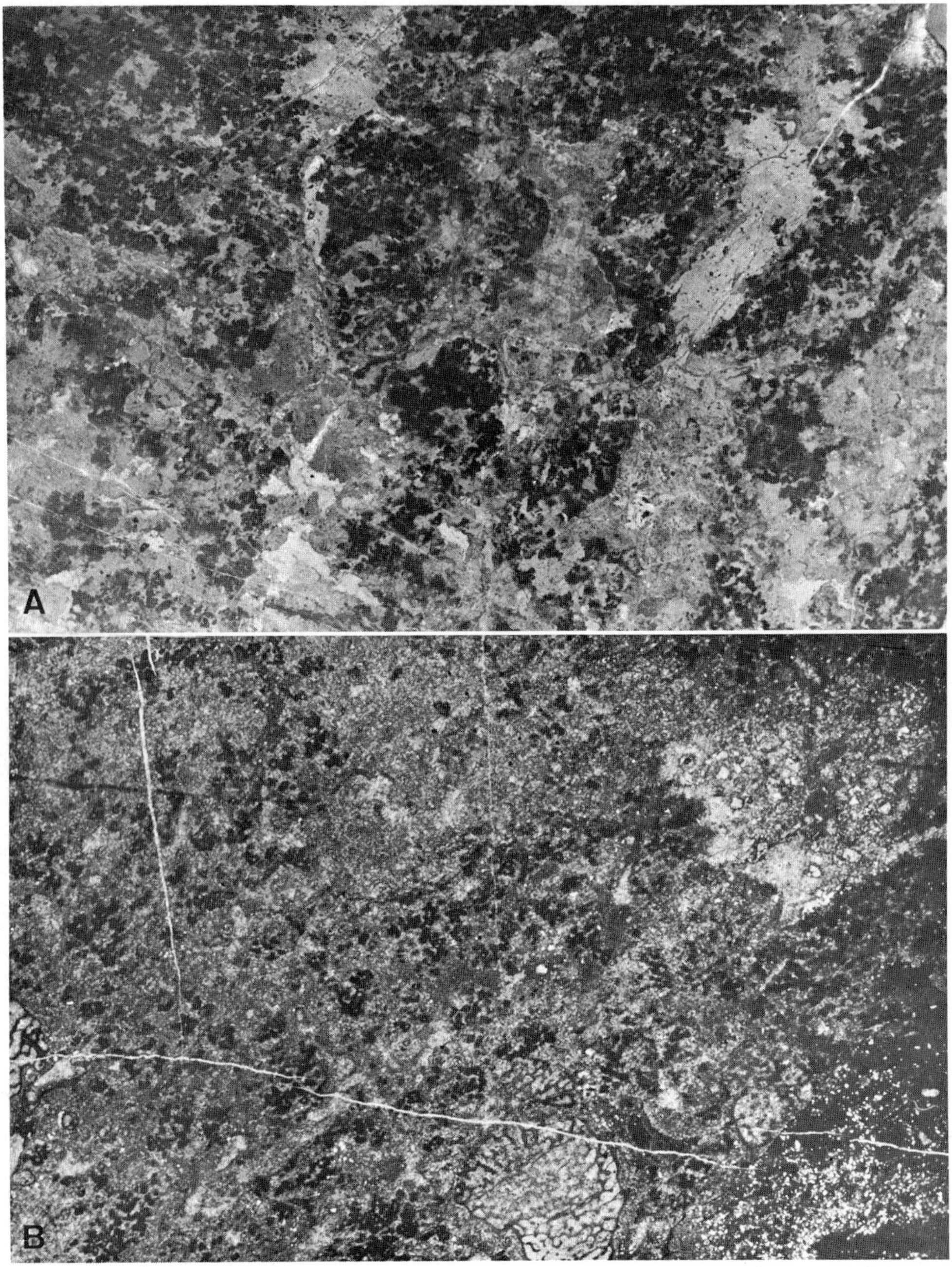

Fig. 3 A and B. Algal carbonate textures.
(A) Clotted macroscopic fabric in algal mound due to *Epiphyton* and *Renalcis*, polished slab, x 0.80
(B) Typical cyclic algal limestone texture where *Renalcis* are predominant while *Epiphyton* and archaeocyathid debris are subordinate, thin-section, x 6

oolite and sandstones beds exhibit cross-stratification and ripple marks, and ooids in the oolitic grainstone suggest active wave and current agitation.

37.3 Conclusions

During deposition of early Cambrian carbonates, southern Spain was a broad shallow marine shelf receiving abundant terrigenous clastic sediments. In this platform, carbonate sedimentation also took place, due chiefly to biogenic accumulation as a result of locally prolific algal growth (formation of algal mounds and cyclic algal limestone).

The most outstanding feature is the association of certain carbonate lithofacies, such as algal carbonates with intraclastic, pelletal, and oolitic grainstones. The faunas of these lithofacies contain archaeocyathid and trilobite debris in varying degrees of abundance.

The lithofacies association suggests depositional environments of different energies, changing from high energy (oolitic, intraclastic lithofacies) to moderate or quiet-water environments (algal carbonates with unbroken *Epiphyton* and *Renalcis*, abundance of mud and lack of bedding). In general the distribution of lithofacies in the sequence indicates a gradual reduction of energy input in the environment: the cyclic algal limestones, and occasionally the algal mounds, rest upon a foundation of oolitic and intraclastic grainstones.

All these features: type of flora and fauna present (*Epiphyton, Renalcis*, archaeocyathids, trilobites, *Chancelloria*) along with the absence of evidence for subaerial exposure (mud-cracks, erosion surfaces, etc.) indicate a subtidal marine environment of deposition.

In this shallow-water environment, algal mound development was due to localized proliferation of the alga *Epiphyton* and the problematical *Renalcis*, which are the dominant biotic constituents, archaeocyathids always being subordinate.

Cyclic algal limestone development appears to have started on local topographic highs (oolitic grainstone) which provided local environmental differences favorable for the flourishing of *Renalcis* and rare *Epiphyton*. Both organisms grew rapidly, stabilized mud and skeletons (archaeocyathids, trilobites, *Chancelloria*), and finally led to the formation of the broad algal limestones. The cyclic character of these carbonates was probably controlled by sedimentary distributive mechanisms. Variation in clastic influx probably was the chief factor controlling algal limestone development.

37.4 Bibliography

Debrenne, F., Zamarreño, I.: Sedimentologie et biologie des constructions organogenes du Cambrien inferieur du Sud de l'espagne. In: 2nd Symp. Intern. Coraux Recifs coralliens fossiles, Paris 1975. Chevallier, J.-P. (ed.). Mem. B.R.G.M. Paris.

Duff, P.M.D., Hallam, A., Walton, E.K.: Cyclic sedimentation. Amsterdam: Elsevier Publ. Co., 1967.

Index of Generic and Species Names

Abbreviations used: F = figure in text; pl = plate

Subject Index

Related Titles

The Study of Trace Fossils
A Synthesis of Principles, Problems, and Procedures in Ichnology
Editor: R.W. Frey
1975. 1 portrait. 250 figures. 18 tables. XVI, 562 pages

Introductory Petrography of Fossils
By A.S. Horowitz and P.E. Potter
Photomicrography by G.R. Ringer
1971. 100 plates. 28 figures. XIV, 302 pages

The Geology of Continental Margins
Editors: C.A. Burk, C.L. Drake
1974. 730 figures. XIII, 1009 pages

Marine Geology and Oceanography of the Arctic Seas
Editor: Y. Herman
1974. 194 figures. X, 416 pages

Depositional Sedimentary Environments
With Reference to Terrigenous Clastics
By H.-E. Reineck and I.B. Singh
Corrected reprint. Springer Study Edition
1975. 579 figures. 23 tables. XV, 439 pages

Springer-Verlag Berlin Heidelberg New York

Related Titles

Petrogenesis of Metamorphic Rocks
English Editor: E. Froese
3rd edition. Springer Study Edition
1974. 92 figures. 15 tables. XI, 320 pages

Recent Sedimentary Carbonates
In two volumes
By J.D. Milliman, G. Müller, and F. Förstner
Part 1: *Marine Carbonates*
By J.D. Milliman
1974. 94 figures. 39 plates. XIII, 375 pages

Sand and Sandstone
By F.J. Pettijohn and P.E. Potter
Springer Study Edition
1973. 258 figures. XIV, 618 pages

The Minor Structures of Deformed Rocks
A Photographic Atlas
By L.E. Weiss
1972. 203 plates. VII, 431 pages

Soil Components
Vol. 1: *Organic Components*
Editor: J.E. Gieseking
1975. 188 figures. IX, 534 pages

Vol. 2: *Inorganic Components*
Editor: J.E. Gieseking
1975. 212 figures. XI, 684 pages